Advanced
Engineering Mathematics

Fourth Edition

Advanced Engineering Mathematics

Fourth Edition

Dr. R. N. Yadava
M.Sc. (BHU Varanasi), Ph.D. (IIT Bombay)
Professor & Director of R&D, AISECT University
&
Former Director Gr. Sc. and Head Resources Management and Modelling
CSIR-Advanced Material and Process Research Institute
Bhopal (M.P.)

Dr. Anil Goyal
M.Sc., Ph.D.
Professor of Mathematics
Dean, Student Welfare &
Dean Applied Sciences
R.G.P.V., Bhopal (M.P.)

Dr. Ramakant Bhardwaj
M.Sc., Ph.D.
Deputy Director (R & D)
TIT Group of Institutions Bhopal
& Head, Department of Mathematics
TIT & Science Bhopal (M.P.)

Dr. Sarvesh Kumar Agrawal
M.Sc., M.Phil, Ph.D.
Associate Professor
Department of Mathematics
TRUBA College of Science & Technology
Bhopal (M.P.)

BSP BS Publications
A unit of **BSP Books Pvt., Ltd.**
4-4-309/316, Giriraj Lane, Sultan Bazar,
Hyderabad - 500 095.

Advanced Engineering Mathematics *Fourth Edition by* Dr. R. N. Yadava, Dr. Anil Goyal, Dr. Ramakant Bhardwaj AND Dr. Sarvesh Kumar Agrawal

First Reprint, 2018

Published by :

BSP BS Publications
A unit of **BSP Books Pvt., Ltd.**

4-4-309/316, Giriraj Lane, Sultan Bazar,
Hyderabad - 500 095.
Phone : 040 - 23445600, 23445688
e-mail : info@bspbooks.net
www.bspbooks.net

Printed at

Aditya Offset Process (I) Pvt., Ltd.

Hyderabad.

Price: 595.00

ISBN: 978-93-5230-091-4 (Paperback)

Dedicated to

Our Parents

Foreword

In the recent years books on engineering mathematics have been published, but the present books by authors fill the gap as a niche that exists between more elementary books and books for advance studies. The authors are well-known experts in engineering mathematics and its applications, and their book will be valuable to students of Master of Engineering of all branches and specialization.

The material in the book is presented in a simple manner, very clear with numerous worked out illustrative examples as per the syllabus of various Technical Universities in India, and need of a textbook on Advanced Engineering Mathematics for post graduate students of engineering has been overcome by the authors with interesting features. The information presented in this book shall be of special value to those engineers who are engaged in industrial consultancy and want to improve their understanding of the subject.

As a textbook on Advanced Engineering Mathematics, this book has no competitors and can be recommended to the serious students of Master of Engineering of all branches and specialization pursuing their studies with Technical Universities in India.

Dr. K. R. Pardasani
Prof, & Former Head
Department of Mathematics
Maulana Azad National Institute of Technology, Bhopal (M.P) India

Message

The Advanced Engineering Mathematics is recognized as an essential part of the mathematical background for Engineers, Scientists, Business Managers and others. The present textbook brought out by the authors provides to post graduate students a simple, clear and logical presentation of the basic concepts for obtaining the solution of problems of heat flow, fluid mechanics, digital electronics, computational mechanics and many other fields of science and engineering in which mathematical methods, statistical methods, and numerical methods and computational methods are employed.

I congratulate the authors for the great efforts they have put into bring out the book and warmly recommend to all postgraduate Engineering and Technical students.

Dr. R. R. Bhargava
Professor
Department of Mathematics
Indian Institute of Technology Roorkee
Roorkee (Uttarakhand) India

Preface

The need for a textbook on Advanced Engineering Mathematics for post graduate students of engineering was first felt by PG students of engineering. Taking into account the fact that there is no comprehensive textbook for the Master of Engineering students at present an attempt has been made to fill a niche that exists between more elementary books and books for advance studies.

Objective of this book on Advanced Engineering Mathematics is to provide to the students of Master of Technology/Engineering a simple, clear and logical presentation of the basic concepts of Vector Spaces, Hermite Polynomials, Functions, Solution of Partial Differential Equations, Numerical Solution of Partial Differential Equations, Fourier Transform, The Discrete Fourier Transform, Wavelet and Harr Transform, Theory of Probability, Theoretical Distribution, Sampling Distribution, Theory of Estimation, Theory of Testing a Hypothesis, Markov Analysis, Queuing Theory, Fuzzy Sets, Decision Theory, Calculus of Variation, Theory of Reliability and Fault Tolerantce, Goal Programming, MATLAB, Hankel and Mellin Transforms and Integral Equation and its application in different branches of engineering.

The book has been prepared according to the syllabus of various Technical Universities and is conceived as a textbook for the course in Master of Engineering of all branches and specializations. The contents of the book have been systematically organized and spread over twenty three chapters. Each chapter begins with the fundamentals and explained with different type of solved examples. The book not only covers the entire scope of the subject but explains the philosophy of the mathematical methods, statistical methods, and numerical methods and computational methods including MATLAB; and makes the understanding of the subject more interesting. Misprints and errors which came to our notice have been corrected; and suggestions from the students and teachers to improve the book have been incorporated.

R. N. Yadava
Anil Goyal
Ramakant Bhardwaj
Sarvesh Kumar Agrawal

Contents

Chapter 2

Hermite Polynomials

Chapter 3

Functions

Chapter 4

Solution of Partial Differential Equations

Chapter 5
Numerical Solution of Partial Differential Equations

Chapter 6
Fourier Transform

Chapter 10

Theoretical Distribution

Chapter 11
Sampling Distribution (Large Samples)

Chapter 12
Theory of Estimation

Chapter 13

Theory of Testing a Hypothesis

Chapter 14

Markov Analysis

Chapter 15

Queuing Theory

Chapter 16

Fuzzy Sets

Chapter 17

Decision Theory

Chapter 18

Calculus of Variation

Chapter 19
Theory of Reliability and Fault Tolerance

Chapter 20
Goal Programming

Chapter 21

MATLAB

Chapter 22

Hankel and Mellin Transforms

Chapter 23

Integral Equation

Chapter - 1

Vector Spaces

Vector Space

Let $(F, +;)$ be a field. Let V be a non empty set whose elements are vectors. Then V is a vector space over the field F, if the following conditions are satisfied:

1. $(V, +)$ is an abelian group

(i) ***Closure property:*** V is closed with respect to addition i.e.,

$$\alpha \in V, \beta \in V \Rightarrow \alpha + \beta \in V$$

(ii) ***Associative:*** $\alpha + (\beta + \Upsilon) = (\alpha + \beta) + \Upsilon, \forall\ \alpha, \beta, \Upsilon \in V$

(iii) ***Existence of identity:*** $\exists$ an elements $0 \in V$ (zero vector) such that

$$\alpha + 0 = \alpha, \forall\ \alpha \in V$$

(iv) ***Existence of inverse:*** To every vector α in V can be associated with a unique vector $-\alpha$ in V called the additive inverse *i.e.,*

$$\alpha + (-\alpha) = 0$$

(v) **Commutative:** $\alpha + \beta = \beta + \alpha, \forall\ \alpha, \beta \in V$

2. V is closed under scalar multiplication *i.e.,*

$$a \in F, \alpha \in V \Rightarrow a\,\alpha \in V$$

3. Multiplication and addition of vector is a distributive property i.e.,

(i) $a(\alpha + \beta) = a\alpha + a\beta, \forall\ a \in F, \alpha, \beta \in V$

(ii) $(a + b)\alpha = a\alpha + b\alpha, \forall\ a, b \in F, \alpha \in V$

(iii) $(ab)\alpha = a(b\alpha), \forall\ a, b \in F, \alpha \in V$

(iv) $1 \cdot \alpha = \alpha, \forall\ \alpha \in V$ and 1 is the unity element in F.

Example 1. The vector space of all ordered n-tuples over a field F.

Proof. Let F be a field. An ordered set $\alpha = (a_1, a_2, ...a_n)$ of n-elements in F is called an n-tuples over F. Let V be the all ordered n-tuple over F. Let $V = \{(a_1, a_2, ...a_n) : a_1, a_2, ...a_n \in F\}$. Now, we will prove that V is a vector space over the field F. For this we define two n-tuples, addition and multiplication of two n-tuples by a scalar as follows.

Equality of two n-tuples : Let $\alpha = (a_1, a_2, ...a_n)$ and $\beta = (b_1, b_2, ...b_n)$ of V. Then $(a_1, a_2, ...a_n) = (b_1, b_2, ...b_n) \Rightarrow a_i = b_i, \forall i = 1, 2, ..., n.$

Addition of n-tuples : we take

$$\alpha + \beta = (a_1 + b_1, a_2 + b_2, ..., a_n + b_n), \forall \alpha = (a_1, a_2, ...a_n) \in V,$$

$$\beta = (b_1, b_2, ...b_n) \in V$$

Since $a_1 + b_1, a_2 + b_2, ..., a_n + b_n$ are all elements of F, therefore, $\alpha + \beta \in V$ and thus V is closed with respect to addition of n-tuples. Scalar multiplication of n-tuples : we define.

$$a\alpha = (aa_1, aa_2, ..., aa_n), \forall a \in F, \alpha = (a_1, a_2, ...a_n) \in V,$$

Since $aa_1, aa_2, ..., aa_n$ are all elements of F, therefore $a\alpha \in V$ and thus V is closed w.r.t. multiplication of n-tuples.

Now, we shall show that V is a vector space for the above two compositions.

1. (i) Associative : Let $(c_1, c_2, ..., c_n) = \Upsilon \in V$

$$\begin{aligned}\alpha + (\beta + \Upsilon) &= (a_1, a_2, ..., a_n) + [(b_1, b_2, ..., b_n) + (c_1, c_2, ..., c_n)]\\ &= (a_1, a_2, ..., a_n) + [b_1 + c_1, b_2 + c_2, ..., b_n + c_n]\\ &= a_1 + (b_1 + c_1), a_2 + (b_2 + c_2), ..., a_n + (b_n + c_n)\\ &= (a_1 + b_1) + c_1, (a_2 + b_2) + c_2, ..., (a_n + b_n) + c_n\\ &= [(a_1, a_2, ..., a_n) + (b_1, b_2, ..., b_n)] + (c_1, c_2, ..., c_n)\\ &= (\alpha + \beta) + \Upsilon\end{aligned}$$

(ii) Commutative: We have

$$\begin{aligned}\alpha + \beta &= (a_1, a_2, ..., a_n) + (b_1, b_2, ..., b_n)\\ &= (a_1 + b_1, a_2 + b_2, ..., a_n + b_n)\\ &= (b_1 + a_1, b_2 + a_2, ..., b_n + a_n)\\ &= (b_1, b_2, ..., b_n) + (a_1, a_2, ..., a_n)\\ &= \beta + \alpha\end{aligned}$$

(iii) Existence of Identify : Let $(0, 0, ..., 0) \in V$ then, we have

$$\begin{aligned}\alpha + 0 &= (a_1, a_2, ..., a_n) + (0, 0, ..., 0)\\ &= (a_1 + 0, a_2 + 0, ..., a_n + 0)\\ &= (a_1, a_2, ..., a_n) = \alpha\end{aligned}$$

(iv) Existence of Inverse : If $\alpha = (a_1, a_2, \ldots, a_n)$ then

$$-\alpha = (-a_1, -a_2, \ldots, -a_n) \in V$$

Then we have

$$\begin{aligned}\alpha + (-\alpha) &= (a_1, a_2, \ldots, a_n) + (-a_1, -a_2, \ldots, -a_n)\\ &= (a_1 - a_1, a_2 - a_2, \ldots, a_n - a_n)\\ &= (0, 0, \ldots, 0)\end{aligned}$$

Hence V is an abelian group under addition.

2. (i) If $a \in F$ and $(a_1, a_2, \ldots, a_n) = \alpha \in V$, $(b_1, b_2, \ldots, b_n) = \beta \in V$ then,

$$\begin{aligned}a(\alpha + \beta) &= a[(a_1, a_2, \ldots, a_n) + (b_1, b_2, \ldots, b_n)]\\ &= a[a_1 + b_1, a_2 + b_2, \ldots, a_n + b_n]\\ &= a(a_1 + b_1), a(a_2 + b_2), \ldots, a(a_n + b_n)\\ &= (aa_1 + ab_1, aa_2 + ab_2, \ldots, aa_n + ab_n)\\ &= (aa_1, aa_2, \ldots, aa_n) + (ab_1, ab_2, \ldots, ab_n)\\ &= a(a_1, a_2, \ldots, a_n) + a(b_1, b_2, \ldots, b_n) = a\alpha + a\beta\end{aligned}$$

(ii) If $a, b \in F$ and $\alpha = (a_1, a_2, \ldots, a_n) \in V$ then

$$\begin{aligned}(a + b)\alpha &= (a + b)(a_1, a_2, \ldots, a_n)\\ &= [(a + b)a_1, (a + b)a_2, \ldots, (a + b)a_n]\\ &= (aa_1 + ba_1, aa_2 + ba_2, \ldots, aa_n + ba_n)\\ &= (aa_1, aa_2, \ldots, aa_n) + (ba_1, ba_2, \ldots, ba_n)\\ &= a(a_1, a_2, \ldots, a_n) + b(a_1, a_2, \ldots, a_n)\\ &= a\alpha + b\alpha\end{aligned}$$

(iii) If $a, b \in F$ and $\alpha = (a_1, a_2, \ldots, a_n) \in V$ then

$$\begin{aligned}(ab)\alpha &= (ab)(a_1, a_2, \ldots, a_n)\\ &= [(ab)a_1, (ab)a_2, \ldots, (ab)a_n]\\ &= [a(ba_1), a(ba_2), \ldots, a(ba_n)]\\ &= a(ba_1, ba_2, \ldots, ba_n)\\ &= a[b(a_1, a_2, \ldots, a_n)]\\ &= a(b\alpha)\end{aligned}$$

(iv) If 1 is the unity element of F and $\alpha = (a_1, a_2, \ldots, a_n) \in V$ then

$$1 \cdot \alpha = 1 \cdot (a_1, a_2, \ldots, a_n) = (a_1, a_2, \ldots, a_n) = \alpha$$

Hence V is a vector space over the field F. The vector space of all ordered n-tuples over F will be denoted by $V_n(F)$.

Example 2. *Prove that the set of all vectors in a plane over the field of real member is a vector space.*

Proof: Let V be the set of all vectors in a plane and R be the field of real numbers.

Then we observe that

1. **$(V, +)$ is abelian group:**

   (i) **Closure property :** Let $\alpha, \beta \in V \Rightarrow \alpha + \beta \in V$

   (ii) **Commutative property :** Let $\alpha, \beta \in V$ then

   $$\alpha + \beta = \beta + \alpha, \forall \alpha, \beta \in V$$

   (iii) **Associative property :** $(\alpha + \beta) + \Upsilon = \alpha + (\beta + \Upsilon), \forall \alpha, \beta, \Upsilon \in V$

   (iv) **Existance of Identity :** Zero vector O in V such that

   $$\alpha + 0 = \alpha, \forall \alpha \in V$$

   (v) **Existance of inverse :** If $\alpha \in V$, then the vector $-\alpha \in V$ such that

   $$\alpha + (-\alpha) = 0$$

2. If $\alpha \in V$ and $m \in R$ (m is any scalar). Then the scalar multiplication

   $$m\alpha \in V$$

3. Scalar multiplication and addition of vectors satisfy the following properties:

   (i) $m(\alpha + \beta) = m\alpha + m\beta, \forall m \in R, \forall \alpha, \beta \in V$

   (ii) $(m+n)\alpha = m\alpha + n\alpha \; \forall m, n \in R, \forall \alpha \in V$

   (iii) $(mn)\alpha = m(n\alpha) \; \forall m, n \in R, \forall \alpha \in V$

   (iv) $1 \cdot \alpha = \alpha, \forall \alpha \in V$ and 1 is the unit element of field R.

   Hence V is a vector space over the field R.

Example 3. *Let R be the field of real numbers and let R_n be the set of all polynomials over the field R. Prove that R_n is a vector space over the field R. Where R_n is of degree at most n.*

Solution. Here R_n is the set of polynomials of degree at most n over the field R. The set R_n is also includes the zero polynomial.

So, $\quad R_n = \{f(x) : f(x) = a_0 + a_1x + a_2x^2 + \ldots . a_nx^n,$

Where $\quad a_0, a_1, a_2, \ldots . a_n \in R\}$

If $\quad f(x) = a_0 + a_1x + a_2x^2 + \ldots .. + a_nx^n$

$\quad g(x) = b_0 + b_1x + b_2x^2 + \ldots .. + b_nx^n$

$\quad r(x) = c_0 + c_1x + c_2x^2 + \ldots .. + c_nx^n$

Then, $\quad f(x) + g(x) = (a_0 + b_0) + (a_1 + b_1)x + \ldots .. + (a_n + b_n)x^n \in R_n$

Because it is also a polynomial of degree at most n over the field R.

Thus R_n is close for addition of polynomials.

∵ Addition of polynomials is commutative as well as associative. The zero polynomial 0 is a member of R_n and is identity for addition of polynomials.

Again if $f(x) = a_0 + a_1x + \dots\dots + a_nx^n \in R_n$

then $-f(x) = -a_0 - a_1x - a_2x^2 \dots\dots - a_nx^n \in R_n$

because it is also a polynomials of degree at most n over the field R.

We have $-f(x) + f(x)$ = zero polynomial.

The polynomial $-f(x)$ is the inverse of $f(x)$ for addition of polynomials.

Hence R_n is an addition group for addition of polynomials.

Now we define scalar multiplication $c\,f(x)$ by the relation.

$$cf(x) = ca_0 + (ca_1)x + (ca_2)x^2 + \dots\dots + (ca_n)x^n$$

Clearly $cf(x) \in R_n$ because it is also a polynomial of degree at most n over the field R. Then R_n is closed for scalar multiplication.

Now if $k_1, k_2 \in R$ and $f(x), g(x) \in R_n$ we have

$$k_1[(f(x) + g(x)] = k_1 f(x) + k_2 g(x)$$

$$(k_1 + k_2) f(x) = k_1 f(x) + k_2 f(x)$$

and $(k_1k_2) f(x) = k_1 [k_2 f(x)]$ can be proved easily.

Also $1 \cdot f(x) = f(x), f(x) \in R_n$

Hence R_n is a vector space over the field R.

General properties of vector spaces: Let $V(f)$ be a vector space over field F and $\bar{0}$ be the zero vector of V· then,

(i) $a.\bar{0} = \bar{0}, \forall a \in F$

(ii) $a\alpha = \bar{0}, \forall a \in V$

(iii) $a(-\alpha) = -(a\alpha), \forall a \in F, \alpha \in V$

(iv) $(-a)\alpha = -(a\alpha), \forall a \in F, \alpha \in V$

(v) $a(\alpha - \beta) = a\alpha - a\beta \quad \forall a \in F, \alpha, \beta \in V$

(vi) $a = \bar{0} \Rightarrow a = 0$ or $\alpha = \bar{0}$

Proof :

(i) We have, $a\bar{0} = a(\bar{0} + \bar{0})$

$= a(\bar{0}) + a(\bar{0})$

$\therefore$ $\bar{0} + a\bar{0} = a\bar{0} + a\bar{0}$

$\because$ V is an abelian group with respect to addition therefore by right cancellalion in V, we get $\bar{0} = a \cdot \bar{0}$

(ii) $0\alpha = (0+0)\alpha$ [$\because 0 + 0 = 0 \in F$, by distributive law]

$= 0\alpha + 0\alpha$

$\bar{0} + 0\alpha = 0\alpha + 0\alpha$

By right cancellation law in V, we get

$\bar{0} = 0\alpha$

(iii) $a[\alpha + (-\alpha)] = a\alpha + a(-\alpha)$

$a \cdot \bar{0} = a\alpha + a(-\alpha)$

$\Rightarrow$ $\bar{0} = a\alpha + a(-\alpha)$

$\Rightarrow$ $a(-\alpha) = -(a\alpha)$

(iv) Now, $[a + (-a)]\alpha = a\alpha + (-a)\alpha$

$\Rightarrow$ $0\alpha = a\alpha + (-a)\alpha$

$\Rightarrow$ $\bar{0} = a\alpha + (-a)\alpha$

$(-a)\alpha$ is the additive inverse of $a\alpha$

$\Rightarrow$ $(-a)\alpha = -(a\alpha)$

(v) We have, $a(\alpha - \beta) = a[\alpha + (-\beta)]$

$= a\alpha + a(-\beta)$

$= a\alpha + [-(a\beta)]$ [$\because a(-\beta) = -(a\beta)$]

$= a\alpha - a\beta$

(vi) Let $a\alpha = \bar{0}$ then we have to prove that either $a = 0$ or $\alpha = \bar{0}$.

Let $a\alpha = \bar{0}$ and $a \neq 0 \in F$, so $\hat{a}$ exists

Then $\hat{a}(a\alpha) = \hat{a} \cdot \bar{0}$

$\Rightarrow$ $(\hat{a}a)\alpha = \bar{0}$

$\Rightarrow$ $1 \cdot \alpha = \bar{0}$

$$\Rightarrow \qquad \alpha = \overline{0}$$

So when $a \neq 0$ then $\alpha = \overline{0}$

again let $a\alpha = \overline{0}$ and we have to prove that $a = 0$. Suppose $a \neq 0$ then $\hat{a}$ exists.

Now, $$a\alpha = \overline{0}$$

$$\hat{a}(a\alpha) = \hat{a} \cdot \overline{0}$$

$$(\hat{a}a)\alpha = \overline{0}$$

$$\Rightarrow \qquad 1 \cdot \alpha = \overline{0}$$

$$\Rightarrow \qquad \alpha = \overline{0}$$

Which is a contradictions that α must be a zero vector. Therefore $a = 0$
Hence $a\alpha = \overline{0}$ then either $a = 0$ or $\alpha = \overline{0}$.

Vector Subspace: Let *V* be a vector space over the field *F* and *W* be a subset of *V*. then *W* is said to be vector subspace of *V* if *W* is also a vector space with scalar multiplication and vector addition over the field *F* as *V*.

Some basic theorems of vector subspaces

***Theorem* 1:** The necessary and sufficient condition for a non empty subset *W* of a vector space *V (f)* to be subspace of *V* is that *W* is closed under vector addition and scalar multiplication.

***Proof: Condition is necessary*.** Let *V* be a vector space and *W* be subspace of *V* over the same field *F*. Since W is vector sub space of *V*, so it is also a vector space under vector addition and scalar multiplications so it is closed. Hence condition is necessary.

The condition is sufficient: Let *V* be a vector space over field *F* and *W* be a non empty subset of *V*, such that *W* is closed under vector addition and scalar multiplication then we have to prove that *W* is subspace of *V*. For this we will prove that it is a vector space over field itself.

Let α∈W, if 1 is the unit element of *F* then -1∈ *F*. Now *W* is closed under scalar multiplication. Therefore,

$$(-1) \in F, \alpha \in W \Rightarrow (-1)\alpha \in W \Rightarrow -(1\alpha) \in W \Rightarrow -\alpha \in W$$

Thus, the additive inverse of each element of *W* is also in *W*. Now, *W* is closed under vector addition. Therefore

$$\alpha \in W \Rightarrow -\alpha \in W \Rightarrow \alpha + (-\alpha) \in W \Rightarrow \overline{0} \in W$$

Where $\overline{0}$ is the zero vector of *V*. Hence the zero vector of *V* is also the zero vector of *W*. Since $W \subseteq V$ therefore vector addition will be commutative as well as associative in W. Hence *W* is an abelian group with respect to vector addition. Also it is given that *W* is closed under scalar multiplication. The remaining properties of a vector space will hold in W. since they hold in *V* of which *W* is a subset. Hence *W* is itself a vector space with respect to vector addition and scalar multiplication as in *V*, so *W* is subspace of *V*. Hence condition in sufficient.

Theorem 2: The necessary and sufficient condition for a non empty subset *W* of a vector space *V(f)* to be a subspace of *V* is

$$a, b \in F, \alpha, \beta \in W \Rightarrow a\alpha + b\beta \in W$$

Proof: ***The condition is necessary:*** let *V* be a vector space over field *F* and *W* is subspace of *V*; then by the definition of subspace, *W* is a vector space over field *F* itself as *V*.

So, $\forall a \in F, \alpha \in W \Rightarrow a\alpha \in W$

$\forall b \in F, \beta \in W \Rightarrow b\beta \in W$

$\forall a\alpha \in W, b\beta \in W$, by vector addition in *W*, $a\alpha + b\beta \in W$

So condition is necessary.

The sufficient condition: Suppose *V* is a vector space over field *F* and *W* is nonempty subset of *V* such that $\forall a, b \in F, \alpha, \beta \in W \Rightarrow a\alpha + b\beta \in W$ then we have to show that *W* is subspace of *V*, for this we will show that *W* is a vector space itself as *V*.

$\because \quad a\alpha + b\beta \in W$

Put $\quad a = b = 1 \in F$

$\Rightarrow \quad 1 \cdot \alpha + 1 \cdot \beta \in W$

$\Rightarrow \quad \alpha + \beta \in W, \forall \alpha, \beta \in W$

So *W* is closed under vector addition.

Now taking $a = 0$, $b = 0$, we see that if

$\alpha \in W$ then

$0\alpha + 0\beta \in W$

$\overline{0} \in W$

Thus the zero vector of *V* belongs to *W*. It will also be the zero vector of *W*.

Now again $1 \in F, -1 \in F$

Taking $a = -1, b = 0$

We get $-1 \cdot \alpha + 0\bar{0} \in W$

$\Rightarrow$ $-\alpha \in W$

Thus the additive inverse of each elements of W is also in W.

Now taking $\beta = \bar{0}$, we see that if $a, b \in F$ and $\alpha \in W$, then $a\alpha + b\bar{0} \in W$ i.e., $a\alpha \in W$

Thus W is closed under scalar multiplication. The remaining properties of a vector space will hold in W since they hold in V of which W is a subset.

Hence W is vector space itself. So by the definition W will be sub space of V.

***Theorem* 3.** The necessary and sufficient conditions for a non-empty subset W of a vector space $V(F)$ to be a subspace of V are

(i) $\alpha \in W, \beta \in W \Rightarrow \alpha - \beta \in W$

(ii) $a \in F, \alpha \in W \Rightarrow a\alpha \in W$

Proof: As theorem 2.

***Theorem* 4.** V be a vector space and W is non empty subset of V then W will be sub space of V if and only if $\forall \alpha, \beta \in W$, $a \in F \Rightarrow a\alpha + \beta \in W$.

Proof: As theorem 2.

Examples on Vector Sub Spaces

***Example* 1.** *The set W of ordered trails* $(k_1, k_2, 0)$ *where* $k_1, k_2 \in F$ *is a subspace of* $V_3(F)$.

Solution. Let $\alpha = (k_1, k_2, 0)$, and $\beta = (l_1, l_2, 0)$ be any two element of W. Where $k_1, k_2, l_1, l_2 \in F$. If a, b be any two elements of F, we have

$$\begin{aligned} a\alpha + b\beta &= a(k_1, k_2, 0) + b(l_1, l_2, 0) \\ &= (ak_1, ak_2, 0) + (bl_1, bl_2, 0) \\ &= (ak_1 + bl_1, ak_2 + bl_2, 0) \end{aligned}$$

$\because$ $ak_1 + bl_1, ak_2 + bl_2 \in F$ so

$a\alpha + b\beta \in W$

Hence W is a subspace of $V_3(F)$.

Example 2. *Prove that the set of all solution (l ,m, n) of the equation l+m+2n=0 is a subspace of the vector space V_3 (R)*

Solution. Let $W = \{(l,m,n) : l,m,n \in R \; and \; l+m+2n=0\}$

To prove that W is a subspace of V_3 (*R*) or R^3.

Let $\alpha = (l_1, m_1, n_1) \; and \; \beta = (l_2, m_2, n_2)$ be any two elements of W. Then

$$l_1 + m_1 + 2n_1 = 0$$
$$l_2 + m_2 + 2n_2 = 0$$

If a, b be any two elements of *R*, we have

$$a\alpha + b\beta = a(l_1, m_1, n_1) + b(l_2, m_2, n_2)$$
$$= (al_1, am_1, an_1) + (bl_2, bm_2, bn_2)$$
$$= (al_1 + bl_2, am_1 + bm_2, an_1 + bn_2)$$

Now $(al_1 + bl_2) + (am_1 + bm_2) + 2(an_1 + bn_2)$

$$= a(l_1 + m_1 + 2n_1) + b(l_2 + m_2 + 2n_2)$$
$$= a \cdot 0 + b \cdot 0 = 0$$

So $a\alpha + b\beta = (al_1 + bl_2, am_1 + bm_2, an_1 + bn_2) \in W$

Thus, $\alpha, \beta \in W and \; a, b \in R \Rightarrow a\alpha + b\beta \in W$.

Hence W is a subspace of $V_3(R)$.

Example **3.** If V is a vector space of all real valued continuous functions over the field of real numbers *R*, then show that the set W of solutions of the differential equation.

$$\frac{d^2y}{dx^2} - 7\frac{dy}{dx} + 12y = 0 \text{ is a subspace of } V.$$

Solution. We have $W = \left\{ y : \frac{d^2y}{dx^2} - 7\frac{dy}{dx} + 12y = 0 \right\}$

It is clear that $y = 0$ satisfies the given differential equation and as such it belongs to W and thus $W \neq \Phi$.

Now let $y_1, y_2 \in W$ then

$$\frac{d^2 y_1}{dx^2} - 7\frac{dy_1}{dx} + 12y_1 = 0 \qquad \dots \text{(i)}$$

$$\frac{d^2 y_2}{dx^2} - 7\frac{dy_2}{dx} + 12y_2 = 0 \qquad \dots \text{(ii)}$$

Let a, b$\in R$. If W is to be subspace then we should show that $ay_1 + by_2$ also belongs to W *i.e.* It is a solution of the given differential equation. We have

$$\frac{d^2}{dx^2}(ay_1 + by_2) - 7\frac{d}{dx}(ay_1 + by_2) + 12(ay_1 + by_2)$$

$$= a\cdot\left(\frac{d^2 y_1}{dx^2} - 7\frac{dy_1}{dx} + 12y_1\right) + b\left(\frac{d^2 y_2}{dx^2} - 7\frac{dy_2}{dx} + 12y_2\right)$$

$$= a\cdot 0 + b\cdot 0$$

Thus a y_1 + by_2 is a solution of the given differential equation and so it belongs to W. Hence W is a subspace of V.

Algebra of subspaces

Theorem 1. The intersection of any two subspaces W_1 and W_2 of a vector space $V(f)$ is also a subspace of $V(f)$.

Proof : Let V be a *vector space* over field F and W_1, W_2 are two subspaces of V. It is clear that $\overline{0} \in W_1$ *and* $\overline{0} \in W_2$ *so* $W_1 \cap W_2 \neq \Phi$

Let, $\quad \alpha, \beta \in W_1 \cap W_2$ *and* $a, b \in F$

$$\alpha \in W_1 \cap W_2 \Rightarrow \alpha \in W_1 \text{ and } \alpha \in W_2$$

$$\beta \in W_1 \cap W_2 \Rightarrow \beta \in W_1 \text{ and } \beta \in W_2$$

$\because W_1$ is subspace of V so

$$\forall a, b \in F, \alpha, \beta \in W_1 \Rightarrow a\alpha + b\beta \in W_1$$

$$\forall a, b \in F, \alpha, \beta \in W_2 \Rightarrow a\alpha + b\beta \in W_2$$

So, $\quad a\alpha + b\beta \in W_1 \cap W_2$

So, $W_1 \cap W_2$ is subspace of V.

Theorem 2. The union of two subspaces is a subspace if and only if one is contained in the other .

Proof: Suppose W_1 and W_2 are two subspaces of V.

Condition is necessary: Let $W_1 \subseteq W_2$ *or* $W_2 \subseteq W_1$, then we will prove that $W_1 \cup W_2$ will be subspace of V.

If $W_1 \subseteq W_2 \Rightarrow W_1 \cup W_2 = W_2$ *and if* $W_2 \subseteq W_1 \Rightarrow W_1 \cup W_2 = W_1$.

But W_1 and W_2 both are subspace of V, so $W_1 \cup W_2$ will be subspace of V.

Condition is sufficient: Let W_1 and W_2 be two subspaces of V such that $W_1 \cup W_2$ be also subspace of V, we have to show that $W_1 \subseteq W_2$ *or* $W_2 \subseteq W_1$. Let us assume that W_1 is not a sub set of W_2 and W_2 is also not a subset of W_1.

$\therefore \quad W_1 \subseteq W_2 \Rightarrow \exists \alpha \in W_1$ such that $\alpha \notin W_2$

and $\quad W_2 \subseteq W_1 \Rightarrow \exists \beta \in W_2$ such that $\beta \notin W_1$

But, $\quad \alpha \in W_1 \cup W_2$

and $\quad \beta \in W_1 \cup W_2$

But $W_1 \cup W_2$ is subspace of V so

$$\alpha + \beta \in W_1 \cup W_2$$

$\Rightarrow \quad \alpha + \beta \in W_1$ *or* $\alpha + \beta \in W_2$

If $\quad \alpha + \beta \in W_1$ *and* $\alpha \in W_1, -\alpha \in W_1$

$\Rightarrow \quad (\alpha + \beta) - \alpha \in W_1 \Rightarrow \beta \in W_1$

Which is a contradiction, again if

$\Rightarrow \quad \alpha + \beta \in W_2$ *and* $-\beta \in W_2$

$\Rightarrow \quad (\alpha + \beta) + (-\beta) \in W_2$

$\Rightarrow \quad \alpha \in W_2$

Again we get a contradiction. Hence either $W_1 \subseteq W_2$ *or* $W_2 \subseteq W_1$.

Theorem 3. Intersection of any family of subspaces of a vector space is a subspace.

Proof: As above

Linear combination : Let *V (f)* be a vector space if $\alpha_1, \alpha_2, ...\alpha_n \in V$ then any vector $\alpha \in V$.

$\alpha = a_1\alpha_1 + a_2\alpha_2 + ... + a_n\alpha_n$ where $a_i \in F$ is called a linear combination of the vectors $\alpha_1, \alpha_2 ..., ...\alpha_n$.

Linear span : Let *V (f)* be a vector space and *S* be any non empty subset of *V*. Then the linear span of *S* is the set of all linear combinations of finite sets of elements of *S* and is denoted by *L*(*S*). Thus we have

$$L(S) = \{a_1\alpha_1 + a_2\alpha_2 + ... + a_n\alpha_n, \alpha_i \in V, a_i \in F\}$$

Linear dependence and linear independence. Let *V* be a vector space over field *F*. A finite set $S = \{\alpha_1, \alpha_2, ...\alpha_n\}$ is said to be linearly dependent if

$$a_1\alpha_1 + a_2\alpha_2 + ... + a_n\alpha_n = \overline{0}$$

Where $\alpha_i \in V, a_i \in F$, and all a_i s may not zero. There will be minimum one $a_i \neq 0$.

A finite set $S = \{\alpha_1, \alpha_2, ...\alpha_n\}$ is said to be linearly independent if $a_1\alpha_1 + a_2\alpha_2 + ... + a_n\alpha_n = \overline{0}$

Where α_i's $\in V$ and a_i's $\in F$ and all a_i's $= 0$

Any infinite setof vectors of *V* is said to be linearly independent if its every finite subset is linearly independent, otherwise it is linearly dependent.

Example 4. Show that the vector *(1, 2, 0), (0, 3, 1), (-1, 0, 1)* forms linearly independent set over field *R*.

Solutions. Let,

$$S = \{\alpha_1, \alpha_2, \alpha_3\}$$

Where,

$$\alpha_1 = (1, 2, 0), \alpha_2 = (0, 3, 1), \alpha_3 = (-1, 0, 1)$$

Let $a_1, a_2, a_3 \in F$ such that

$$a_1\alpha_1 + a_2\alpha_2 + a_3\alpha_3 = \overline{0}$$

Then *S* will be linearly independent if all $a_1 = a_2 = a_3 = 0$

Now

$$a_1\alpha_1 + a_2\alpha_2 + a_3\alpha_3 = \overline{0}$$

$$a_1(1, 2, 0) + a_2(0, 3, 1) + a_3(-1, 0, 1) = (0, 0, 0)$$

$$(a_1, 2a_1, 0) + (0, 3a_2, a_2) + (-a_3, 0, a_3) = (0,0,0)$$

$$(a_1 - a_3, 2a_1 + 3a_2, a_2 + a_3) = (0,0,0)$$

$$a_1 + 0a_2 - a_3 = 0$$

$$2a_1 + 3a_2 + 0a_3 = 0$$

$$0a_1 + a_2 + a_3 = 0$$

Coefficient matrix

$$A = \begin{bmatrix} 1 & 0 & -1 \\ 2 & 3 & 0 \\ 0 & 1 & 1 \end{bmatrix}, |A| = \begin{vmatrix} 1 & 0 & -1 \\ 2 & 3 & 0 \\ 0 & 1 & 1 \end{vmatrix}$$

$$= 1[3] - 0 - 1[2]$$

$$|A| = 1 \neq 0$$

$\therefore$ **Rank A = 3, hence there will be only one solution $a_1 = a_2 = a_3 = 0$**

Hence $S = \{\alpha_1, \alpha_2, \alpha_3\}$ is linearly independent.

Example 5. Show that $S = \{\alpha_1, \alpha_2, \alpha_3\}$ is linearly dependent over field ***R***. Where $\alpha_1 = (1,3,2), \alpha_2 = (1,-7,-8), \alpha_3 = (2,1,-1)$.

Proof: Let $a_1, a_2, a_3 \in R$

Now, $a_1\alpha_1 + a_2\alpha_2 + a_3\alpha_3 = \overline{0}$

$$\Rightarrow \quad a_1(1,3,2) + a_2(1,-7,-8) + a_3(2,1,-1) = (0,0,0)$$

$$\Rightarrow \quad (a_1 + a_2 + 2a_3, 3a_1 - 7a_2 + a_3, 2a_1 - 8a_2 - a_3) = (0,0,0)$$

$$a_1 + a_2 + 2a_3 = 0$$

$$3a_1 - 7a_2 + a_3 = 0$$

$$2a_1 - 8a_2 - a_3 = 0$$

Coefficient matrix

$$A=\begin{bmatrix}1 & 1 & 2\\ 3 & -7 & 1\\ 2 & -8 & -1\end{bmatrix}$$

$$|A|=\begin{vmatrix}1 & 1 & 2\\ 3 & -7 & 1\\ 2 & -8 & -1\end{vmatrix}$$

$$=1[7+8]-1[-3-2]+2[-24+14]$$

$$=15+5-20=0$$

$$|A|=0$$

So rank of A<3

Rank of A<number of unknowns

So there is minimum one $a_i \neq 0$

So $S=\{\alpha_1,\alpha_2,\alpha_3\}$ is linearly dependent.

Example 6. If α,β,γ are linearly independent vectors of $V(f)$ where F is any sub field of complex numbers than prove that $\alpha+\beta,\beta+\gamma,\gamma+\alpha$ are also linearly independent.

Solution. Let a_1,a_2,a_3 be scalar then

$$a_1(\alpha+\beta)+a_2(\beta+\gamma)+a_3(\gamma+\alpha)=\overline{0}$$

$$(a_1+a_3)\alpha+(a_1+a_2)\beta+(a_2+a_3)\gamma=\overline{0} \quad \text{...(i)}$$

But α,β,γ are linearly independent. Therefore (i) implies

$$a_1+0a_2+a_3=0$$

$$a_1+a_2+0a_3=0$$

$$0a_1+a_2+a_3=0$$

The coefficient matrix A of these equations is

$$A = \begin{bmatrix} 1 & 0 & 1 \\ 1 & 1 & 0 \\ 0 & 1 & 0 \end{bmatrix}$$

$$|A| = 1[0] + 0 + 1[1-0] = 1 \neq 0$$

Rank $A = 3$ = number of unknowns

There is only one solution

$$a_1 = a_2 = a_3 = 0$$

So $\alpha+\beta, \beta+\gamma, \gamma+\alpha$ are also linearly independent.

Basic of a vector space : A subset $S = \{\alpha_1, \alpha_2, ..., \alpha_n\}$ of a vector space $V(f)$ is said to be a basis of $V(f)$ if

(i) S consists of linearly independent vectors *i.e.,*

$$a_1\alpha_1 + a_2\alpha_2 + a_3\alpha_3 = \overline{0}$$

all a_i's are zero, $\forall a_i \in F, \alpha_i \in V$

(ii) $L(S) = V(f)$ *i.e.,* every element of V can be written as linear combination of element of S.

Example 7. Show $S = \{(1,2,1),(2,1,0),(1,-1,2)\}$ forms a basis of R^3.

Proof : Since $\dim R^3 = 3$

So $L(S) = V(R^3)$

Now we only to prove that S is linearly independent

Let $a_1, a_2, a_3 \in F$ such that

$$a_1\alpha_1 + a_2\alpha_2 + a_3\alpha_3 = \overline{0}$$

We will prove that $a_1 = a_2 = a_3 = 0$

Now $a_1(1,2,1) + a_2(2,1,0) + a_3(1,-1,2) = (0,0,0)$

$$(a_1 + 2a_2 + a_3, 2a_1 + a_2 - a_3, a_1 + 0a_2 + 2a_3) = (0,0,0)$$

$$a_1 + 2a_2 + a_3 = 0$$

$$2a_1 + a_2 - a_3 = 0$$

$$a_1 + 0a_2 + 2a_3 = 0$$

Coefficient matrix

$$A=\begin{bmatrix}1 & 2 & 1\\ 2 & 1 & -1\\ 1 & 0 & 2\end{bmatrix}$$

$$|A|=\begin{vmatrix}1 & 2 & 1\\ 2 & 1 & -1\\ 1 & 0 & 2\end{vmatrix}$$

$= 1\,[2 - 0] - 2\,[4 + 1] + 1\,[0 - 1]$

$= 2 - 10 - 1$

$\neq 0$

So rank of $A = 3$ = no of unknown

So there is only one solution $a_1 = a_2 = a_3 = 0$

So $S = \{\alpha_1, \alpha_2, \alpha_3\}$ is linearly independent

So S forms the basis of R^3.

Example 8. Select a basis of R^3 (R) from the set $S = \{\alpha_1, \alpha_2, \alpha_3, \alpha_4\}$ where

$$\alpha_1 = (1, -3, 2), \alpha_2 = (2, 4, 1), \alpha_3 = (3, 1, 3), \alpha_4 = (1, 1, 1)$$

Solution. If any three vectors in S are linearly independent, then they will form a basis of the vector space R^3 (R).

First we take $S_1 = \{\alpha_1, \alpha_2, \alpha_3\}$

For this we take $a_1, a_2, a_3 \in R$ such that $a_1\alpha_1 + a_2\alpha_2 + a_3\alpha_3 = \overline{0}$

i.e., $a_1(1,-3,2) + a_2(2,4,1) + a_3(3,1,3) = (0,0,0)$

$$(a_1 + 2a_2 + 3a_3, -3a_1 + 4a_2 + a_3, 2a_1 + a_2 + 3a_3) = (0,0,0)$$

$$a_1 + 2a_2 + 3a_3 = 0$$

$$-3a_1 + 4a_2 + a_3 = 0$$

$$2a_1 + a_2 + 3a_3 = 0$$

Coefficient matrix

$$A=\begin{bmatrix}1 & 2 & 3\\ -3 & 4 & 1\\ 2 & 1 & 3\end{bmatrix}$$

$$|A|=0$$

So rank of A<3

i.e., rank of A < no. of unknowns

So $S_1=\{\alpha_1,\alpha_2,\alpha_3\}$ are linearly dependent

Now we take $S_2=\{\alpha_1,\alpha_2,\alpha_4\}$

Then we get

$$|A|=\begin{vmatrix}1 & 2 & 1\\ -3 & 4 & 1\\ 2 & 1 & 1\end{vmatrix}=0$$

So rank of A = No. of unknown

So $[\alpha_1,\alpha_2,\alpha_4]$ is linearly independent

So $S_2=\{\alpha_1,\alpha_2,\alpha_4\}$ forms the basis of R^3 (*R*).

Linear Transformation: Let *U*(*f*) and *V* (*f*) be two vector space over the same field $F.T:U\to V$ is said to be linear transformation if

$$T(a\alpha+b\beta)=aT(\alpha)+bT(\beta) \quad \ldots\text{(i)}$$

$\forall\,\alpha,\beta$ in *U* and *a*, *b* in *F*

in another way the properly (i) can be defined in two ways

(i) $T(\alpha+\beta)=T(\alpha)+T(\beta)$ and

(ii) $T(a\alpha)=aT(\alpha),\forall\, a\in F,\ \alpha,\beta\in U$

Example 9. The function $T:V_3(R)\to V_2(R)$ defined by $T(a,b,c)=(a,b),\forall\, a,b,c\in R,$ is a linear transformation from V_3 (*R*) into V_2 (*R*).

Proof: If $T:V_3(R)\to V_2(R)$ will be linear transformation then

$$T(a\alpha+b\beta)=aT(\alpha)+bT(\beta)$$

Let $\alpha=(a_1,b_1,c_1)\in V_3(R)$

$\beta=(a_2,b_2,c_2)\in V_3(R)$ and $a,b\in R$

Now $a\alpha+b\beta=a(a_1,b_1,c_1)+b(a_2,b_2,c_2)$

$$=(aa_1,ab_1,ac_1)+(ba_2,bb_2,bc_2)$$

$$=(aa_1+ba_1,ab_2+bb_2,ac_1+bc_2)$$

L.H.S. $=T(a\alpha+b\beta)$

$$=T(aa_1+ba_2,ab_1+bb_2,ac_1+bc_2)$$

$$=(aa_1+ba_2,ab_1+bb_2),\ \text{by def. of } T$$

$$=(aa_1,ab_1)+(ba_2,bb_2)$$

$$=a(a_1,b_1)+b(a_2,b_2)$$

$$=aT(a_1,b_1,c_1)+bT(a_2,b_2,c_2)$$

$$=aT(\alpha)+bT(\beta)=R.H.S.$$

So T is linear transformation from $V_3(R)\to V_2(R)$

Range and Null space of a linear transformation

Let U and V be two vector spaces over the same field F and let T be a linear transformation from U into V.

Then the range of T is written as $R(T)$ and it is the set of all vectors β is V such that $\beta=T(\alpha)$ for some $\alpha\in U$

$$\text{Range } T=\{T(\alpha)=\beta:\alpha\in U,\beta\in V\}$$

Null space of a linear transformation

Let U and V be two vector spaces over same field F and let T be a linear transformation from U into V. Then the null space of T is written as $N(T)$ and it is the set of all vectors α in U such that $T(\alpha)=\bar{0}$ for some $\alpha\in U$

That is $N(T)=\{\alpha\in U:T(\alpha)=\bar{0}\in V\}$

It is to be noted that if we take T as vector space homomorphism of U into V, then the null space of T is also called the karnel of T.

Rank and nullity of a linear transformation

Let U and V be two vector spaces over the same field F and T be a linear transformation from U to V, with U as finite dimensional.

The rank of T is denoted by $\rho(T)$ and it is the dimension of the range of T *i.e.*,

$$\rho(T) = \dim R(T)$$

The nullity of T is denoted by $v(T)$ and it is the dimension of null space of T *i.e.*, $v(T) = \dim N(T)$

Note: Let $T: U(F) \to V(F)$ be a linear transformation from U into V. Suppose that U is finite dimensional. Then rank T + nullity T = dim U.

Example 10. Show that the mapping $T: V_2(R) \to V_3(R)$ defined as $T(a,b) = (a+b, a-b, b)$ is a linear transformation from $V_2(R) \to V_3(R)$. Find the range, rank, null space and nullity of T.

Solution: Let $\alpha = (a_1, b_1)$, $\beta = (a_2, b_2)$ be arbitrary elements of $V_2(R)$. Then $T: V_2(R) \to V_3(R)$ will be a linear transformation if

$$T(a\alpha + b\beta) = aT(\alpha) + bT(\beta), \forall a, b \in R$$

$\alpha, \beta \in V_2(R), a, b \in R$ then $a\alpha + b\beta \in V_2(R)$

Now, $T(a\alpha + b\beta) = T\ [a(a_1, b_1) + b(a_2, b_2)]$

$$= T\ [(aa_1, ab_1) + (ba_2, bb_2)]$$

$$= T\left[(aa_1 + ba_2, ab_1 + bb_2)\right]$$

$$= \left[(aa_1 + ba_2) + (ab_1 + bb_2), (aa_1 + ba_2) - (ab_1 + bb_2), (ab_1 + bb_2)\right]$$

$$= \left[a(a_1 + b_1) + b(a_2 + b_2), a(a_1 - b_1) + b(a_2 - b_2), (ab_1 + bb_2)\right]$$

$$= a(a_1 + b_1, a_1 - b_1, b_1) + b(a_2 + b_2, a_2 - b_2, b_2)$$

$$= aT(a_1, b_1) + bT(a_2, b_2)$$

$$= aT(\alpha) + bT(\beta)$$

So $T(a\alpha + b\beta) = aT(\alpha) + bT(\beta)$

So T is a linear transformation from $V_2(R)$ into $V_3(R)$.

Now $\{(1,0),(0,1)\}$ is a basis for $V_2(R)$

We have, $T(1, 0) = (1 + 0, 1 - 0, 0) = (1, 1, 0)$

$T(0, 1) = (0 + 1, 0 - 1, 0) = (1, -1, 1)$

The vectors $T(1, 0)$, $T(0, 1)$ span the range of T.

Thus the range of T is sub space of $V_3(R)$ spanned by the vectors $(1, 1, 0)$ and $(1, -1, 1)$.

Now the vectors $(1,1,0),(1,-1,1) \in V_3(R)$ are linearly independent if $x, y \in R$,

Then

$$x(1,1,0) + y(1,-1,1) = (0,0,0)$$

$$\Rightarrow \quad (x+y, x-y, y) = (0,0,0)$$

$$\Rightarrow \quad x+y=0, x-y=0, y=0$$

$$\Rightarrow \quad x=0, y=0$$

The vectors $(1, 1, 0)$, $(1, -1, 1)$ form a basis for range of T.

Hence rank T = dim of range of $T = 2$

Nullity of T = dim of $V_2(R)$ – rank of $T = 2 - 2 = 0$

Null space of T must be the zero subspace of $V_2(R)$

Otherwise $(a,b) \in$ null space of T

$$\Rightarrow \quad T(a,b) = (0,0,0)$$

$$(a+b, a-b, b) = (0,0,0)$$

$$a + b = 0$$

$$a - b = 0$$

$$b = 0$$

$$\Rightarrow \quad a = 0, b = 0$$

$\therefore$ (0,0) is the only element of $V_2(R)$ which belong to null space of T.

$\therefore$ Null space of T is the zero subspace of $V_2(R)$.

Representation of transformation by matrices

Let U be an n-dimensional vector space over the field F and let V be an m-dimensional vector space over the field F.

We take two ordered basis

$$\beta=\{\alpha_1,\alpha_2,......\alpha_n\} \text{ and } \beta'=\{\beta_1,\beta_2,......\beta_m\}$$

for U and V respectively

Let $T:U\rightarrow V$ be a linear operator: since T is completely determined by its action on the vectors α_j belonging to a basis for U. Each of the n vectors $T(\alpha_j)$ is uniquely expressible as a linear combination of $\beta_1,\beta_2,......\beta_m$. For $j = 1, 2,……n$.

Then, $$T(\alpha_j)=a_{1j}\beta_1+a_{2j}\beta_2+......+a_{mj}\beta_m=\sum_{i=1}^{m}a_{ij}\beta_i$$

The scalars $a_{1j}, a_{2j},......a_{mj}$ are the co-ordinates of $T(\alpha_j)$ in the ordered basis β'.

The m x n matrix whose j^{th} column (j = 1, 2, ……n) consists of these co-ordinates is called the matrix of the linear transformation T relative to the pair of ordered basis β and β`. It is denoted by the symbol $[T:\beta:\beta']$ or simply by $[T]$ if the basis is understood. Thus,

$[T]=[T:\beta:\beta']$ = matrix of T relative to ordered basis β and $\beta'=\left[a_{ij}\right]_{m\times n}$

and $$T(\alpha_j)=\sum_{i=1}^{m}a_{ij}\beta_i,\ \forall\ j=1,2,......n$$

Example 11. Find the matrix of the linear transformation T on V_3 (R) defined as T (x, y, z) = $(2y + z, x - 4y, 3x)$ with respect to the ordered basis β and also with respect to the ordered basis β` where

(i) β = {(1, 0, 0), (0, 1, 0), (0, 0, 1)}

(ii) β` = {(1, 1, 1), (1, 1, 0), (1, 0, 0)}

Solution.

(i) We have

$T(1, 0, 0) = (0, 1, 3) = 0\,(1, 0, 0) + 1\,(0, 1, 0) + 3\,(0, 0, 1)$

$T(0, 1, 0) = (2 - 4, 0) = 2(1, 0, 0) - 4\,(0, 1, 0) + 0\,(0, 0, 1)$

and $T(0, 0, 1) = (1, 0, 0) = 1\,(1, 0, 0) + 0\,(0, 1, 0) + 0\,(0, 0, 1)$

so by def of matrix of T, with respect to β , we have

$$[T]_\beta = \begin{bmatrix} 0 & 2 & 1 \\ 1 & -4 & 0 \\ 3 & 0 & 0 \end{bmatrix}$$

(ii) We have $T(1, 1, 1) = (3, -3, 3)$

We have to express (3, -3, 3) as a linear combination of vectors in β`.

Let $(a,b,c) = x_1(1,1,1) + y_1(1,1,0) + z_1(1,0,0)$

$$= (x_1 + y_1 + z_1,\ x_1 + y_1, x_1)$$

$$x_1 + y_1 + z_1 = a,\ x_1 + y_1 = b, x_1 = c$$

So $\quad x_1 = c, y_1 = b - c, z_1 = a - b$

For (3, -3, 3), putting $a = 3, b = -3, c = 3$

$$x_1 = 3, y_1 = -6 \, and \, z_1 = 6 \qquad \ldots \text{(i)}$$

So, $T(1, 1, 1) = (3, -3, 3) = 3\,(1, 1, 1) - 6\,(1, 1, 0) + 6\,(1, 0, 0)$

Also, $\quad T(1, 1, 0) = (2, -3, 3)$

Putting $a = 2$, $b = -3$ and $c = 3$ in (i) we get

$T(1, 1, 0) = (2, -3, 3) = 3\,(1, 1, 1) - 6\,(1, 1, 0) + 6\,(1, 0, 0)$

Similarly, $T(1, 0, 0) = (0, 1, 3)$

So, $\quad a = 0, b = 1, c = 3$

$T(1, 0, 0) = (0, 1, 3) = 3\,(1, 1, 1) - 2\,(1, 1, 0) - 1\,(1, 0, 0)$

So, $$[T]_{\beta'} = \begin{bmatrix} 3 & 3 & 3 \\ -6 & -6 & -2 \\ 6 & 5 & -1 \end{bmatrix}$$

Practice Problems

1. Suppose R be the field of real numbers. Which of the following are subspace of $V_3(R)$:

   (i) $\{(a, 2b, 3c) : a, b, c \in R\}$, (ii) $\{(a, a, a) : a \in \mathrm{R}\}$

   (iii) $\{(a, b, c)$: a, b, c are rational numbers$\}$

   Ans. (i) and (ii)

2. In $V_3(R)$, where R is the field of real numbers, examines each of the following sets of vectors for linear independence/ dependence.

(i) {(2, 1, 2), (8, 4, 8)} (ii) {(-1, 2, 1), (3, 0, -1), (-5, 4, 3)}

(iii) {(2, 3, 5), (4, 9, 25)} (iv) {(1, 2, 1), (3, 1, 5), (, -4, 7)}

Ans. (i) Dependent, (ii) Dependent (iii) Independent (iv) Dependent.

3. Show that the three vectors (1, 1, -1), (2, -3, 5) and (-2, 1, 4) of R^3 are linearly independent.
4. Determine if the set {(2, -1, 0), (3, 5, 1), (1, 1, 2)} is a basis of $V_3(R)$.
5. Show that the vectors $\alpha_1 = (1,0,-1), \alpha_2 = (1,2,1), \alpha_3 = (0,-3,2)$ form a basis of $V_3(R)$. Express each of the standard basis vectors as a linear combination of $\alpha_1, \alpha_2, \alpha_3$.
6. Show that the set {(1, i, 0), (2i, 1, 1), (0, 1+i, 1-i)} is a basis for $V_3(c)$.
7. Let $T : V_3(R) \to V_3(R)$ defined by

$$T(x_1, x_2, x_3) = (3x_1 + x_3, -2x_1 + x_2, -x_1 + 2x_2 + 4x_3).$$

What is the matrix of T in the ordered basis $\{\alpha_1, \alpha_2, \alpha_3\}$ where $\alpha_1 = (1,1,0), \alpha_2 = (-1,2,1), \alpha_3 = (2,1,1)$.

$$\textbf{Ans. } T = \frac{1}{4}\begin{bmatrix} 17 & 35 & 22 \\ -3 & 15 & -6 \\ 2 & -14 & 0 \end{bmatrix}$$

Chapter - 2

Hermite Polynomials

We define the hermite polynomials $H_n(x)$ by means of the relation

$$e^{2xt-t^2} = \sum_{n=0}^{\infty} \frac{H_n(x)t^n}{n!} \text{ ,for all finite } x \text{ and } t.$$

§ : Prove that the series representation of Hermite Polynomial

$$H_n(x) = \sum_{k=0}^{\left[\frac{n}{2}\right]} \frac{(-1)^k n!(2x)^{n-2k}}{k!(n-2k)!}$$

Proof We know by the basic definition of Hermite Polynomial

$$e^{2xt-t^2} = \sum_{n=0}^{\infty} \frac{H_n(x)t^n}{n!}$$

$$\sum_{n=0}^{\infty} \frac{H_n(x)t^n}{n!} = e^{2xt} \cdot e^{-t^2}$$

or

$$= \sum_{n=0}^{\infty} \frac{(2x)^n t^n}{n!} \cdot \sum_{k=0}^{\infty} \frac{(-1)^k t^{2k}}{k!}$$

$$= \sum_{n=0}^{\infty} \sum_{k=0}^{\infty} \frac{(2x)^n (-1)^k t^{n+2k}}{n!k!} \qquad \text{..... (i)}$$

Since the relation $\sum_{n=0}^{\infty} \sum_{k=0}^{\infty} A(k,n) = \sum_{n=0}^{\infty} \sum_{k=0}^{\left[\frac{n}{2}\right]} A(k,n-2k)$

Then equation (i) becomes $\sum_{n=0}^{\infty} \frac{H_n(x) t^n}{n!} = \sum_{n=0}^{\infty} \sum_{k=0}^{\left[\frac{n}{2}\right]} \frac{(2x)^{n-2k} (-1)^k t^n}{(n-2k)!\, k!}$

Equating the coefficient of t^n on both the sides, we have.

$$H_n(x) = \sum_{k=0}^{\left[\frac{n}{2}\right]} \frac{(-1)^k\, n!(2x)^{n-2k}}{k!(n-2k)!}$$

§ : Show that $H_n(x)$ is an even function of x for even n and an odd function of x for odd n.

Proof Since the series representation of Hermite Polynomial is

$$H_n(x) = \sum_{k=0}^{\left[\frac{n}{2}\right]} \frac{(-1)^k\, n!(2x)^{n-2k}}{k!\,(n-2k)!} \quad \text{... (ii)}$$

Putting $x = -x$ in (ii), we have

$$H_n(-x) = \sum_{k=0}^{\left[\frac{n}{2}\right]} \frac{(-1)^k\, n!(-2x)^{n-2k}}{k!\,(n-2k)!}$$

$$= (-1)^n \sum_{k=0}^{\left[\frac{n}{2}\right]} \frac{(-1)^k\, n!(2x)^{n-2k}}{k!\,(n-2k)!\,(-1)^{2k}}$$

$$= (-1)^n \sum_{k=0}^{\left[\frac{n}{2}\right]} \frac{(-1)^k\, n!(2x)^{n-2k}}{k!\,(n-2k)!}$$

$$H_n(-x) = (-1)^n H_n(x) \quad \text{..... (iii)}$$

Now, if n is even then (iii) becomes

$$H_{2n}(-x) = H_{2n}(x)$$

Thus, $H_n(x)$ is an even function of x.

If n is odd then (iii) becomes

$$H_{2n+1}(-x) = (-1)^{2n+1} H_{2n+1}(x)$$

$$= -H_{2n+1}(x)$$

Thus, $H_n(x)$ is an odd function of x.

Recurrence Relations

(i) $xH_n'(x) = n\,H_{n-1}'(x) + n\,H_n(x)$

Proof From Hermite Polynomial, we know that

$$e^{2xt-t^2} = \sum_{n=0}^{\infty} \frac{H_n(x)\cdot t^n}{n!} \qquad \text{..... (A)}$$

Differentiating both sides w.r.t. t, we get

$$e^{2xt-t^2}(2x-2t) = \sum_{n=0}^{\infty} \frac{H_n(x)\cdot nt^{n-1}}{n!} \qquad \text{.....(iv)}$$

Further, differentiating (A) w.r.t. x, we get

$$e^{2xt-t^2}\cdot 2t = \sum_{n=0}^{\infty} \frac{H_n'(x)t^n}{n!}$$

i.e

$$e^{2xt-t^2} = \sum_{n=0}^{\infty} \frac{H_n'(x)t^{n-1}}{2\cdot n!}$$

Multiplying both sides by $(2x - 2t)$, this takes the form

$$e^{2xt-t^2}(2x-2t) = 2(x-t)\sum_{n=0}^{\infty} \frac{H_n'(x)t^{n-1}}{2\cdot n!}$$

or

$$e^{2xt-t^2}(2x-2t) = (x-t)\sum_{n=0}^{\infty} \frac{H_n'(x)t^{n-1}}{n!} \qquad \text{..... (v)}$$

Comparing (iv) and (v), we find that

$$\sum_{n=0}^{\infty} \frac{H_n(x)nt^{n-1}}{n!} = x\sum_{n=0}^{\infty} \frac{H_n'(x)t^{n-1}}{n!} - \sum_{n=0}^{\infty} \frac{H_n'(x)t^n}{n!}$$

Equating the coefficients of t^{n-1} from both sides, we get

$$nH_n(x) = x\,H_n'(x) - nH_{n-1}'(x)$$

or

$$xH_n'(x) = n\,H_{n-1}'(x) + nH_n(x)$$

(ii) $H_n'(x) = 2n\,H_{n-1}(x)$ and hence show that

$$D^s H_n(x) = \frac{2^s\, n!\, H_{n-s}(x)}{(n-s)!}, D = \frac{d}{dx}$$

Proof We know that

$$e^{2xt-t^2} = \sum_{n=0}^{\infty} \frac{H_n(x)t^n}{n!}$$

Differentiating both sides w.r.t. x, we have

$$2te^{2xt-t^2} = \sum_{n=0}^{\infty} \frac{H_n'(x)t^n}{n!}$$

or
$$2t\sum_{n=0}^{\infty} \frac{H_n(x)t^n}{n!} = \sum_{n=0}^{\infty} \frac{H_n'(x)t^n}{n!}$$

or
$$2\sum_{n=0}^{\infty} \frac{H_n(x)t^{n+1}}{n!} = \sum_{n=0}^{\infty} \frac{H_n'(x)t^n}{n!}$$

Equating the coefficient of t^n on both sides. We get

$$\frac{2H_{n-1}(x)}{(n-1)!} = \frac{H_n'(x)}{n!}$$

or
$$H_n'(x) = \frac{2n!\,H_{n-1}(x)}{(n-1)!} \qquad \text{..... (vi)}$$

or
$$H_n'(x) = 2n\,H_{n-1}(x) \qquad \text{..... (vii)}$$

Now, differentiating (vi) w.r.t. x, we have

$$H_n''(x) = \frac{2\cdot n!}{(n-1)!} H_{n-1}'(x)$$

from (vi)
$$= \frac{2.n!}{(n-1)!} \frac{2(n-1)!}{(n-2)!} H_{n-2}(x),$$

$$= \frac{2^2\, n!}{(n-2)!} H_{n-2}(x)$$

Continuing like this for S times, this takes the form

$$D^s H_n(x) = \frac{2^s\, n!}{(n-s)!} H_{n-s}(x), \text{ where } D = \frac{d}{dx}$$

(iii) $H_n(x) = 2x\,H_{n-1}(x) - H_{n-1}'(x)$

Proof From recurrence formula (i), we have

$$xH_n'(x) = n\,H_{n-1}'(x) + n\,H_n(x)$$

Also from recurrence formula (ii), we have

$$H_n'(x) = 2n\,H_{n-1}(x)$$

Combination of both the recurrence formula, we have

$$H_n(x) = 2xH_{n-1}(x) - H_{n-1}'(x)$$

(iv) $H_n(x) = 2xH_{n-1}(x) - 2(n-1)\,H_{n-2}(x)$

Proof We have, $e^{2xt-t^2} = \sum_{n=0}^{\infty} \frac{H_n(x)t^n}{n!}$

Differentiating both sides w.r.t. t, we get,

$$e^{2xt-t^2}(2x-2t) = \sum_{n=0}^{\infty} \frac{H_n(x)nt^{n-1}}{n!}$$

or $$2(x-t)\sum_{n=0}^{\infty} \frac{H_n(x)t^n}{n!} = \sum_{n=0}^{\infty} \frac{H_n(x)nt^{n-1}}{n!}$$

or $$2x\sum_{n=0}^{\infty} \frac{H_n(x)t^n}{n!} - 2\sum_{n=0}^{\infty} \frac{H_n(x)t^{n+1}}{n!} = \sum_{n=0}^{\infty} \frac{H_n(x)nt^{n-1}}{n!}$$

Comparing the coefficients of t^n on both the sides, we have

$$2x\,H_n(x) - 2n\,H_{n-1}(x) = H_{n+1}(x) \quad .....(viii)$$

This relation is a pure recurrence relation, which holds for $n \geq 0$, using equation (vii), we get

$$2x\,H_n(x) - H_n'(x) = H_{n+1}(x)$$

replacing n by $(n - 1)$, we have

$$2x\,H_{n-1}(x) - H_{n-1}'(x) = H_n(x) \quad .....(ix)$$

Again replacing n by n-1in (viii), we get

$$2x\,H_{n-1}(x) - 2(n-1)H_{n-2}(x) = H_n(x)$$

(v) $H_n''(x) - 2x\,H_n'(x) + 2n\,H_n(x) = 0$

Proof From recurrence formula (ii)

$$H_n'(x) = 2n\,H_{n-1}(x)$$

Differentiating both sides w.r.t. x, we get

$$H_n''(x) = 2n\,H_{n-1}'(x) \qquad \text{..... (x)}$$

Substituting the values of $H_{n-1}'(x)$ from (x) and $H_{n-1}(x)$ from (vii) in (ix), we have

$$2x \cdot \frac{1}{2n} H_n'(x) - \frac{1}{2n} H_n''(x) = H_n(x)$$

or $$2x \cdot H_n'(x) - H_n''(x) = 2n\,H_n(x)$$

or $$H_n''(x) - 2x\,H_n'(x) + 2n\,H_n(x) = 0$$

The Rodrigues formula for $H_n(x)$

To prove that $$H_n(x) = (-1)^n e^{x^2} D_n\left\{e^{-x^2}\right\}$$

or $$H_n(x) = (-1)^n \exp(x^2) \frac{d^n}{dx^n}\{\exp(-x^2)\}$$

Proof The Hermite Polynomial is

$$e^{2xt-t^2} = \sum_{n=0}^{\infty} \frac{H_n(x)t^n}{n!} \qquad \text{..... (i)}$$

Let $f(t)$ be a function of t, then by maclaurin's theorem

$$f(t) = f(0) + t\,f'(0) + \frac{t^2}{2!} f''(0) + \ldots\ldots$$
1

$$= \sum_{n=0}^{\infty} \frac{f^n(0)t^n}{n!} \qquad \text{..... (ii)}$$

Then from (i) and (ii), we have

$$H_n(x) = \left[\frac{d^n}{dt^n} e^{2xt-t^2}\right]_{t=0}$$

Since e^{-x^2} is independent of *t*, so we may write

$$e^{-x^2} H_n(x) = \left[e^{-x^2} \frac{d^n}{dt^n} e^{2xt-t^2} \right]_{t=0}$$

$$e^{-x^2} H_n(x) = \left[\frac{d^n}{dt^n} e^{-(x-t)^2} \right]_{t=0} = \frac{d^n V}{dt^n} \qquad \text{..... (iii)}$$

Where, $e^{-(x-t)^2} = V$ (Say)

Putting x-t = w, so that $\frac{dw}{dt} = -1$

Then,
$$\frac{dV}{dt} = \frac{dV}{dw} \cdot \frac{dw}{dt} = -\frac{dV}{dw}$$

$$\frac{d^2V}{dt^2} = (-1)^2 \frac{d^2V}{dw^2}$$

$$\frac{d^nV}{dt^n} = (-1)^n \frac{d^nV}{dw^n}$$

Thus, (iii) becomes

$$e^{-x^2} H_n(x) = \left[(-1)^n \frac{d^nV}{dw^n} \right]_{t=0}$$

$$= \left[(-1)^n \frac{d^n}{dw^n} e^{-(x-t)^2} \right]_{t=0}$$

$$= (-1)^n \frac{d^n}{dw^n} e^{-x^2}$$

But it is rediculous to differentiate with respect to *w*, a function of *w* alone and afterward to put *w*=*x*. The *w* is superfluous. Therefore we write

$$e^{-x^2} H_n(x) = (-1)^n \frac{d^n}{dx^n} e^{-x^2}$$

or
$$H_n(x) = (-1)^n e^{x^2} \frac{d^n}{dx^n} e^{-x^2}$$

Example 1 Find first Hermite polynomials.

Solution By using Rodrigues formula for $H_n(x)$

$$H_n(x) = (-1)^n e^{x^2} \frac{d^n}{dx^n} e^{-x^2} \quad \text{..... (i)}$$

Putting $n = 0$, we get

$$H_0(x) = (-1)^0 e^{x^2} \cdot e^{-x^2} = 1$$

Putting $n = 1$, we have

$$H_1(x) = (-1)^1 e^{x^2} \frac{d}{dx}(e^{-x^2})$$

$$= -e^{x^2} \cdot e^{-x^2} \cdot (-2x) = 2x$$

Putting $n = 2$, to get

$$H_2(x) = (-1)^2 e^{x^2} \frac{d^2}{dx^2}(e^{-x^2})$$

$$= e^{x^2} \cdot \frac{d}{dx}(-e^{-x^2} \cdot 2x)$$

$$= -e^{x^2} \left[2e^{-x^2} + 2x \cdot e^{-x^2}(-2x)\right]$$

$$= -e^{x^2} \cdot e^{-x^2}\left[2 - 4x^2\right] = 4x^2 - 2$$

Putting $n = 3$, we have

$$H_3(x) = (-1)^3 e^{x^2} \frac{d^3}{dx^3}(e^{-x^2})$$

$$= -e^{x^2} \frac{d^2}{dx^2}(-2xe^{-x^2})$$

$$= 2e^{x^2} \frac{d}{dx}(e^{-x^2} - 2x^2 e^{-x^2}),$$

$$= 2e^{x^2}\left[e^{-x^2} \cdot (-2x) - 2\left\{2xe^{-x^2} - 2x^3 e^{-x^2}\right\}\right]$$

$$= 2e^{x^2} e^{-x^2}\left[-2x - 4x + 4x^3\right]$$

$$= 8x^3 - 12x$$

Putting $n = 4$, we have

$$H_4(x) = e^{x^2} \frac{d}{dx}\left[e^{-x^2}(12x - 8x^3)\right]$$

$$= e^{x^2}\left[e^{-x^2}(-2x)(12x - 8x^3) + e^{-x^2}(12 - 24x^2)\right]$$

$$= e^{x^2} \cdot e^{-x^2}\left[-24x^2 + 16x^4 + 12 - 24x^2\right]$$

$$= 16x^4 - 48x^2 + 12$$

Similarly, $H_6(x) = 64x^6 - 480x^3 + 720x^2 - 120$

$$H_5(x) = 32x^5 - 160x^2 + 120x$$

Brafman Generating Function

To prove, $$\sum_{n=0}^{\infty} \frac{(c)_n H_n(x) t^n}{n!} = (1 - 2xt)^{-c} \, {}_2F_0\left[\begin{matrix}\frac{1}{2}c, \frac{1}{2}c + \frac{1}{2}; \frac{-4t^2}{(1-2xt)^2} \\ -; \end{matrix}\right]$$

Proof Consider the sum

$$\sum_{n=0}^{\infty} \frac{(c)_n H_n(x) t^n}{n!} = \sum_{n=0}^{\infty} \frac{(c)_n}{n!} \sum_{k=0}^{[n/2]} \frac{(-1)^k (2x)^{n-2k} n! t^n}{k!(n-2k)!}$$

$$= \sum_{n=0}^{\infty} \sum_{k=0}^{[n/2]} \frac{(-1)^k (c)_n (2x)^{n-2k} t^n}{k!(n-2k)!}$$

Since we know that,

$$\sum_{n=0}^{\infty} \sum_{k=0}^{[n/2]} A(k,n) = \sum_{n=0}^{\infty} \sum_{k=0}^{\infty} A(k, n+2k)$$

Using this results in above equation, we have

$$\sum_{n=0}^{\infty} \frac{(c)_n H_n(x) t^n}{n!} = \sum_{n=0}^{\infty} \sum_{k=0}^{\infty} \frac{(-1)^k (c)_{(n+2k)} (2x)^n t^{n+2k}}{k! n!}$$

To change the order of summation, we have

$$= \sum_{k=0}^{\infty} \sum_{n=0}^{\infty} \frac{(-1)^k (c)_{(n+2k)} (2x)^n t^{n+2k}}{k! n!}$$

$$= \sum_{k=0}^{\infty} \left[\sum_{n=0}^{\infty} \frac{(c+2k)_n (2xt)^n}{n!} \right] \frac{(-1)^k (c)_{2k} t^{2k}}{k!}$$

$$= \sum_{k=0}^{\infty} (1-2xt)^{-(c+2k)} \frac{(-1)^k 2^{2k} \left(\frac{c}{2}\right)_k \left(\frac{c}{2}+\frac{1}{2}\right)_k t^{2k}}{k!}$$

$$\left[\because \text{using the relation} (c)_{2k} = 2^{2k} \left(\frac{c}{2}\right)_k \left(\frac{c}{2}+\frac{1}{2}\right)_k \right]$$

$$= (1-2xt)^{-c} \sum_{k=0}^{\infty} \frac{(-1)^k 2^{2k} \left(\frac{c}{2}\right)_k \left(\frac{c}{2}+\frac{1}{2}\right)_k t^{2k}}{(1-2xt)^{2k} \cdot k!}$$

$$= (1-2xt)^{-c} \sum_{k=0}^{\infty} \frac{\left(\frac{c}{2}\right)_k \left(\frac{c}{2}+\frac{1}{2}\right)_k}{k!} \left\{ \frac{-4t^2}{(1-2xt)^2} \right\}^k$$

$$\sum_{n=0}^{\infty} \frac{(c)_n H_n(x) t^n}{n!} = (1-2xt)^{-c} {}_2F_0 \left[\begin{matrix} \frac{c}{2}, \frac{c}{2}+\frac{1}{2}; \frac{-4t^2}{(1-2xt)^2} \\ -; \end{matrix} \right]$$

The Hermite Polynomial as a ${}_2F_0$

The series representation of Hermite Polynomial is

$$H_n(x) = \sum_{k=0}^{[n/2]} \frac{(-1)^k n!(2x)^{n-2k}}{k!(n-2k)!}$$

$$= (2x)^n \sum_{k=0}^{[n/2]} (-1)^k \frac{(-n)_{2k} \cdot x^{-2k}}{2^{2k} \cdot k!}$$

$$= (2x)^n \sum_{k=0}^{[n/2]} \frac{\left(\frac{-n}{2}\right)_k \left(\frac{-n}{2}+\frac{1}{2}\right)_k (-1)^k}{k! x^{2k}}$$

$$= (2x)^n \, {}_2F_0 \left[\begin{matrix} \frac{-n}{2}, \frac{-n}{2}+\frac{1}{2}; -\frac{1}{x^2} \\ -; \end{matrix} \right]$$

Intergral Representation

Let the Hermite Polynomial be

$$e^{(2xt-t^2)} = \sum_{n=0}^{\infty} \frac{H_n(x)t^n}{n!}$$

By Maclaurin's Theorem, we have

$$f(t) = f(0) + t\,f'(0) + \frac{t^2}{2!} f''(0) + ... + \frac{t^n}{n!} f^n(0) + ...$$

Where $f^n(0)$ represents the value of nth partial derivative of e^{2xt-t^2} w.r.t. t at $t=0$. It follows that

$$H_n(x) = \left[\frac{d^n}{dt^n} e^{(2xt-t^2)}\right]_{t=0}$$

But by the complex variable theory,

$$f'(0) = \frac{n!}{2\pi i} \int_c u^{-n-1} f(u)du$$

Where c is the contour encircling the origin in positive direction. Hence we obtain

$$H_n(x) = \frac{n!}{2\pi i} \int_c u^{-n-1} e^{(2xu-u^2)} du$$

Curzon's Integral for $P_n(x)$

To prove, $$P_n(x) = \frac{2}{n!\sqrt{\pi}} \int_0^{\infty} e^{-t^2} \cdot t^n\, H_n(xt)\,dt$$

Proof: The integral on the R.H. S. as

$$\frac{2}{n!\sqrt{\pi}} \int_0^{\infty} e^{-t^2}\, t^n\, H_n(xt)\,dt = \frac{2}{n!\sqrt{\pi}} \int_0^{\infty} e^{-t^2} \cdot t^n \sum_{k=0}^{[n/2]} \frac{(-1)^k\, n!(2xt)^{n-2k}}{k!(n-2k)!} dt$$

$$= \frac{2}{n!\sqrt{\pi}} \sum_{k=0}^{[n/2]} \frac{(-1)^k\, n!(2x)^{n-2k}}{k!(n-2k)!} \int_0^{\infty} e^{-t^2}\, t^{2(n-k)}\,dt \qquad ..... \text{(i)}$$

Now, let $$\int_0^{\infty} e^{-t^2}\, t^{2(n-k)}\,dt$$

Putting, $t^2=u$, so that $2t\,dt = du \Rightarrow dt = \frac{1}{2\sqrt{u}} du$

$$\therefore \quad \int_0^\infty e^{-t^2} t^{2(n-k)} dt = \frac{1}{2}\int_0^\infty e^{-u} u^{n-k} \frac{du}{\sqrt{u}}$$

$$= \frac{1}{2}\int_0^\infty e^{-u} u^{n-k-\frac{1}{2}} du$$

$$= \frac{1}{2}\int_0^\infty e^{-u} u^{\left(n-k+\frac{1}{2}\right)-1} du$$

$$= \frac{1}{2}\Gamma\left(n-k+\frac{1}{2}\right)$$

$$= \frac{1}{2}\left(\frac{1}{2}\right)_{n-k} \Gamma\frac{1}{2}$$

$$= \frac{\sqrt{\pi}}{2}\left(\frac{1}{2}\right)_{n-k} \qquad \left[\because \Gamma\frac{1}{2} = \sqrt{\pi}\right]$$

Putting this value in (i), we have

$$\frac{2}{n!\sqrt{\pi}}\int_0^\infty e^{-t^2} t^n H_n(xt)\, dt = \frac{2}{\sqrt{\pi}}\sum_{k=0}^{[n/2]} \frac{(-1)^k (2x)^{n-2k}}{k!(n-2k)!}\left(\frac{1}{2}\right)_{n-k} \frac{\sqrt{\pi}}{2}$$

$$= \sum_{k=0}^{[n/2]} \frac{(-1)^k (2x)^{n-2k}}{k!(n-2k)!}\left(\frac{1}{2}\right)_{n-k}$$

$$= P_n(x)$$

(By the definition of legendre polynomial)

Orthogonal Property

Prove that

(i) $\int_{-\infty}^{\infty} e^{-x^2} H_n(x) H_m(x)\, dx = 0, m \neq n$ (ii) $\int_{-\infty}^{\infty} e^{-x^2} H_n^2(x)\, dx = 2^n\, n!\sqrt{\pi}$

Proof (i) We have already obtained by Hermite differential equation (v) *i.e.*,

$$H_n''(x) - 2x H_n'(x) + 2nH_n(x) = 0$$

Multiply this by e^{-x^2}, we have

$$e^{-x^2} H_n''(x) - 2x e^{-x^2} H_n'(x) + 2n e^{-x^2} H_n(x) = 0$$

or
$$\left[e^{-x^2} H_n'(x)\right]^1 + 2n e^{-x^2} H_n(x) = 0 \qquad \text{..... (i)}$$

Similarly, for the Hermite polynomial of *m*th degree,

$$\left[e^{-x^2} H_m'(x)\right]^1 + 2m e^{-x^2} H_m(x) = 0 \qquad \text{..... (ii)}$$

Multiply (i) by $H_m(x)$ and (ii) by $H_n(x)$ and subtracting, we have

$$2(n-m)e^{-x^2} H_n(x) H_m(x) = H_n(x)\left[e^{-x^2} H_m'(x)\right]'$$

$$-H_m(x)\left[e^{-x^2} H_m'(x)\right]'$$

$$= \left[e^{-x^2}\{H_n(x) H_m'(x) - H_m(x) H_n'(x)\}\right]'$$

Now, on integrating the equation form the limits -∞ to ∞ w.r.t. *x*, we get

$$2(n-m)\int_{-\infty}^{\infty} e^{-x^2} H_n(x) H_m(x)\, dx$$

$$= \int_{-\infty}^{\infty} \left\{[H_n(x) H_m'(x) - H_m(x) H_n'(x)]e^{-x^2}\right\}' dx$$

$$= \left[e^{-x^2}\{H_n(x) H_m'(x) - H_m(x) H_n'(x)\}\right]_{-\infty}^{\infty}$$

Since the product of any polynomial in *x* by $e^{-x^2} \to 0$ as $x \to \infty$ or as $x \to -\infty$, we may conclude that

$$2(n-m)\int_{-\infty}^{\infty} e^{-x^2} H_n(x) H_m(x)\, dx = 0$$

$$\Rightarrow \qquad \int_{-\infty}^{\infty} e^{-x^2} H_n(x) H_m(x)\, dx = 0, m \neq n$$

(ii) By the definition of Hermite Polynomial

$$e^{2xt-t^2} = \sum_{m=0}^{\infty} \frac{H_m(x) t^m}{m!} \qquad \text{..... (iii)}$$

and $$e^{2xt-t^2} = \sum_{k=0}^{\infty} \frac{H_k(x)t^k}{k!} \qquad \text{..... (iv)}$$

we obtain from (iii) and (iv),

$$e^{4xt-2t^2} = \sum_{m=0}^{\infty}\sum_{k=0}^{\infty} \frac{H_m(x)\, H_k(x) t^{m+k}}{m!\, k!}$$

$$\sum_{n=0}^{\infty}\sum_{k=0}^{\infty} A(k,n) = \sum_{n=0}^{\infty}\sum_{k=0}^{\infty} A(k,n-k)$$

Since,

Thus, the above equation becomes

$$e^{4xt-2t^2} = \sum_{m=0}^{\infty}\sum_{k=0}^{\infty} \frac{H_{m-k}(x)\, H_k(x) t^m}{(m-k)!\, k!}$$

Multiplying both sides by e^{-x^2} and integrating from -∞ to ∞ w.r.t. x, we have

$$\int_{-\infty}^{\infty} e^{-x^2+4xt-2t^2}\, dx = \sum_{m=0}^{\infty}\sum_{k=0}^{\infty} \int_{-\infty}^{\infty} \frac{e^{-x^2}\; H_{m-k}(x)\, H_k(x) t^m}{(m-k)!k!} dx$$

or $$\int_{-\infty}^{\infty} e^{-x^2+4xt-4t^2+2t^2}\, dx = \sum_{m=0}^{\infty}\sum_{k=0}^{\infty} \int_{-\infty}^{\infty} \frac{e^{-x^2}\; H_{m-k}(x)\, H_k(x) t^m}{(m-k)!k!} dx$$

$$e^{2t^2} \int_{-\infty}^{\infty} e^{-(x-2t)^2}\, dx = \sum_{m=0}^{\infty}\sum_{k=0}^{\infty} \frac{t^m}{(m-k)!k!}$$

$$\int_{-\infty}^{\infty} e^{-x^2}\; H_{m-k}(x) H_k(x)\, dx \qquad \text{..... (v)}$$

Now, Let $$\int_{-\infty}^{\infty} e^{-(x-2t)^2}\, dx$$

Putting, $x - 2t = V$ so that $dx = dV$

and $$\int_{-\infty}^{\infty} e^{-V^2}\, dV$$

32 Again putting $V^2 = P$ so that $2VdV = dP \Rightarrow dV = \frac{1}{2\sqrt{p}}dP$

Thus, $$\int_{-\infty}^{\infty} e^{-(x-2t)^2} dx = \int_{-\infty}^{\infty} e^{-V^2} dV = 2\int_{0}^{\infty} e^{-V^2} dV$$

$$= 2\int e^{-P} \frac{dP}{2\sqrt{P}} = \int_{0}^{\infty} e^{-P} \cdot P^{-1/2} dP$$

$$= \int_{0}^{\infty} e^{-P} \cdot P^{\frac{1}{2}-1} dP = \Gamma\frac{1}{2} = \sqrt{\pi}$$

Hence equation (v) becomes

$$e^{2t^2}\sqrt{\pi} = \sum_{m=0}^{\infty}\sum_{k=0}^{m} \frac{t^m}{(m-k)!k!} \int_{-\infty}^{\infty} e^{-x^2} H_{m-k}(x) H_k(x) dx$$

When $m - k \neq k$ then from first part integration become vanish.

So, we assume $m - k = k$ so that $m = 2k$.

and when $k = n$, then $m = 2n$.

Thus, the above equation becomes

$$\sqrt{\pi}\sum_{n=0}^{\infty} \frac{2^n t^{2n}}{n!} = \sum_{n=0}^{\infty} \frac{t^{2n}}{n!n!} \int_{-\infty}^{\infty} e^{-x^2} H_n(x) H_n(x) dx$$

or $$\sqrt{\pi}\sum_{n=0}^{\infty} \frac{2^n t^{2n}}{n!} = \sum_{n=0}^{\infty} \frac{t^{2n}}{(n!)^2} \int_{-\infty}^{\infty} e^{-x^2} H_n^{\,2}(x) dx$$

Equating the coefficient of t^{2n} on both sides, we obtain.

$$\sqrt{\pi}\, 2^n = \frac{1}{n!} \int_{-\infty}^{\infty} e^{-x^2} H_n^{\,2}(x) dx$$

Expansion of Polynomials

We obtain the expansion of x^n in terms of Hermite Polynomial. We know that

$$e^{(2xt-t^2)} = \sum_{n=0}^{\infty} \frac{H_n(x)t^n}{n!}$$

or $$e^{2xt}=e^{t^2}\sum_{n=0}^{\infty}\frac{H_n(x)t^n}{n!}$$

or $$\sum_{n=0}^{\infty}\frac{2^n x^n t^n}{n!}=\sum_{k=0}^{\infty}\frac{t^{2k}}{k!}\sum_{n=0}^{\infty}\frac{H_n(x)t^n}{n!}$$

$$=\sum_{n=0}^{\infty}\sum_{k=0}^{\infty}\frac{H_n(x)t^{n+2k}}{k!n!}=\sum_{n=0}^{\infty}\sum_{k=0}^{[n/2]}\frac{H_{n-2k}(x)t^n}{k!(n-2k)!}$$

Equating the coefficient of t^n on both the sides, we have

$$x^n=\sum_{k=0}^{[n/2]}\frac{n!H_{n-2k}(x)}{k!(n-2k)!\cdot 2^n}$$

More Generating Function : Prove that

$$\sum_{n=0}^{\infty}\frac{H_{n+k}(x)t^n}{n!}=e^{(2xt-t^2)}H_k(x-t)$$

Proof: Consider

$$\sum_{k=0}^{\infty}\sum_{n=0}^{\infty}\frac{H_{n+k}(x)t^n v^k}{k!n!}=\sum_{n=0}^{\infty}\sum_{k=0}^{n}\frac{H_n(x)t^{n-k}v^k}{k!(n-k)!}$$

$$=\sum_{n=0}^{\infty}\frac{H_n(x)\cdot(t+v)^n}{n!}$$

$$\left[\because\ (t+v)^n=\sum_{k=0}^{\infty}{}^nc_k t^{(n-k)}v^k\right]$$

$$=e^{[2x(t+v)-(t+v)^2]}$$

$$\left[\because\ \sum_{n=0}^{\infty}\frac{H_n(x)t^n}{n!}=e^{(2xt-t^2)}\right]$$

$$=e^{(2xt-t^2)}\cdot e^{[2(x-t)v-v^2]}$$

$$=e^{(2xt-t^2)}\sum_{k=0}^{\infty}\frac{H_k(x-t)v^k}{k!}$$

By equating coefficient of v^k, we find that

$$\sum_{n=0}^{\infty}\frac{H_{n+k}(x)t^n}{n!}=e^{(2xt-t^2)}\cdot H_k(x-t) \qquad \text{..... (A)}$$

Bilinear Generating Function : Prove that

$$\sum_{n=0}^{\infty}\frac{H_n(x)H_n(y)t^n}{n!}=(1-4t^2)^{-1/2}\,e^{\left[y^2-\frac{(y-2xt)^2}{1-4t^2}\right]}$$

Proof : Let $\displaystyle\sum_{n=0}^{\infty}\frac{H_n(x)H_n(y)t^n}{n!}=\sum_{n=0}^{\infty}\sum_{k=0}^{[n/2]}\frac{(-1)^k(2x)^{n-2k}}{k!(n-2k)!}H_n(y)t^n$

$$=\sum_{n=0}^{\infty}\sum_{k=0}^{\infty}\frac{(-1)^k(2x)^n H_{n+2k}(y)t^{n+2k}}{k!\,n!}$$

$$=\sum_{k=0}^{\infty}\sum_{n=0}^{\infty}\frac{H_{n+2k}(y)(2xt)^n}{n!}\frac{(-1)^k t^{2n}}{k!}$$

Using the result (A), we obtain

$$\sum_{n=0}^{\infty}\frac{H_n(x)H_n(y)t^n}{n!}=\sum_{k=0}^{\infty}\frac{e^{(4xyt-4x^2t^2)}H_{2k}(y-2xt)(-1)^k t^{2k}}{k!} \quad \text{..... (i)}$$

We know that,

$$H_n(x)=\sum_{k=0}^{[n/2]}\frac{(-1)^k\,n!(2x)^{n-2k}}{k!(n-2k)!}$$

$$\Rightarrow \quad H_{2k}(V)=\sum_{s=0}^{k}\frac{(-1)^s(2k)!(2V)^{2k-2s}}{s!(2k-2s)!}$$

$$\Rightarrow \quad H_{2k}(y-2xt)=\sum_{s=0}^{k}\frac{(-1)^s(2k)!(2y-4xt)^{2k-2s}}{s!(2k-2s)!} \quad \text{..... (ii)}$$

and $$(2k)!=2^{2k}\,k!\left(\frac{1}{2}\right)_k \quad \text{.....(iii)}$$

From (i), (ii) and (iii), we have

$$\sum_{n=0}^{\infty}\frac{H_n(x)H_n(y)t^n}{n!}=\sum_{k=0}^{\infty}\sum_{s=0}^{k}\frac{(-1)^{k+s}2^{2k}\left(\frac{1}{2}\right)_k(2y-4xt)^{2k-2s}\cdot t^{2k}}{s!(2k-2s)!}\cdot e^{(4xyt-4x^2t^2)}$$

$$=e^{(4xyt-4x^2t^2)}\sum_{k=0}^{\infty}\sum_{s=0}^{\infty}\frac{(-1)^k 2^{2k+2s}\left(\frac{1}{2}\right)_{k+s}(2y-4xt)^{2k}\cdot t^{2k+2s}}{s!(2k)!}$$

$$= e^{(4xyt-4x^2t^2)} \sum_{k=0}^{\infty}\sum_{s=0}^{\infty} \frac{(-1)^k\, 2^{2s} \left(\frac{1}{2}\right)_{k+s} \quad (2y-4xt)^{2k} \cdot t^{2k+2s}}{s!\,k! \left(\frac{1}{2}\right)_k}$$

$$= e^{(4xyt-4x^2t^2)} \sum_{k=0}^{\infty}\sum_{s=0}^{\infty} \frac{\left(\frac{1}{2}+k\right)_s \; 2^{2s}\, t^{2s}}{s!} \;\; \frac{(-1)^k\, t^{2k} (2y-4xt)^{2k}}{k!}$$

$$= e^{(4xyt\ 4x^2t^2)} \sum_{k=0}^{\infty} \frac{(-1)^k\, t^{2k}\, (2y-4xt)^{2k}}{k!(1-4t^2)^{\frac{1}{2}+k}}$$

$$\left[\because\ (1-x)^{-a} = \sum_{n=0}^{\infty} \frac{a_n z^n}{n!}\right]$$

$$= (1-4t^2)^{-\frac{1}{2}}\, e^{(4xyt-4x^2t^2)} \cdot e^{\left[\frac{-4t^2(y-2xt)^2}{1-4t^2}\right]}$$

Hence, we obtain

$$\sum_{n=0}^{\infty} \frac{H_n(x) H_n(y) t^n}{n!} = (1-4t^2)^{-1/2}\, e^{\left[y^2 - \frac{(y-2xt)^2}{1-4t^2}\right]}$$

Chapter - 3

Functions

I. Hash Function

A hash function is any well-defined procedure or mathematical function which converts a large, possibly variable-sized amount of data in a small datum. Usually a single integer that may serve as an index into an array. The valued returned by a hash function are called hash values, hash codes, hash sum. Hash function are used to data comparison tasks – such as finding items in a data base, similar records in a large file and so on. A hash function may map two or more keys to the same hash value. Hash function are related to checksums, fingerprints, randomization functions, error correcting codes.

Applications of Hash function

1. **Hash tables:** Hash function are primarily used in hash tables, to quickly locate a data record. The hash function is used to map the search key to the hash. Hash tables are used to implement associative arrays and dynamic sets. In general, a function may map several different keys to the same index. Therefore, each slot of a hash table is associated with a set of records, rather than a single record. For this reason, each slot of a hash table is often called a bucket and hash values are called bucket indices.
2. **Caches :** Hash function are also used to build caches for large data sets stored in slow media. A cache is a generally simpler than a hashed search table, since any collision can be resolved by discarding or writing back the older of the two colliding items.
3. **Bloom Filters :** Hash function are an essential ingredient of the Bloom filter, a compact data structure that provides an enclosing approximation to a set of keys.
4. **Finding Duplicate Records :** To find duplicate records in a large unsorted file, one may use a hash function to map each file record to an index in a table P and collect in each bucket $P[i]$ a list of the numbers of all records with the some hash value i. Once the table is complete, any two duplicate records with end up in the same bucket. The duplicates can then be found by scanning every bucket $P[i]$ which contains two or more numbers, fetching those records and comparing them.

5. **Finding Similar Records :** Hash functions can also be used to locate table records whose key is similar, but not identical, to a given key. For that purpose, one needs a hash function that maps similar keys to hash values that differ by at most *n*, where *n* is a small integer. If one builds a table of *P* of all record numbers, using hash function, then similar records will end up in the same bucket. Then one need only check the records in each bucket $P[i]$ against those in bucket $P[i + j]$, where *j* ranges between $-n$ and *n*.
6. **Geometric hashing :** This principle is used in computer graphics and many other disciplines, to solve many proximity problems in the plane or in 3-D space, such as finding closet pairs in a set of points, similar images in an image database and so on.

In these applications, the set of all inputs is some sort of metric space and the hashing function can be interpreted as a partition of that space into a grid of cells. The table is often an array with two or more indices (called a grid file), and the hash function returns an index tuple. This special case of hashing is called as geometric hashing or the grid method.

Properties of Hash Function

Hash functions are usually required to satisfy certain properties listed below.

1. **Low cost:** The cost of computing a hash function must be small enough to make a hashing – based solution. For instance, binary search can locate an item in a sorted value of *m* item with $\log_2 m$ key comparisons. Therefore, a hash-table solution will be more efficient than binary search only if computing the hash function for one key cost less than performing $\log_2 m$ key comparisons.
2. **Determinism:** The meaning of determinism is that for given input value it must always generate the same hash value. In other words, it must be a function of the hashed data, in the mathematical sense of the term. This requirement excludes hash functions that depend on external variable parameters, such as pseudo-random number generators that depend on the time of day.
3. **Uniformity :** If some hash values are more likely to occur than others, a larger fraction of the lookup operations will have to search through a larger set of colliding table entries. This criterian only requires the value to be uniformly distributed.

   In other words, if a typical set of *n* records is hashed to *m* table slots, the probability of a bucket receiving many more than $\frac{n}{m}$ records should be vanishingly small. In particular, if *n* is less than *m*, very few buckets should have more than one or two records.

   When testing a hash function, the uniformity of the distribution of hash values can be evaluated by the Chi-square test (χ^2).

4. **Continuity :** A hash function is used to search for similar data must be an continuous as possible, two inputs that differ by a little should be mapped to equal or nearly equal hash values. Continuity is desirable for hash functions only in some applications, such as hash tables that use linear search.

5. **Variable range :** The range of hash values may be different for each run of the program or may change along the same run. In those cases, one needs a hash function which takes two parameters the input data *z* and the number *n* of allowed hash values.

A common solution is to compute a fixed hash function with a very large range, divide the result by *n* and use the division's remainder.

Perfect hashing

A hash function *i.e.*, injective maps, each valid input to a different hash values is said to be perfect. With such a function one can directly locate the desired entry in a hash table, without any additional searching.

Perfect hash functions are effective only in the situations where the inputs are fixed and entirely known in advance, such as mapping month names to the integer 0 to 11 or words to the entries of a dictionary. A perfect hash function for a given number of set of n keys, suitable for use in hash table, can be found in time proportional to *n*, can be represented in less than 3 * *n* bits and can be evaluated in a constant number of operations. There are generators that will emit optimized executable code to evaluate a perfect hash for a given input set.

Minimal perfect hashing

A perfect hash function for *n* keys is said to be minimal if its range consists of *n* consecutive integers, usually from 0 to $n - 1$. Besides providing single step, lookup, a minimal perfect hash function also yields a compact hash table, without any vacant slots. Minimal perfect hash functions are much harder to find than perfect ones with a wider range.

Origin of the term "Hash"

"Hash Peter Luhn" of IBM appears to have been the first to use the concept in 'January 1953', that the term "hash" comes by way of analogy with its non-technical meaning, to "chop and mix". Indeed typical hash functions, like the mod operation, "chop" the input domain into many sub domains that get "mixed" into the output range to improve the unformity of the key distribution.

II. Heaviside Step Functions

The Heaviside step function, *H,* is also called the unit step function. It is a discontinuous function whose value is zero for negative argument and one for positive argument.

The function is used in the mathematics of control theory and signal processing to represent a signal that switches on at a specified time and stays switched on, indefinitely. It was named after the English polymath oliver Heaviside. It is the cumulative distribution function of a random variable which is almost surdyo.

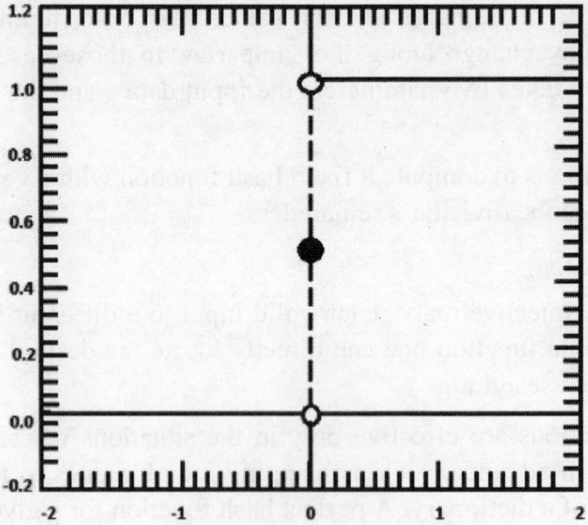

The Heaviside function is the integral of the Dirac delta function : $H' = \delta$. This can also be denoted as

$$H(x) = \int_{-\infty}^{x} \delta(t)\,dt$$

Although this expansion may not hold, for $x = 0$

Discrete form. We can also define an alternative form of the unit step as a function of a discrete variable n.

$$H(x)\begin{cases} 0, & n < 0 \\ 1, & n \geq 0 \end{cases}$$

Where n is an integer.

The discrete time unit impulse is the first difference of the discrete time step

$$\delta(n) = H(n) - H(n-1)$$

This function is the cumulative summation of the Kronecker delta.

$$H(n) = \sum_{k=-\infty}^{\infty} \delta(k)$$

where $\delta(k) = \delta_{k_i}$ is the discrete unit impulse function.

Remark I. The value of the function at 0 can be defined as $H(0)=0, H(0)=\frac{1}{2}$ or H (0) = 1. H(0) = $\frac{1}{2}$ is the most consistent choice used since it maximizes the symmetry of the function and becomes completely consistent with the sign function. This makes for a more general definition.

$$H(x)=\frac{H\ sgn\ (x)}{2}=\begin{cases}0, & x<0\\ \frac{1}{2}, & x=0\\ 1, & x>0\end{cases}$$

To remove the ambiguity of which value to use for $H(0)$, a subscript specifying the value may be used

$$H_a(x)==\begin{cases}0, & x<0\\ a, & x=0\\ 1, & x>0\end{cases}$$

Remark II. The antiderivative of the Heavisible step function is known as ramp function and defined as

$$R(x)=\int_{-\infty}^{x} H(\xi)d\xi = x\,H(x)$$

The derivative of the Heavisible step function is the Dirac delta function: $\frac{dH(x)}{dx}=\delta x$

III. Error Function

The error function is a special function of sigmoid shape which occurs in probability, statistics, materials science and partial differential equations. It is defined as

$$\text{erf}(x)=\frac{2}{\sqrt{\pi}}\int_0^x e^{-t^2}dt$$

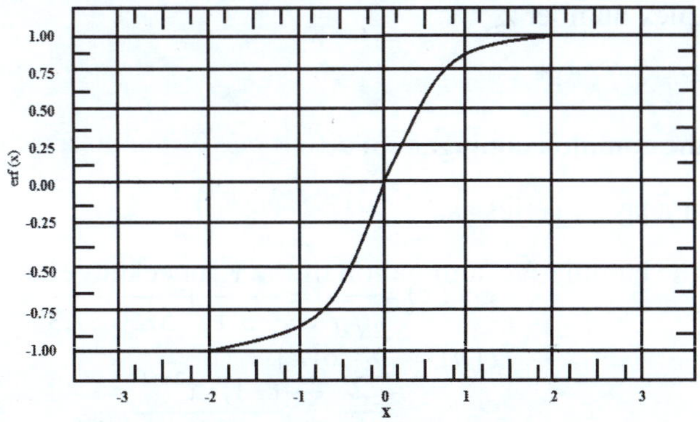

Note: The error function's value at plus/minus infinity is equal to plus/minus 1.

The complementary Error function: It is denoted by erfc, is defined in terms of the error function:

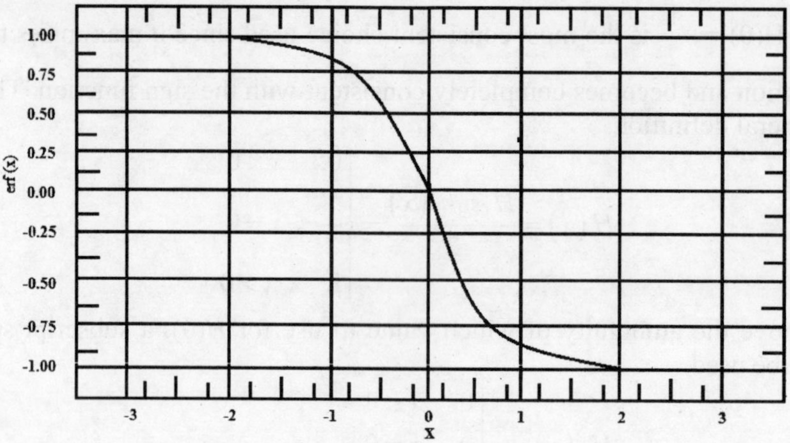

$$\text{erfc}\left(x\right) = 1 - \text{erf}\ (x)$$

$$= \frac{2}{\sqrt{\pi}} \int_x^{\infty} e^{-t^2}\, dt$$

The complex error function : It is denoted by w(x) is also defined in terms of the error function.

$$w(x) = e^{-x^2}\, erfc(-ix)$$

It is known as the Faddeeva function.

Properties

(i) The error function is odd i.e.,

$$\text{erf}\left(-x\right) = -erf\,(x)$$

(ii) For, any complex number z,

$$\text{erf}\left(z^*\right) = -erf\,(z)^*$$

Where z* is the complex conjugate of z.

(iii) Error function's taylor series as :

$$erf\,(x) = \frac{2}{\sqrt{\pi}}\left(x - \frac{x^2}{3.1!} + \frac{x^5}{5.2!} + \frac{x^7}{7.3!} + \ldots\ldots\right)$$

$$= \frac{2}{\sqrt{\pi}} \sum_{n=0}^{\infty} \frac{(-1)^n\, x^{2n+1}}{(2n+1)\cdot n!}$$

Complementary error function's Taylor series as :

$$erfc(x) = 1 - \frac{2}{\sqrt{\pi}}\left(x - \frac{x^2}{3.1!} + \frac{x^5}{5.2!} + \frac{x^7}{7.3!} + ...\right)$$

$$= 1 - \frac{2}{\sqrt{\pi}}\sum_{n=0}^{\infty}\frac{(-1)^n x^{2n+1}}{(2n+1)\cdot n!}$$

(iv) The derivative of the error function is

$$\frac{d}{dx}erf(z) = \frac{2}{\sqrt{\pi}}e^{-x^2}$$

(v) An antiderivative of the error function is

$$x\,erf(x) + \frac{e^{-x^2}}{\sqrt{\pi}}$$

Remarks : Some useful results are

(i) erf (0) = 0 (ii) erfc (0) =1

(ii) erf (∞) =1 (iv) erfc (∞) = 0

Inverse Error Function : The inverse error function is defined as :

$$erf^{-1}(x) = \sum_{k=0}^{\infty}\frac{a_k}{2k+1}\left(\frac{\sqrt{\pi}}{2}x\right)^{2k+1}$$

Where $a_0 = 1$ and

$$a_k = \sum_{n=0}^{k-1}\frac{a_n a_{k-1-n}}{(n+1)(2n+1)} = \left\{1, 1, \frac{7}{6}, \frac{127}{90}, ...\right\}$$

Therefore we have the series expansion

$$erf^{-1}(x) = \frac{\sqrt{\pi}}{2}\left(x + \frac{\pi}{12}x^3 + \frac{7\pi^2}{480}x^5 + \frac{127\pi^3}{40320}x^7 + ...\right)$$

Applications : Mainly two types of applications are as under :

(i) Asymptotic expansion : A useful exapansion of the complementary error function (also of the error function) for large x is

$$erfc(x)=\frac{e^{-x^2}}{x\sqrt{\pi}}\left(1+\sum_{n=1}^{\infty}(-1)^n\frac{1\cdot3\cdot5......(2n-1)}{(2x^2)^n}\right)$$

$$=\frac{e^{-x^2}}{x\sqrt{\pi}}\sum_{n=0}^{\infty}(-1)^n\frac{(2n)!}{(2x)^{2n}\cdot n!}$$

This series diverges for every finite x. However, only the first few terms of this expansion are needed to obtain a good approximation of $erfc(x)$.

(ii) Approximation with elementary function : One approximation is given by

$$erf^2(x)\approx 1-\exp\left(-x^2\frac{\frac{4}{\pi}+ax^2}{1+ax^2}\right)$$

$$a=-\frac{8(\pi-3)}{3\pi(\pi-4)}$$

Where,

Such a fit gives at least one correct decimal digit of function erf in vicinity of the real axis.

Table of values

x	$erf(x)$	$erfc(x)$	x	$erf(x)$	$erfc(x)$
0.00	0.0000000	1.0000000	1.30	0.9340079	0.0659921
0.05	0.0563720	0.9436280	1.40	0.9522851	0.0477149
0.10	0.1124629	0.8875371	1.50	0.9661051	0.0338949
0.15	0.1679960	0.8320040	1.60	0.9763484	0.0236516
0.20	0.2227026	0.7772974	1.70	0.9837905	0.0162095
0.25	0.2763264	0.7236736	1.80	0.9890905	0.0109095
0.30	0.3286268	0.6713732	1.90	0.9927904	0.0072096
0.35	0.3793821	0.6206179	2.00	0.9953223	0.0046777
0.40	0.4283924	0.5716076	2.10	0.9970205	0.0029795
0.45	0.4754817	0.5245183	2.20	0.9981372	0.0018628
0.50	0.5204999	0.4795001	2.30	0.9988568	0.0011432
0.55	0.5633234	0.4366766	2.40	0.9993115	0.0006885
0.60	0.6038561	0.3961439	2.50	0.9995930	0.0004070
0.65	0.6420293	0.3679707	2.60	0.9997640	0.0002360
0.70	0.6778012	0.3221988	2.70	0.9998657	0.0001343

0.75	0.7111556	0.2888444	2.80	0.9999250	0.0000750
0.80	0.7421010	0.2578990	2.90	0.9999589	0.0000411
0.85	0.7706681	0.2293319	3.00	0.9999779	0.0000221
0.90	0.7969082	0.2030918	3.10	0.9999884	0.0000116
0.95	0.8208908	0.1791092	3.20	0.9999940	0.0000060
1.00	0.8427008	0.1572992	3.30	0.9999969	0.0000031
1.10	0.8802051	0.1197949	3.40	0.9999985	0.0000015
1.20	0.9103140	0.0896860	3.50	0.9999993	0.0000007

IV. Modular Mathematics

Introduction

Modular mathematics is similar to division. But some numbers are divide evenly, while some numbers do not. When a number does not divide evenly, a remainder is left over.

For example : 9 divided by 4 is 2 with a remainder of 1.

Modular mathematics uses these remainders. Consider the above example 9 divided by 4. Our number is 9 and 4 is called our mod. The mod is the number that we divided by

2R1
4 | 9
  | −8
  1

Clock method : There is another way to think about modular maths.

For example : What time is it three hours after 11:00? It is 2:00 because to continue past the 12:00 to the 1:00 on the clock. For this, draw the face of a clock. Instead of using the number 12. Put a zero. Now, move in clockwise direction to the next number as you count to 14. You should stop at 2 and hence 14 is 2 in mod 12.

1R2
   | 14
12 | −12
  2

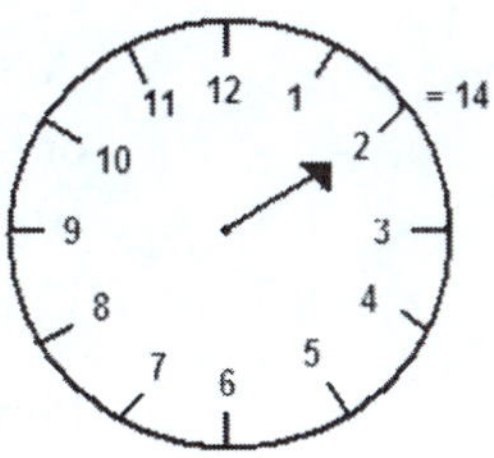

Since 12 divide by 12 gives a remainder 0. 13 divided by 12 gives a remainder 1. 14 divided by 12 gives a remainder 2. It shows that no any number divided by 12 gives a remainder 0. Hence we use 0 instead of 12.

Simultaneous equations : The simultaneous solutions to sets of equations in the form

$$x \equiv a_1 \pmod{m_1}$$
$$\vdots$$
$$x \equiv a_k \pmod{m_k}$$

There are of two types:

(i) Successive substitution, and

(ii) The chinese remainder theorem.

(i) **Successive substitution :** The method of successive substitution is that where we use the definition of the modulus to rewrite these simultaneous equations and then successively make substitutions.

Example 1 Solve $3x \equiv 10 \pmod{19}$, and $x \equiv 19 \pmod{21}$ using successive substitution.

Solution Let $3x \equiv 10 \pmod{19}$

Find the inverse of 3 in Z_{19}; $3^{-1}=-6$

Then, $x \equiv -60 \pmod{19}$

$$x \equiv 16 \pmod{19}$$

$$x \equiv 16 + 19j, \exists j \in Z \qquad \ldots\text{(i)}$$

Substitute in the equation $x \equiv 19 \pmod{21}$

$\Rightarrow$ $(16+19j) \equiv 19 \pmod{21}$

or $19j \equiv 3 \pmod{21}$

Find the inverse of 19 in Z_{21} ; $19^{-1} = 10$

$\therefore$ $j = 30 \pmod{21}$

or $j = 9 \pmod{21}$

We write in the equivalent form

$$j = 9 + 21k, \exists k \in Z \qquad \ldots\text{(ii)}$$

Putting the value of j from (ii) in (i), we have

$$x \equiv 16 + 19(9 + 21k)$$

$$x \equiv 187 + 399k$$

Writing back in the first form

$$x \equiv 187 \pmod{399}$$

Which is the required solution.

(ii) Chinese remainder theorem : The C.R.T. is a method for solving simultaneous linear congruences when the moduli are coprime. Let the equation.

$$x \equiv a_k \pmod{m_k}$$

$$\vdots$$

$$x \equiv a_1 \pmod{m_1}$$

Multiply the moduli together, i.e. $N = m_1 m_2 \ldots\ldots m_k$

$$\Rightarrow \qquad n_1 = \frac{N}{m_1},\ n_2 = \frac{N}{m_2}, \ldots\ldots,\ n_k = \frac{N}{m_k},$$

Then we set y_i be the inverse of n_i mod $m_i, \forall i$

So that $\qquad = 1 \bmod m_i$.

Thus, the solution will be $x = a_1 y_1 n_1 + a_2 y_2 n_2 + \ldots\ldots + a_k y_k n_k \pmod{N}$

The term $a_k y_k n_k \bmod m_k$ becomes equal to a_k, as $y_k n_k = 1 \bmod m_k$ and all other terms becomes equal to zero as when $j \neq k, m_k$ is a factor of n_j.

Powers and roots

Let $a^b \pmod{m}$ be a power of a number.

First we calculate a^b and then taking the modulus. It would take an exorbitant amount of time.

If $a^b \pmod{m} = 1$ which is known as fermat's theorem.

Which states that

If p is prime and $gcd(a, p) = 1$ then $a^{p-1} = 1$ in Z_p.

Primitive elements : If p is prime and a be an element. Then there is an element $a \in Z_p^*$, such that every element of Z_p^* is a power of a.

For example : For Z_3.

Here, $\qquad 2^0 = 1, 2^1 = 2, 2^2 = 1,$

Hence, the element $\{1, 2\}$ constitute in Z_3^*.

Orders : Let a be an element and order of $a \in Z_n^*$ by the smallest integer k such that

$$a^k = 1 \text{ in } Z_n$$

The order of an element is denoted by $O_n(a)$.

For example : $O_n(-1) = 1$, for all n except 2.

Since, $(-1)^2 = 1$

except when $n = 2$, since in that field $-1 = 1$ and thus order is 1.

Remarks. When $gcd(a, n) \neq 1, i.e., a \notin Z_n^*$. Then order is not defined.

Chapter - 4

Solution of Partial Differential Equations

Introduction

Partial differential equations generally arise from the mathematical formulation of physical problems. Subject to certain given conditions known as boundary conditions is known as solving a barndany value problem. We first find out the general solution of the ordinary differential equation and determine the particular solution with the help of given conditions. Method of separation of variables is employed to solve the applied partial differential equation.

Method of Separation of Variable

We consider the solution to be the product of two functions, each of which involves only one of the variables. The following examples explains the method :

Example 1 Solve $\frac{\partial u}{\partial x} = 2\frac{\partial u}{\partial t} + u$ by the method of separation of variables, where $u(x,0) = 6e^{-3x}$.

Solution The given equation is $\frac{\partial u}{\partial x} = 2\frac{\partial u}{\partial t} + u$ ...(i)

Assume that the solution of (i) be $u(x,t) = X(x)T(t)$ ...(ii)

$$\therefore \quad \frac{\partial u}{\partial x} = X'T, \ \frac{\partial u}{\partial t} = XT'$$

Substituting these values in (i), we get

$$\therefore \quad X'T = 2XT' + XT$$

$$\therefore \quad (X' - X)T = 2XT'$$

$\therefore \qquad \dfrac{X'-X}{2X}=\dfrac{T'}{T}=\lambda \quad$ (Say)

Now $\qquad \dfrac{X'-X}{2X}=\lambda$

$\therefore \qquad X'-X=2\lambda X$

$X'=(1+2\lambda)X$

Integrating, we get $\log X=(1+2\lambda)x+\log C_1$

$\therefore \qquad X=C_1e^{(1+2\lambda)x}$ ...(iii)

and $\qquad \dfrac{T'}{T}=\lambda$

$\therefore \qquad \log T=\lambda t+\log C_2$

$\therefore \qquad T=C_2\,e^{\lambda t}$ ...(iv)

Putting values of X and T in (ii), we get

$u(x,t)=C_1C_2\,e^{(1+2\lambda)x}\cdot e^{\lambda t}$ ...(v)

Given $\qquad u(x,0)=6e^{-3x}$ ...(vi)

$\therefore$ From (v) and (vi), we get

$6e^{-3x}=C_1C_2\,e^{(1+2\lambda)x}$

$\therefore \qquad C_1C_2=6, 1+2\lambda=-3$

$\therefore \qquad C_1C_2=6, \lambda=-2$

Putting these values in equation (v), we get

$u(x,t)=6\,e^{-3x}\,e^{-2t}$

$=6\,e^{-(3x+2t)}$

Example 2 Use the separation of variables to solve $3u_x+2u_y=0$ with $u(x,0)=4e^{-x}$

Solution The given equation is $3\dfrac{\partial u}{\partial x}+2\dfrac{\partial u}{\partial y}=0$ ...(i)

Assume the solution $u(x,y)=X(x)Y(y)$ ...(ii)

$\therefore \qquad \dfrac{\partial u}{\partial x}=X'Y, \dfrac{\partial u}{\partial y}=XY'$

Substituting the values in (i), we get

$$3X'Y+2XY'=0$$

or $$\frac{X'}{X}=\frac{-2}{3}\frac{Y'}{Y}=\lambda \text{ (say)}$$

Now $$\frac{X'}{X}=\lambda$$

$\Rightarrow$ $$X=C_1e^{\lambda x} \qquad \text{...(iii)}$$

and $$\frac{-2}{3}\frac{Y'}{Y}=\lambda$$

or $$\frac{Y'}{Y}=\frac{-3\lambda}{2}$$

$\Rightarrow$ $$Y=C_2\,e^{\frac{-3}{2}\lambda y} \qquad \text{...(iv)}$$

Putting values of X and Y in (ii), we get

$$u(x,y)=C_1C_2\,e^{\lambda x}\,e^{\frac{-3}{2}\lambda y} \qquad \text{...(v)}$$

Given, $$u(x,0)=4e^{-x} \qquad \text{...(vi)}$$

$\therefore$ from (v) and (vi), we get

$$4e^{-x}=C_1C_2\,e^{\lambda x}$$

$\Rightarrow$ $$C_1C_2=4, \lambda=-1$$

$\therefore$ from equation (v), we get

$$u(x,y)=4e^{-x}\cdot e^{\frac{-3}{2}y}$$

$$=4e^{\frac{-1}{2}(2x-3y)}$$

Example 3 Using the method of separation of variable. Solve $\frac{\partial^2 z}{\partial x^2}-2\frac{\partial z}{\partial x}+\frac{\partial z}{\partial y}=0$.

Solution The given equation is $$\frac{\partial^2 z}{\partial x^2}-2\frac{\partial z}{\partial x}+\frac{\partial z}{\partial y}=0 \qquad \text{...(i)}$$

Assume that the solution of (i) is

$$Z(x,y)=X(x)Y(y) \quad \text{...(ii)}$$

$$\therefore \quad \frac{\partial z}{\partial x}=X'Y, \frac{\partial^2 z}{\partial x^2}=X''Y, \frac{\partial z}{\partial y}=XY'$$

Substituting these values in (i), we get

$$X''Y-2X'Y+XY'=0$$

or $$\frac{X''-2X'}{X}=\frac{-Y'}{Y}=\lambda\,(say)$$

Now $$\frac{X''-2X'}{X}=\lambda$$

$$\Rightarrow \quad X''-2X'-\lambda X=0$$

The auxiliary equation is $m^2-2m-\lambda=0 \Rightarrow m=1\pm\sqrt{1+\lambda}$

$$\therefore \quad X=C_1e^{(1+\sqrt{1+\lambda})x}+C_2e^{(1-\sqrt{1+\lambda})x} \quad \text{...(iii)}$$

and $$\frac{-Y'}{Y}=\lambda$$

$$\Rightarrow \quad Y=C_3e^{-\lambda y} \quad \text{...(iv)}$$

Substituting the value of *X* and *Y* in (ii), we get

$$z(x,y)=\left[C_1e^{(1+\sqrt{1+\lambda})x}+C_2e^{(1-\sqrt{1+\lambda})x}\right]\cdot C_3e^{-\lambda y}$$

Which is the required solution of the given equation.

Example 4 Use the method of separation of variables to solve the equation

$$\frac{\partial^2 u}{\partial x^2}=\frac{\partial u}{\partial t}$$

given that $u = 0$, when $t \to \infty$ as well as $u = 0$ at $x = 0$ and $x = l$.

Solution The given partial differential equation be

$$\frac{\partial^2 u}{\partial x^2}=\frac{\partial u}{\partial t} \quad \text{...(i)}$$

Assume the solution of equation (i) is

$$u(x_1t)=X(x)T(t) \quad \text{...(ii)}$$

$\therefore \qquad \frac{\partial u}{\partial x} = X'T, \ \frac{\partial^2 u}{\partial x^2} = X''T$

and $\qquad \frac{\partial u}{\partial t} = X\ T'$

Putting the values in equation (i), we get

$$X''T = X\ T'$$

$\Rightarrow \qquad \frac{X''}{X} = \frac{T'}{T} = -\lambda^2$ (let)

$\therefore \qquad \frac{X''}{X} = -\lambda^2$ and $\frac{T'}{T} = -\lambda^2$

'or' $\qquad X'' + \lambda^2 X = 0 \qquad$ ...(iii)

and $\qquad \frac{T'}{T} = -\lambda^2 \qquad$ ...(iv)

From equation (iii), we have

$$X'' + \lambda^2 X = 0$$

The A.E. is $m^2 + \lambda^2 = 0$

$\Rightarrow \qquad m = \pm\ \lambda i$

The solution of equation (iii) is

$$X = C_1 \cos\ \lambda\ x + C_2\ \sin \lambda\ x$$

Again from equation (iv), we have

$$\frac{T'}{T} = -\lambda^2$$

Integrating, we get

$$\log\ T = -\ \lambda^2\ t + \log C_3$$

$\Rightarrow \qquad T = C_3\ e^{-\lambda^2 t}$

Putting the values of X and T in equation (ii), we get

$$u(x_1 t) = (C_1\ \cos \lambda x + C_2 \sin \lambda x) C_3\ e^{-\lambda^2 t} \qquad ...$$

Case 1: Putting $u = 0$ at $x = 0$ in equation (v), we get

$$0 = C_1\, C_3\, e^{-\lambda^2 t}$$

But $C_3 \neq 0$ then $C_1 = 0$

Case 2: Putting $u = 0$ at $x = l$ in equation (v), we get

$$0 = C_2 \sin \lambda l \times C_3\, e^{-\lambda^2 t}$$

But $C_3 \neq 0$ and $C_2 \neq 0$, then

$$\sin \lambda l = 0 \Rightarrow \lambda l = n\pi$$

$$\therefore \quad \lambda = \frac{n\pi}{l}$$

Putting the value of C_1 and λ in equation (v), we have

$$u(x,t) = C_2 C_3 \sin \frac{n\pi x}{l} e^{\frac{-n^2\pi^2}{l^2}t}$$

$$= b_n \sin\left(\frac{n\pi x}{l}\right) e^{-\left(\frac{n^2\pi^2}{l^2}\right)t}, \text{ where } b_n = C_2\, C_3$$

Hence, for all values of n, the required solution is

$$u(x,t) = \sum_{n=1}^{\infty} b_n \sin\left(\frac{n\pi x}{l}\right) e^{-\left(\frac{n^2\pi^2}{l^2}\right)t}$$

Example 5 Solve

$$\frac{\partial^2 u}{\partial x^2} = h^2 \frac{\partial u}{\partial t}$$

given that $u(0,t) = 0,\ u\ (l,\ t) = 0$ and $u\ (x,\ 0) = \sin\left(\frac{\pi x}{l}\right)$

by the method of separation of variables

Solution The given partial differential equation be

$$\frac{\partial^2 u}{\partial x^2} = h^2 \frac{\partial u}{\partial t} \qquad \text{...(i)}$$

Assume the solution of equation (i) is

$$u(x,t) = X(x) \times T(t)$$

$\therefore \qquad \frac{\partial u}{\partial x} = X'T, \ \frac{\partial^2 u}{\partial x^2} = X''T$

and $\qquad \frac{\partial u}{\partial t} = X\ T'$

Putting the values in equation (i), we get

$$X''T = h^2\ X\ T'$$

or $\qquad \frac{X''}{X} = h^2 \frac{T'}{T} = -\lambda^2 \text{ (let)}$

$\therefore \qquad \frac{X''}{X} = -\lambda^2 \text{ and } h^2 \frac{T'}{T} = -\lambda^2$

or $\qquad X'' + \lambda^2 X = 0 \qquad \text{...(iii)}$

and $\qquad \frac{T'}{T} = \frac{-\lambda^2}{h^2} \qquad \text{...(iv)}$

From equation (iii), we get

The A.E. is $m^2 + \lambda^2 = 0$

$\Rightarrow \qquad m = \pm\ \lambda i$

The solution of equation (iii) is

$$X = C_1 \cos \lambda x + C_2 \sin \lambda x$$

Again from equation (iv), we have

$$\frac{T'}{T} = \frac{-\lambda^2}{h^2}$$

Integrating, we get

$$\log T = \frac{-\lambda^2}{h^2} t + \log C_3$$

$\Rightarrow \qquad T = C_3\ e^{\frac{-\lambda^2}{h^2} t}$

Putting the value of X and T in equation (ii), we get

$$u(x_1 t) = (C_1 \cos \lambda x + C_2 \sin \lambda x) C_3\ e^{\frac{-\lambda^2}{h^2} t} \qquad \text{...(v)}$$

Case 1: Putting $u(0,t) = 0$ i.e., $u = 0$ at $x = 0$ in equation (v), we get

$$0 = C_1 C_3\ e^{\frac{-\lambda^2}{h^2} t}$$

But $\qquad C_3 \neq 0, \text{ then } C_1 = 0$

Case 2: Putting $u(l,t)=0$ i.e., $u=0$ at $x=l$, we get

$$0=C_2C_3\sin\lambda l\; e^{\frac{-\lambda^2}{h^2}t}$$

But $\quad C_3\neq 0,\; C_2\neq 0,$ then

$$\sin\lambda l=0 \;\Rightarrow\; \lambda l=n\pi$$

$$\Rightarrow\quad \lambda=\frac{n\pi}{l}$$

Putting the values of C_1 and λ in equation (v), we get

$$u(x_1t)=C_2C_3\sin\left(\frac{n\pi x}{l}\right)e^{-\left(\frac{n^2\pi^2}{h^2l^2}\right)t}$$

'or' $$u(x_1t)=b_n\sin\left(\frac{n\pi x}{l}\right)e^{-\left(\frac{n^2\pi^2}{h^2l^2}\right)t}, \text{ where } b_n=C_2C_3 \qquad \text{...(vi)}$$

Hence, for all values of n, the required solution is

$$u(x_1t)=\sum_{n=1}^{\infty}b_n\sin\left(\frac{n\pi x}{l}\right)e^{-\left(\frac{n^2\pi^2}{h^2l^2}\right)t} \qquad \text{...(vii)}$$

Now, putting $\quad u(x,0)=\sin\dfrac{\pi x}{l}$

$$\therefore\quad \sin\frac{\pi x}{l}=\sum_{n=1}^{\infty}b_n\sin\left(\frac{n\pi x}{l}\right)$$

Compare both sides for $n=1$, we get

$$b_1=1$$

Hence the required solution for $n=1$ is

$$u(x,t)=\sin\left(\frac{\pi x}{l}\right)e^{-\left(\frac{\pi^2}{h^2l^2}\right)t}$$

One dimensional wave equation: $\dfrac{\partial^2 y}{\partial t^2}=c^2\dfrac{\partial^2 y}{\partial x^2}$

Solution of one dimensional wave equation: The vibration of an elastic string are governed by the one dimensional wave equation

$$\frac{\partial^2 y}{\partial t^2}=c^2\frac{\partial^2 y}{\partial x^2} \qquad \text{...(i)}$$

Where $y(x,t)$ is the deflection of the string stretched between the fixed points (0, 0) and (l,0). Let $f(x)$ be the initial deflection and $g(x)$ the initial velocity of the string. The boundary conditions are given by

$$\left.\begin{aligned} y(0,t)&=0 \\ y(l,t)&=0 \end{aligned}\right\} \text{for all } t \geq 0$$

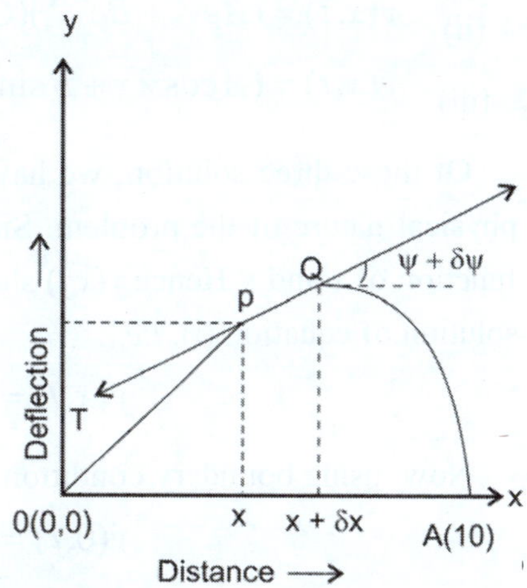

and initial conditions are given by

$$y(x,0)=f(x)$$

and $$\left(\frac{\partial y}{\partial t}\right)_{t=0}=g(x)$$

'irst method (solution by separation of variables)

Since $y(x, t)$ is a function of x and t then

$$y(x,t)=X(x)T(t)$$

then $$\frac{\partial^2 y}{\partial x^2}=X''T, \frac{\partial^2 y}{\partial t^2}=XT''$$

Substituting in (i), we get

$$XT''=c^2TX''$$

or $$\frac{X''}{X}=\frac{1}{c^2T}T''$$

Since x and t are independent variables. Thus we have three possibilities given by

$$\frac{X''}{X}=\frac{1}{c^2}\frac{T''}{T}=0 \text{ or } \lambda^2 \text{ or } -\lambda^2$$

Hence, in the three cases, we have

$$\frac{X''}{X}=0, \qquad \frac{1}{c^2}\frac{T''}{T}=0$$

$$X''-\lambda^2X=0, \qquad T'-c^2\lambda^2T=0$$

$$X''+\lambda^2X=0, \qquad T'+c^2\lambda^2T=0$$

The general solutions are given by

(a) $X=Ax+B, T=Ct+D$

(b) $X=Ae^{\lambda x}+Be^{-\lambda x}, T=Ce^{+c\lambda t}+De^{-c\lambda t}$

(c) $X=A\cos\lambda x+B\sin\lambda x, T=C\cos\lambda ct+D\sin\lambda ct$

Substituting in (ii), we have

(i) $y(x,t) = (Ax + B)(Ct + D)$

(ii) $y(x,t) = (Ae^{\lambda x} + Be^{-\lambda x})(Ce^{c\lambda t} + De^{-c\lambda t})$

(iii) $y(x,t) = (A\cos\lambda x + B\sin\lambda x)(C\cos c\lambda t + D\sin c\lambda t)$

Of these three solution, we have to choose that solution which is consistent with the physical nature of the problem. Since the problem is on variations, y must be a periodic function of x and t. Hence $y(x,t)$ should invalve trigonometric terms. Hence (iii) must be a solution of equation (i). *i.e.,*

$$y(x,t) = (A\cos\lambda x + B\sin\lambda x)(C\cos c\lambda t + D\sin c\lambda t) \quad \text{...(iii)}$$

Now, using boundary conditions, we

$$y(0,t) = A(C\cos c\lambda t + D\sin c\lambda t) = 0 \Rightarrow A = 0$$

and $$y(l,t) = B\sin\lambda l(C\cos c\lambda t + D\sin c\lambda t) = 0$$

which will be satisfied if $\sin\lambda l = 0$

or $$\sin\lambda l = \sin n\pi$$

$$\Rightarrow \quad \lambda l = n\pi$$

$$\Rightarrow \quad \lambda = \frac{n\pi}{l}; \quad n = 1, 2, 3, \ldots\ldots$$

Hence, $$y_n(x,t) = \left(C_n\cos\frac{n\pi ct}{l} + D_n\sin\frac{n\pi ct}{l}\right)\sin\frac{n\pi x}{l} \quad \text{...(iv)}$$

which is called eigen function and $\lambda_n = \dfrac{n\pi c}{\lambda}$ is called eigen values.

Now, using initial conditions, we get

$$y(x,0) = C_n\sin\frac{n\pi x}{l} = f(x)$$

and $$\left(\frac{\partial y}{\partial t}\right)_{t=0} = \frac{n\pi c}{l}D_n\sin\frac{n\pi x}{l} = g(x)$$

The sum of all solutions, we have

$$y(x,t)=\sum_{n=1}^{\infty}\left(C_n\cos\frac{n\pi ct}{l}+D_n\sin\frac{n\pi ct}{l}\right)\sin\frac{n\pi x}{l} \qquad \text{...(v)}$$

$$y(x,0)=\sum_{n=1}^{\infty}C_n\sin\frac{n\pi x}{l}=f(x)$$

and $$\left(\frac{\partial y}{\partial t}\right)_{t=0}=\sum_{n=1}^{\infty}\frac{n\pi c}{l}D_n\sin\frac{n\pi x}{l}=g(x)$$

The L.H.S. can be considered as a fourier sine expansion of R.H.S. Hence we have

$$C_n=\frac{2}{l}\int_0^l f(x)\sin\frac{n\pi x}{l}dx \qquad \text{...(vi)}$$

and $$\frac{n\pi c}{l}D_n=\frac{2}{l}\int_0^l g(x)\sin\frac{n\pi x}{l}dx \qquad \text{...(vii)}$$

Hence $y(x,t)$ is the solution of wave equation with the coefficient given by (vi) and (vii).

Particular cases

If the initial velocity $g(x)=0$

Then, $$\left(\frac{\partial y}{\partial t}\right)_{t=0}=0\Rightarrow D_n=0$$

Then (v) become $$y(x,t)=\sum_{n=1}^{\infty}C_n\cos\frac{n\pi ct}{l}\sin\frac{n\pi x}{l}$$

$$=\frac{1}{2}\sum_{n=1}^{\infty}C_n\cdot 2\cos\lambda_n t\sin\frac{n\pi x}{l},\lambda_n=\frac{n\pi c}{l}$$

$$=\frac{1}{2}\sum_{n=1}^{\infty}C_n\left[\sin\frac{n\pi}{l}(x-ct)+\sin\frac{n\pi}{l}(x+ct)\right]$$

$$=\frac{1}{2}\sum_{n=1}^{\infty}C_n\sin\frac{n\pi}{l}(x-ct)+\frac{1}{2}\sum_{n=1}^{\infty}\sin\frac{n\pi}{l}(x+ct)$$

The two series $\sum_{n=1}^{\infty} C_n \sin\frac{n\pi}{l}(x-ct)$ and $\sum_{n=1}^{\infty} C_n \sin\frac{n\pi}{l}(x+ct)$ are determined by replacing $(x - ct)$ and $(x + ct)$.

Hence, $$y(x,t)=\frac{1}{2}\left[F(x-ct)+F(x+ct)\right]$$

where F is an odd periodic expansion of f with the period $2l$.

Second method (De-Almbert's solution)

Consider the new independent variables

$$u=x+ct, v=x-ct$$

then, $$\frac{\partial y}{\partial x}=\frac{\partial y}{\partial u}+\frac{\partial y}{\partial v}$$

and $$\frac{\partial^2 y}{\partial x^2}=\frac{\partial}{\partial x}\left(\frac{\partial y}{\partial x}\right)$$

$$=\frac{\partial}{\partial x}\left(\frac{\partial y}{\partial u}+\frac{\partial y}{\partial v}\right)$$

$$=\left(\frac{\partial}{\partial u}+\frac{\partial}{\partial v}\right)\left(\frac{\partial y}{\partial u}+\frac{\partial y}{\partial v}\right)$$

$$=\frac{\partial^2 y}{\partial u^2}+2\frac{\partial^2 y}{\partial u\partial v}+\frac{\partial^2 y}{\partial v^2}$$

Similarly, $$\frac{\partial^2 y}{\partial t^2}=c^2\left(\frac{\partial^2 y}{\partial u^2}-2\frac{\partial^2 y}{\partial u\partial v}+\frac{\partial^2 y}{\partial v^2}\right)$$

Substituting these values in (i), we have

$$c^2\left(\frac{\partial^2 y}{\partial u^2}-2\frac{\partial^2 y}{\partial u\partial v}+\frac{\partial^2 y}{\partial v^2}\right)=c^2\left(\frac{\partial^2 y}{\partial u^2}+2\frac{\partial^2 y}{\partial u\partial v}+\frac{\partial^2 y}{\partial v^2}\right)$$

or $$\frac{\partial^2 y}{\partial u\partial v}=0$$

integrating *w.r.t.*, *v*, we get $\frac{\partial y}{\partial u}=f(u)$

Again, integrating *w.r.t., u,* we get

$$y = \int f(u)du + \phi(v)$$

or $$y = \psi(u) + \phi(v)$$

$$\left[\because \int f(u)\,du = \psi(u)\right]$$

where $\Psi(u)$ and $\Phi(v)$ are an arbitrary function of u and v.

$$\therefore \quad y(x,t) = \psi(x+ct) + \phi(x-ct) \quad \text{...(ii)}$$

Which is D'Alembert's solution of the wave equation.

Let the initial conditions $u(x,0) = f(x)$ and $\left(\dfrac{\partial u}{\partial t}\right)_{t=0} = 0$.

Differentiating (ii) *w.r.t. t,* we have

$$\frac{\partial y}{\partial t} = c\psi'(x+ct) - c\phi'(x-ct)$$

At $t = 0, y(x,0) = \psi(x) + \phi(x) = f(x)$ ...(iii)

and $$\left(\frac{\partial y}{\partial t}\right)_{t=0} = c\left[\psi'(x) - \phi'(x)\right] = 0$$

$$\Rightarrow \quad \psi'(x) = \phi'(x)$$

Integrating, we get $\psi(x) = \phi(x) + \lambda$ ...(iv)

Thus from (iii) and (iv), we get

$$2\phi(x) + \lambda = f(x)$$

or $$\phi(x) = \frac{1}{2}\left[f(x) - \lambda\right]$$

and $$\psi(x) = \frac{1}{2}\left[f(x) + \lambda\right]$$

Hence from (ii), we have

$$y(x,t) = \frac{1}{2}\left[f(x+ct) + f(x-ct)\right] \quad \text{...(v)}$$

This is called D'Alembert's solution of wave equation.

Example 6 A string is stretched between the fixed points *(0,0)* and *(1,0)* and released from rest from the position $y=A\ \sin\pi x$. Find the formula for its subsequent displacement $y(x,t)$.

Solution The vibration of the string is given by

$$\frac{\partial^2 y}{\partial t^2}=c^2\frac{\partial^2 y}{\partial x^2} \qquad \text{...(i)}$$

The boundary and initial conditions are

$$y(0,t)=0$$

$$y(1,t)=0$$

and $$y(x,0)=A\sin\pi x$$

$$\left(\frac{\partial y}{\partial t}\right)_{t=0}=0$$

The displacement $y(x,t)$ of the string is given by

$$y(x,t)=\sum_{n=1}^{\infty}C_n\cos(nc\pi t)\cdot\sin n\pi x \qquad \text{...(ii)}$$

where, $C_n=2\int_0^1 A\sin\pi x\cdot\sin n\pi x\,dx,\ n=1,2,3,....$

Here all the coefficient $C_2,C_3,.....$ vanish except C_1 because $\int_0^1 \sin m\pi x\sin n\pi x\ dx=0$ if $m\neq n$

Hence, $C_1=2A\int_0^1 \sin^2\pi x\ dx$

$$=A\int_0^1(1-\cos 2\pi x)dx$$

$$=A\left[x-\frac{\sin 2\pi x}{2\pi}\right]_0^1$$

$$=A[1-0]=A$$

Thus (ii) becomes

$$y(x,t)=C_1\cos c\pi t\cdot\sin\pi x+C_2\cos 2c\pi t\sin 2\pi x+.....$$

or $$y(x,t)=A\cos c\pi t\sin\pi x$$

Example 7 A string is stretched and fastened to two points l apart. Motion is started by displacing the string in the form $y = a \sin\frac{\pi x}{l}$ from which it is released at time $t = 0$. Shown that the displacing of any point at a distance x from one end at time t is given by

$$y(x,t) = a\sin\frac{\pi x}{l}\cos\frac{\pi ct}{l}$$

Solution The equation of string is

$$\frac{\partial^2 y}{\partial t^2} = c^2\frac{\partial^2 y}{\partial x^2} \quad \ldots(i)$$

Since the string stretched between two fixed points (0, 0) and (l, 0). Hence the displacement of the string at these points will be zero i.e., (i) $y(0, t) = 0$ (ii) $y(l, t) = 0$

Since the string is released from rest. Then its velocity will be zero.

∴ The initial conditions are

(iii) $y(x,0) = f(x) = a\sin\frac{\pi x}{l}$ (iv) $\frac{\partial y}{\partial t} = 0$ at $t = 0$

We know that the solution of wave equation if length of string be l and it is in released position

$$y(x,t) = \sum_{n=1}^{\infty} a_n \cos\left(\frac{n\pi ct}{l}\right)\sin\left(\frac{n\pi x}{l}\right) \quad \ldots(ii)$$

where $$a_n = \frac{2}{l}\int_0^l f(x)\sin\left(\frac{n\pi x}{l}\right)dx \quad \ldots(iii)$$

∴ $$a_n = \frac{2}{l}\int_0^l a\sin\left(\frac{\pi x}{l}\right)\sin\left(\frac{n\pi x}{l}\right)dx$$

$$= \frac{a}{l}\int_0^l \left[\cos\frac{(n-1)\pi x}{l} - \cos\frac{(n+1)\pi x}{l}\right]dx$$

$$= \frac{a}{l}\left[\frac{l}{(n-1)\pi}\sin\frac{(n-1)\pi x}{l} - \frac{l}{(n+1)\pi}\sin\frac{(n+1)\pi x}{l}\right]_0^l$$

$$= \frac{a}{l}\left[\left\{\frac{l}{(n-1)\pi}\sin(n\pi - \pi) - \frac{l}{(n+1)\pi}\sin(n\pi + \pi)\right\}\right]$$

$$= 0,\ n \neq 0 \quad \left[\begin{array}{l}\therefore \sin(n\pi - \pi) = 0 \\ \sin(n\pi + \pi) = 0\end{array}\right]$$

Now, for $n = 1$, we have

$$a_1 = \frac{2}{l}\int_0^l a\,\sin^2\left(\frac{\pi x}{l}\right)dx$$

$$= \frac{a}{l}\int_0^l \left[1 - \cos\left(\frac{2\pi x}{l}\right)\right]dx$$

$$= \frac{a}{l}\left[x - \frac{l}{2\pi}\sin\left(\frac{2\pi x}{l}\right)\right]_0^l$$

$$= \frac{a}{l}\left[l - \frac{l}{2\pi}\sin 2\pi\right]$$

$\therefore \quad a_1 = a$

Hence solution for n = 1 from equation (ii), we get

$$y(x,t) = a_1 \cos\left(\frac{\pi ct}{l}\right)\ \sin\left(\frac{\pi x}{l}\right)$$

Example 8 A tightly stretched string with fixed end points $x=0$ and $x=l$ is initially given a shape defined by $F(x) = \mu x(l-x)$, where μ is a constant and then released. Find the displacement of any point x of the string at any time $t > 0$.

Solution The equation of vibrating string is given by

$$\frac{\partial^2 y}{\partial t^2} = c^2 \frac{\partial^2 y}{\partial x^2} \qquad \text{...(i)}$$

The boundary condition are $y(0,t) = 0$

$y(l,t) = 0$

and the initial condition are $y(x,0) = \mu x(l - x)$

and $\quad \left(\dfrac{\partial y}{\partial t}\right)_{t=0} = 0$

The solution of equation (i) is of the form

$$y(x,t) = (A\cos\lambda x + B\sin\lambda x)(C\cos\lambda ct + D\sin\lambda ct) \qquad \text{...(ii)}$$

For the boundary condition

$$y(0,t) = A(C\cos\lambda ct + D\sin\lambda ct) = 0$$

$\Rightarrow \quad A = 0$

$\therefore \quad y(x,t) = B\sin\lambda x(C\cos\lambda ct + D\sin\lambda ct)$

when, $\quad y(l,t) = B\sin\lambda l(C\cos\lambda ct + D\sin\lambda ct) = 0$

which gives $\sin\lambda l = 0 = \sin n\pi$

$\Rightarrow \quad \lambda = \frac{n\pi}{l}$, n be an integer

Thus, $$y(x,t) = B\sin\frac{n\pi x}{l}\left(C\cos\frac{n\pi ct}{l} + D\sin\frac{n\pi ct}{l}\right) \qquad \ldots\text{(iii)}$$

Now, $$\frac{\partial y}{\partial t} = \left(B\sin\frac{n\pi x}{l}\right)\frac{n\pi c}{l}\left(-C\sin\frac{n\pi ct}{l} + D\cos\frac{n\pi ct}{l}\right)$$

For the initial condition,

$$\left(\frac{\partial y}{\partial t}\right)_{t=0} = \left(B\sin\frac{n\pi x}{l}\right)\frac{n\pi c}{l}\cdot D = 0$$

$\Rightarrow \quad D = 0$

Thus, from (iii), we have

$$y(x,t) = BC\sin\frac{n\pi x}{l}\cos\frac{n\pi ct}{l},$$

The general solution of (i) is $y(x,t) = \sum_{n=1}^{\infty} C_n\sin\frac{n\pi x}{l}\cdot\cos\frac{n\pi ct}{l} \qquad \ldots\text{(iv)}$

$(\because BC = C_n)$

For the initial condition, $y(x,0) = \sum_{n=1}^{\infty} C_n\sin\frac{n\pi x}{l} = \mu x(l-x)$

where, $$C_n = \frac{2}{l}\int_0^l \mu x(l-x)\sin\frac{n\pi x}{l}\,dx$$

$$= \frac{2\mu}{l}\left[\left\{-(lx-x^2)\frac{\cos\frac{n\pi x}{l}}{\frac{n\pi}{l}}\right\}_0^l - \int_0^l\left\{-(l-2x)\frac{\cos\frac{n\pi x}{l}}{\frac{n\pi}{l}}\right\}dx\right]$$

$$= \frac{2\mu}{l} \cdot \frac{l}{n\pi} \left\{ \int_0^l (l-2x) \cos \frac{n\pi x}{l}\, dx \right\}$$

$$= \frac{2\mu}{l} \left[\left\{ (l-2x) \frac{\sin \frac{n\pi x}{l}}{\frac{n\pi}{l}} \right\}_0^l + 2 \int_0^l \frac{\sin \frac{n\pi x}{l}}{\frac{n\pi}{l}}\, dx \right]$$

$$= \frac{2\mu}{n\pi} \cdot \frac{2l}{n\pi} \left\{ \frac{-\cos \frac{n\pi x}{l}}{\frac{n\pi}{l}} \right\}_0^l$$

$$= \frac{4\mu l^2}{n^3 \pi^3} \{1 - (-1)^n\}$$

Hence, the solution of (iv) is

$$y(x,t) = \frac{4\mu l^2}{\pi^3} \sum_{n=1}^{\infty} \frac{1-(-1)^n}{n^3} \sin \frac{n\pi x}{l} \cos \frac{n\pi ct}{l}$$

$$= \frac{8\mu l^2}{\pi^3} \sum_{\lambda=1}^{\infty} \frac{1}{(2\lambda-1)^3} \sin \frac{(2\lambda-1)\pi x}{\lambda} \cos \frac{(2\lambda-1)\pi ct}{l}$$

Example 9 A tightly stretched string of length l with fixed ends is initially in equilibrium position. It is set vibrating by giving each point a velocity $y_0 \sin^3 \frac{\pi x}{l}$. Find the displacement $y(x, t)$.

Solution The vibrating string equation is given by

$$\frac{\partial^2 y}{\partial t^2} = c^2 \frac{\partial^2 y}{\partial x^2} \quad \text{...(i)}$$

The boundary and initial conditions are $y(0,t) = 0,\ y(l,t) = 0$

and $y(x,0) = 0, \left(\frac{\partial y}{\partial t} \right)_{t=0} = y_0 \sin^3 \frac{\pi x}{l}$

The displacement $y(x,t)$ of the string is given by

$$y(x,t) = \sum_{n=1}^{\infty} \left(C_n \cos \frac{n\pi ct}{\lambda} + D_n \sin \frac{n\pi ct}{l} \right) \sin \frac{n\pi x}{l} \quad \text{...(ii)}$$

where $C_n = \frac{2}{l}\int_0^l f(x)\sin\frac{n\pi x}{l}\,dx = 0$

$$[\because f(x) = y(x,0) = 0]$$

and $\frac{n\pi c}{l} D_n = \frac{2}{l}\int_0^l g(x)\sin\frac{n\pi x}{l}\,dx$

$$= \frac{2}{l}\int_0^l y_0 \sin^3\frac{\pi x}{l}\sin\frac{n\pi x}{l}\,dx$$

$$D_n = \frac{2y_0}{n\pi c}\int_0^l \frac{1}{4}\left(3\sin\frac{\pi x}{l} - \sin\frac{3\pi x}{l}\right)\sin\frac{n\pi x}{l}\,dx$$

$$= \frac{y_0}{2n\pi c}\left[\int_0^l 3\sin\frac{\pi x}{l}\sin\frac{n\pi x}{l}\,dx - \int_0^l \sin\frac{3\pi x}{l}\sin\frac{n\pi x}{l}\,dx\right]$$

Here D_n vanishes for all n except when $n = 1$ and $n = 3$.

Now, $D_1 = \frac{3y_0}{2\pi c}\int_0^l \sin^2\frac{\pi x}{l}\,dx = \frac{3y_0}{2\pi c}\int_0^l\left(1 - \cos\frac{2\pi x}{l}\right)dx = \frac{3ly_0}{4\pi c}$

and $D_3 = \frac{-y_0}{6\pi c}\int_0^l \sin^2\frac{3\pi x}{l}\,dx = \frac{-y_0}{12\pi c}\int_0^l\left(1 - \cos\frac{6\pi x}{l}\right)dx = \frac{-ly_0}{12\pi c}$

Thus, the solution of (i) becomes

$$y(x,t) = D_1\sin\frac{\pi ct}{l}\sin\frac{\pi x}{l} + D_3\sin\frac{3\pi ct}{l}\sin\frac{3\pi x}{l}$$

$$= \frac{ly_0}{12\pi c}\left[9\sin\frac{\pi ct}{l}\sin\frac{\pi x}{l} - \sin\frac{3\pi ct}{l}\sin\frac{3\pi x}{l}\right]$$

***Solution of one dimensional heat flow equation*:** One dimensional heat flow equation is

$$\frac{\partial u}{\partial t} = c^2\frac{\partial^2 u}{\partial x^2} \qquad \text{...(i)}$$

Let the solution of (i) is of the form

$$u(x,t) = X(x)T(t)$$

$$\therefore \qquad \frac{\partial u}{\partial t} = XT', \frac{\partial^2 u}{\partial x^2} = X''T$$

Substituting in (i), we have

$$XT' = c^2 X''T$$

or
$$\frac{X''}{X} = \frac{1}{c^2}\frac{T'}{T}$$

Since L.H.S. is a function of x and R.H.S. is a function of t alone. Since x and t are independent variables. Then the two cannot be equal to each other unless both sides equal to a constant values. We have

$$\frac{X''}{X} = \frac{1}{c^2}\frac{T'}{T} = 0 \ or \ \lambda^2 \ or \ -\lambda^2$$

We have,
$$X'' = 0, T' = 0$$

$$X'' - \lambda^2 X = 0,\ T' - c^2\lambda^2 T = 0$$

$$X'' + \lambda^2 X = 0,\ T' + c^2\lambda^2 T = 0$$

The general solution of the above three cases are

(i) $X = Ax + B, T = C$

(ii) $X = Ae^{\lambda x} + Be^{-\lambda x},\ T = Ce^{\lambda^2 c^2 t}$

(iii) $X = A\cos\lambda x + B\sin\lambda x,\ T = Ce^{-\lambda^2 c^2 t}$

Thus, the possible solution of (i) becomes

(a) $u(x,t) = (Ax + B)C$

(b) $u(x,t) = (Ae^{\lambda x} + Be^{-\lambda x})Ce^{\lambda^2 c^2 t}$

(c) $u(x,t) = (A\cos\lambda x + B\sin\lambda x)Ce^{-\lambda^2 c^2 t}$

We are dealing with problems on heat conduction, it must be a transient solution *i.e.*, $u(x,t)$ is to decrease with the increase of time t. Hence $u(x,t) = (A\cos\lambda x + B\sin\lambda x)Ce^{-\lambda^2 c^2 t}$ is the only suitable solution of the heat equation.

Case I When both ends of the bar are kept at zero temperature

In this case, boundary conditions $u\ (0, t) = 0$

$$u\ (l, t) = 0$$

are initial conditions $u(x,0) = f(x) \qquad 0 < x < l$

Since temperature $u(x,t)$ decreases with time t. Therefore the solution is

$$u(x,t) = (A\cos\lambda x + B\sin\lambda x)e^{-\lambda^2 c^2 t} \qquad \text{...(i)}$$

Now, using boundary conditions

$$u(0,t) = Ae^{-\lambda^2 c^2 t} = 0 \Rightarrow A = 0$$

and $$u(l,t) = B\sin\lambda l e^{-\lambda^2 c^2 t} = 0$$

$\Rightarrow$ $$\sin\lambda l = 0$$

$\Rightarrow$ $$\lambda l = n\pi$$

$\Rightarrow$ $$\lambda = \frac{n\pi}{l}$$

$\therefore$ $$u_n(x,t) = B_n \sin\frac{n\pi x}{l} e^{\frac{-n^2\pi^2c^2t}{l^2}} \quad \text{...(ii)}$$

$\therefore$ $$u(x,t) = \sum_{n=1}^{\infty} B_n \sin\frac{n\pi x}{l} e^{\frac{-n^2\pi^2c^2t}{l^2}} \quad \text{...(iii)}$$

Now, using initial conditions,

We get $$u(x,0) = \sum_{n=1}^{\infty} B_n \sin\frac{n\pi x}{l} = f(x)$$

$\Rightarrow$ $$B_n = \frac{2}{l}\int_0^l f(x)\sin\frac{n\pi x}{l}dx \quad \text{...(iv)}$$

by fourier sine series. Thus equation (iii) is the solution of heat equation with coefficient (iv)

Case II. One face of the bar is kept at temperature u_0:

In this case, Boundary conditions

$$u(0,t) = 0$$

$$u(l,t) = u_0$$

and initial condition $$u(x,0) = v_0$$

Let the solution be $$u(x,t) = u_p(x) + u_t(x,t) \quad \text{...(v)}$$

When the end $x = 0$ of the rod is kept at zero temperature and the end $x = l$ at temperature u_0.

Then, $$u_p(x) = \frac{u_0}{l}x$$

$\therefore$ equation (v) becomes

$$u(x,t) = \frac{u_0}{l}x + u_t(x,t) \qquad \text{...(vi)}$$

using Boundary conditions $u(0,t) = u_t(0,t) = 0$

$$u(l,t) = u_t(0,t) = 0$$

Initial conditions $\quad u(x,0) = v_0 = \frac{u_0 x}{l} + u_t(x,0)$

$$\Rightarrow \quad u_t(x,0) = v_0 - \frac{u_0}{l}x \qquad \text{...(vii)}$$

Since $u_t(x,t)$ is a solution of the type (c), so

$$u_t(x,t) = (A'\cos\lambda x + B'\sin\lambda x)e^{-\lambda^2 c^2 t} \qquad \text{...(viii)}$$

for $\quad u_t(0, t), A' = 0$

for $\quad u_t(l,t), \sin\lambda l = 0 \Rightarrow \lambda = \frac{n\pi}{l}, n = 1, 2, 3, ...$

$$\therefore \quad u_t(x,t) = B_n \sin\frac{n\pi x}{l} e^{\frac{-n^2\pi^2 c^2 t}{l}}$$

Adding the solution for all n, the general solution is

$$u_t(x,t) = \sum_{n=1}^{\infty} B_n \sin\frac{n\pi x}{l} e^{\frac{-n^2\pi^2 c^2 t}{l}} \qquad \text{...(ix)}$$

at $t = 0$, $\quad v_0 - \frac{u_0}{l}x = \sum_{n=1}^{\infty} B_n \sin\frac{n\pi x}{l}$

By fourier sine series, we have

$$B_n = \frac{2}{l}\int_0^l \left(v_0 - \frac{u_0}{l}x\right)\sin\frac{n\pi x}{l}dx$$

$$= \frac{2}{n\pi}\left[v_0 - (-1)^n\left(v_0 - u_0\right)\right]$$

Hence equation (vi) becomes

$$u(x,t) = \frac{u_0}{l}x + \frac{2}{\pi}\sum_{n=1}^{\infty}\frac{1}{n}\left[v_0 - (-1)^n(v_0 - u_0)\right]\sin\frac{n\pi x}{l} e^{\frac{-n^2\pi^2 c^2 t}{l}} \qquad \text{...(x)}$$

Case III When the bar is infinitely long

In this case, no boundary condition is given. The initial condition is $u(x,0)=f(x)$.

The solution is $u(x,t)=\left(A\cos\lambda x+B\sin\lambda x\right)e^{-\lambda^2c^2t}$

Let A = $A(P)$ and B = $B(P)$ are periodic function of P, where λ = P and since heat equation is linear homogeneous, then

$$u(x,t)=\int_0^\infty\left[A(P)\cos Px+B(P)\sin Px\right]e^{-P^2c^2t}dP \qquad \text{...(xi)}$$

Now, using initial condition, we get

$$\int_0^\infty\left[A(P)\cos Px+B(P)\sin Px\right]dP=f(x)$$

So, $$A(P)=\frac{1}{\pi}\int_{-\infty}^\infty f(y)\cos Py\,dy$$

and $$B(P)=\frac{1}{\pi}\int_{-\infty}^\infty f(y)\sin Py\,dy$$

∴ $$u(x,0)=\frac{1}{\pi}\int_0^\infty\left[\int_{-\infty}^\infty f(y)\cos(Px-Py)\,dy\right]dP$$

Hence (xi) becomes,

$$u(x,t)=\frac{1}{\pi}\int_0^\infty\left[\int_{-\infty}^\infty f(y)\cos(Px-Py)e^{-c^2P^2t}\,dy\right]dP$$

$$=\frac{1}{\pi}\int_{-\infty}^\infty f(y)\left[\int_0^\infty e^{-c^2P^2t}\cos(x-y)P\,dP\right]dy \qquad \text{...(xii)}$$

Now, let $v=\int_0^\infty e^{-c^2P^2t}\cos(x-y)P\,dP$

Putting $cP\sqrt{t}=S\Rightarrow dP=\dfrac{dS}{c\sqrt{t}}$, we have

$$v=\frac{1}{c\sqrt{t}}\int_0^\infty e^{-S^2}\cos\frac{(x-y)}{2c\sqrt{t}}2S\,dS$$

$$=\frac{1}{c\sqrt{t}}\frac{\sqrt{\pi}}{2}e^{-\left(\frac{(x-y)}{2c\sqrt{t}}\right)^2}=\frac{\sqrt{\pi}}{2c\sqrt{t}}e^{\frac{-(x-y)^2}{4c^2t}}$$

$$\left[\because\int_0^\infty e^{-S^2}\cos 2asdS=\frac{\sqrt{\pi}e^{-a^2}}{2}\right]$$

Putting in equation (xii) we get

$$u(x,t) = \frac{1}{\pi}\int_{-\infty}^{\infty} f(y)\frac{\sqrt{\pi}}{2c\sqrt{t}}\, e^{\frac{-(x-y)^2}{4c^2t}}\, dy$$

$$= \frac{1}{2c\sqrt{\pi t}}\int_{-\infty}^{\infty} f(y)\, e^{\frac{-(x-y)^2}{4c^2t}}\, dy$$

which is the required solution

Case IV Bar with insulated ends: In this case, the boundary conditions are $u_x(0,t) = 0,\ u_x(l,t) = 0$

The initial conditions are $u(x,0) = f(x),\ 0 < x < l.$

Now assume that the solution of heat equation is

$$u(x,t) = (Ax + B)C = A'x + B',\ \text{(Say)}$$

$$\therefore \qquad u_x(x,t) = A'$$

using boundary conditions, we get $u_x(0,t) = A' = 0$

$\therefore$ Solution of heat equation is $u(x, t) = B'$ (constant) ...(i)

Again assume that the solution of heat equation is

$$u(x,t) = (C'\cos\lambda x + D'\sin\lambda x)e^{-\lambda^2c^2t} \qquad \text{...(ii)}$$

using boundary conditions, we get

$$u_x(0,t) = D'\lambda e^{-\lambda^2c^2t} = 0 \Rightarrow D' = 0$$

and

$$u_x(l,t) = \sin\lambda l = 0$$

$$\Rightarrow \qquad \lambda = \frac{n\pi}{l},\ n = 1, 2, 3, ...$$

Thus another solution of heat equation from (ii) is

$$u_n(x,t) = A_n\cos\left(\frac{n\pi x}{l}\right)e^{-\left(\frac{n^2\pi^2c^2}{l^2}\right)t} \qquad \text{...(iii)}$$

Hence the general solution is given by [from (i) and (iii)]

$$u(x,t) = B' + \sum_{n=1}^{\infty} A_n\cos\frac{n\pi x}{l}\cdot e^{-\left(\frac{n\pi c}{l}\right)^2 t} \qquad \text{...(iv)}$$

Applying initial condition, we get

$$u(x,0) = B' + \sum_{n=1}^{\infty} A_n \cos\frac{n\pi x}{l} = f(x)$$

where, $$A_n = \frac{2}{l}\int_0^l f(x)\cos\frac{n\pi x}{l}\,dx,$$

and $$B' = \frac{1}{l}\int_0^l f(x)\cos\frac{0\cdot\pi x}{l}\,dx \Rightarrow B' = \frac{1}{2}A_0$$

Hence the complete solution of one dimensional heat equation is given by

$$u(x,t) = \frac{1}{2}A_0 + \sum_{n=1}^{\infty} A_n \cos\frac{n\pi x}{l}\, e^{-\left(\frac{n\pi c}{l}\right)^2 t}$$

Example 10 Find the temperature $u(x, t)$ in a bar of silver (length 10cm, constant cross section of area 1cm^2 , density 10.6 gm/cm^3, thermal conductivity 1.04 cal/cm, deg. Sec. specific heat 0.056 cal/gm deg.) which is perfectly insulated laterally, whose ends are kept at temperature 0°C and whose initial temperature in $f(x)$, where

(i) $f(x) = \begin{cases} x, & 0 < x < 5 \\ 10 - x, & 5 < x < 10 \end{cases}$.

(ii) $f(x) = x(10 - x)$.

Solution The equation of one dimensional heat equation,

$$\frac{\partial u}{\partial t} = c^2 \frac{\partial^2 u}{\partial t^2}, \text{ where } c^2 = \frac{k}{\sigma\rho},$$

σ is the specific heat, k is thermal conductivity, ρ is the density

Now, the temperature $u(x, t)$ is given by

$$u(x,t) = \sum_{n=1}^{\infty} B_n \sin\frac{n\pi x}{l}\, e^{\frac{-n^2\pi^2c^2t}{l}}$$

where, $$B_n = \frac{2}{l}\int_0^l f(x)\sin\frac{n\pi x}{l}\,dx$$

Given, $l = 10\text{cm}$ and $c^2 = \dfrac{1.04}{0.056\times 10.6} = 1.752$

(i) $B_n = \frac{2}{10}\int_0^{10} f(x)\sin\frac{n\pi x}{l}dx$

$$= \frac{1}{5}\left[\int_0^5 x\sin\frac{n\pi x}{10}dx + \int_5^{10}(10-x)\sin\frac{n\pi x}{10}dx\right]$$

$$= \frac{40}{n^2\pi^2}\sin\frac{n\pi}{2}$$

$\therefore$ When n is even, $B_n = 0$

When $n = 1, 5, 9, \ldots\ldots, B_n = \frac{40}{n^2\pi^2}$

and when $n = 3, 7, 11, \ldots\ldots, B_n = \frac{-40}{n^2\pi^2}$

$\therefore$ Solution becomes

$$u(x,t) = \frac{40}{\pi^2}\left[\sin\frac{\pi x}{10}e^{-0.01752\pi^2 t} - \frac{1}{3^2}\sin\frac{3\pi x}{10}e^{-0.01752(3^2\pi^2)t} + \ldots\right]$$

(ii) $B_n = \frac{2}{10}\int_0^{10} x(10-x)\sin\frac{n\pi x}{10}dx$

$$= \frac{1}{5}\left[\left\{-x(10-x)\frac{10}{n\pi}\cos\frac{n\pi x}{10}\right\}_0^{10} + \int_0^{10}(10-2x)\frac{10}{n\pi}\cos\frac{n\pi x}{10}dx\right]$$

$$= \frac{1}{5}\cdot\frac{20}{n\pi}\int_0^{10}(5-x)\cos\frac{n\pi x}{10}dx$$

$$= \frac{400}{n^3\pi^3}(1-\cos n\pi)$$

When, n is even, $B_n = 0$

When n is odd, $B_n = \frac{800}{n^3\pi^3}$

Hence, solution becomes

$$u(x,t) = \frac{800}{\pi^3}\left[\frac{1}{1^3}\sin\frac{\pi x}{10}e^{-0.01752\pi^2 t} + \frac{1}{3^2}\sin\frac{3\pi x}{10}e^{-0.01752(3^2\pi^2)t} + \ldots\right]$$

Example 11 Solve $\frac{\partial u}{\partial t}=c^2\frac{\partial^2 u}{\partial x^2}$, $0<x<\pi,\ t>0.\ u_x(0,t)=0,\ u_x(\pi,t)=0$ and $u(x,0)=\sin x$.

Solution The temperature distribution of the given equation is

$$u(x,t)=\frac{1}{2}A_0+\sum_{n=1}^{\infty}A_n\cos\frac{n\pi x}{\pi}e^{-\left(\frac{n\pi c}{\pi}\right)^2 t}$$

$$=\frac{1}{2}A_0+\sum_{n=1}^{\infty}A_n\cos nx\ e^{-n^2c^2t} \quad \text{...(i)}$$

where, $A_n=\frac{2}{\pi}\int_0^{\pi}\sin x\cos nx\,dx$

$$=\frac{1}{\pi}\int_0^{\pi}\left[\sin(n+1)x-\sin(n-1)x\right]dx$$

$$=\frac{1}{\pi}\left[-\frac{\cos(n+1)\pi}{n+1}+\frac{\cos(n-1)\pi}{n-1}+\frac{1}{n+1}-\frac{1}{n-1}\right]$$

when n is odd, $A_n=0$

when n is even, let $n=2m$, then

$$A_n=\frac{2}{\pi}\left[\frac{1}{2m+1}-\frac{1}{2m-1}\right]=\frac{-4}{\pi(4m^2-1)}$$

and $A_0=\frac{2}{\pi}\int_0^{\pi}\sin x\,dx=\frac{4}{\pi}$

Hence the required solution of the given equation becomes

$$u(x,t)=\frac{2}{\pi}-\frac{4}{\pi}\sum_{m=1}^{\infty}(4m-1)^{-1}\cos 2mx\,e^{-4m^2c^2t}$$

***Two Dimensional wave equation (vibrating membrane)*:** The deflection $u(x,y,t)$ for a vibrating membrane is given by the two dimensional wave equation

$$\frac{\partial^2 u}{\partial t^2}=C^2\left(\frac{\partial^2 u}{\partial x^2}+\frac{\partial^2 u}{\partial y^2}\right) \quad \text{...(i)}$$

Solution of two dimensional wave equation (Rectangular Membrane):

The boundary conditions are

$$\left.\begin{array}{l} u(0,y,t)=0 \\ u(a,y,t)=0 \\ u(x,0,t)=0 \\ u(x,b,t)=0 \end{array}\right\} \text{for all } t \geq 0$$

and initial conditions are $u(x,y,0)=f(x,y)$

$$\left(\frac{\partial u}{\partial t}\right)_{t=0}=g(x,y)$$

Thus, assume that a solution of (i) is of the form

$$u(x,y,t)=X(x)Y(y)T(t)$$

Substituting its differential forms in (i), we have

$$\frac{1}{C^2T}T''=\frac{X''}{X}+\frac{Y''}{Y}$$

This can hold good if each number is a constant.

Therefore, three possibilities arise:

(i) $\quad \frac{T''}{C^2T}=0, \frac{X''}{X}=0, \frac{Y''}{Y}=0$

(ii) $\quad \frac{T''}{C^2T}=\lambda^2, \frac{X''}{X}=\lambda_1^2, \frac{Y''}{Y}=\lambda_2^2$, where $\lambda^2=\lambda_1^2+\lambda_2^2$

(iii) $\quad \frac{T''}{C^2T}=-\lambda^2, \frac{X''}{X}=-\lambda_1^2, \frac{Y''}{Y}=-\lambda_2^2$

The general solution becomes

(a) $\quad X=A_1x+B_1, Y=A_2y+B_2, T=A_3t+B_3$

(b) $\quad X=A_1e^{\lambda_1 x}+B_1e^{-\lambda_1 x}, Y=A_2e^{\lambda_2 y}+B_2e^{-\lambda_2 y}, T=A_3e^{C\lambda t}+B_3e^{-C\lambda t}$

(c)

$$X=A_1\cos\lambda_1 x+B_1\sin\lambda_1 x, Y=A_2\cos\lambda_2 y+B_2\sin\lambda_2 y, T=A_3\cos C\lambda t+B_3\sin C\lambda t$$

Since $u(x, y, t)$ contains trigonometric terms. Hence the solution becomes

$$y(x,y,t) = (A_1 \cos \lambda_1 x + B_1 \sin \lambda_1 x)(A_2 \cos \lambda_2 y + B_2 \sin \lambda_2 y)(A_3 \cos C\lambda t + B_3 \sin C\lambda t) \quad \ldots\text{(ii)}$$

Now, using boundary conditions, we have

$$u(0, y, t) = A_1 (A_2 \cos \lambda_2 y + B_2 \sin \lambda_2 y)(A_3 \cos \lambda Ct + B_3 \sin \lambda Ct) = 0$$

$$\Rightarrow \qquad A_1 = 0$$

and

$$u(a, y, t) = (0 + B_1 \sin \lambda_1 a)(A_2 \cos \lambda_2 y + B_2 \sin \lambda_2 y)(A_3 \cos \lambda Ct + B_3 \sin \lambda Ct = 0)$$

$$\Rightarrow \qquad \sin \lambda_1 a = 0 \Rightarrow \lambda_1 = \frac{m\pi}{a}, m = 1, 2, 3, \ldots..$$

Similarly, $A_2 = 0$ and $\lambda_2 = \frac{n\pi}{b}, n = 1, 2, 3, \ldots..$

Putting in (ii), we have

$$u_{mn}(x, y, t) = (A_{mn} \cos C\lambda_{mn} t + B_{mn} \sin C\lambda_{mn} t) \sin \frac{m\pi x}{a} \sin \frac{n\pi y}{b} \quad \ldots\text{(iii)}$$

where $$\lambda^2 = \lambda_{mn}^2 = \pi^2 \left(\frac{m^2}{a^2} + \frac{n^2}{b^2} \right)$$

Now, using initial conditions, we get

$$u(x, y, 0) = \sum_{m=1}^{\infty} \sum_{n=1}^{\infty} A_{mn} \sin \frac{m\pi x}{a} \sin \frac{n\pi y}{b} = f(x, y)$$

Which is a double fourier sine series of $f(x, y)$. Then

$$A_{mn} = \frac{2}{a} \cdot \frac{2}{b} \int_0^a \int_0^b f(x, y) \sin \frac{m\pi x}{a} \cdot \sin \frac{n\pi y}{b} \, dydx \quad \ldots\text{(iv)}$$

Differentiating (iii) w.r.t. t, we get

$$\left(\frac{\partial u}{\partial t} \right)_{t=0} = \sum_{m=1}^{\infty} \sum_{n=1}^{\infty} C\lambda_{mn} B_{mn} \sin \frac{m\pi x}{a} \sin \frac{n\pi y}{b} = g(x, y)$$

$$\therefore \qquad B_{mn} = \frac{4}{abc\lambda_{mn}} \int_0^a \int_0^b g(x, y) \sin \frac{m\pi x}{a} \sin \frac{n\pi y}{b} \, dydx \quad \ldots\text{(v)}$$

Hence the solution of wave equation is given by (iii)

Particular case

When $g(x,y)=0 \Rightarrow \left(\frac{\partial u}{\partial t}\right)_{t=0} = g(x,y)=0$

Then, $B_{mn}=0$

and hence $u(x,y,t)=\sum_{m=1}^{\infty}\sum_{n=1}^{\infty} A_{mn}\cos C\lambda_{mn}t\sin\frac{m\pi x}{a}\sin\frac{n\pi y}{b}$

where $A_{mn}=\frac{4}{ab}\int_0^a\int_0^b f(x,y)\sin\frac{m\pi x}{a}\sin\frac{n\pi y}{b}\,dydx$

Solution of two dimensional wave equation (circular membrane): The deflection $u(x,y,t)$ for a vibrating membrane is given by two dimensional wave equation

$$\frac{\partial^2 u}{\partial t^2}=C^2\left(\frac{\partial^2 u}{\partial x^2}+\frac{\partial^2 u}{\partial y^2}\right)$$

Changing it into Polar form by putting $x=r\cos\theta, y=r\sin\theta$, we have

$$\frac{\partial^2 u}{\partial t^2}=C^2\left(\frac{\partial^2 u}{\partial r^2}+\frac{1}{r}\frac{\partial u}{\partial r}+\frac{1}{r^2}\frac{\partial^2 u}{\partial \theta^2}\right)$$

For a radially symmetric membrane (u does not depend on θ), the above equation becomes

$$\frac{\partial^2 u}{\partial t^2}=C^2\left(\frac{\partial^2 u}{\partial r^2}+\frac{1}{r}\frac{\partial u}{\partial r}\right) \qquad \text{...(i)}$$

The boundary condition is $u(1,t)=0,$ for $t\geq 0$

The initial conditions are $u(r,0)=f(r)$

and $\left(\frac{\partial u}{\partial t}\right)_{t=0}=0$

Now, let the solution of (i) $u(r,t)=R(r)\cdot T(t)$ ...(ii)

Substituting its differential forms in (i), we have

$$\frac{1}{C^2}\frac{T''}{T}=\frac{1}{R}\left(R''+\frac{1}{r}R'\right)=-\lambda^2 \text{ (Say)}$$

which gives $$T''+\lambda^2C^2T=0 \quad \text{...(iii)}$$

and $$R''+\frac{1}{r}R'+\lambda^2R=0 \quad \text{...(iv)}$$

Putting $S=\lambda r$, then $R''+\frac{1}{S}R'+R=0$ *i.e.*, $\frac{\partial^2R}{\partial S^2}+\frac{1}{S}\frac{\partial R}{\partial S}+R=0$.

Which is Bessel's equation. Its general solution $R=aJ_0(S)+bY_0(S)$

Where J_0 and Y_0 are Bessel's function of the first and second kind of order zero. Since the deflection of the membrane is always finite. We must have $b = 0$ and taking $a = 1$, we get $R(r)=J_0(S)=J_0(\lambda r)$

On the boundary of the circular membrane, we must have $J_0(\lambda)=0, \lambda=k_m, m=1,2,...$ thus the solution of (iv) are $R(r)=J_0(k_mr)$, $m=1,2,.....$ and the solution of (iii) is $T(t)=A_m\cos C\lambda_mt+B_m\sin C\lambda_mt$

Hence the general solution of (i) is

$$u(r,t)=(A_m\cos C\lambda_mt+B_m\sin C\lambda_mt)J_0(k_mr)$$

For initial condition, $u(r,0)=\sum_{m=1}^{\infty}A_mJ_0(k_mr)=f(r)$

where $$A_m=\frac{2}{J_1^2(k_m)}\int_0^1 rf(r)J_0(k_mr)\,dr$$

and $$\left(\frac{\partial u}{\partial t}\right)_0=\sum_{m=1}^{\infty}C\lambda_mB_mJ_0(k_mr)=0$$

$\Rightarrow$ $$B_m=0$$

and hence the general solution of the wave equation (i) is given by

$$u(r,t)=\sum_{m=1}^{\infty}A_mJ_0(k_mr)\cos Ck_mt.$$

Example 12: Find the deflection $u(x, y, t)$ of the square membrane with $a = b = 1$ and $c = 1$ if the initial velocity is zero and the initial deflection is $f(x, y) = A \sin\pi x \sin 2\pi y$.

Solution: The vibration of the square membrane are governed by two dimensional wave equation

$$\frac{\partial^2 u}{\partial t^2} = C^2\left(\frac{\partial^2 u}{\partial x^2} + \frac{\partial^2 u}{\partial y^2}\right)$$

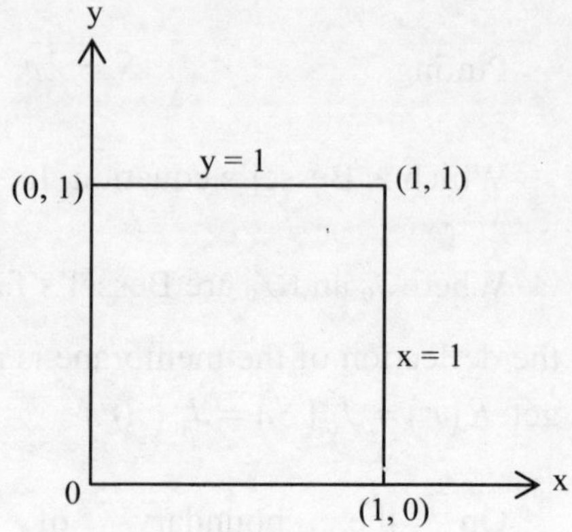

The boundary conditions are

$$u(0, y, t) = 0$$

$$u(1, y, t) = 0$$

$$u(x, 0, t) = 0$$

$$u(x, 1, t) = 0$$

and the initial conditions are

$$u(x, y, 0) = f(x, y) = A \sin \pi x \sin 2\pi y$$

$$\left(\frac{\partial u}{\partial t}\right)_{t=0} = 0$$

Let the solution of (i) becomes

$$u(x, y, t) = \sum_{n=1}^{\infty}\sum_{m=1}^{\infty} A_{mn} \cos \lambda_{mn} t \sin m\pi x \sin n\pi y \quad \text{...(i)}$$

where $\lambda_{mn}^2 = \pi^2 (m^2 + n^2)$

and $A_{mn} = 4\int_0^1\int_0^1 f(x, y) \sin m\pi x \sin n\pi y \, dxdy$

$$= 4A\int_0^1\int_0^1 \sin \pi x \sin 2\pi y \sin m\pi x \sin n\pi y \, dxdy$$

Obviously $A_{m1} = A_{m3} = \ldots\ldots = 0$

But, $A_{m2} = 4A\int_0^1\int_0^1 \sin \pi x \sin m\pi x \sin^2 2\pi y \, dxdy$

Now, $A_{mn} = A\int_0^1 2\sin \pi x \sin m\pi x \, dx \cdot \int_0^1 2\sin 2\pi y \sin n\pi y \, dy$

$$= A\int_0^1 \left[\cos \pi x(m-1) - \cos \pi x(m+1)\right] dx$$

$$\therefore \int_0^1 \left[\cos(n-2)\pi y - \cos(n+2)\pi y\right] dy$$

$$= A\left[\frac{\sin\pi(m-1)x}{\pi(m-1)} - \frac{\sin\pi(m+1)x}{\pi(m+1)}\right]_0^1 \cdot$$

$$\left[\frac{\sin\pi(n-2)y}{\pi(n-2)} - \frac{\sin\pi(n+2)y}{\pi(n+2)}\right]_0^1$$

$$= A\left[\frac{\sin\pi(m-1)}{\pi(m-1)} - \frac{\sin\pi(m+1)}{\pi(m+1)}\right] \cdot$$

$$\left[\frac{\sin\pi(n-2)}{\pi(n-2)} - \frac{\sin\pi(n+2)}{\pi(n+2)}\right]$$

$$= 0 \text{ if } m \neq 1, n \neq 2$$

Now, $$A_{12} = 4A\int_0^1\int_0^1 \sin^2 \pi x \sin^2 2\pi y \, dx dy$$

$$= A\int_0^1 2\sin^2 \pi x \, dx \int_0^1 2\sin^2 2\pi y \, dy$$

$$= A\int_0^1 (1-\cos 2\pi x) \, dx \int_0^1 (1-\cos 4\pi y) \, dy$$

$$= A$$

Hence the required solution of (i) is

$$u(x, y, t) = A_{12} \cos \lambda_{12} t \sin \pi x \sin 2\pi y$$

where $$\lambda_{12}^2 = \pi^2\left(1^2 + 2^2\right) = 5\pi^2 \Rightarrow \lambda_{12} = \pi\sqrt{5}$$

$\therefore$ $$u(x, y, t) = A\cos\left(\pi\sqrt{5}\right)t \sin \pi x \sin 2\pi y$$

Solution of Laplace Equation: The two dimensional heat equation is

$$\frac{\partial u}{\partial t} = C^2\left[\frac{\partial^2 u}{\partial x^2} + \frac{\partial^2 u}{\partial y^2}\right]$$

If the heat flow is steady, then $\frac{\partial u}{\partial t} = 0$, and the heat equation reduces to Laplace's equation:

$$\nabla^2 u = \frac{\partial^2 u}{\partial x^2} + \frac{\partial^2 u}{\partial y^2} = 0 \qquad \ldots\text{(i)}$$

Assume the solution of (i) is

$$u(x,y)=X(x)Y(y) \qquad \text{...(ii)}$$

Then $$\frac{\partial^2 u}{\partial x^2}=X''Y \text{ and } \frac{\partial^2 u}{\partial y^2}=X\,Y''$$

Putting the values in equation (i), we get

$$X''Y+X\,Y''=0$$

or $$\frac{X''}{X}=\frac{-Y''}{Y}=\lambda^2,\ 0,-\lambda^2 \text{ (let)} \qquad \text{...(iii)}$$

Solution of equation (iii) be

(i) when $\dfrac{X''}{X}=\lambda^2$ and $\dfrac{-Y''}{Y}=\lambda^2$

Then $X=C_1\,e^{\lambda x}+C_2\,e^{-\lambda x}$ and $Y=C_3\cos\lambda y+C_4\sin\lambda y$

(ii) when $\dfrac{X''}{X}=0$ and $\dfrac{Y''}{Y}=0$

Then $X=C_1\,x+C_2$ and $Y=C_3\,y+C_4$

(iii) when $\dfrac{X''}{X}=-\lambda^2$ and $\dfrac{-Y''}{Y}=-\lambda^2$

Then $X=C_1\,\cos\lambda x+C_2\,\sin\lambda x$ and $Y=C_3\,e^{\lambda y}+C_4\,e^{-\lambda y}$

Thus the various possible solutions of equation (i) are

$$u(x,y)=\left(C_1\,e^{\lambda x}+C_2\,e^{-\lambda x}\right)\left(C_3\cos\lambda y+C_4\sin\lambda y\right)$$

$$u(x,y)=\left(C_1\,x+C_2\right)\left(C_3 y+C_4\right)$$

$$u(x,y)=\left(C_1\,\cos\lambda x+C_2\sin\lambda x\right)\left(C_3\,e^{\lambda y}+C_4\,e^{-\lambda y}\right)$$

of these we take that solution which is consistent with the given boundary conditions

Example 13 Solve $$\frac{\partial^2 u}{\partial x^2}+\frac{\partial^2 u}{\partial y^2}=0 \text{ for } 0<x<\pi,\ 0<\text{y}<\pi$$

$$u(x,0)=x^2, u(x,\pi)=0,\ \frac{\partial u}{\partial x}(0,y)=\frac{\partial u}{\partial x}(\pi,y)=0$$

Solution: The given partial differential equation is

$$\frac{\partial^2 u}{\partial x^2} + \frac{\partial^2 u}{\partial y^2} = 0 \quad \text{...(i)}$$

Assume the solution of equation (i) be

$$u(x, y) = X(x)Y(y) \quad \text{...(ii)}$$

$$\therefore \quad \frac{\partial^2 u}{\partial x^2} = X''Y \quad \text{and} \quad \frac{\partial^2 u}{\partial y^2} = XY''$$

Putting the values, in (i), we get

$$X''Y + XY'' = 0$$

'or' $$\frac{X''}{X} = \frac{-Y''}{Y} = 0 \text{ or } -\lambda^2 (\text{let})$$

The general solution in the two cases are given by

(i) $X = Ax + B, \quad Y = (y + 1)$

(ii) $X = A \cos \lambda x + B \sin \lambda x, y = C \cos h \lambda Y + D \sin h \lambda y$

Now taking the solution by (i), we get

$$u (x, y) = (Ax + B) (Cy + D) \quad \text{...(iii)}$$

given $u (x, \pi) = 0$

then from (iii), we get

$$u (x, \pi) = (Ax + B) (C\pi + D) = 0 \Rightarrow D = - C\pi$$

Putting in (iii), we get

$$u (x, y) = (Ax + B) C (y - \pi)$$

$$u(x, y) = (A'x + B')(y - \pi), \quad (\text{where } AC = A', BC = B')$$

Then using condition $\frac{\partial u}{\partial x}(0, y) = 0$

$$\frac{\partial u}{\partial x}(0, y) = A'(y - \pi) = 0 \Rightarrow A' = 0$$

$\therefore$ One solution of (i) may be taken as

$$u(x, y) = B'(y - \pi) \quad \text{...(iv)}$$

Again taking the solutions given by (ii), we get

$$u(x,y) = (A\cos\lambda x + B\sin\lambda x)(C\cos h\,\lambda y + D\sin h\,\lambda y) \quad \text{...(v)}$$

given $u(x, \pi) = 0$ then from (v)

$$u(x,\pi) = (A\cos\lambda x + B\sin\lambda x)(C\cos h\,\lambda\pi + D\sin h\,\lambda\pi) = 0$$

$$\Rightarrow \quad D = -\frac{C\cos h\,\lambda\pi}{\sin h\,\lambda\pi}$$

then from (v), we get

$$u(x,y) = (A'\cos\lambda x + B'\sin\lambda x)\left(\cos h\lambda y - \frac{\cos h\lambda y}{\sin h\lambda\pi}\sin h\lambda y\right)$$

$$u(x,y) = (A'\cos\lambda x + B'\sin\lambda x)\frac{\sin h\lambda(\pi - y)}{\sin h\,\lambda\,\pi} \quad \text{...(vi)}$$

when $\frac{\partial u}{\partial x}(0,y) = 0$ then $B' = 0$

and then using $\frac{\partial u}{\partial x}(\pi,y) = 0$, we get

$$\frac{\partial u}{\partial x}(\pi,y) = -A'\lambda\sin\lambda\pi\;\frac{\sin h\lambda\,(\pi - y)}{\sin h\,\lambda\pi}$$

$$\Rightarrow \quad \sin\lambda\pi = 0 \Rightarrow \lambda\pi = n\pi$$

$$\Rightarrow \lambda = n,\ n = 1, 2, 3, \ldots$$

Thus for each value of n, the solution of (i) is

$$u(x,y) = A_n\cos nx\;\frac{\sin hn\,(\pi - y)}{\sin h\,n\pi}$$

Hence the complete solution of (i) is

$$u(x,y) = B'(y - \pi) + \sum_{n=1}^{\infty} A_n\cos nx\;\frac{\sin hn\,(\pi - y)}{\sin h\,n\pi} \quad \text{...(vii)}$$

Now, when $u(x, 0) = x^2$, then

$$x^2 = -B'\pi + \sum_{n=1}^{\infty} A_n\cos nx \quad \text{...(viii)}$$

Multiplying both sides by cos $n\,x$ and integrating between the limits 0 to π, we get

$$A_n = \frac{2}{\pi}\int_0^{\pi} x^2 \cos nx\, dx$$

$$= \frac{4\pi}{x^2\pi}\cos n\pi$$

$$= \frac{4}{n^2}(-1)^n$$

Again integrating equation (viii), we get

$$-B'\pi^2 = \int_0^{\pi} x^2 dx$$

$\therefore$ $$B' = \frac{-\pi}{3}$$

Hence the complete solution of (i) is

$$u(x,y) = \frac{\pi}{3}(\pi - y) + 4\sum_{n=1}^{\infty}\frac{(-1)^n \sin\, hn\,(\pi - y)\cos nx}{n^2 \sin\, h\, n\pi}$$

Laplace equation in three Dimensions: In the steady state, the three dimensional heat flow equation reduces to

$$\nabla^2 u = \frac{\partial^2 u}{\partial x^2} + \frac{\partial^2 u}{\partial y^2} + \frac{\partial^2 u}{\partial z^2} = 0$$

Which is the Laplace's equation in three dimensions.

Solution of three dimensional Laplace's equation:

(i) ***Cartesion form*** : The equation is $\frac{\partial^2 u}{\partial x^2} + \frac{\partial^2 u}{\partial y^2} + \frac{\partial^2 u}{\partial z^2} = 0$ ...(i)

Let $$u(x,y,z) = X(x)Y(y)Z(z) \quad \text{...(ii)}$$

Substituting (ii) in (i) and divided by XYZ, we have

$$\frac{1}{X}\frac{\partial^2 X}{\partial x^2} + \frac{1}{Y}\frac{\partial^2 Y}{\partial y^2} + \frac{1}{Z}\frac{\partial^2 Z}{\partial z^2} = 0 \quad \text{...(iii)}$$

which is of the form $\xi_1(x) + \xi_2(y) + \xi_3(z) = 0$

But x, y, z are independent variables. Then ξ_1, ξ_2, ξ_3 are constant functions. Let us assume that constants are $k^2, l^2 - (k^2 + l^2)$ respectively. Then equation (iii) becomes

$$\frac{d^2X}{dx^2} - k^2 X = 0, \frac{d^2Y}{dy^2} - l^2 Y = 0, \frac{d^2Z}{dz^2} + (k^2 + l^2) Z = 0$$

Their solutions are $X = Ae^{kx} + Be^{-kx}, Y = Ce^{ly} + De^{-ly}$,

$$Z = E\cos\sqrt{k^2 + l^2}z + F\sin\sqrt{k^2 + l^2}z$$

Substituting these values in (ii), we have

$$u(x, y, z) = \left(Ae^{kx} + Be^{-kx}\right)\left(Ce^{ly} + De^{-ly}\right)\left(E\cos\sqrt{k^2 + l^2}z + F\sin\sqrt{k^2 + l^2}z\right)$$

If we take three constants to be $-k^2, -l^2, k^2 + l^2$, then an alternative solution of (i) becomes

$$u(x, y, z) = (A\cos kx + B\sin kx)(C\cos ly + D\sin ly)\left(Ee^{\sqrt{k^2+l^2}z} + Fe^{-\sqrt{k^2+l^2}z}\right)$$

(ii) ***Cylindrical form***: The Laplace equation in cylindrical form is

$$\frac{\partial^2 u}{\partial r^2} + \frac{1}{r}\frac{\partial u}{\partial r} + \frac{1}{r^2}\frac{\partial^2 u}{\partial \theta^2} + \frac{\partial^2 u}{\partial z^2} = 0 \qquad \text{...(i)}$$

Let
$$u(r, \theta, z) = R(r)H(\theta)Z(z) \qquad \text{...(ii)}$$

be a solution of (i)

Substituting its differential form in (i) and divide by RHZ, we get

$$\frac{1}{R}\left(\frac{d^2R}{dr^2} + \frac{1}{r}\frac{dR}{dr}\right) + \frac{1}{r^2 H}\frac{d^2H}{d\theta^2} + \frac{1}{Z}\frac{d^2Z}{dz^2} = 0 \qquad \text{...(iii)}$$

Assuming that $\frac{d^2H}{d\theta^2} = -n^2 H$ and $\frac{d^2Z}{dz^2} = k^2 Z$ ...(iv)

Then equation (iii) becomes

$$\frac{1}{R}\left(\frac{d^2R}{dr^2} + \frac{1}{r}\frac{dR}{dr}\right) - \frac{n^2}{r^2} + k^2 = 0$$

or
$$r^2\frac{d^2R}{dr^2} + r\frac{dR}{dr} + (k^2r^2 - n^2)R = 0$$

which is Bessel's equation and its solution is

$$R = C_1 J_n(kr) + C_2 Y_n(kr)$$

Also the solution of (iv) are

$$H = C_3 \cos n\theta + C_4 \sin n\theta, Z = C_5 e^{kz} + C_6 e^{-kz}$$

Thus a solution of (i) is

$$u(r,\theta,z) = \left[C_1 J_n(kr) + C_2 Y_n(kr)\right]\left[C_3 \cos n\theta + C_4 \sin n\theta\right]\left[C_5 e^{kz} + C_6 e^{-kz}\right]$$

(iii) ***Spherical form***: The Laplace's equation in spherical form is given by

$$\frac{\partial^2 u}{\partial r^2} + \frac{2}{r}\frac{\partial u}{\partial r} + \frac{1}{r^2}\frac{\partial^2 u}{\partial \theta^2} + \frac{\cot\theta}{r^2}\frac{\partial u}{\partial \theta} + \frac{1}{r^2 \sin^2\theta}\frac{\partial^2 u}{\partial \phi^2} = 0 \quad \text{...(i)}$$

Let $u(r,\theta,\phi) = R(r)F(\theta)H(\phi)$ ...(ii)

Substituting its differential form in (i), we get

$$\frac{1}{R}\left(r^2 \frac{d^2 R}{dr^2} + 2r\frac{dR}{dr}\right) + \frac{1}{F}\left(\frac{d^2 F}{d\theta^2} + \cot\theta \frac{dF}{d\theta}\right) + \frac{1}{H \sin^2\theta}\frac{d^2 H}{d\phi^2} = 0$$

...(iii)

Putting $\frac{1}{R}\left(r^2 \frac{d^2 R}{dr^2} + 2r\frac{dR}{dr}\right) = n(n+1)$ ...(iv)

and $\frac{1}{H}\frac{d^2 H}{d\phi^2} = -m^2$ ...(v)

Then equation (iii) becomes

$$\frac{d^2 F}{d\theta^2} + \cot\theta \frac{dF}{d\theta} + \left[n(n+1) - m^2 \operatorname{cosec}^2\theta\right]F = 0 \quad \text{...(vi)}$$

Now differentiating the Legendre's equation

$$\left(1-\mu^2\right)y'' - 2\mu y' + n(n+1)y = 0$$

Differentiating m times w.r.t. μ and write $u = \frac{d^m y}{d\mu^m}$, we get

$$\left(1-\mu^2\right)u'' - 2(m+1)\mu u' + (n-m)(n+m+1)u = 0 \quad \text{...(vii)}$$

Now putting $F=\left(1-\mu^2\right)^{m/2} u$ in (vii), we get

$$\left(1-\mu^2\right)\frac{d^2F}{d\mu^2}-2\mu\frac{dF}{d\mu}+\left[n(n+1)-\frac{m^2}{1-\mu^2}\right]F=0 \quad \text{...(viii)}$$

Putting $\mu=\cos\theta$ in (viii), it reduces to (vi) and its solution is

$$F=C_1P_n^m\cos\theta+C_2Q_n^m\cos\theta$$

$\therefore$ The solution of (v) is $H=C_3\cos m\phi+C_4\sin m\phi$

From equation (iv), Put $R=r^k$, so that

$$k(k-1)+2k=n(n+1) \text{ which gives } k=n \text{ or } -(n+1)$$

Thus $$R=C_5r^n+C_6r^{-(n+1)}$$

Hence the general solution of (i) is

$$u(r,\theta,\phi)=\sum_{n=0}^{\infty}\sum_{m=0}^{\infty}\left[C_1P_n^m\cos\theta+C_2Q_n^m\cos\theta\right]\left[C_3\cos m\phi+C_4\sin m\phi\right]\left[C_5r^n+C_6r^{-(n+1)}\right]$$

Practice Problems

1. Solve the following equations by the method of separation of variables:

(i) $u\dfrac{\partial u}{\partial x}+\dfrac{\partial u}{\partial y}=3u$, given $u=3e^{-y}-e^{-5y}$ when $x=0$

[**Ans.** $u=3e^{x-y}-e^{2x-5y}$]

(ii) $py^3+qx^2=0$

[**Ans.** $z=Ce^{4kx^3}\cdot e^{-3ky^4}, c_1c_2=C, k=\dfrac{\lambda}{12}$]

(iii) $\dfrac{\partial^2 u}{\partial x^2}=h^2\dfrac{\partial u}{\partial t}, u(0,t)=u(l,t)=0$ and $u(x,0)=\sin\dfrac{\pi x}{l}$

[**Ans.** $u(x,t)=\sin\dfrac{\pi x}{l}e^{\frac{-\pi^2 t}{h^2l^2}}$]

2. A string is stretched between the fixed points (0,0) and (1,0) and released from rest from the position $y(x,0)=A\sin 2\pi x$. Find the displacement $y(x,t)$.

[**Ans.** $y(x,t)=A\cos 2\pi ct\sin 2\pi x$]

3. Find the deflection $y(x, t)$ of the vibrating string ($l = \pi$) and $c = 1$ corresponding to zero initial velocity and initial deflection $f(x)$ given by $f(x) = Ax(\pi - x)$.

$$\textbf{[Ans. } y(x,t) = \frac{8A}{\pi}\left[\cos t \sin x + \frac{1}{3^3}\cos 3t \sin 3x + \frac{1}{5^3}\cos 5t \sin 5x + ...\right]\textbf{]}$$

4. Solve the wave equation $\frac{\partial^2 y}{\partial t^2} = c^2\frac{\partial^2 y}{\partial x^2}$ subject to the condition $y(0,t) = 0,\ y(l,t) = 0, y(x,0) = f(x)$ and $\left(\frac{\partial y}{\partial t}\right)_{t=0} = 0$.

$$\textbf{[Ans. } y(x,t) = \sum_{n=1}^{\infty} C_n \sin\frac{n\pi x}{l}\cos\frac{n\pi ct}{l} \text{ where } C_n = \frac{2}{l}\int_0^l f(x)\sin\frac{n\pi x}{l}dx\ \textbf{]}$$

5. Find the temperature $u(x,t)$ in a bar of length l, which is perfectly insulated already whose ends are kept at temperature zero and initial temperature is

$$f(x) = \begin{cases} x, & 0 < x < \frac{l}{2} \\ l - x, & \frac{l}{2} < x < l \end{cases}$$

$$\textbf{[Ans. } u(x,t) = \frac{4l}{\pi^2}\left[\frac{1}{1^2}\sin\frac{\pi x}{l}e^{\frac{-c^2\pi^2 t}{l^2}} - \frac{1}{3^2}\sin\frac{3\pi x}{l}e^{\frac{-3^2\pi^2 c^2 t}{l^2}} + ...\right]\textbf{]}$$

6. A bar10cm long have the temperature at 30°C and 80°C until steady state prevails. The temperature of the ends are changed to 40°C and 60°C respectively. Find the temperature distribution in the bar at time t.

$$\textbf{[Ans. } u(x,t) = u_s(x) + u_t(x,t) = 40 + 2x - \frac{20}{\pi}\sum_{n=1}^{\infty}\left[\left(\frac{2(-1)^n + 1}{n}\right)\sin\frac{n\pi x}{10}e^{\frac{-c^2 n^2\pi^2 t}{100}}\right]\textbf{]}$$

7. Solve the equation $\frac{\partial^2 v}{\partial x^2} = \frac{\partial v}{\partial t}$ given that $v = 0$ when $t = \infty$ and when $x = 0$ and when $x = l$.

$$\textbf{[Ans. } u(x,t) = \sum_{n=1}^{\infty} B_n \sin\frac{n\pi x}{l}e^{\frac{-n^2\pi^2 t}{l^2}}\ \textbf{]}$$

8. A rod of length l with insulated sides is initially at a uniform temperature u_0. Its ends are suddenly cooled to 0°C and are kept at that temperature. Find the temperature function $u(x, t)$.

$$\textbf{[Ans. } u(x,t) = \frac{4u_0}{\pi}\sum_{m=1}^{\infty}\frac{1}{2m-1}\sin\frac{(2m-1)\pi x}{l}e^{\frac{-c^2(2m-1)^2\pi^2 t}{l^2}}, n = 2m-1]$$

9. A homogeneous rod of conducting material of length 100cm has its ends kept at zero temperature and the initially temperature is

$$u(x,0) = \begin{cases} x, & 0 \le x \le 50 \\ 100 - x, & 50 \le x \le 100 \end{cases}$$

Find the temperature $u(x, t)$ at any time.

$$\textbf{[Ans. } u(x,t) = 25 + \sum_{n=\text{even}}^{\infty}\left(\frac{-400}{n^2\pi^2}\right)\cos\frac{n\pi x}{100}e^{\frac{-c^2n^2\pi^2 t}{10000}}\ ]$$

Chapter - 5

Numerical Solution of Partial Differential Equations

Classification of Second order partial differential equation: The general partial differential equations of the second order in two independent variables x and y is of the form

$$A\frac{\partial^2 u}{\partial x^2}+B\frac{\partial^2 u}{\partial x\partial y}+C\frac{\partial^2 u}{\partial y^2}+F\left(x,y,u,\frac{\partial u}{\partial x},\frac{\partial u}{\partial y}\right)=0 \quad .....(i)$$

The equation (i) is said to be

(i) elliptic if $B^2-4AC<0$,

(ii) Parabolic if $B^2-4AC=0$.

(iii) Hyperbolic if $B^2-4AC>0$.

Example 1 *Classify the following equations.*

(i) $$\frac{\partial^2 u}{\partial x^2}+2\frac{\partial^2 u}{\partial x\partial y}+3\frac{\partial^2 u}{\partial y^2}=0$$

(ii) $$x^2\frac{\partial^2 u}{\partial x^2}+(1-y^2)\frac{\partial^2 u}{\partial y^2}=0,-\infty<x<\infty,-1<y<1$$

(iii) $$x\frac{\partial^2 u}{\partial x\partial y}-y\frac{\partial^2 u}{\partial y^2}=\frac{\partial u}{\partial y}$$

(iv) $$\frac{\partial^2 u}{\partial x^2}+u\frac{\partial^2 u}{\partial x\partial y}+u\frac{\partial^2 u}{\partial y^2}-\frac{\partial u}{\partial x}+2\frac{\partial u}{\partial y}=0$$

(v) $$(1+x^2)\frac{\partial^2 u}{\partial x^2}+(5+2x^2)\frac{\partial^2 u}{\partial t\partial x}+(4+x^2)\frac{\partial^2 u}{\partial t^2}=0$$

Solution Comparing the given equations with general form of partial differential equation

$$A\frac{\partial^2 u}{\partial x^2}+B\frac{\partial^2 u}{\partial x \partial y}+C\frac{\partial^2 u}{\partial y^2}+F\left(x, y, u, \frac{\partial u}{\partial x}, \frac{\partial u}{\partial y}\right)=0$$

We have,

(i) $A=1, B=2, C=3$

$\therefore \quad B^2-4AC=4-12=-8<0$

Hence the given equation is elliptic.

(ii) Here, $A=x^2$, $B=0$, $C=1-y^2$

$\therefore \quad B^2-4AC=0-4x^2(1-y^2)$

$\qquad\qquad = 4x^2(y^2-1)$

For all x between $-\infty$ and ∞, x^2 is positive

For all y between -1 and 1, $y^2<1$

$\therefore \quad B^2-4AC=x^2$

Hence the equation is elliptic.

(iii) Here, $A=0$, $B=x$, $C=-y$

$\therefore \quad B^2-4AC=x^2$

for all x between $-\infty$ to ∞, x^2 is positive

$\therefore \quad B^2-4AC>0$

Hence, the equation is Hyperbolic,

(iv) Here, $A=1$, $B=4$, $C=4$

$\therefore \quad B^2-4AC=16-16=0$

Hence, the equation is parabolic.

(v) Here, $A=1+x^2$, $B=5+2x^2$, $C=4+x^2$

$\therefore \quad B^2-4AC=(5+2x^2)^2-4(1+x^2)(4+x^2)$

$\qquad\qquad = 25+4x^4+20x^2-16-4x^4-20x^2$

$\qquad\qquad = 9>0$

Hence, the equation is hyperbolic.

Finite difference approximations to partial derivatives : Let us consider a rectangular region R in xy-plane. We divide this region R into a small rectangular regions of sides $\Delta x = h$ and $\Delta y = k$.

The points of intersection of the lines are called mesh point (nodal points or grid points). (fig.1)

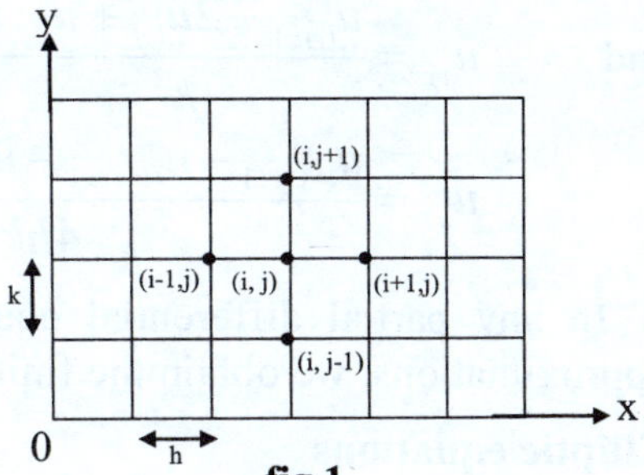

fig.1

Then finite difference approximation for the partial derivatives in x-direction.

$$\frac{\partial u}{\partial x} = \frac{u(x+h, y) - u(x, y)}{h},$$

(neglecting higher terms)

Which is forward difference approximation.

$$\frac{\partial u}{\partial x} = \frac{u(x, y) - u(x-h, y)}{h}$$

Which is back ward difference approximation and the central difference approximation is

$$\frac{\partial u}{\partial x} = \frac{u(x+h, y) - u(x-h, y)}{2h}$$

and $$\frac{\partial^2 u}{\partial x^2} = \frac{u(x-h, y) - 2u(x, y) + u(x+h, y)}{h^2}$$

If (x, y) is any mesh point then we can write $u(x, y) = u(ih, jk)$ or $u_{i,j}$ the above equations may be written as

$$u_x = \frac{u_{i+1,j} - u_{i,j}}{h}$$

or $$= \frac{u_{i,j} - u_{i-1,j}}{h}$$

or $$= \frac{u_{i+1,j} - u_{i-1,j}}{2h}$$

and $$u_{xx} = \frac{u_{i-1,j} - 2u_{i,j} + u_{i+1,j}}{h^2}$$

Similarly, in y-direction,

$$u_y = \frac{u_{i,j+1} - u_{i,j}}{k}$$

or $$= \frac{u_{i,j} - u_{i,j-1}}{k}$$

or $$= \frac{u_{i,j+1} - u_{i,j-1}}{2k}$$

and $$u_{yy} = \frac{u_{i,j-1} - 2u_{i,j} + u_{i,j+1}}{k^2}$$

$$u_{xy} = \frac{u_{i+1,j+1} - u_{i+1,j-1} - u_{i-1,j+1} + u_{i-1,j-1}}{4hk}$$

In any partial differential equation, we substitute corresponding above difference approximations, we obtain the finite difference analogues of the given equation.

Elliptic equations

The examples of elliptic partial differential equations are

(i) Laplace's equation $\nabla^2 u = \frac{\partial^2 u}{\partial x^2} + \frac{\partial^2 u}{\partial y^2} = 0$

(ii) Poisson's equation $\frac{\partial^2 u}{\partial x^2} + \frac{\partial^2 u}{\partial y^2} = f(x, y)$

Solution of Laplace Equation

$$\nabla^2 u = \frac{\partial^2 u}{\partial x^2} + \frac{\partial^2 u}{\partial y^2} = 0 \qquad \text{.....(i)}$$

The solution of $u(x, y)$ of Laplace equation is satisfied at every point of the rectangular region R, which is known as boundary conditions.

We divide this region into a square mesh of side h in xy-plane. (for square mesh $h = k$).

Substituting the values of derivatives of (i) by their above difference approximations,

We have

$$\frac{1}{h^2}\left[u_{i-1,j} - 2u_{i,j} + u_{i+1,j}\right] + \frac{1}{h^2}\left[u_{i,j-1} - 2u_{i,j} + u_{i,j+1}\right] = 0$$

or $$u_{i,j} = \frac{1}{4}\left[u_{i-1,j} + u_{i+1,j} + u_{i,j-1} + u_{i,j+1}\right] \qquad \text{.....(ii)}$$

The values of $u_{i,j}$ is the average of its values at four neighboring points to the left, right, below and above. The equation (ii) is said to be standard 5- point formula. It is also called Liebman's averaging procedure. (fig.2) Some times a formula similar to equation (ii) is used

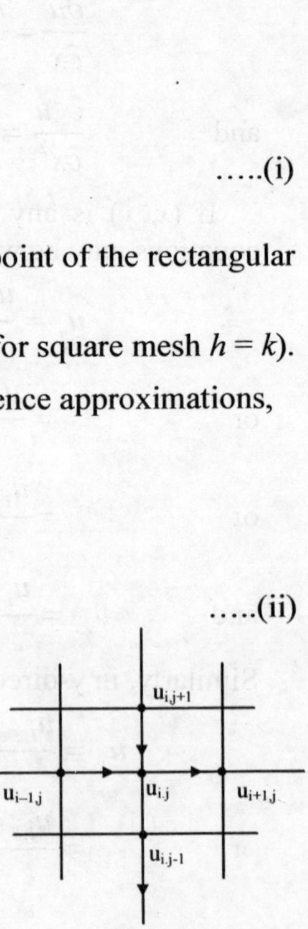

fig.2

which is given by

$$u_{i,j} = \frac{1}{4}\left[u_{i-1,j+1} + u_{i+1,j-1} + u_{i+1,j+1} + u_{i-1,j-1}\right] \quad \ldots\text{(iii)}$$

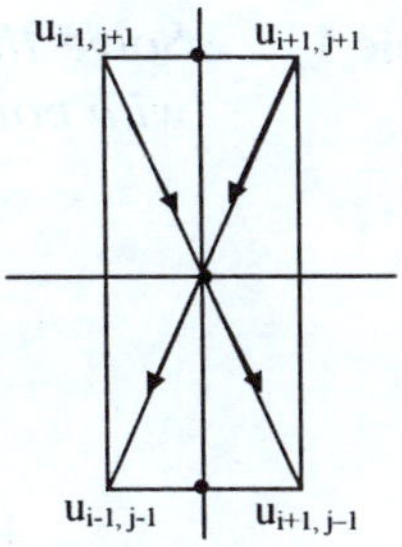

fig.3

The formula (iii) is called the diagonal 5-point formula. (fig.3) which shows the values of $u_{i,j}$ is the average of its values at the four neighboring diagonal mesh point. Although equation (iii) is less accurate than (ii).

Procedure

Let the given boundary values are $a_1, a_2, \ldots\ldots\ldots\ldots\ldots a_{16}$. Now we compute the $u(x, y)$ at the interior mesh point of the region. Since the value u_5 at the center and u_1, u_3, u_7, u_9 are computed by using diagonal five point formula. Thus we have,

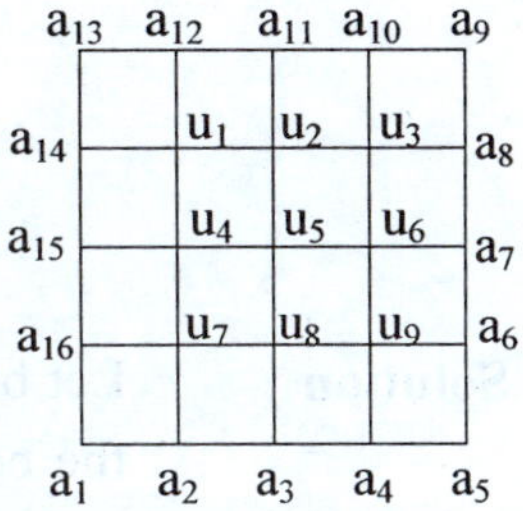

$$u_5 = \frac{1}{4}(a_1 + a_5 + a_9 + a_{13})$$

$$u_1 = \frac{1}{4}(a_{15} + u_5 + a_{11} + a_{13}),\; u_3 = \frac{1}{4}(u_5 + a_7 + a_9 + a_{11})$$

$$u_7 = \frac{1}{4}(a_1 + a_3 + u_5 + a_{15}),\; u_9 = \frac{1}{4}(a_3 + a_5 + a_7 + u_5)$$

The values at the remaining interior points *i.e.*, values of u_2, u_4, u_6, u_8 are computed by standard five point formula. Thus, we have

$$u_2 = \frac{1}{4}(u_1 + u_3 + u_5 + a_{11}),\; u_4 = \frac{1}{4}(u_1 + u_5 + u_7 + a_{15})$$

$$u_6 = \frac{1}{4}(u_3 + u_5 + u_9 + a_7),\; u_8 = \frac{1}{4}(u_5 + u_7 + a_3 + u_9)$$

The accuracy of $u_1, u_2, \ldots\ldots\ldots\ldots u_9$ are improved by any one of the following iterative methods:

(i) Jacobi's iteration method :

$$u_{i,j}^{(n+1)} = \frac{1}{4}\left[u_{i-1,j}^{(n)} + u_{i+1,j}^{(n)} + u_{i,j+1}^{(n)} + u_{i,j-1}^{(n)}\right] \quad \ldots\text{(iv)}$$

Where $u_{i,j}^{(n)}$ is the nth iterative value of $u_{i,j}$.

(ii) Gauss seidel interation method :

$$u_{i,j}^{(n+1)} = \frac{1}{4}\left[u_{i-1,j}^{(n+1)} + u_{i+1,j}^{(n)} + u_{i,j+1}^{(n+1)} + u_{i,j-1}^{(n)}\right] \quad \ldots\text{(v)}$$

Example 1 *Solve the elliptic equation $u_{xx} + u_{yy} = 0$ for the following square mesh with boundary values as shown.*

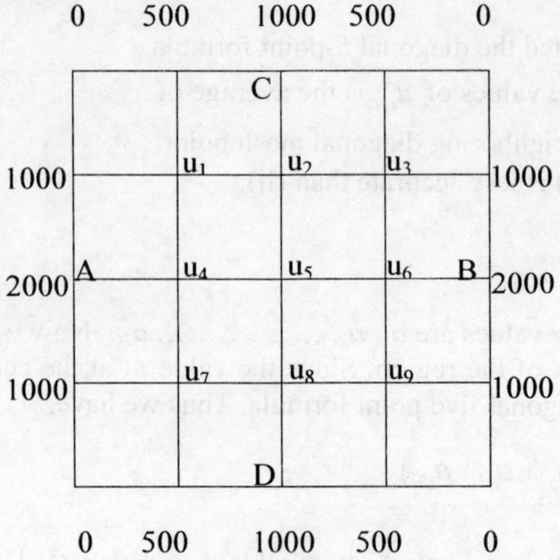

Solution Let be $u_1, u_2, \ldots\ldots\ldots\ldots u_9$ the values of u at the interior mesh points. Since the boundary values of u are symmetrical about the horizontal (*AB*) and vertical (*CD*) lines. Therefore,

$$u_1 = u_3 = u_7 = u_9,\ u_2 = u_8,\ u_4 = u_6$$

Now, we determine their initial values in the following order.

By standard formula, $u_5 = \frac{1}{4}(2000 + 2000 + 1000 + 1000) = 1500$

By diagonal formula, $u_1 = \frac{1}{4}(0 + u_5 + 1000 + 2000) = 1125$

$$u_1 = u_3 = u_7 = u_9 = 1125$$

By standard formula, $u_2 = \frac{1}{4}(u_1 + u_3 + 1000 + u_5) = 1188$

$$u_2 = u_8 = 1188$$

$$u_4 = \frac{1}{4}(2000 + u_5 + u_1 + u_7) = 1438, u_4 = u_6 = 1438$$

Now, we carry out the Gauss-seidel formula for successive iterations are

$$u_1^{(n+1)} = \frac{1}{4}\left(1000 + u_2^{(n)} + 500 + u_4^{(n)}\right)$$

$$u_2^{(n+1)} = \frac{1}{4}\left(u_1^{(n+1)} + u_3^{(n)} + 1000 + u_5^{(n)}\right)$$

$$u_4^{(n+1)} = \frac{1}{4}\left(2000 + u_5^{(n)} + u_1^{(n+1)} + u_7^{(n)}\right)$$

$$u_5^{(n+1)} = \frac{1}{4}\left(u_4^{(n+1)} + u_6^{(n)} + u_2^{(n+1)} + u_8^{(n)}\right)$$

Ist iteration : Putting $n = 0$,

$$u_1^{(1)} = \frac{1}{4}(1000 + 1188 + 500 + 1438) = 1032$$

$$u_2^{(1)} = \frac{1}{4}(1032 + 1125 + 1000 + 1500) = 1164,$$

$$u_4^{(1)} = \frac{1}{4}(2000 + 1500 + 1032 + 1125) = 1414,$$

$$u_5^{(1)} = \frac{1}{4}(1414 + 1438 + 1164 + 1188) = 1301,$$

IIst iteration : Putting $n = 1$,

$$u_1^{(2)} = \frac{1}{4}(1000 + 1164 + 500 + 1414) = 1020,$$

$$u_2^{(2)} = \frac{1}{4}(1020 + 1032 + 1000 + 1301) = 1088,$$

$$u_4^{(2)} = \frac{1}{4}(2000 + 1301 + 1020 + 1032) = 1338,$$

$$u_5^{(2)} = \frac{1}{4}(1338 + 1414 + 1088 + 1164) = 1251,$$

IIIrd iteration : Putting $n = 2$,

$$u_1^{(3)} = \frac{1}{4}(1000 + 1088 + 500 + 1338) = 982,$$

$$u_2^{(3)} = \frac{1}{4}(982 + 1020 + 1000 + 1251) = 1063,$$

$$u_4^{(3)} = \frac{1}{4}(2000 + 1251 + 982 + 1020) = 1313,$$

$$u_5^{(3)} = \frac{1}{4}(1313 + 1338 + 1063 + 1088) = 1201,$$

Similarly, $u_1^{(4)} = 969, u_2^{(4)} = 1038, u_4^{(4)} = 1288, u_5^{(4)} = 1176$

$$u_1^{(5)} = 957, u_2^{(5)} = 1026, u_4^{(5)} = 1276, u_5^{(5)} = 1157$$

$$u_1^{(6)} = 951, u_2^{(6)} = 1016, u_4^{(6)} = 1266, u_5^{(6)} = 1146$$

$$u_1^{(7)} = 946, u_2^{(7)} = 1011, u_4^{(7)} = 1260, u_5^{(7)} = 1138$$

$$u_1^{(8)} = 943, u_2^{(8)} = 1007, u_4^{(8)} = 1257, u_5^{(8)} = 1134$$

$$u_1^{(9)} = 941, u_2^{(9)} = 1005, u_4^{(9)} = 1255, u_5^{(9)} = 1131$$

$$u_1^{(10)} = 940, u_2^{(10)} = 1003, u_4^{(10)} = 1253, u_5^{(10)} = 1129$$

$$u_1^{(11)} = 939, u_2^{(11)} = 1002, u_4^{(11)} = 1252, u_5^{(11)} = 1128$$

$$u_1^{(12)} = 939, u_2^{(12)} = 1001, u_4^{(12)} = 1251, u_5^{(12)} = 1126$$

The values obtained in the 11th and 12th iteralions.

Hence, $u_1 = 939, u_2 = 1001, u_4 = 1251, u_5 = 1126$

Example 2 *Solve the elliptic equation* $\dfrac{\partial^2 u}{\partial x^2} + \dfrac{\partial^2 u}{\partial y^2} = 0$ *for the square mesh with boundary values as shown in figure.*

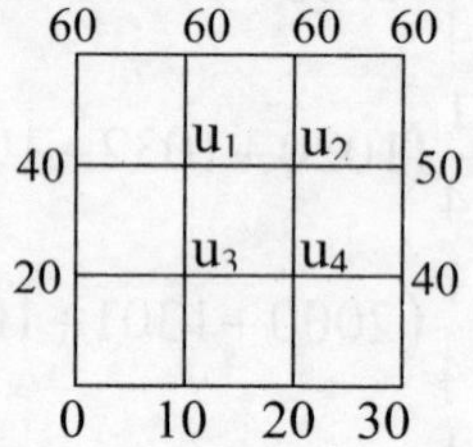

Solution We shall determine the initial values of u_1, u_2, u_3, u_4. Suppose $u_4 = 0$, then by diagonal formula,

$$u_1 = \frac{1}{4}(60 + 0 + 60 + 20) = 35,$$

by standard formula,

$$u_2 = \frac{1}{4}(35 + 50 + 60 + 0) = 36.25,$$

$$u_3 = \frac{1}{4}(20 + 0 + 35 + 10) = 16.25,$$

$$u_4 = \frac{1}{4}(16.25 + 40 + 36.25 + 20) = 28.125,$$

Now, we shall find successive iterations by Gauss-seidel approximation.

$$u_1^{(n+1)} = \frac{1}{4}\left(40 + u_2^{(n)} + 60 + u_3^{(n)}\right)$$

$$u_2^{(n+1)} = \frac{1}{4}\left(u_1^{(n+1)} + 50 + 60 + u_4^{(n)}\right)$$

$$u_3^{(n+1)} = \frac{1}{4}\left(20 + u_4^{(n)} + u_1^{(n+1)} + 10\right)$$

$$u_4^{(n+1)} = \frac{1}{4}\left(u_3^{(n+1)} + 40 + u_2^{(n+1)} + 20\right)$$

Ist iteration : Putting $n = 0$,

$$u_1^{(1)} = \frac{1}{4}(40 + 36.25 + 60 + 16.25) = 38.125$$

$$u_2^{(1)} = \frac{1}{4}(38.125 + 50 + 60 + 28.125) = 44.062,$$

$$u_3^{(1)} = \frac{1}{4}(20 + 28.125 + 38.125 + 10) = 24.0625,$$

$$u_4^{(1)} = \frac{1}{4}(24.0625 + 40 + 44.0625 + 20) = 32.0313,$$

IIst iteration : Putting $n = 1$,

$$u_1^{(2)} = \frac{1}{4}(40 + 40.0625 + 60 + 24.0625) = 42.0313$$

$$u_2^{(2)} = \frac{1}{4}(42.0313 + 50 + 60 + 32.0313) = 46.015,$$

$$u_3^{(2)} = \frac{1}{4}(20 + 32.0313 + 42.0313 + 10) = 26.0157,$$

$$u_4^{(2)} = \frac{1}{4}(26.0157 + 40 + 46.0157 + 20) = 33.0079,$$

IIIrd iteration : Putting $n = 2$,

$$u_1^{(3)} = \frac{1}{4}(40 + 46.0157 + 60 + 26.0157) = 43.0079,$$

$$u_2^{(3)} = \frac{1}{4}(43.0079 + 50 + 60 + 33.0079) = 46.5040,$$

$$u_3^{(3)} = \frac{1}{4}(20 + 33.0079 + 43.0079 + 10) = 26.5040,$$

$$u_4^{(3)} = \frac{1}{4}(26.5040+40+46.5040+20) = 33.2520,$$

IVth iteration : Putting $n = 3$,

$$u_1^{(4)} = \frac{1}{4}(40+46.5040+60+26.5040) = 43.2520,$$

$$u_2^{(4)} = \frac{1}{4}(43.2520+50+60+33.2520) = 46.6260,$$

$$u_3^{(4)} = \frac{1}{4}(20+33.2520+32.2520+10) = 26.6260,$$

$$u_4^{(4)} = \frac{1}{4}(26.6260+40+46.6260+20) = 33.3130,$$

Vth iteration : Putting $n = 4$,

$$u_1^{(5)} = \frac{1}{4}(40+46.6260+60+26.6260) = 43.3130,$$

$$u_2^{(5)} = \frac{1}{4}(50+33.3130+43.3130+60) = 46.6565,$$

$$u_3^{(5)} = \frac{1}{4}(20+33.3130+43.3130+10) = 26.6565,$$

$$u_3^{(5)} = \frac{1}{4}(26.6565+40+46.6565+20) = 33.3283,$$

Since, there is very small difference between 4th and 5th iteration.

Hence, $\quad u_1 = 43.31, u_2 = 46.66, u_3 = 26.66, u_4 = 33.33$

Example 3 *Solve the elliptic equation $u_{xx} + u_{yy} = 0$ for the following square mesh with boundary values as shown:*

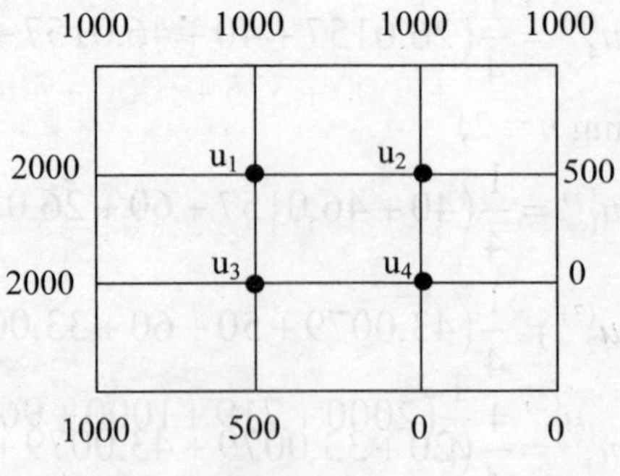

Solution To get the initial values of u_1, u_2, u_3, u_4 we assume that $u_4 = 0$ then by diagonal formula

$$u_1 = \frac{1}{4}[1000 + 0 + 1000 + 2000] = 1000$$

by standard formula

$$u_2 = \frac{1}{4}[1000 + 500 + 1000 + 0] = 625$$

$$u_3 = \frac{1}{4}[2000 + 0 + 1000 + 500] = 875$$

$$u_4 = \frac{1}{4}[875 + 0 + 625 + 0] = 375$$

First method (by Jacobi's method)

$$u_1^{(n+1)} = \frac{1}{4}\left[2000 + u_2^{(n)} + 1000 + u_3^{(n)}\right]$$

$$u_2^{(n+1)} = \frac{1}{4}\left[u_1^{(n)} + 500 + 1000 + u_4^{(n)}\right]$$

$$u_3^{(n+1)} = \frac{1}{4}\left[2000 + u_4^{(n)} + u_1^{(n)} + 500\right]$$

$$u_4^{(n+1)} = \frac{1}{4}\left[u_3^{(n)} + 0 + u_2^{(n)} + 0\right]$$

First iteration : (Put $n = 0$)

$$u_1^{(1)} = \frac{1}{4}(2000 + 625 + 1000 + 875) = 1125$$

$$u_2^{(1)} = \frac{1}{4}(1000 + 500 + 1000 + 375) = 719$$

$$u_3^{(1)} = \frac{1}{4}(2000 + 375 + 1000 + 500) = 969$$

$$u_4^{(1)} = \frac{1}{4}(875 + 0 + 625 + 0) = 375$$

Second iteration : (Put $n = 1$)

$$u_1^{(2)} = \frac{1}{4}(2000 + 719 + 1000 + 969) = 1172$$

$$u_2^{(2)} = \frac{1}{4}(1125 + 500 + 1000 + 375) = 750$$

$$u_3^{(2)} = \frac{1}{4}(2000 + 375 + 1125 + 500) = 1000$$

$$u_4^{(2)} = \frac{1}{4}(969 + 0 + 719 + 0) = 422$$

Similarly $u_1^{(3)} = 1188,\quad u_2^{(3)} = 774,\quad u_3^{(3)} = 1024,\quad u_4^{(3)} = 438$

$$u_1^{(4)} = 1200,\quad u_2^{(4)} = 782,\quad u_3^{(4)} = 1032,\quad u_4^{(4)} = 450$$

$$u_1^{(5)} = 1204,\quad u_2^{(5)} = 788,\quad u_3^{(5)} = 1038,\quad u_4^{(5)} = 454$$

$$u_1^{(6)} = 1206,\quad u_2^{(6)} = 790,\quad u_3^{(6)} = 1040,\quad u_4^{(6)} = 456.5$$

$$u_1^{(7)} = 1208,\quad u_2^{(7)} = 791,\quad u_3^{(7)} = 1041,\quad u_4^{(7)} = 458$$

$$u_1^{(8)} = 1208,\quad u_2^{(8)} = 791.5,\quad u_3^{(8)} = 1041.5,\quad u_4^{(8)} = 458$$

Hence $u_1 = 1208,\quad u_2 = 792,\quad u_3 = 1042,\quad u_4 = 458$

Second method (by Gauss seidal formula)

$$u_1^{(n+1)} = \frac{1}{4}\left[2000 + u_2^{(n)} + 1000 + u_3^{(n)}\right]$$

$$u_2^{(n+1)} = \frac{1}{4}\left[u_1^{(n+1)} + 500 + 1000 + u_4^{(n)}\right]$$

$$u_3^{(n+1)} = \frac{1}{4}\left[2000 + u_4^{(n)} + u_1^{(n+1)} + 500\right]$$

$$u_4^{(n+1)} = \frac{1}{4}\left[u_3^{(n+1)} + 0 + u_2^{(n+1)} + 0\right]$$

First iteration: (Put $n = 0$)

$$u_1^{(1)} = \frac{1}{4}[2000 + 625 + 1000 + 875] = 1125$$

$$u_2^{(1)} = \frac{1}{4}[1125 + 500 + 1000 + 375] = 750$$

$$u_3^{(1)} = \frac{1}{4}[2000 + 375 + 1125 + 500] = 1000$$

$$u_4^{(1)} = \frac{1}{4}[1000 + 0 + 750 + 0] = 438$$

Second iteration: (Put $n = 1$)

$$u_1^{(2)} = \frac{1}{4}[2000 + 750 + 1000 + 1000] = 1188$$

$$u_2^{(2)} = \frac{1}{4}[1188 + 500 + 1000 + 438] = 782$$

$$u_3^{(2)} = \frac{1}{4}[2000 + 438 + 1188 + 500] = 1032$$

$$u_4^{(2)} = \frac{1}{4}[1032 + 0 + 782 + 0] = 454$$

Similarly $\quad u_1^{(3)} = 1204,\ u_2^{(3)} = 789,\quad u_3^{(3)} = 1040,\quad u_4^{(3)} = 458$

$$u_1^{(4)} = 1207,\ u_2^{(4)} = 791,\quad u_3^{(4)} = 1041,\quad u_4^{(4)} = 458$$

$$u_1^{(5)} = 1208,\ u_2^{(5)} = 791.5,\ u_3^{(5)} = 1041.5,\ u_4^{(5)} = 458.25$$

Hence $\quad u_1 = 1208,\quad u_2 = 792,\quad u_3 = 1042,\quad u_4 = 458$

Solution of Poisson's equation

$$\nabla^2 u = \frac{\partial^2 u}{\partial x^2} + \frac{\partial^2 u}{\partial y^2} = f(x, y) \qquad \text{...(i)}$$

Its method of solution is similar to that of the Laplace equation. The standard 5- point formula for (i) is given by

$$u_{i-1,j} + u_{i+1,j} + u_{i,j+1} + u_{i,j-1} - 4u_{i,j} = h^2 f(ih, jh) \qquad \text{.....(ii)}$$

We apply (ii) at each mesh point and obtain linear equations which can be solved by Gauss-seidel method.

Example 4 *Solve the equation* $\nabla^2 u = -10\,(x^2 + y^2 + 10)$ *over the square with sides* $x = 0 = y,\ x = 3 = y$ *with* $u = 0$ *on the boundary and mesh length* $= 1$.

Solution Here, mesh length $h = 1$.

$\therefore$ The standard five – point formula for the given equation is

$$u_{i-1,j} + u_{i+1,j} + u_{i,j+1} + u_{i,j-1} - 4u_{i,j} = -10(i^2 + j^2 + 10) \quad \ldots(i)$$

for u_1 ($i = 1, j = 2$), equation (i) gives

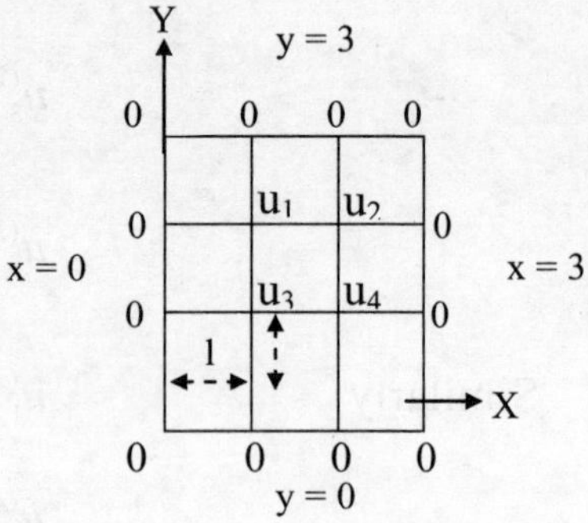

$$0 + u_2 + 0 + u_3 - 4u_1 = -10\,(1 + 4 + 10)$$

$$\therefore\ u_1 = \frac{1}{4}(u_2 + u_3 + 150) \quad \ldots.(ii)$$

for u_2 ($i = 2, j = 2$), equation (i) gives

$$\therefore\ u_2 = \frac{1}{4}(u_1 + u_4 + 180) \quad \ldots(iii)$$

for u_3 ($i = 1, j = 1$), equation (i) gives

$$\therefore\ u_3 = \frac{1}{4}(u_1 + u_4 + 120) \quad \ldots(iv)$$

for u_4 ($i = 2, j = 1$), equation (i) gives

$$\therefore\ u_4 = \frac{1}{4}(u_2 + u_3 + 150) \quad \ldots(v)$$

Equation (ii) and (v) are similar. It shows $u_1 = u_4$.

Thus the above equations reduces to

$$u_1 = \frac{1}{4}(u_2 + u_3 + 150)$$

$$u_2 = \frac{1}{4}(2u_1 + 180)$$

$$u_3 = \frac{1}{4}(2u_1 + 120)$$

Now, for solving these equations, we apply Gauss – seidel iteration method.

Ist iteration : Putting $u_2 = u_3 = 0$,

$$u_1^{(1)} = \frac{1}{4}(0+0+150) = 37.5$$

$$u_2^{(1)} = \frac{1}{4}(2\times 37.5+180) = 63.75,$$

$$u_3^{(1)} = \frac{1}{4}(2\times 37.5+120) = 48.75,$$

IInd iteration :

$$u_1^{(2)} = \frac{1}{4}(63.75+48.75+150) = 65.63$$

$$u_2^{(2)} = \frac{1}{4}(2\times 65.63+180) = 77.82,$$

$$u_3^{(2)} = \frac{1}{4}(2\times 65.63+120) = 62.82,$$

IIIrd iteration :

$$u_1^{(3)} = \frac{1}{4}(77.82+62.82+150) = 74.66$$

$$u_2^{(3)} = \frac{1}{4}(2\times 74.66+180) = 81.33,$$

$$u_3^{(3)} = \frac{1}{4}(2\times 74.66+120) = 66.33,$$

IVth iteration :

$$u_1^{(4)} = \frac{1}{4}(81.33+66.33+150) = 74.42$$

$$u_2^{(4)} = \frac{1}{4}(2\times 74.42+180) = 82.21,$$

$$u_3^{(4)} = \frac{1}{4}(2\times 74.42+120) = 67.21,$$

Vth iteration :

$$u_1^{(5)} = \frac{1}{4}(28.21+67.21+150) = 74.86$$

$$u_2^{(5)} = \frac{1}{4}(2\times 74.86+180) = 82.43,$$

$$u_3^{(5)} = \frac{1}{4}(2\times 74.86+120) = 67.43,$$

VIth iteration :

$$u_1^{(6)} = \frac{1}{4}(82.43 + 67.43 + 150) = 74.97$$

$$u_2^{(6)} = \frac{1}{4}(2 \times 74.97 + 180) = 82.48,$$

$$u_3^{(6)} = \frac{1}{4}(2 \times 74.97 + 120) = 67.49,$$

VIIth iteration :

$$u_1^{(7)} = \frac{1}{4}(82.48 + 67.49 + 150) = 74.99$$

$$u_2^{(7)} = \frac{1}{4}(2 \times 74.99 + 180) = 82.50,$$

$$u_3^{(7)} = \frac{1}{4}(2 \times 74.99 + 120) = 67.50,$$

VIIIst iteration :

$$u_1^{(8)} = \frac{1}{4}(82.50 + 67.50 + 150) = 75$$

$$u_2^{(8)} = \frac{1}{4}(2 \times 75 + 180) = 82.50,$$

$$u_3^{(8)} = \frac{1}{4}(2 \times 75 + 120) = 67.50,$$

Since iteration 7th and 8th are same.

Hence, $$u_1 = 75, u_2 = 82.5, u_3 = 67.5, u_4 = 75$$

Example 5 *Solve the partial differential equation $u_{xx} + u_{yy} = 8\,x^2\,y^2$ for the square mesh of the following figure with u (x, y) = 0 on the boundary and mesh length = 1.*

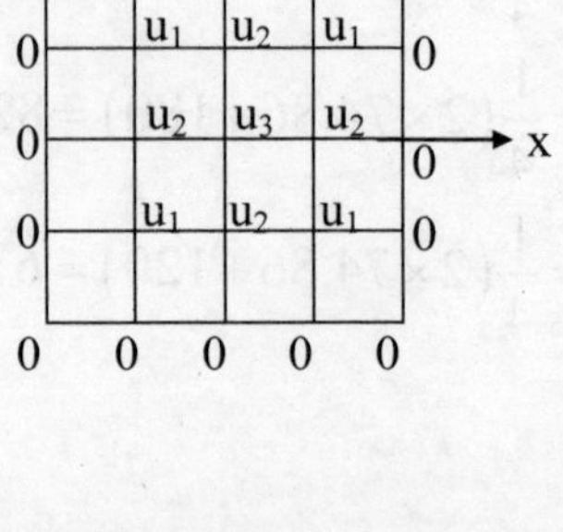

Solution Here, mesh length = 1 *i.e.,* h = 1

The standard five – point formula is

$$u_{i-1,j} + u_{i+1,j} + u_{i,j+1} + u_{i,j-1} - 4u_{i,j} = 8\, i^2 j^2 \quad \text{.....(i)}$$

In the given figure, u_3 is the center and so for u_3, $i = 0, j = 0$

Now, for u_1 ($i = -1, j = 1$) equation (i) gives

$$0 + u_2 + u_2 + 0 - 4u_1 = 8\,(-1)^2\,(1)^2$$

$$\Rightarrow \quad 4u_1 = 2u_2 - 8 \quad \text{.....(ii)}$$

for u_2 ($i = 0, j = 1$) equation (i) gives

$$u_1 + u_1 + u_3 + 0 - 4u_2 = 8\,(0)^2\,(1)^2$$

$$\Rightarrow \quad 4u_2 = 2u_1 + u_3 \quad \text{.....(iii)}$$

for u_3 ($i = 0, j = 0$) equation (i) gives

$$u_2 + u_2 + u_2 + u_2 - 4u_3 = 8\,(0)^2\,(0)^2$$

$$\Rightarrow \quad 4u_3 = 4u_2$$

$$\Rightarrow \quad u_3 = u_2 \quad \text{.....(iv)}$$

by solving (ii), (iii) and (iv), we have

$$u_1 = -3,\ u_2 = u_3 = -2$$

Solution of elliptic by relaxation method

Consider the Laplace equation $\nabla^2 u = \dfrac{\partial^2 u}{\partial x^2} + \dfrac{\partial^2 u}{\partial y^2} = 0$ ...(i)

We divide it into a square mesh size h. Let the value of u at A be u_0 and its values at the four adjacent point u_1, u_2, u_3, u_4. Then

$$\frac{\partial^2 u}{\partial x^2} = \frac{u_1 + u_3 - 2u_0}{h^2}$$

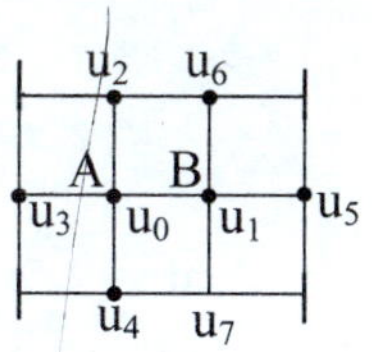

and $$\frac{\partial^2 u}{\partial y^2} = \frac{u_2 + u_4 - 2u_0}{h^2}$$

If equation (i) is satisfied at A, then

$$\frac{u_1 + u_3 - 2u_0}{h^2} + \frac{u_2 + u_4 - 2u_0}{h^2} = 0$$

or $$u_1 + u_2 + u_3 + u_4 - 4u_0 = 0$$

If r_0 be the residual at the mesh point A, then

$$r_0 = u_1 + u_2 + u_3 + u_4 - 4u_0 \quad \text{...(ii)}$$

Similarly, the residual at point B is,

$$r_1 = u_0 + u_5 + u_6 + u_7 - 4u_1 \text{ and so on} \qquad \ldots\text{(iii)}$$

The relaxation method is to reduce all the residuals to zero by making then as small as possible step by step.

If u_0 is given an increment 1, then (a) equation (ii) shows that r_0 is changed by –4 and (b) equation (iii) shows that r_1 is changed by 1.

Working Procedure

(1) We find the initial values of u at the interior mesh points by standard and diagonal methods.

(2) Calculate the residuals at each of these points by (ii).

(3) Write the residuals at a mesh point on the right of this point and the value of u on its left side.

(4) Obtain the solution by reducing the residuals to zero, one by one. At each step, we reduce the largest residual to zero and write the increment of u on the left and the modified residual on the right.

(5) When relaxation is completed, the value of u and increments are added at each point. If some residuals are large, liquidate these again.

(6) Stop the relaxation process, when the present values of the residuals are very small. The solution will be the present value of u at each of the nodes.

Example 6 *Solve by relaxation method, the laplace equation* $\dfrac{\partial^2 u}{\partial x^2} + \dfrac{\partial^2 u}{\partial y^2} = 0$ *inside the square boundary by the lines x = 0, x = 4, y = 0, y = 4, given that* $u = x^2 y^2$ *on the boundary.*

Solution Taking mesh length $h = 1$, the boundary form $u = x^2 y^2$.

at $x = 1, y = 4, u = 16$

at $x = 2, y = 4, u = 64$

at $x = 3, y = 4, u = 144$

at $x = 4, y = 4, u = 256$

The initial values of u at the nine mesh points are estimated as :

by standard formula,

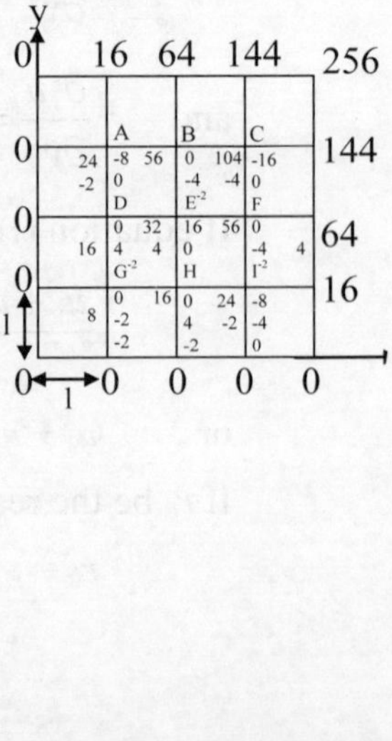

$$E = \frac{1}{4}\left(0 + 64 + 64 + 0\right) = 32,$$

by diagonal formula,

$$A = \frac{1}{4}(0 + 32 + 64 + 0) = 24,$$

$$C = \frac{1}{4}(64 + 64 + 32 + 256) = 104,$$

$$G = \frac{1}{4}(0 + 0 + 32 + 0) = 8,$$

$$I = \frac{1}{4}(32 + 0 + 64 + 0) = 24,$$

by standard formula,

$$B = \frac{1}{4}(24 + 104 + 64 + 32) = 56,$$

$$D = \frac{1}{4}(0 + 32 + 24 + 8) = 16,$$

$$F = \frac{1}{4}(32 + 64 + 104 + 24) = 56,$$

$$H = \frac{1}{4}(8 + 24 + 32 + 0) = 16,$$

Now, residual at A,

$r_A = 0 + 56 + 16 + 16 - 4 \times 24 = -8$ [from equ. (ii)]

Similarly, $r_B = 0$, $r_C = -16$, $r_D = 0$, $r_E = 16$, $r_F = 0$, $r_G = 0$, $r_H = 0$, $r_I = -8$.

(i) The largest residual is 16 at E. To liquidate it, we increase u by 4, so the residual becomes zero and the residuals at neighboring nodes get increased by 4.

(ii) The next largest residual is –16 at C. We increase u by –4, so the residual becomes zero and neighboring nodes are increased by –4.

(iii) Now, the largest residual is –8 at A and I. We increase u by –2 So the residual becomes zero and neighboring nodes are increased by –2.

(iv) Now, we stop the relaxation process.

Hence the final values of u are

$u_A = 22$, $u_B = 56$, $u_C = 100$, $u_D = 16$, $u_E = 36$, $u_F = 56$, $u_G = 8$, $u_H = 16$, $u_I = 22$.

Example 7 *Given the values of u on the boundary of the square in the figure. Evaluate by relaxation method, the function u (x, y) satisfying the laplace equation at the*

$\nabla^2 u = 0$ at the pivotal points.

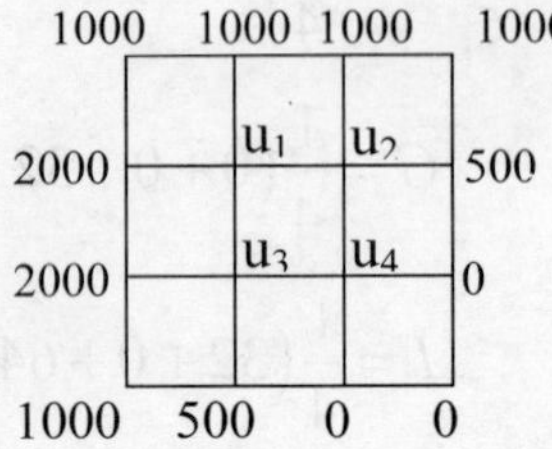

Solution The initial values of u_1, u_2, u_3, u_4 are estimated as:

by standard formula, $u_1 = \frac{1}{4}(2000 + 500 + 1000 + 500) = 1000,$

$$u_2 = \frac{1}{4}(1000 + 0 + 1000 + 500) = 625,$$

$$u_3 = \frac{1}{4}(2000 + 0 + 1000 + 500) = 875,$$

$$u_4 = \frac{1}{4}(875 + 0 + 625 + 0) = 375,$$

(i) Now, residual at u_1,

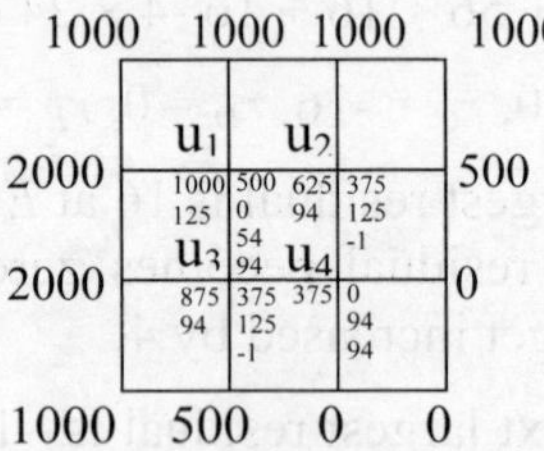

(i)

$r_{u1} = 2000 + 625 + 1000 + 875 - 4 \times 1000 = 500$
$r_{u2} = 1000 + 500 + 1000 + 375 - 4 \times 625 = 375$
$r_{u3} = 2000 + 375 + 1000 + 500 - 4 \times 875 = 375$
$r_{u4} = 875 + 0 + 625 + 0 - 4 \times 375 = 0$
To liquidate r_{u1}, increase by 125.
To liquidate r_{u2}, increase by 94.
To liquidate r_{u2}, increase by 94.

(ii) The modified values of u_2, u_3, u_4 are 1125, 719, 969, 375.

Since the values are very large. Then we find residuals again, we get

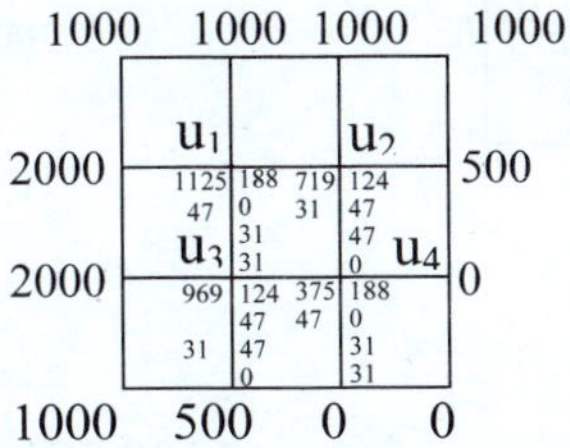

(ii)

$$r_{u1} = 188, r_{u2} = 124, r_{u3} = 124, r_{u4} = 188$$

To liquidate $r_{u1}, r_{u2}, r_{u3}, r_{u4}$ increased by 47,47,31,31.

(iii) Revised values of u_1, u_2, u_3, u_4 are 1172, 750, 1000, 422 residuals are

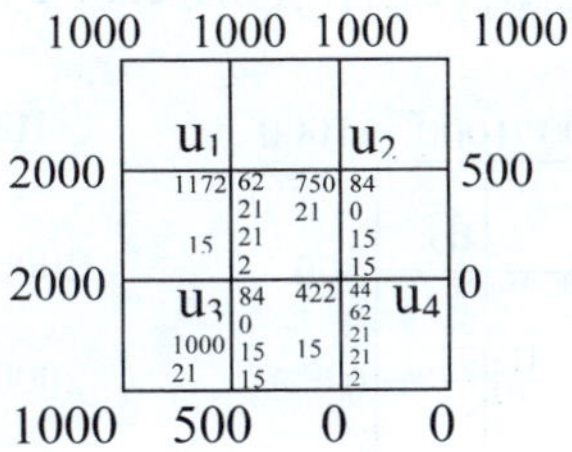

(iii)

Residuals are

$$r_{u1} = 62, r_{u2} = 84, r_{u3} = 84, r_{u4} = 62$$

To liquidate $r_{u1}, r_{u2}, r_{u3}, r_{u4}$ increased by 15, 21, 21, 15.

(iv) The values of u_1, u_2, u_3, u_4 are 1187,771,1021,437.

The residuals are

$$r_{u1} = 44, r_{u2} = 40, r_{u3} = 40, r_{u4} = 44$$

To liquidate $r_{u1}, r_{u2}, r_{u3}, r_{u4}$ increased by 11,10,10,11.

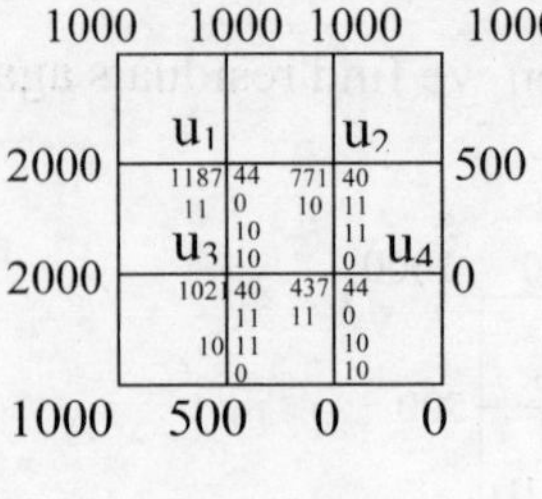

(iv)

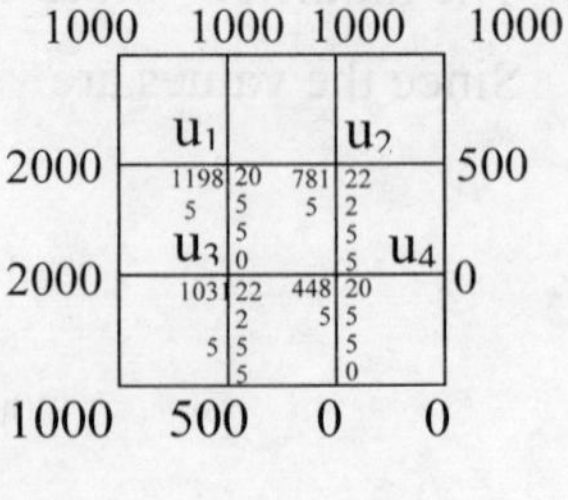

(v)

(v) Modified values of u_1, u_2, u_3, u_4 are 1198,781,1031,448

The residuals are $r_{u1} = 20, r_{u2} = 22, r_{u3} = 22, r_{u4} = 20$

To liquidate $r_{u1}, r_{u2}, r_{u3}, r_{u4}$ increase by 5,5,5,5.

(vi) Revised values of u_1, u_2, u_3, u_4 are 1203,786,1036,453.

The residuals are $r_{u1} = 10, r_{u2} = 12, r_{u3} = 12, r_{u4} = 10$

To liquidate $r_{u1}, r_{u2}, r_{u3}, r_{u4}$ increase by 2,3,3,2.

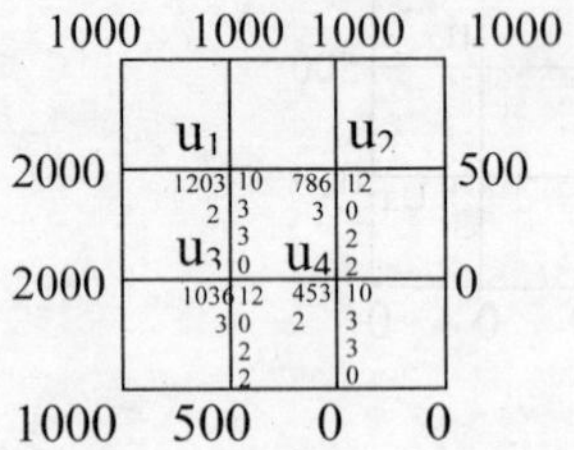

(vi)

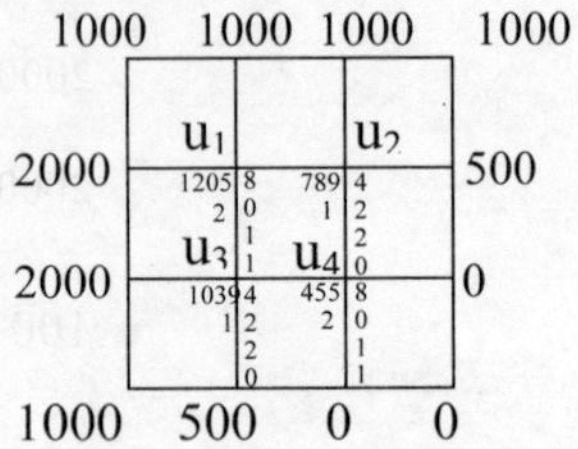

(vii)

(vii) Modified values of u_1, u_2, u_3, u_4 are 1205,789,1039,455.

The residuals are $r_{u1} = 8, r_{u2} = 4, r_{u3} = 4, r_{u4} = 8$.

To liquidate $r_{u1}, r_{u2}, r_{u3}, r_{u4}$ increase by 2,1,1,2.

(viii) Finally the current residuals being 1,0,0,1. We stop the relaxation method.

Hence the values are $u_1 = 1207, \ u_2 = 790$

$$u_3 = 1040, \ u_4 = 457$$

Parabolic Equations (Heat Equation)

Solution of one dimensional heat conduction equation

$$\frac{\partial u}{\partial t} = C^2 \frac{\partial^2 u}{\partial x^2}, \text{ where } C^2 = \frac{k}{S\ell} \text{ is the diffusivity of the substance.} \quad \ldots(i)$$

(i) Schmidt method : We consider a rectangular mesh in the x-t plane of sides h along x direction and k along t direction. We have $\frac{\partial u}{\partial t} = \frac{u_{i,j+1} - u_{i,j}}{k}$ and

$$\frac{\partial^2 u}{\partial x^2} = \frac{u_{i-1,j} - 2u_{i,j} + u_{i+1,j}}{h^2}$$

Substituting these values in (i), we have

$$u_{i,j+1} - u_{i,j} = \frac{Kc^2}{h^2}[u_{i-1,j} - 2u_{i,j} + u_{i+1,j}]$$

or $$u_{i,j+1} = \alpha u_{i-1,j}(1-2\alpha)u_{i,j} + \alpha u_{i+1,j} \quad \ldots(ii)$$

Where $\alpha = \frac{Kc^2}{h^2}$ is the mesh ratio parameter.

It is a 2 – level formula. Hence equation (ii) is called the Schmidt explicit formula which is valid only for $0 < \alpha \le \frac{1}{2}$.

When, $\alpha = \frac{1}{2}$, then (ii) becomes

$$u_{i,j+1} = \frac{1}{2}(u_{i+1,j} + u_{i-1,j}) \quad \ldots(iii)$$

Which is known as Bendre-Schimidt recurrence relation.

(ii) Crank-Nicolson method : According to this method, $\frac{\partial^2 u}{\partial x^2}$ is replaced by the average of its central-difference approximation on the *j*th and (*j+1*) th rows. Thus equation (i) is reduced to

$$\frac{u_{i,j+1} - u_{i,j}}{K} = C^2 . \frac{1}{2}\left[\frac{u_{i-1,j} - 2u_{i,j} + u_{i+1,j}}{h^2} + \frac{u_{i-1,j+1} - 2u_{i,j+1} + u_{i+1,j+1}}{h^2}\right]$$

or $$-\alpha u_{i-1,j+1} + (2+2\alpha)u_{i,j+1} - \alpha u_{i+1,j+1}$$
$$= \alpha u_{i-1,j} + (2-2\alpha)u_{i,j} + \alpha u_{i+1,j} \quad \ldots(iv)$$

where $$\alpha = \frac{Kc^2}{h^2}$$

Equation (iv) is a 2-level implicit relation and is known as Crank-Nicolson formula.

(iii) Iterative method : From equation (iv), we have

$$(1+\alpha)u_{i,j+1} = \frac{1}{2}\alpha(u_{i-1,j+1} + u_{i+1,j+1})$$

$$+u_{i,j} + \frac{1}{2}\alpha(u_{i-1,j} - 2u_{i,j} + u_{i+1,j}) \quad \ldots(v)$$

Putting $b_i = u_{i,j} + \frac{1}{2}\alpha(u_{i-1,j} - 2u_{i,j} + u_{i+1,j})$

And dropping j's from (v), we have

$$u_i = \frac{\alpha}{2(1+\alpha)}(u_{i-1} + u_{i+1}) + \frac{bi}{1+\alpha}$$

The iteration formula is

$$u_i^{(n+1)} = \frac{\alpha}{2(1+\alpha)}[u_{i-1}^{(n)} + u_{i+1}^{(n)}) + \frac{bi}{1+\alpha} \quad \ldots(vi)$$

which is known as **Jacobi's** Iteration formula.

Equation (vi) may be written as

$$u_i^{(n+1)} = \frac{\alpha}{2(1+\alpha)}[u_{i-1}^{(n+1)} + u_{i+1}^{(n)}) + \frac{bi}{1+\alpha} \quad \ldots(vii)$$

which is known as Gauss-Seidel iteration formula.

(iv) Du-fort and frankel method : If we replace the derivatives in equation (i) by the central-difference approximations,

$$\frac{\partial u}{\partial t} = \frac{u_{i,j+1} - u_{i,j-1}}{2k}$$

and
$$\frac{\partial^2 u}{\partial x^2} = \frac{u_{i-1,j} - 2u_{i,j} + u_{i+1,j}}{h^2}$$

we have,
$$u_{i,j+1} - u_{i,j-1} = \frac{2kc^2}{h^2}[u_{i-1,j} - 2u_{i,j} + u_{i+1,j}]$$

or
$$u_{i,j+1} = u_{i,j-1} + 2\alpha[u_{i-1,j} - 2u_{i,j} + u_{i+1,j}] \quad \ldots(viii)$$

where $\alpha = \frac{kc^2}{h^2}$. This equation (viii) is called the Richardson scheme which is a 3-level method.

If $\quad u_{i,j} = \frac{1}{2}(u_{i,j-1} + u_{i,j+1})$ then equation (viii) becomes

$$u_{i,j+1} = u_{i,j-1} + 2\alpha[u_{i-1,j} - u_{i,j-1} - u_{i,j+1} + u_{i+1,j}]$$

or $$u_{i,j+1} = \frac{1-2\alpha}{1+2\alpha}u_{i,j-1} + \frac{2\alpha}{1+2\alpha}[u_{i-1,j} + u_{i+1,j}] \quad \text{...(ix)}$$

which is known as *Du* fort-frankel method and it is a 3-level explicit method.

Example 8 *Solve the equation* $\frac{\partial u}{\partial t} = \frac{\partial^2 u}{\partial x^2}$ *subject to the conditions* $u(x,0) = \sin \pi x, 0 \le x \le 1; u(0,t) = u(1,t) = 0,$ *using (a) Schmidt method (b) crank-Nicolson method (c) Du fort-frankel method. Carryout computations for two levels, taking* $h = \frac{1}{3}, k = \frac{1}{36}$.

Solution Here, $\quad c^2 = 1, h = \frac{1}{3}, k = \frac{1}{36}$

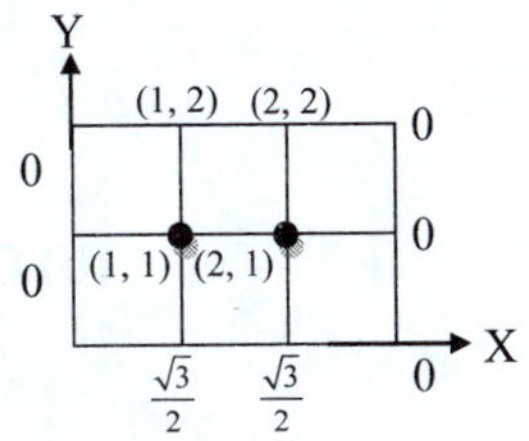

$$\therefore \ \alpha = \frac{Kc^2}{h^2} = \frac{1}{4}$$

Also $u_{i,o} = \sin(ih\pi)$

$$u_{0,0} = 0, u_{1,0} = \sin\frac{\pi}{3} = \frac{\sqrt{3}}{2}, u_{2,0} = \sin\frac{2\pi}{3} = \frac{\sqrt{3}}{2},$$

$u_{3,0} = \sin\frac{3\pi}{3} = 0, u_{0,1} = 0, u_{0,2} = 0$ are the boundary values.

(a) **Schmidt method :** $u_{i,j+1} = \alpha u_{i-1,j} + (1-2\alpha)u_{i,j} + \alpha\, u_{i+1,j}$

Put $\alpha = \frac{1}{4}, u_{i,j+1} = \frac{1}{4}u_{i-1,j} + \left(1 - \frac{2}{4}\right)u_{i,j} + \frac{1}{4}u_{i+1,j}$

Or $$u_{i,j+1} = \frac{1}{4}[u_{i-1,j} + 2u_{i,j} + u_{i+1,j}]$$

For $i = 1, j = 0,$

$$u_{1,1} = \frac{1}{4}[u_{0,0} + 2u_{1,0} + u_{2,0}] = \frac{1}{4}\left(0 + 2 \times \frac{\sqrt{3}}{2} + \frac{\sqrt{3}}{2}\right)$$

$$= \frac{1}{4} \times \frac{3\sqrt{3}}{2} = 0.65$$

for $i=2, j=0$,

$$u_{2,1}=\frac{1}{4}[u_{1,0}+2u_{2,0}+u_{3,0}]$$

$$=\frac{1}{4}\left(\frac{\sqrt{3}}{2}+2\times\frac{\sqrt{3}}{2}+0\right)=\frac{1}{4}\times\frac{3\sqrt{3}}{2}=0.65$$

for $i=1, j=1$,

$$u_{1,2}=\frac{1}{4}[u_{0,1}+2u_{1,1}+u_{2,1}]$$

$$=\frac{1}{4}(0+2\times 0.65+0.65)=0.49$$

for $i=2, j=1$,

$$u_{2,2}=\frac{1}{4}[u_{1,1}+2u_{2,1}+u_{3,1}]$$

$$=\frac{1}{4}(0.65+2\times 0.65+0)=0.49$$

(b) Crank-Nicolson method :

$$-\alpha u_{i-1,j+1}+(2+2\alpha)u_{i,j+1}-\alpha u_{i+1,j+1}=\alpha u_{i-1,j}+(2-2\alpha)u_{i,j}+\alpha u_{i+1,j}$$

Put $\alpha=\frac{1}{4}$,

$$-\frac{1}{4}u_{i-1,j+1}+\frac{10}{4}u_{i,j+1}-\frac{1}{4}u_{i+1,j+1}$$

$$=\frac{1}{4}u_{i-1,j}+\frac{6}{4}u_{i,j}+\frac{1}{4}u_{i+1,j}$$

or $\quad -u_{i-1,j+1}+10u_{i,j+1}-u_{i+1,j+1}=u_{i-1,j}+6u_{i,j}+u_{i+1,j}$

for $\quad i=1, j=0$,

$$-u_{0,1}+10u_{1,1}-u_{2,1}=u_{0,0}+6u_{1,0}+u_{2,0}$$

or $\quad -0+10u_{1,1}-u_{2,1}=0+6\frac{\sqrt{3}}{2}+\frac{\sqrt{3}}{2}$

or $\quad 10u_{1,1}-u_{2,1}=\frac{7\sqrt{3}}{2} \qquad \ldots$

for $i = 2, j = 0$

$$-u_{1,1} + 10u_{2,1} - u_{3,1} = u_{1,0} + 6u_{2,0} + u_{3,0}$$

or $$-u_{1,1} + 10u_{2,1} - 0 = \frac{\sqrt{3}}{2} + 6\frac{\sqrt{3}}{2} + 0$$

or $$-u_{1,1} + 10u_{2,1} = \frac{7\sqrt{3}}{2} \qquad \text{...(ii)}$$

Solving (i) and (ii), we get

$$u_{1,1} = 0.67, u_{2,1} = 0.67$$

for $i = 1, j = 1,$

$$-u_{0,2} + 10u_{1,2} - u_{2,2} = u_{0,1} + 6u_{1,1} + u_{2,1}$$

or $$0 + 10u_{1,2} - u_{2,2} = 0 + 6 \times 0.67 + 0.67$$

or $$10u_{1,2} - u_{2,2} = 4.69 \qquad \text{...(iii)}$$

for $i = 2, j = 1,$

$$-u_{1,2} + 10u_{2,2} - u_{3,2} = u_{1,1} + 6u_{2,1} + u_{3,1}$$

or $$-u_{1,2} + 10u_{2,2} - 0 = 0.67 + 6 \times 0.67 + 0$$

or $$-u_{1,2} + 10u_{2,2} = 4.69 \qquad \text{...(iv)}$$

Solving (iii) and (iv), we get

$$u_{1,2} = 0.52, u_{2,2} = 0.52$$

(c) Du-fort-frankel method :

$$u_{i,j+1} = \frac{1-2\alpha}{1+2\alpha} u_{i,j-1} + \frac{2\alpha}{1+2\alpha} \{u_{i-1,j} + u_{i+1,j}\}$$

Put $\alpha = \frac{1}{4}, u_{i,j+1} = \frac{1}{3}[u_{i,j-1} + u_{i-1;j} + u_{i+1,j}]$

To start the calculation, we take $u_{1,1} = u_{2,1} = 0.65$ from Schmidt method.

Now, for $i = 1, j = 1$,

$$u_{1,2} = \frac{1}{3}[u_{1,0} + u_{0,1} + u_{2,1}] = \frac{1}{3}\left[\frac{\sqrt{3}}{2} + 0 + 0.65\right] = 0.5$$

for $i = 2, j = 1$,

$$u_{2,2} = \frac{1}{3}[u_{2,0} + u_{1,1} + u_{3,1}] = \frac{1}{3}\left[\frac{\sqrt{3}}{2} + 0.65 + 0\right] = 0.5$$

Example 9 *Solve the boundary value problem* $\frac{\partial u}{\partial t} = \frac{\partial^2 u}{\partial x^2}$ *under the conditions* $u\ (0,\ t) = u\ (1,\ t) = 0$ *and* $u\ (x,\ 0) = \sin \pi x,\ 0 \le x \le 1$ *using schimidt method, taking* $h = 0.2$ *and* $k = 0.02$.

Solution Here $C^2 = 1, h = 0.2, k = 0.02$

$$\alpha = \frac{C^2 k}{h^2} = \frac{1}{2}$$

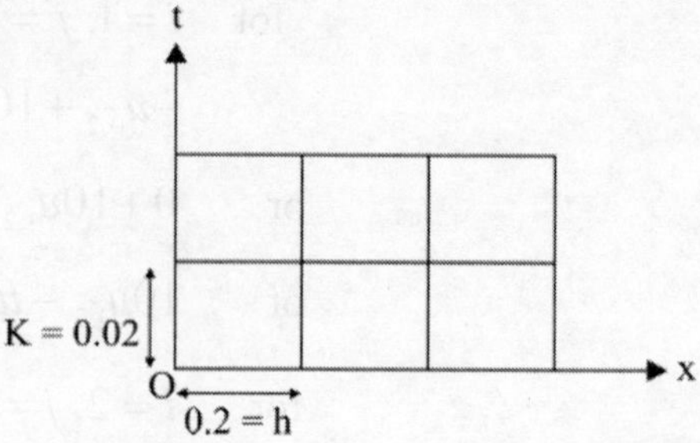

Hence by schimidt formula

$$u_{i,j+1} = \frac{1}{2}\left(u_{i-1,j} + u_{i+1,j}\right) \quad \text{...(i)}$$

Now, $u\ (0, t) = u\ (1, t) = 0$

$\Rightarrow$ $u\ (0, t) = u\ (5 \times 0.2, t) = 0$

$\Rightarrow$ $u\ (0, t) = u\ (5h, t) = 0$

$\Rightarrow$ $u_{0,j} = u_{5,j} = 0$ for all j ...(ii)

Also the initial conditions is $u\ (x, 0) = \sin \pi x,\ 0 \le x \le 1$

$\therefore$ $u_{i,0} = \sin\ (ih\ \pi),\ i = 0, 1, 2, 3, 4, 5$ at $t = 0$

$\therefore$ $u_{0,0} = 0,\ u_{1,0} = \sin\ (0.2\ \pi) = 0.59,\ u_{2,0} = \sin\ (0.4\ \pi) = 0.95,$

$u_{3,0} = \sin\ (0.6\ \pi) = 0.95,\ u_{4,0} = \sin\ (0.8\ \pi) = 0.59,\ u_{5,0} = \sin \pi = 0$

These are the entries of the first row

Putting $j = 0$ in equation (i), we get

$$u_{i,1} = \frac{1}{2}\left(u_{i-1,0} + u_{i+1,0}\right) \quad \text{...(iii)}$$

Putting $i = 1, 2, 3, 4, 5$, we have

$$u_{1,1} = \frac{1}{2}\left(u_{0,0} + u_{2,0}\right) = 0.475,\ u_{2,1} = \frac{1}{2}\left(u_{1,0} + u_{3,0}\right) = 0.77,$$

$$u_{3,1} = \frac{1}{2}\left(u_{2,0} + u_{4,0}\right) = 0.77,\ u_{4,1} = \frac{1}{2}\left(u_{3,0} + u_{5,0}\right) = 0.475,\ u_{5,1} = 0$$

These are the entries of the second row

Putting $j = 1$, we have

$$u_{i,2} = \frac{1}{2}\left(u_{i-1,1} + u_{i+1,1}\right) \qquad \text{...(iv)}$$

Putting $i = 1, 2, 3, 4, 5$ in (iv), we get

$$u_{1,2} = \frac{1}{2}\left(u_{0,1} + u_{2,1}\right) = 0.385,\ u_{2,2} = \frac{1}{2}\left(u_{1,1} + u_{3,1}\right) = 0.623$$

$$u_{3,2} = \frac{1}{2}\left(u_{2,1} + u_{4,1}\right) = 0.623,\ u_{4,2} = \frac{1}{2}\left(u_{3,1} + u_{5,1}\right) = 0.385,\ u_{5,2} = 0$$

These are the entries of the third row

Putting $j = 2$ in (i), we get

$$u_{i,3} = \frac{1}{2}\left(u_{i-1,2} + u_{i+1,2}\right) \qquad \text{...(v)}$$

$$u_{1,3} = \frac{1}{2}\left(u_{0,2} + u_{2,2}\right) = 0.312,\ u_{2,3} = \frac{1}{2}\left(u_{1,2} + u_{3,2}\right) = 0.504,$$

$$u_{3,3} = \frac{1}{2}\left(u_{2,2} + u_{4,2}\right) = 0.504,\ u_{4,3} = \frac{1}{2}\left(u_{3,2} + u_{5,2}\right) = 0.032, u_{5,3} = 0$$

These are entries of the fourth row

Similarly for $j = 4$,

$$u_{i,4} = \frac{1}{2}\left(u_{i-1,3} + u_{i+1,3}\right) \qquad \text{...(vi)}$$

$$u_{1,4} = 0.252,\ u_{2,4} = 0.408,\ u_{3,4} = 0.408,\ u_{4,4} = 0.252,\ u_{5,4} = 0$$

For $j = 5$,

$$u_{i,5} = \frac{1}{2}\left(u_{i-1,4} + u_{i+1,4}\right) \qquad \text{...(vii)}$$

$$u_{1,5} = 0.204,\ u_{2,5} = 0.33,\ u_{3,5} = 0.33,\ u_{4,5} = 0.204,\ u_{5,5} = 0$$

These are the entries of fifth and sixth rows

Hence the solution is given in tabular form as follows:

j↓/i→	0	1	2	3	4	5
0	0	0.59	0.95	0.95	0.59	0
1	0	0.475	0.77	0.77	0.475	0
2	0	0.385	0.623	0.623	0.385	0
3	0	0.312	0.504	0.504	0.312	0
4	0	0.252	0.408	0.408	0.252	0
5	0	0.204	0.330	0.330	0.204	0

Example 10 *Find the values of u(x, t) satisfying the parabolic equation* $\frac{\partial u}{\partial t} = 4\frac{\partial^2 u}{\partial x^2}$ *and the boundary conditions u (0, t) = 0 = u (8, t) and* $u(x,0) = 4x - \frac{1}{2}x^2$ *at the points* $x = i:\ i = 0, 1, 2, ...7$ and $t = \frac{1}{8}j:\ j = 0, 1, 2, ...5$

Solution Here $C^2 = 4,\ h = 1,\ k = \frac{1}{8}$

Then $\alpha = \frac{C^2 k}{h^2} = \frac{1}{2}$

Hence by Schmidt formula

$$u_{i,j+1} = \frac{1}{2}\left(u_{i-1,j} + u_{i+1,j}\right) \quad \text{...(i)}$$

Now since *u (0, t)* = 0 = *u* (8, *t)*

$\therefore u_{0,j} = 0$ and $u_{8,j} = 0$ for all *j* i.e., the entries in the first and last column are zero

Since $u(x,0) = 4x - \frac{1}{2}x^2$

$$\therefore \quad u_{i,0} = 4i - \frac{1}{2}i^2$$

$$= 0, 3.5, 6, 7.5, 8, 7.5, 6, 3.5$$

For $i = 0, 1, 2, 3, 4, 5, 6, 7$ at $t = 0$

These are entries of first row

Putting $j = 0$ in (i), we have

$$u_{i,1} = \frac{1}{2}\left(u_{i-1,0} + u_{i+1,0}\right)$$

Taking $i = 1, 2, \ldots .7$, we get

$$u_{1,1} = \frac{1}{2}\left(u_{0,0} + u_{2,0}\right) = 3,\ u_{2,1} = \frac{1}{2}\left(u_{1,0} + u_{3,0}\right) = 5.5$$

$$u_{3,1} = \frac{1}{2}\left(u_{2,0} + u_{4,0}\right) = 7,\ u_{4,1} = 7.5,\ u_{5,1} = 7,\ u_{6.1} = 5.5,\ u_{7,1} = 3$$

These are entries in the second row

Putting $j = 1$ in (i), we have

$$u_{i,2} = \frac{1}{2}\left(u_{i-1,1} + u_{i+1,1}\right)$$

Similarly putting $i = 2, 3, \ldots .. 7$. The entries of the third, fourth, fifth and sixth rows are obtained

Hence the values of $u_{i,2}$ are as given in the following table:

j↓/i→	0	1	2	3	4	5	6	7	8
0	0	3.5	6	7.5	8	7.5	6	3.5	0
1	0	3	5.5	7	7.5	7	5.5	3	0
2	0	2.75	5	6.5	7	6.5	5	2.75	0
3	0	2.5	4.625	6	6.5	6	4.625	2.5	0
4	0	2.3125	4.25	5.5625	6	5.5625	4.25	2.3125	0
5	0	2.125	3.9375	5.125	5.5625	5.125	3.9375	2.125	0

Solution of two dimensional heat equation : The equation becomes

$$\frac{\partial u}{\partial t} = C^2\left(\frac{\partial^2 u}{\partial x^2} + \frac{\partial^2 u}{\partial y^2}\right) \qquad \ldots(i)$$

If h be the step size then a mesh point $(x,y,t) = (ih,jh,nl)$ may be represented as (i,j,n).

By finite difference approximation, equation (i) becomes.

$$\frac{u_{i,j,n+1} - u_{i,j,n}}{l} = \frac{c^2}{h^2}[(u_{i-1,j,n} - 2u_{i,j,n} + u_{i+1}, j, n)$$
$$+ (u_{i,j-1,n} + u_{i,j+1,n} - 2u_{i,j,n})]$$

or $$u_{i,j,n+1} = u_{i,j,n} + \alpha(u_{i-1,j,n} + u_{i+1,j,n} + u_{i,j-1,n} + u_{i,j+1,n} - 4u_{i,j,n}) \quad \ldots\text{(ii)}$$

where $\alpha = \frac{lc^2}{h^2}$. This equation needs the five points available on the nth plane. This method is known as *ADE* (Alternating Direction Explicit) method.

Example 11 *Solve the equation* $\frac{\partial u}{\partial t} = \frac{\partial^2 u}{\partial x^2} + \frac{\partial^2 u}{\partial y^2}$ *subject to the initial conditions* $u(x,y,0) = \sin 2\pi x \sin 2\pi y, 0 \le x,y \le 1$ *and the conditions* $u(x,y,t) = 0, t>0$ *on the boundaries, using ADE method with* $h = \frac{1}{3}, \alpha = \frac{1}{8}$.

Solution The equation of two dimensional heat equation becomes.

$$u_{i,j,n+1} = u_{i,j,n} + \frac{1}{8}(u_{i-1,j,n} + u_{i+1,j,n} + u_{i,j+1,n} + u_{i,j-1,n} - 4u_{i,j,n})$$

$$= \frac{1}{2}u_{i,j,n} + \frac{1}{8}(u_{i-1,j,n} + u_{i+1,j,n} + u_{i,j+1,n} + u_{i,j-1,n}) \quad \ldots\text{(i)}$$

At the zero level (n=0), the initial and boundary conditions are

$$u_{i,j,0} = u_{i,j,.0} = \sin\frac{2\pi i}{3}\sin\frac{2\pi j}{3}$$

and $u_{i,0,0} = u_{0,j,0} = u_{3,j,0} = u_{0,3,0} = 0, i, j = 0,1,2,3,\ldots.$

Now, we determine the mesh values at the first level:

For $n = 0$, equation (i) becomes

$$u_{i,j,1} = \frac{1}{2}u_{i,j,0} + \frac{1}{8}(u_{i-1,j,0} + u_{i+1,j,0} + u_{i,j+1,0} + u_{i,j-1,0}) \quad \ldots\text{(ii)}$$

Putting $i = 1, j = 1$,

$$u_{1,1,1} = \frac{1}{2}u_{1,1,0} + \frac{1}{8}(u_{0,1,0} + u_{2,1,0} + u_{1,2,0} + u_{1,0,0})$$

$$= \frac{1}{2}\left(\sin\frac{2\pi}{3}\right)^2 + \frac{1}{8}\left(0 + \sin\frac{4\pi}{3}\sin\frac{2\pi}{3} + \sin\frac{2\pi}{3}.\sin\frac{4\pi}{3} + 0\right)$$

$$= \frac{3}{8} + \frac{1}{8}\left(-\frac{\sqrt{3}}{2}\times\frac{\sqrt{3}}{2} - \frac{\sqrt{3}}{2}\times\frac{\sqrt{3}}{2}\right) = \frac{3}{16}$$

Putting $i = 2, j = 1$,

$$u_{2,1,1} = \frac{1}{2}u_{2,1,0} + \frac{1}{8}(u_{1,1,0} + u_{3,1,0} + u_{2,2,0} + u_{2,0,0})$$

$$= \frac{1}{2}\sin\frac{4\pi}{3}\sin\frac{2\pi}{3} + \frac{1}{8}\left[\left(\sin\frac{2\pi}{3}\right)^2 + 0 + \left(\sin\frac{4\pi}{3}\right)^2 + 0\right]$$

$$= -\frac{1}{2}\left(\frac{\sqrt{3}}{2}\right)^2 + \frac{1}{8}\left[\left(\frac{\sqrt{3}}{2}\right)^2 + \left(-\frac{\sqrt{3}}{2}\right)^2\right] = \frac{-3}{16}$$

Putting $i = 1, j = 2$,

$$u_{1,2,1} = \frac{1}{2}u_{1,2,0} + \frac{1}{8}[u_{0,2,0} + u_{2,2,0} + u_{1,3,0} + u_{1,1,0})]$$

$$= \frac{1}{2}\sin\frac{2\pi}{3}\sin\frac{4\pi}{3} + \frac{1}{8}\left[0 + \left(\sin\frac{4\pi}{3}\right)^2 + 0 + \left(\sin\frac{2\pi}{3}\right)^2\right]$$

$$= \frac{-3}{8} + \frac{1}{8}\left(\frac{\sqrt{3}}{2}\times\frac{\sqrt{3}}{2} + \frac{\sqrt{3}}{2}\times\frac{\sqrt{3}}{2}\right) = \frac{-3}{16}$$

Putting $i = 2, j = 2$,

$$u_{2,2,1} = \frac{1}{2}u_{2,2,0} + \frac{1}{8}(u_{1,2,0} + u_{3,2,0} + u_{2,3,0} + u_{2,1,0})$$

$$= \frac{1}{2}\left(\sin\frac{4\pi}{3}\right)^2 + \frac{1}{8}\left[\sin\frac{2\pi}{3}\sin\frac{4\pi}{3} + 0 + 0 + \sin\frac{4\pi}{3}\sin\frac{2\pi}{3}\right]$$

$$= \frac{3}{8} + \frac{1}{8}\left(-\frac{3}{4} - \frac{3}{4}\right) = \frac{3}{16}$$

Similarly, the mesh values at the second and higher level can be calculated.

Hyperbolic equations

Hyperbolic equations are connected with oscillating system. For example, the wave equation in one dimension

$$\frac{\partial^2 u}{\partial t^2} = C^2\frac{\partial^2 u}{\partial x^2} \qquad \text{...(i)}$$

Solution of wave equation

Subject to the initial conditions :

$$u = f(x), \frac{\partial u}{\partial t} = g(x), \qquad 0 \le x \le 1 \text{ at } t = 0$$

and boundary conditions are $u(0,t) = \phi(t), u(1,t) = \psi(t)$

We consider a rectangular region R in the x-t plane and divide it into rectangular mesh of sides h along x-direction and k along t-direction. We represent a mesh point $(x,t) = (ih, jk)$, we have

$$\frac{\partial^2 u}{\partial x^2} = \frac{u_{i-1,j} - 2u_{i,j} + u_{i+1,j}}{h^2}$$

and

$$\frac{\partial^2 u}{\partial t^2} = \frac{u_{i,j-1} - 2u_{i,j} + u_{i,j+1}}{h^2}$$

Putting these values in (i), we have

$$u_{i,j-1} - 2u_{i,j} + u_{i,j+1} = \frac{c^2k^2}{h^2}(u_{i-1,j} - 2u_{i,j} + u_{i+1,j})$$

or

$$u_{i,j+1} = 2(1-\alpha^2c^2)u_{i,j} + \alpha^2c^2(u_{i-1,j} + u_{i+1,j}) - u_{i,j-1} \qquad \ldots\text{(ii)}$$

where

$$\alpha = \frac{k}{h}$$

Example 12 *Evaluate the pivotal values of the equation* $u_{tt} = 16u_{xx,}$ taking $\Delta x = 1$ *upto* t*=1.25. The boundary conditions are* $u(0,t) = u(5,t) = 0, u_t(x,0) = 0$ and $u(x,0) = x^2(5-x)$.

Solution Here, $c^2 = 16$,

The difference equation is

$$u_{i,j+1} = 2(1-c^2\alpha^2)u_{i,j} + c^2\alpha^2(u_{i-1,j} + u_{i+1,j}) - u_{i,j-1}$$

or

$$u_{i,j+1} = 2(1-16\alpha^2)u_{i,j} + 16\alpha^2(u_{i-1,j} + u_{i+1,j}) - u_{i,j-1} \qquad \ldots\text{(i)}$$

where $\alpha = \frac{k}{h}$

Given h= 1, we choose k so that $16\alpha^2 = 1$

$$\Rightarrow \quad 16\frac{k^2}{h^2} = 1 \Rightarrow k^2 = \frac{1}{16} \Rightarrow k = \frac{1}{4}$$

$\therefore$ equation (i) reduces to

$$u_{i,j+1} = u_{i-1,j} + u_{i+1,j} - u_{i,j-1} \quad \text{...(ii)}$$

Now, the boundary conditions are $u(0,t) = u(5,t) = 0$

$\therefore \quad u_{0,j} = 0$ and $u_{5,j} = 0$, for all j

i.e., the entries in the first and last columns are zero.

Since $u(x,0) = x^2(5-x)$

$\therefore \quad u_{i,o} = i^2(5-i)$

$= 4, 14, 18, 16$, for $i = 1,2,3,4$ at $t = 0$

i.e., $\quad u_{1,0} = 4, u_{2,0} = 12, u_{3,0} = 18, u_{4,0} = 16, u_{5,0} = 0$

These entries are for the first row.

Finally, since $u_t(x,0) = 0$ becomes

$$\frac{u_{i,j+1} - u_{i,j-1}}{2k} = 0$$

$$\Rightarrow \quad u_{i,j+1} = u_{i,j-1} \Rightarrow u_{i,1} = u_{i,-1}, \text{ for j= 0} \quad \text{...(iii)}$$

Putting $j = 0$ in equation (ii), we have

$$u_{i,1} = u_{i-1,0} + u_{i+1,0} - u_{i,-1}$$

$$= u_{i-1,0} + u_{i+1,0} - u_{i,1} \quad \text{using (iii)}$$

or $\quad u_{i,1} = \frac{1}{2}(u_{i-1,0} + u_{i+1,0}) \quad \text{...(iv)}$

Putting $i = 1,2,3,4$ we get

$$u_{1,1} = \frac{1}{2}(u_{0,0} + u_{2,0}) = \frac{1}{2}(0 + 12) = 6$$

$$u_{2,1} = \frac{1}{2}(u_{1,0} + u_{3,0}) = \frac{1}{2}(4 + 18) = 11$$

$$u_{3,1} = \frac{1}{2}(u_{2,0} + u_{4,0}) = \frac{1}{2}(12 + 16) = 14$$

$$u_{4,1} = \frac{1}{2}(u_{3,0} + u_{5,0}) = \frac{1}{2}(18 + 0) = 9$$

These are the entries of second row.

Putting $j = 1$ (ii), we get

$$u_{i,2} = u_{i-1,1} + u_{i+1,1} - u_{i,0}$$

Taking $i = 1,2,3,4$ we get

$u_{1,2} = u_{0,1} + u_{2,1} - u_{1,0} = 0 + 11 - 4 = 7$ $u_{2,2} = u_{1,1} + u_{3,1} - u_{2,0} = 6+14-12=8$

$u_{3,2} = u_{2,1} + u_{4,1} - u_{3,0} = 11 + 9 - 18 = 2$

$u_{4,2} = u_{3,1} + u_{5,1} - u_{4,0} = 14 + 0 - 16 = -2$

These are entries of the 3rd row.

Similarly, putting j = 2, 3, 4 in (ii), the entries of the 4th, 5th and 6th rows are obtained. Hence, the solution u (x, t) is given in table from :

j↓/i→	0	1	2	3	4	5
0	0	4	12	18	16	0
1	0	6	11	14	9	0
2	0	7	8	2	-2	0
3	0	2	-2	-8	-7	0
4	0	-9	-14	-11	-6	0
5	0	-16	-18	-12	-4	0

Example 13 *The transverse displacement u of the point at a distance x from one end and any time t of a vibrating string satisfies the equation* $\frac{\partial^2 u}{\partial t^2} = 4\frac{\partial^2 u}{\partial x^2}$, *with boundary conditions u =0 at x = 0, t > 0 and u = 0 at x = 4, t > 0 and initial conditions u = x (4 – x) and*

$\frac{\partial u}{\partial t} = 0, 0 \le x \le 4.$ *Solve this equation numberically for one half period of virbation, taking* $h = 1$ *and* $k = \frac{1}{2}$.

Solution Here, h = 1, k = $\frac{1}{2}$, $\frac{h}{k}$ =2 = c

The difference equation is

$$u_{i,j+1} = u_{i-1,j} + u_{i+1,j} - u_{i,j-1} \quad \text{...(i)}$$

Since, the boundary conditions are $u(0,t) = u(4,t) = 0$

$\therefore \quad u_{0,j} = 0$ and $u_{4,j} = 0$, for all j

$\therefore$ the entries in the first and last columns are zero.

Since $u(x,0) = x(4-x)$

$\therefore \quad u_{i,o} = i(4-i)$

for $i = 1, 2, 3$ at $t = 0$, $u_{1,0} = 3$, $u_{2,0} = 4$, $u_{3,0} = 3$

These are entries of the first row.

Also $u_t(x,0) = 0$ becomes

$$\frac{u_{i,j+1} - u_{i,j-1}}{2k} = 0$$

$$\Rightarrow \quad u_{i,j+1} = u_{i,j-1} \Rightarrow u_{i,1} = u_{i,-1}, \text{ for j= 0} \qquad \text{...(ii)}$$

Putting $j = 0$ in equation (i), we have

$$u_{i,1} = u_{i-1,0} + u_{i+1,0} - u_{i,-1}$$

$$= u_{i-1,0} + u_{i+1,0} - u_{i,1} \qquad \text{using (ii)}$$

or $$u_{i,1} = \frac{1}{2}(u_{i-1,0} + u_{i+1,0}) \qquad \text{...(iii)}$$

Putting $i = 1,2,3$ we have

$$u_{1,1} = \frac{1}{2}(u_{0,0} + u_{2,0}) = \frac{1}{2}(0+4) = 2$$

$$u_{2,1} = \frac{1}{2}(u_{1,0} + u_{3,0}) = \frac{1}{2}(3+3) = 3$$

$$u_{3,1} = \frac{1}{2}(u_{2,0} + u_{4,0}) = \frac{1}{2}(4+0) = 2$$

These are the entries of second row.

Putting $j = 1$ in equation (i) we get

$$u_{i,2} = u_{i-1,1} + u_{i+1,1} - u_{i,0}$$

Taking $i = 1,2,3$, we get

$u_{1,2} = u_{0,1} + u_{2,1} - u_{1,0} = 0+3-3=0$ $\quad u_{2,2} = u_{2,1} + u_{3,1} - u_{2,0} = 2+2-4=0$

$u_{3,2} = u_{3,1} + u_{4,1} - u_{3,0} = 3+0-3=0$

These are entries of the 3rd row.

Now, the equation of the vibrating string of length l is $u_{tt} = c^2 u_{xx}$.

$\therefore$ Its period of vibration $= \dfrac{2l}{c}$ $(\because l = 4, c = 2) = \dfrac{2\times 4}{2} = 4\,\text{sec}$

Which shows that we have to determine $u(x,t)$ upto $t = 2$, i.e., we obtain the value of $u_{i,2}$ and $u_{i,3}$ (*i.e.* 5th row). Hence the values of $u(x, t)$ are given in table from:

j↓/i→	0	1	2	3	4
0	0	3	4	3	0
1	0	2	3	2	0
2	0	0	0	0	0
3	0	-2	-3	-2	0
4	0	-3	-4	-3	0

Example 14 *Solve* $\dfrac{\partial^2 u}{\partial t^2} = \dfrac{\partial^2 u}{\partial x^2}$ *with conditions* $u(0,t) = u(1,t) = 0,\ u(x,0) = \dfrac{1}{2}x(1-x)$ and $u_x(x,0) = 0$ taking $h = k = 0.1$ for $0 \le t \le 0.4$.

Solution Here $C^2 = 1$, $h = k = 0.1$

The difference equation is

$$u_{i,j+1} = 2\left(1 - C^2\alpha^2\right) u_{i,j} + C^2\alpha^2\left(u_{i-1,j} + u_{i+1,j}\right) - u_{i,j-1} \qquad \ldots\text{(i)}$$

where $\alpha = \dfrac{k}{h} = \dfrac{0.1}{0.1} = 1$

equation (i) reduces to

$$u_{i,j+1} = u_{i-1,j} + u_{i+1,j} - u_{i,j-1} \qquad \ldots\text{(ii)}$$

Boundary conditions are $u\,(0, t) = u\,(1, t) = 0$

$$\Rightarrow\ u_{0,j} = u_{1,j} = 0$$

Since $u(x,0) = \dfrac{1}{2}x(1-x)$

$$\Rightarrow\ u(i,0) = \frac{1}{2}i(1-i)$$

$= 0.045, 0.08, 0.105, 0.120, 0.125, 0.120, 0.105$

For $i = 0.1, 0.2, 0.3, 0.4, 0.5, 0.6, 0.7$ at $t = 0$.

These are entries in first row

Since $u_x(x,0)=0$

$$\therefore \quad \frac{u_{i,j+1}-u_{i,j}}{k}=0 \text{ for } j=0 \text{ at } t=0,\ u_{i,1}=u_{i,0}$$

Putting $j = 0$ in (ii), we get

$$u_{i,1}=u_{i-1,0}+u_{i+1,0}-u_{i,-1}$$

$$\therefore u_{i,1}=u_{i-1,0}+u_{i+1,0}-u_{i,1} \qquad (\because u_{i,-1}=u_{i,1})$$

$$\therefore 2u_{i,1}=u_{i-1,0}+u_{i+1,0}$$

$$\Rightarrow \quad u_{i,1}=\frac{1}{2}\left[u_{i-1,0}+u_{i+1,0}\right]$$

For i = 1, 2, 3, 4, 5, 6, we get

$u_{1,1} = 0.040$, $u_{2,1} = 0.075$, $u_{3,1} = 0.100$, $u_{4,1} = 0.115$, $u_{5,1} = 0.120$, $u_{6,1} = 0.115$

These are entries in the second row

Putting $j = 1$ in (ii), we get

$u_{i,2} = u_{i-1,1} + u_{i+1,1} - u_{i,0}$

Similarly putting i = 1, 2, 3, 4, 5. The entries of the third, fourth, fifth rows are obtained.

Hence the values of $u_{i,j}$ are as given in the following table:

j↓/i→	0	1	2	3	4	5	6
0	0	0.045	0.080	0.105	0.120	0.125	0.120
1	0	0.040	0.075	0.100	0.115	0.120	0.115
2	0	0.030	0.060	0.085	0.100	0.105	
3	0	0.020	0.040	0.060	0.075	0.080	
4	0	0.010	0.020	0.030	0.040	0.048	

Practice Problem

1. Classify the equation $y^2\frac{\partial^2 u}{\partial x^2}-2y\frac{\partial^2 u}{\partial x\partial y}+\frac{\partial^2 u}{\partial y^2}-\frac{\partial u}{\partial y}=8y.$

[**Ans.** Parabolic]

2. In which part of the xy- plane, the equation $\frac{\partial^2 u}{\partial x^2}+\frac{\partial^2 u}{\partial x \partial y}+\left(x^2+4y^2\right)\frac{\partial^2 u}{\partial x^2} = 2$ sin(xy) is elliptic.

[**Ans**. The equation is elliptic outside the elliptic $\frac{x^2}{\left(\frac{1}{2}\right)^2}+\frac{y^2}{\left(\frac{1}{4}\right)^2}=1$]

3. Solve $\frac{\partial^2 u}{\partial x^2}+\frac{\partial^2 u}{\partial y^2} = 0$ for the square mesh with boundary values as shown in the figure. Iterate until the maximum difference between successive values at any point is less than 0.005.

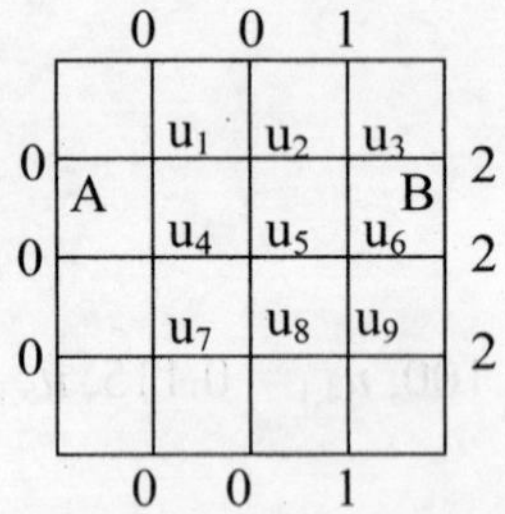

[**Ans**. $u_1 = 0.187$, $u_2 = 0.5$, $u_3 = 1.188$, $u_4 = 0.25$, $u_5 = 0.625$, $u_6 = 1.25$]

4. Given the values of $u(x, y)$ on the boundary of the square in the figure, evaluate the function $u(x, y)$ satisfying the laplace equation $\nabla^2 u = 0$ at the pivotal points of this figure by (i) Jacobi's method (ii) Gauss-Seidel method.

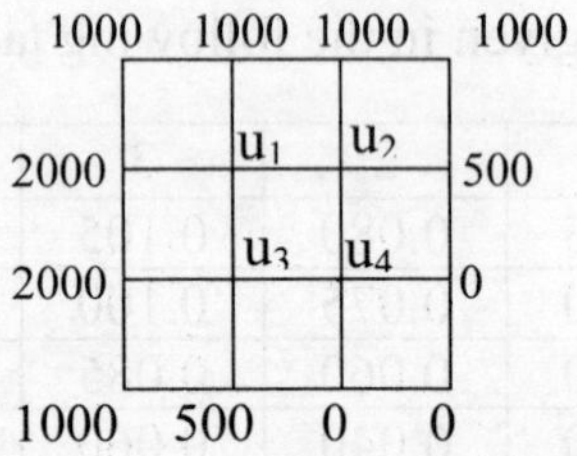

[**Ans.** $u_1 = 1208$, $u_2 = 792$, $u_3 = 1042$, $u_4 = 458$]

5. Solve the laplace equation $u_{xx} + u_{yy} = 0$ in the domain of given figure by

(i) Jacobi's method (ii) Gauss – Seidel method.

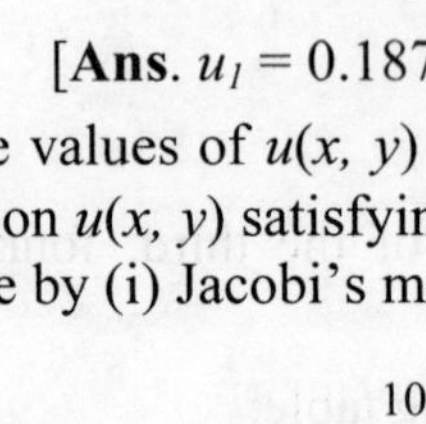

[**Ans.** $u_1 = 0.126$, $u_2 = 0.126$, $u_3 = 0.376$, $u_4 = 0.376$]

6. Given that $u(x, y)$ satisfies the equation $\nabla^2 u = 0$ and the boundary conditions are $u(0, y) = 0$, $u(4, y) = 8+2y$, $u(x, 0) = \frac{1}{2}x^2$, $u(x, 4) = x^2$, find the values $u(i, j)$, $i = 1,2,3; j = 1,2,3$ by relaxation method.

[**Ans.** $u_1 = 1.9$, $u_2 = 4.9$, $u_3 = 9.1$, $u_4 = 2.1$, $u_5 = 4.7$, $u_6 = 8.4$, $u_7 = 1.6$, $u_8 = 3.9$, $u_9 = 6.7$]

7. Solve by relaxation method, the equation $\nabla^2 u = 0$ in the square region with square meshes in the given figure starting with the initial values $u_1 = u_2 = u_3 = u_4 = 1$.

[**Ans.** $u_1 = 1$, $u_2 = 1.3$, $u_3 = 0.7$, $u_4 = 1$]

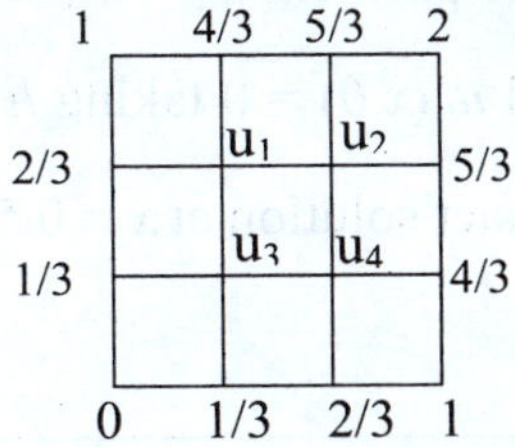

8. Find the values of $u(x, t)$ satisfying the parabolic equation $\frac{\partial u}{\partial t} = 4\frac{\partial^2 u}{\partial x^2}$ and the boundary conditions $u(0, t) = 0 = u(8, t)$ and $u(x, 0) = 4x - \frac{1}{2}x^2$ at the point $x = i$, $i = 1,2,\ldots\ldots.7$ and $t = \frac{1}{8}j$; $j = 0,1\ldots\ldots\ldots\ldots.5$,

[**Ans.**

j↓/i→	0	1	2	3	4	5	6	7	8
0	0	3.5	6	7.5	8	7.5	6	3.5	0
1	0	3	5.5	7	7.5	7	5.5	3	0
2	0	2.75	5	6.5	7	6.5	5	2.75	0
3	0	2.5	4.625	6	6.5	6	4.625	2.5	0
4	0	2.3125	4.25	5.5625	6	5.5625	4.25	2.3125	0
5	0	2.125	3.9375	5.125	5.5625	5.125	3.9375	2.125	0

9. Solve the boundary value problem $u_t = u_{xx}$ under the conditions $u(0, t) = u(1, t) = 0$ and $u(x, 0) = \sin \pi x$, $0 \le x \le 1$, by the Gauss-seidal method (take $h = 0.2$ and $\alpha = \frac{1}{2}$)

[**Ans.** $u_1 = 0.4853$, $u_2 = 0.7854$, $u_3 = 0.7854$, $u_4 = 0.4853$]

10. Solve the 2-dimensional heat equation $\frac{\partial u}{\partial t} = \frac{\partial^2 u}{\partial x^2} + \frac{\partial^2 u}{\partial y^2}$

Satisfying the initial conditions $u(x,y,0) = \sin\pi x \sin\pi y, 0 \le x, y \le 1$ and the boundary conditions $u = 0$ at $x = 0$ and $x = 1$ for $t > 0$. Obtain the solution upto two time level with $h = \frac{1}{3}$ and $\alpha = \frac{1}{8}$

[**Ans.** $u_{1,1,1} = u_{2,1,1} = u_{1,2,1} = u_{2,2,1} = \frac{9}{16}$, $u_{1,1,2} = u_{2,1,2} = u_{1,2,2} = u_{2,2,2} = \frac{27}{64}$,]

11. Solve the boundary value problem $u_{tt} = u_{xx}$ with the conditions $u(0, t) = u(1, t) = 0$, $u(x, 0) = \frac{1}{2}x(1-x)$ and $u_t(x,0) = 0$ taking $h = 0.1 = k$, for $0 \le t \le 0.4$. Compare your solution with the exact solution at $x = 0.5$ and $t = 0.3$.

[**Ans.**

$t = 0.3, x =$	0.1	0.2	0.3	0.4	0.5
Num.Sol.u =	0.02	0.04	0.06	0.075	0.08
Exact Sol.u =	0.02	0.04	0.06	0.075	0.08

12. Solve the heat equation $u_t = u_{xx}$ subject to the conditions $u(0, t) = u(1, t) = 0$, and $u(x, 0) = 2x$, for $0 \le x \le \frac{1}{2}$ and $2(1-x)$, for $\frac{1}{2} \le x \le 1$

Take $h = \frac{1}{4}$ and k according to Bandre- Schmidt equation.

[**Ans.**

j↓/i→	0	1	2	3	4
0	0	0.5	1	0.5	0
1	0	0.5	0.5	0.5	0
2	0	0.25	0.5	0.25	0
3	0	0.25	0.25	0.25	0

Chapter - 6

Fourier Transform

Fourier Integral Formula

A function, $f(x)$, which is pairwise continuous in every finite interval - $a \le x \le b$ and at every point of discontinuity as $\frac{1}{2}[f(x+o)+f(x-o)]$, further let $\int_{-\infty}^{\infty}|f(x)|dx$ converges *i.,e.,* $f(x)$ is absolutely integrable in $-\infty \le x \le \infty$, then

$$f(x)=\frac{1}{2\pi}\int_{-\infty}^{\infty}dv\int_{-\infty}^{\infty}f(t)\cos v(t-x)dt \qquad ...(i)$$

which is known as fourier integral formula.

Another form :

We have,

$$0=\frac{1}{2\pi}\int_{-\infty}^{\infty}f(t)dt\int_{-\infty}^{\infty}\sin v(t-x)dv \qquad ...(ii)$$

On adding Eqs. (i) and (ii) , we get

$$f(x)=\frac{1}{2\pi}\int_{-\infty}^{\infty}f(t)dt\int_{-\infty}^{\infty}e^{iv(t-x)}dv \qquad ...(iii)$$

which is the complex form of fourier integral.

Fourier Sine and Cosine Integrals

From Eqs. (i), expanding cos v $(x - t)$

$$f(x)=\frac{1}{2\pi}\int_{-\infty}^{\infty}\cos vx\int_{-\infty}^{\infty}f(t)\cos vt\,dt\,dv+\frac{1}{2\pi}$$

$$\int_{-\infty}^{\infty}\sin vx\int_{-\infty}^{\infty}f(t)\sin vt\,dt\,dv$$

If *f (x)* is an odd function, then

$$f(x)=\frac{1}{2\pi}\int_{-\infty}^{\infty}\sin vx\int_{-\infty}^{\infty}f(t)\sin vt\,dt\,dv$$

$$=\frac{2}{\pi}\int_{0}^{\infty}\sin vx\int_{0}^{\infty}f(t)\sin vt\,dt\,dv \quad \text{...(iv)}$$

which is known as fourier sine integral formula.

Similarly if *f (x)* is an even function. Then

$$f(x)=\frac{2}{\pi}\int_{0}^{\infty}\cos vx\int_{0}^{\infty}f(t)\cos vt\,dt\,dv \quad \text{...(v)}$$

Which is known as fourier cosine integral formula.

Fourier Transforms

Let *f(x)* be a function defined on interval $(-\infty,\infty)$ and be pairwise continuous in every finite interval and absolutely integrable in $(-\infty,\infty)$, then from Eq (iii)

$$f(x)=\frac{2}{\pi}\int_{-\infty}^{\infty}e^{-isx}ds\int_{-\infty}^{\infty}f(t)\,e^{ist}dt$$

if $$\text{F}(s)=\int_{-\infty}^{\infty}f(t)\,e^{ist}dt \quad \text{...(vi)}$$

then $$f(x)=\frac{1}{2\pi}\int_{-\infty}^{\infty}\text{F}(s)\,e^{-isx}ds \quad \text{...(vii)}$$

Eq. (vi) is called the fourier transform of *f (x)* and Eq. (vii) is called the inverse fourier transform.

Fourier Sine and Cosine Integrals

From Eqs. (iv).

$$F_s(s)=\int_{0}^{\infty}f(x)\sin sx\,dx \quad \text{...(viii)}$$

then $$f(x)=\frac{2}{\pi}\int_0^{\infty} F_s(s)\sin sx\, ds \qquad \text{...(ix)}$$

The function F_s(s) us called the fourier sine transform and *f(x)* is known as the inverse fourier sine transform.

Similary, from Eq. (v),

$$F_c(s)=\int_0^{\infty} f(x)\cos sx\, dx \qquad \text{...(x)}$$

then $$f(x)=\frac{2}{\pi}\int_0^{\infty} F_c(s)\cos sx\, ds \qquad \text{...(xi)}$$

The function $F_c(s)$ is called the fourier cosine transform and $f(x)$ is called the inverse fourier cosine transform.

Properties of Fourier Transform

1. **Linear property:** if *F(s)* and G (*s)* are fourier transforms of *F(x)* and g (*x)*respectively. Then,

$$F\{af(x)+bg(x)\}=a\,F(s)+b\,G(s)$$, where *a* and *b* are constants.

Proof. We have,

$$F(s)=\int_{-\infty}^{\infty} f(x)e^{isx}dx$$

and $$G(s)=\int_{-\infty}^{\infty} g(x)e^{isx}dx$$

$$\therefore \quad F\{af(x)+b\,g(x)\}=\int_{-\infty}^{\infty}\{a\,f(x)+b\,g(x)\}\,e^{isx}dx$$

$$=a\int_{-\infty}^{\infty} f(x)\,e^{isx}dx+b\int_{-\infty}^{\infty} g(x)\,e^{isx}dx$$

$$=a\,F(s)+bG(s)$$

2. **Change of Scale Property:** Let F (s) is the complex fourier transform of f (x). Then

$$F\{f(ax)\}=\frac{1}{a}F\left(\frac{s}{a}\right)$$

Proof. We have,

$$F(s)=\int_{-\infty}^{\infty} f(x)\,e^{isx}dx$$

$$\therefore \quad F\{f(ax)\}=\int_{-\infty}^{\infty} f(ax)e^{isx}dx$$

put $\quad ax = t, \Rightarrow a\,dx = dt$

$$= \frac{1}{a}\int_{-\infty}^{\infty} f(t)\, e^{\frac{ist}{a}}\, dt$$

$$= \frac{1}{a} F\left(\frac{s}{a}\right)$$

Remarks.

(i) For fourier sine transform, $F_s\{f(ax)\} = \frac{1}{a} F_s\left(\frac{s}{a}\right)$

(ii) For fourier cosine transform, $F_c\{f(ax)\} = \frac{1}{a} F_c\left(\frac{s}{a}\right)$

3. **Shifting Property :** Let $F(s)$ is the complex fourier transform of $f(x)$ Then
$F\{f(x-a)\} = F(s)\, e^{isa}$

Proof. We have, $F(s) = \int_{-\infty}^{\infty} f(x)\, e^{isx}\, dx$

$$\therefore \quad F\{f(x-a)\} = \int_{-\infty}^{\infty} f(x-a)\, e^{isx}\, dx$$

put $\quad x - a = t \Rightarrow dx = at$

$$= \int_{-\infty}^{\infty} f(t)\, e^{is(a+t)}\, dt$$

$$= e^{isa} \int_{-\infty}^{\infty} f(t)\, e^{ist}\, dt$$

$$= e^{isa} F(s)$$

4. **Modulation Property:** If $F(s)$ is the complex fourier transform of $f(x)$. then

$$F\{f(x)\cos ax\} = \frac{1}{2}\left[F(s+a) + F(s-a)\right]$$

Proof: We have,

$$F(s) = \int_{-\infty}^{\infty} f(x)\, e^{isx}\, dx$$

$$\therefore \quad F\{f(x)\cos ax\} = \int_{-\infty}^{\infty} f(x)\cos ax\, e^{isx}\, dx$$

$$= \int_{-\infty}^{\infty} e^{isx} f(x) \frac{e^{isx} + e^{-iax}}{2}\, dx \quad \left(\because \cos ax = \frac{e^{iax} + e^{-iax}}{2}\right)$$

$$=\frac{1}{2}\left[\int_{-\infty}^{\infty} f(x)e^{i(s+a)x}dx+\int_{-\infty}^{\infty} f(x)e^{i(s-a)x}dx\right]$$

$$=\frac{1}{2}\left[F(s+a)+F(s-a)\right]$$

Inversion Theorem for Complex Fourier Transform

Let $F(s)$ is the fourier transform of $f(x)$ defind on interval $(-l,l)$ and be pairwise continuous in every finite interval and further, if $\int_{-\infty}^{\infty}|f(x)|dx$ is convergent. Then

$$f(x)=\frac{1}{2\pi}\int_{-\infty}^{\infty} F(s)e^{-isx}ds$$

Proof. From fourier integral formula, we have

$$f(x)=\frac{1}{2\pi}\int_{-\infty}^{\infty} f(t)dt\int_{-\infty}^{\infty} e^{iv(t-x)}dv, \qquad \text{[from (iii)]}$$

$$=\frac{1}{2\pi}\int_{-\infty}^{\infty} e^{-ivx}dv\int_{-\infty}^{\infty} f(t)e^{ivt}dt,$$

Putting $v = s$ so that $dv = ds$

$$\therefore \quad f(x)=\frac{1}{2\pi}\int_{-\infty}^{\infty} e^{-isx}ds\int_{-\infty}^{\infty} f(t)e^{ist}dt,$$

$$=\frac{1}{2\pi}\int_{-\infty}^{\infty} F(s)e^{-isx}ds \qquad \text{[from (vi)]}$$

Example 1 Find the fourier transform of f(x) defined by $f(x)=\begin{cases}1 & ,|x|<1\\ 0 & ,|x|>1\end{cases}$ and hence evaluate (i) $\int_{-\infty}^{\infty}\frac{\sin s\cos sx}{s}ds$ (ii) $\int_{-\infty}^{\infty}\frac{\sin s}{s}ds$

Solution. We have,

$$F(s)=\int_{-\infty}^{\infty} f(x)e^{isx}dx$$

$$=\int_{-1}^{1} 1.e^{isx}dx$$

$$=\left(\frac{e^{isx}}{is}\right)_{-1}^{1}=\frac{e^{is}-e^{-is}}{is}$$

$$\therefore \quad F(s) = \frac{2\sin s}{s} \qquad \left[\because \sin x = \frac{e^{ix} - e^{-ix}}{2i}\right]$$

(i) By inversion formula

$$f(x) = \frac{1}{2\pi}\int_{-\infty}^{\infty} F(s)e^{-isx}\,ds$$

$$= \frac{1}{2\pi}\int_{-\infty}^{\infty} \frac{2\sin s}{s} e^{-isx}\,ds$$

$$= \frac{1}{\pi}\int_{-\infty}^{\infty} \frac{\sin s\cos sx}{s}\,ds - \frac{i}{\pi}\int_{-\infty}^{\infty} \frac{\sin s\sin sx}{s}\,ds,$$

$$[\because e^{-ix} = \cos x - i\sin x]$$

$$= \frac{1}{\pi}\int_{-\infty}^{\infty} \frac{\sin s\cos sx}{s}\,ds$$

(since the integrad of other integral is an odd function)

$$\therefore \quad \int_{-\infty}^{\infty} \frac{\sin s\cos sx}{s}\,ds = \pi f(x) = \begin{cases} \pi & ,|x| < 1 \\ 0 & ,|x| > 1 \end{cases}$$

(ii) Putting $x = 0$ in (i), then

$$\int_{-\infty}^{\infty} \frac{\sin s}{s}\,ds = \pi$$

$$\Rightarrow \quad 2\int_{0}^{\infty} \frac{\sin s}{s}\,ds = \pi$$

$$\Rightarrow \quad \int_{0}^{\infty} \frac{\sin s}{s}\,ds = \frac{\pi}{2}$$

Example 2 *Find the fourier sine transform of* $\frac{e^{-ax}}{x}$.

Solution If $f(x) = \frac{e^{-ax}}{x}$, then fourier sine transform

$$F_s(s) = \int_{0}^{\infty} f(x)\sin sx\,dx$$

$$= \int_{0}^{\infty} \frac{e^{-ax}}{x}\sin sx\,dx$$

Differentiating both sides *w.r.t. s*, we get

$$\frac{d}{ds}F_s(s) = \int_0^\infty \frac{e^{-\alpha x}}{x} x \cos sx \, dx$$

$$= \int_0^\infty e^{-\alpha x} \cos sx \, dx$$

$$= \frac{\alpha}{s^2 + \alpha^2}$$

Integrating *w.r.t.* s, we get

$$F_s(s) = \int \frac{\alpha}{s^2 + \alpha^2} ds$$

$$= \tan^{-1} \frac{s}{\alpha} + c$$

When $s = 0, F_s(s) = 0 \therefore c = 0$

Hence, $F_s(s) = \tan^{-1} \frac{s}{\alpha}$

Example 3 Find the fourier cosine transform of the function f(x), if

$$f(x) = \begin{cases} \cos x & ,0 < x < a \\ 0 & ,x > a \end{cases}$$

Solution. We have

$$F_c(s) = \int_0^\infty f(x) \cos sx dx$$

$$= \int_0^a \cos x \cos sx \, dx$$

$$= \frac{1}{2} \int_0^a [\cos (1+s)\, x + \cos (1-s)\, x] dx$$

$$= \frac{1}{2} \left[\frac{\sin(1+s)x}{1+s} + \frac{\sin(1-s)x}{1-s} \right]_0^a$$

$$= \frac{1}{2} \left[\frac{\sin(1+s)a}{1+s} + \frac{\sin(1-s)a}{1-s} \right]$$

Example 4 Find the fourier transform of f(x) = $\begin{cases} 1-x^2 & ,|x| \le 1 \\ 0 & ,|x| > 1 \end{cases}$, Hence evaluate

$$\int_0^\infty \frac{x \cos x - \sin x}{x^3} . \cos \frac{x}{2} dx$$

Solution: We have, $F(s) = \int_{-\infty}^{\infty} f(x)\, e^{isx}\, dx$

$$= \int_{-\infty}^{-1} 0.\, e^{isx}\, dx + \int_{-1}^{1} (1-x^2)\, e^{isx}\, dx + \int_{1}^{\infty} 0.\, e^{isx}\, dx$$

$$= \int_{-1}^{1} (1-x^2)\, e^{isx}\, dx$$

$$= \left[\frac{(1-x^2)\, e^{isx}}{is} - \frac{2\,x\, e^{isx}}{(is)^2} - \frac{2\, e^{isx}}{(is)^3}\right]_{-1}^{1}$$

$$= \frac{-2\, e^{is}}{-s^2} - \frac{2\, e^{is}}{-is^3} - \frac{2\, e^{-is}}{-s^2} + \frac{2\, e^{-is}}{-is^3}$$

$$= -2\left(\frac{e^{is} + e^{-is}}{-s^2}\right) - 2\left(\frac{e^{is} - e^{-is}}{-is^3}\right)$$

$$= \frac{-4}{s^3}(s\cos s - \sin s)$$

By inversion formula, we get

$$f(x) = \frac{1}{2\pi}\int_{-\infty}^{\infty} F(s)\, e^{-isx}\, ds$$

$$= \frac{1}{2\pi}\int_{-\infty}^{\infty}\left(\frac{-4}{s^3}\right)(s\cos s - \sin s)\, e^{-isx}\, ds = \begin{cases} 1-x^2 & , |x| \le 1 \\ 0 & , |x| > 1 \end{cases}$$

Putting, $x = \frac{1}{2}$ we have

$$-\frac{2}{\pi}\int_{-\infty}^{\infty}\frac{1}{s^3}(s\cos s - \sin s)e^{\frac{-is}{2}}\, ds = \frac{3}{4}$$

or $$\int_{-\infty}^{\infty}\frac{s\cos s - \sin s}{s^3}\left(\cos\frac{s}{2} - i\sin\frac{s}{2}\right)ds = \frac{-3\pi}{8}$$

or

$$\int_{-\infty}^{\infty}\frac{s\cos s - \sin s}{s^3}\cos\frac{s}{2}\, ds - i\int_{-\infty}^{\infty}\frac{s\cos s - \sin s}{s^3}\sin\frac{s}{2}\, ds = \frac{-3\pi}{8}$$

or $$\int_{-\infty}^{\infty}\frac{s\cos s - \sin s}{s^3}\cos\frac{s}{2}\, ds = \frac{-3\pi}{8}$$

(since integrad of other integral is an odd function)

or, $$\int_0^\infty \frac{x\cos x - \sin x}{x^3}\cos\frac{x}{2}\,dx = \frac{-3\pi}{16}$$

Example 5 Find the fourier cosine transform of $f(x) = \dfrac{1}{1+x^2}$ and hence find fourier sine transform of $f(x) = \dfrac{x}{1+x^2}$.

Solution: Fourier cosine transform is

$$F_c(s) = \int_0^\infty f(x)\cos sx\,dx$$

$$= \int_0^\infty \frac{\cos sx}{1+x^2}\,dx \qquad \text{...(i)}$$

Differentiating w.r.t. s, we have

$$\frac{d}{ds}F_c(s) = -\int_0^\infty \frac{x\sin sx}{1+x^2}\,dx \qquad \text{...(ii)}$$

$$= -\int_0^\infty \frac{x^2 \sin sx}{x(1+x^2)}\,dx$$

$$= -\int_0^\infty \frac{\left[(1+x^2)-1\right]\sin sx}{x(1+x^2)}\,dx$$

$$= -\int_0^\infty \frac{\sin sx}{x}\,dx + \int_0^\infty \frac{\sin sx}{x(1+x^2)}\,dx$$

$$= -\frac{\pi}{2} + \int_0^\infty \frac{\sin sx}{x(1+x^2)}\,dx \qquad \text{...(iii)}$$

Differentiating again w.r.t. s, we get

$$\frac{d^2}{ds^2}F_c(s) = \int_0^\infty \frac{x\cos sx}{x(1+x^2)}\,dx = F_c(s)$$

or $(D^2\text{-}1)\,F_c(s) = 0$

Whose general solution is

$$F_c(s) = C_1e^{s} + C_2e^{-s} \qquad \text{...(iv)}$$

$$\therefore \quad \frac{d}{ds}F_c(s) = C_1e^{s} - C_2e^{-s} \qquad \text{...(v)}$$

when $s = 0$, eqs. (i) and (iv), gives $C_1 + C_2 = \int_0^\infty \frac{1}{1+x^2}\,dx = \frac{\pi}{2}$

Also, when $s = 0$, eqs. (iii) and (v), gives $C_1 - C_2 = \dfrac{-\pi}{2}$

By solving these equations, we get $C_1 = 0, C_2 = \dfrac{\pi}{2}$

$\therefore$ From (iv), $F_c(s) = \dfrac{\pi}{2} e^{-s}$

Now, $$F_s\{f(x)\} = \int_0^\infty \frac{x \sin sx}{1+x^2} dx$$

$$= -\frac{d}{ds} F_c(s) \qquad \text{[from Eq. (ii)]}$$

$$= -\frac{d}{ds}\left(\frac{\pi}{2} e^{-s}\right)$$

$$= \frac{\pi}{2} e^{-s}$$

Example 6. *Show that the fourier transform of* $f(x) = e^{\frac{-x^2}{2}}$ *is* $e^{\frac{-s^2}{2}}.\sqrt{2\pi}$.

Solution. We have,

$$F(s) = \int_{-\infty}^{\infty} f(x) e^{isx} dx$$

$$= \int_{-\infty}^{\infty} e^{\frac{-x^2}{2}} e^{isx} dx$$

$$= \int_{-\infty}^{\infty} e^{\frac{-1}{2}(x-is)^2} . e^{\frac{-s^2}{2}} dx$$

$\therefore$ Putting $\dfrac{x - is}{\sqrt{2}} = t$

$\Rightarrow$ $dx = \sqrt{2} dt$

$$= e^{\frac{-s^2}{2}} \int_{-\infty}^{\infty} e^{-t^2} . \sqrt{2} dt$$

$$= \sqrt{2} e^{\frac{-s^2}{2}} \int_{-\infty}^{\infty} e^{-t^2} dt$$

$$= \sqrt{2\pi} e^{\frac{-s^2}{2}} \qquad \left(\because \int_{-\infty}^{\infty} e^{-t^2} dt = \sqrt{\pi}\right)$$

Example 7 Find the fourier sine transfer of $e^{-|x|}$. Hence, show that

$$\int_0^\infty \frac{x\sin mx}{1+x^2}dx = \frac{\pi e^{-m}}{2}, m > 0.$$

Solution. In the interval $(0,\infty)$, x is positive.

Hence, $e^{-|x|} = e^{-x}$

$\therefore$ Fourier sine transform of $f(x)$ is

$$F_s(s) = \int_0^\infty f(x)\sin sx\, dx$$

$$= \int_0^\infty e^{-x}\sin sx\, dx$$

$$= \left[\frac{e^{-x}}{1+s^2}(-\sin sx - s\cos sx)\right]_0^\infty$$

$$= \frac{s}{1+s^2}$$

Using inversion formula, we get

$$f(x) = \frac{2}{\pi}\int_0^\infty F_s(s)\sin sx\, ds$$

or, $$e^{-x} = \frac{2}{\pi}\int_0^\infty \frac{s}{1+s^2}\sin sx\, ds$$

Putting $x = m$, we get

$$e^{-m} = \frac{2}{\pi}\int_0^\infty \frac{s\sin ms}{1+s^2}ds$$

$$\therefore \quad \int_0^\infty \frac{x\sin mx}{1+x^2}dx = \frac{\pi}{2}e^{-m}$$

Multiple Fourier Transforms

Let $f(x, y)$ be a function of two variables x and y, as a function of x, is its fourier transform is $F(s, y) = y \int_{-\infty}^\infty f(x,y)e^{isx}\, dx$

$F(s, y)$, as a function of y, its fourier transform is

$$F(q,y) = \int_{-\infty}^\infty F(s,y)e^{iqy}\, dy$$

$$\therefore \quad F(s,q) = \frac{1}{2\pi}\int_{-\infty}^\infty \int_{-\infty}^\infty f(x,y)e^{i(sx+qy)}\, dx\, dy$$

which is fourier transform of *f(x, y)*

and $$f(x,y)=\frac{1}{2\pi}\int_{-\infty}^{\infty}\int_{-\infty}^{\infty}F(s,q)e^{-i(sx+qy)}\,dx\,dy$$

which is the inversion formula of *f(x, y)*.

Convolution

The convolution of two functions *f(x)* and *g(x)* over the interval (-∞,∞) is defined as

$$f*g=\int_{-\infty}^{\infty}f(u)g(x-u)du$$

Convolution theorem for fourier transform (Falting theorem): The fourier transform of the convolution of *f(x)* and *g(x)* is the product of their fourier transforms *i.e.,*

$$F\{f(x)*g(x)\}=F\{f(x)\}.F\{g(x)\}$$

Proof. We have, $F\{f(x)*g(x)\}=F\left\{\int_{-\infty}^{\infty}f(u)g(x-u)du\right\}$

$$=\int_{-\infty}^{\infty}\left[\int_{-\infty}^{\infty}f(u)g(x-u)du\right]e^{isx}dx$$

$$=\int_{-\infty}^{\infty}f(u)\left[\int_{-\infty}^{\infty}g(x-u)e^{isx}dx\right]du$$

$$=\int_{-\infty}^{\infty}f(u)\left[\int_{-\infty}^{\infty}e^{is(x-u)}g(x-u)d(x-u)\right]e^{isu}du$$

Putting $x - u = t$

$$=\int_{-\infty}^{\infty}e^{isu}f(u)\left[\int_{-\infty}^{\infty}e^{ist}g(t)dt\right]du$$

$$=\int_{-\infty}^{\infty}e^{isu}f(u)\,du.\,F\{g(t)\}$$

$$=\int_{-\infty}^{\infty}e^{isx}f(x)\,dx.\,F\{g(x)\}$$

$$=F\{f(x)\}.F\{g(x)\}$$

Perseval's Identity for Fourier Transforms

If the fourier transform of *f(x)* and *g(x)* are *F(s)* and *G(s)* respectively, then

(i) $\frac{1}{2\pi}\int_{-\infty}^{\infty}F(s)G^{*}(s)\,ds=\int_{-\infty}^{\infty}f(x)\,g^{*}(x)dx$

(ii) $$\frac{1}{2\pi}\int_{-\infty}^{\infty}|F(s)|^2 ds = \int_{-\infty}^{\infty}|f(x)|^2 dx$$

Where'*' implies the complex conjugate.

Proof. We have,

(i) $$\int_{-\infty}^{\infty} f(x)g^*(x)\,dx = \int_{-\infty}^{\infty} f(x)\left[\frac{1}{2\pi}\int_{-\infty}^{\infty} G(s)e^{isx}ds\right]dx$$

$$= \frac{1}{2\pi}\int_{-\infty}^{\infty} G^*(s)\left[\int_{-\infty}^{\infty} f(x)e^{isx}dx\right]ds$$

$$= \frac{1}{2\pi}\int_{-\infty}^{\infty} G^*(s)\,F(s)ds$$

(ii) Taking $g(x) = f(x)$, we have

$$\frac{1}{2\pi}\int_{-\infty}^{\infty} F(s)F^*(s)ds = \int_{-\infty}^{\infty} f(x)f^*(x)dx$$

or, $$\frac{1}{2\pi}\int_{-\infty}^{\infty}|F(s)|^2 ds = \int_{-\infty}^{\infty}|f(x)|^2 dx$$

Relationship between Fourier and Laplace Transforms

Let us assume the function $f(t) = \begin{cases} e^{-xt}g(t), & t > 0 \\ 0 \quad , & t < 0 \end{cases}$ ...(i)

$\therefore$ The fourier transform of $f(t)$ is given by

$$F\{f(t)\} = \int_{-\infty}^{\infty} f(t)e^{ist}dt$$

$$= \int_{-\infty}^{0} 0.e^{ist}dt + \int_{0}^{\infty} e^{-xt}g(t)e^{ist}dt$$

$$= \int_{0}^{\infty} g(t)e^{(is-x)t}dt$$

Putting, $is - x = p$ $$= \int_{0}^{\infty} g(t)e^{-pt}dt$$

$$= L\{g(t)\}$$

Hence the fourier transform of the function $f(t)$ defined by (i) is the laplace transform of the function g(t).

Fourier Transform of the Derivatives of a Function

The Fourier Transform of $f'(x)$, the derivative of f(x) is –is F(s), where F(s) is the Fourier Transform of f(x).

Proof. We have,

$$F\{f'(x)\} = \int_{-\infty}^{\infty} f'(x)e^{isx}dx$$

$$= \int_{-\infty}^{\infty} \lim_{\delta x\to 0} \frac{f(x+\delta x)-f(x)}{\delta x}.e^{isx}dx$$

$$= \lim_{\delta x\to 0}\int_{-\infty}^{\infty} \frac{f(x+\delta x)}{\delta x}e^{isx}dx - \lim_{\delta x\to 0}\int_{-\infty}^{\infty}\frac{f(x)}{\delta x}e^{isx}dx$$

$$= \lim_{\delta x\to 0}\int_{-\infty}^{\infty} \frac{f(x+\delta x)}{\delta x}e^{is(x+\delta x)}.e^{-is\delta x}d(x+\delta x) - \lim_{\delta x\to 0}\frac{F(s)}{\delta x}$$

$$= \lim_{\delta x\to 0} e^{-is\delta x}\int_{-\infty}^{\infty}\frac{f(y)}{\delta x}e^{isy}dy - \lim_{\delta x\to 0}\frac{F(s)}{\delta x}, \text{Putting } x+\delta x = y$$

$$= \lim_{\delta x\to 0} e^{-is\delta x}\frac{F(s)}{\delta x} - \lim_{\delta x\to 0}\frac{F(s)}{\delta x}$$

$$= F(s)\lim_{\delta x\to 0}\frac{e^{-is\delta x}-1}{\delta x}$$

$$= -isF(s)$$

In general, the fourier transform of *n*th derivatives of *f*(x) is

$$F\{f^n(x)\} = (-is)^n F(s)$$

Example 8 If $f(x) = \begin{cases} 1\ , |x| < a \\ 0\ , |x| > a \end{cases}$ and $F(s) = \dfrac{2\sin sa}{s}$ then prove that

$$\int_0^{\infty}\frac{\sin^2 ax}{x^2}dx = \frac{\pi a}{2}$$

Solution: By using Perseval's identity

$$\int_{-\infty}^{\infty}|f(x)|^2 dx = \frac{1}{2\pi}\int_{-\infty}^{\infty}[F(s)]^2 ds$$

or $$\int_{-a}^{a} 1^2\, dx = \frac{1}{2\pi}\int_{-\infty}^{\infty}\left(\frac{2\sin sa}{s}\right)^2 ds$$

or $$2a = \frac{2}{\pi}\int_{-\infty}^{\infty}\frac{\sin^2 as}{s^2}ds$$

or $$\int_{-\infty}^{\infty}\frac{\sin^2 as}{s^2}ds = a\pi$$

or $$\int_{0}^{\infty}\frac{\sin^2 as}{s^2}ds = \frac{a\pi}{2}$$

$\therefore$ $$\int_{0}^{\infty}\frac{\sin^2 ax}{x^2}dx = \frac{a\pi}{2}$$

Example 9 *Solve the integral equation* $\int_0^{\infty} f(x)\cos sx\, dx = \begin{cases} 1-s & ,0 \le s \le 1 \\ 0 & ,s > 1 \end{cases}$. *Hence deduce that* $\int_0^{\infty}\frac{\sin^2 t}{t^2}dt = \frac{\pi}{2}$

Solution Let $\int_0^{\infty} F(x)\cos sx\, dx = f(s)$

$\therefore$ $$f(s) = \begin{cases} 1-s & ,0 \le s \le 1 \\ 0 & ,s > 1 \end{cases}$$

$\therefore$ By fourier cosine inversion formula, we have

$$F(x) = \sqrt{\frac{2}{\pi}}\int_0^{\infty} f_c(s)\cos sx\, ds$$

$$= \frac{2}{\pi}\int_0^{\infty} f(s)\cos sx\, ds$$

$$= \frac{2}{\pi}\int_0^{1}(1-s)\cos sx\, ds + \frac{2}{\pi}\int_1^{\infty} 0\cos sx\, ds$$

$$= \left[\frac{2}{\pi}.\frac{1-s}{x}\sin sx\right]_0^1 + \frac{2}{\pi x}\int_0^1 \sin sx\, ds$$

$$= \frac{2}{\pi x^2}(-\cos sx)_0^1$$

$$= \frac{2(1-\cos x)}{\pi x^2}$$

Now, we have, $\int_0^\infty F(x)\cos sx\,dx = \begin{cases} 1-s & ,0\le s\le 1 \\ 0 & ,s>1 \end{cases}$

where $\quad F(x) = \dfrac{2(1-\cos x)}{\pi x^2}$

$$\therefore \quad \frac{2}{\pi}\int_0^\infty \frac{1-\cos x}{x^2}\cos sx\,dx = \begin{cases} 1-s & ,0\le s\le 1 \\ 0 & ,s>1 \end{cases}$$

Letting $s \to 0$, we get

$$\frac{2}{\pi}\int_0^\infty \frac{1-\cos x}{x^2}\,dx = 1$$

or, $$\int_0^\infty \frac{2\sin^2\frac{x}{2}}{x^2}\,dx = \frac{\pi}{2}$$

Putting $x = 2t, dx = 2dt$

$$\therefore \quad \int_0^\infty \frac{\sin^2 t}{t^2}\,dt = \frac{\pi}{2}$$

Finite Fourier Sine Transforms

Let $f(x)$ be a function over some finite interval $(0, l)$ of the variable x. The finite fourier sine transform of $f(x)$ is defined as

$F_s(s) = \int_0^l f(x)\sin\frac{\pi sx}{l}\,dx$, where s is an integer and Inversion formula of sine transform is

$$f(x) = \frac{2}{l}\sum_{s=1}^{\infty} F_s(s)\sin\frac{\pi sx}{l}$$

Finite Fourier Cosine Transforms

Let $f(x)$ be a function over some finite interval $(0,l)$ of the variable x. The finite fourier cosine transform of $f(x)$ is defined as

$F_c(s) = \int_0^l f(x)\cos\frac{\pi sx}{l}\,dx$, where s is an integer.

and Inversion formula of cosine transform is

$$f(x)=\frac{1}{l}F_c(0)+\frac{2}{l}\sum_{s=1}^{\infty}F_c(s)\cos\frac{\pi sx}{l}$$

where $F_c(0)=\int_0^l f(x)dx$

Example 10 *Find the finite fourier sine and cosine transforms of the function $f(x)=2x, 0<x<4$.*

Solution. We have,

$$F_s(s)=\int_0^l f(x)\sin\frac{\pi sx}{l}dx$$

$$=\int_0^4 2x\sin\frac{\pi sx}{4}dx$$

$$=\left(\frac{-2x\cos\frac{\pi sx}{4}}{\frac{\pi s}{4}}\right)_0^4+2\int_0^4\frac{\cos\frac{\pi sx}{4}}{\frac{\pi s}{4}}dx$$

$$=\frac{-32}{s\pi}\cos\pi s+\frac{8}{\pi s}\left(\frac{\sin\frac{\pi sx}{4}}{\frac{\pi s}{4}}\right)_0^4$$

$$=\frac{-32}{\pi s}\cos\pi s$$

and $F_c(s)=\int_0^l f(x)\cos\frac{\pi sx}{l}dx$

$$=\int_0^4 2x\cos\frac{\pi sx}{4}dx$$

$$=\left(\frac{-2x\sin\frac{\pi sx}{4}}{\frac{\pi s}{4}}\right)_0^4-2\int_0^4\frac{\sin\frac{\pi sx}{4}}{\frac{\pi s}{4}}dx$$

$$=\frac{-8}{\pi s}\left(\frac{-\cos\frac{\pi sx}{4}}{\frac{\pi s}{4}}\right)_0^4$$

$$= \frac{32}{\pi^2 s^2}(\cos \pi s - 1), \text{ if } s > 0$$

If $s = 0$, then $F_c(s) = \int_0^4 2x.1dx = 16$

Example 11 *Find the finite cosine transform of* $\left(1 - \frac{x}{\pi}\right)^2$

Solution. We have,

$$F_c(s) = \int_0^{\pi} \left(1 - \frac{x}{\pi}\right)^2 \cos sx \, dx$$

$$\left[\left(1 - \frac{x}{\pi}\right)^2 \frac{\sin sx}{s}\right]_0^{\pi} + \frac{2}{\pi s}\int_0^{\pi}\left(1 - \frac{x}{\pi}\right)\sin sx \, dx$$

$$= \frac{2}{\pi s}\left[-\left(1 - \frac{x}{\pi}\right)\frac{\cos sx}{s}\right]_0^{\pi} - \frac{2}{\pi s}.\frac{1}{\pi s}\int_0^{\pi}\cos sx \, dx$$

$$= \frac{2}{\pi s^2} - \frac{2}{\pi^2 s^2}\left(\frac{\sin sx}{s}\right)_0^{\pi}$$

$$= \frac{2}{\pi s^2}, \text{ if } s > 0$$

and if $s = 0$, then

$$F_c(s) = \int_0^{\pi}\left(1 - \frac{x}{\pi}\right)^2 dx$$

$$= \left[\frac{-\pi}{3}\left(1 - \frac{x}{\pi}\right)^3\right]_0^{\pi}$$

$$= \frac{\pi}{3}$$

Example 12 *Find the finite fourier sine transform of f(x), if* $f(x) = \begin{cases} -x, & x<c \\ \pi - x, & x>c \end{cases}$, *where* $0 \le c \le \pi$.

Solution. We have

$$F_s(s) = \int_0^{\pi} f(x)\sin sx \, dx$$

$$= \int_0^c -x \sin sx \, dx + \int_c^\pi (\pi - x) \sin sx \, dx$$

$$= \left(\frac{x \cos sx}{s} \right)_0^c - \frac{1}{s} \int_0^c \cos sx \, dx - \left(\frac{(\pi - x) \cos sx}{x} \right)_c^\pi - \frac{1}{s} \int_c^\pi \cos sx \, dx$$

$$= \frac{c \cos sc}{s} - \frac{1}{s^2} (\sin sx)_0^c + \left(\frac{\pi - c}{s} \right) \cos sc - \frac{1}{s^2} (\sin sx)_c^\pi$$

$$= \frac{c \cos sc}{s} - \frac{\sin sc}{s^2} + \frac{(\pi - c) \cos sc}{s} - \frac{\sin s\pi}{s^2} + \frac{\sin sc}{s^2}$$

$$= \frac{\pi}{s} \cos sc$$

Application of Fourier Transforms to Boundary Value Problems

To find the required solution, we apply the following methods :

(i) If $u(x,t)_{x=0}$ is given, then we use infinite sine transform to remove $\frac{\partial^2 u}{\partial x^2}$ from the differential equation.

(ii) If $\left[\frac{\partial u(x,t)}{\partial x} \right]_{x=0}$ is given, then we use infinite cosine transform to remove $\frac{\partial^2 u}{\partial x^2}$ from the differential equation.

(iii) If $u(0,t)$ and $u(l,t)$ are given, then we use finite sine transform to remove $\frac{\partial^2 u}{\partial x^2}$ from the differential equation.

(iv) If $\left(\frac{\partial u}{\partial x} \right)_{x=0}$ and $\left(\frac{\partial u}{\partial x} \right)_{x=l}$ are given, then we use finite cosine transform to remove $\frac{\partial^2 u}{\partial x^2}$ from the differential equation.

Example 13 *The temperature V in the semi-infinite rod $0 \le x < \infty$ is determined by the differential equation*

$$\frac{\partial V}{\partial t} = k \frac{\partial^2 V}{dx^2}$$

Subject to the condition

(i) $V = 0$, *when* $t = 0, x \ge 0$

(ii) $\frac{\partial V}{\partial x} = -\mu$ (a constant), when $x = 0$ and $t > 0$.

Making use of cosine transform, show that

$$V(x,t) = \frac{2\mu}{\pi}\int_0^\infty \frac{\cos sx}{s^2}\left(1-e^{-ks^2t}\right)ds$$

Solution Taking the fourier cosine transform on both the sides, we get

$$\sqrt{\frac{2}{\pi}}\int_0^\infty \frac{\partial V}{\partial t}\cos sx\,dx = k\sqrt{\frac{2}{\pi}}\int_0^\infty \frac{\partial^2 V}{\partial x^2}\cos sx\,dx$$

or, $$\sqrt{\frac{2}{\pi}}\frac{d}{dt}\int_0^\infty V\cos sx\,dx = k\sqrt{\frac{2}{\pi}}\left[\frac{\partial V}{\partial x}\cos sx\right]_0^\infty + ks\sqrt{\frac{2}{\pi}}\int_0^\infty \frac{\partial V}{\partial x}\sin sx\,dx$$

or, $$\frac{d}{dt}V_c = -k\sqrt{\frac{2}{\pi}}\left(\frac{\partial V}{\partial x}\right)_{x=0} + ks\sqrt{\frac{2}{\pi}}(V\sin sx)_0^\infty - ks^2\sqrt{\frac{2}{\pi}}\int_0^\infty V\cos sx\,dx$$

Initially, when $x = 0$, $\left(\frac{\partial V}{\partial x}\right)_{x=0} = -\mu$ and if $x \to \infty$, then $V \to 0$

$$\therefore \frac{dV_c}{dt} = \sqrt{\frac{2}{\pi}}k\mu - ks^2V_c \quad \text{or,} \quad \frac{dV_c}{dt} + ks^2V_c = \sqrt{\frac{2}{\pi}}k\mu$$

which is linear differential equation of first order, then

$$\text{I.F.} = e^{\int ks^2 dt} = e^{ks^2t}$$

∴ It solution is

$$e^{ks^2t}V_c = \sqrt{\frac{2}{\pi}}k\mu\int e^{ks^2t}dt + A = \sqrt{\frac{2}{\pi}}\frac{\mu}{s^2}e^{ks^2t} + A$$

when $t = 0, V_c = 0$

$$\therefore \quad 0 = \sqrt{\frac{2}{\pi}}.\frac{\mu}{s^2} + A$$

or, $$A = -\frac{\mu}{s^2}\sqrt{\frac{2}{\pi}}$$

$$\therefore \quad V_c = \sqrt{\frac{2}{\pi}}.\frac{\mu}{s^2}\left(+1-e^{-ks^2t}\right)$$

Applying the inverse fourier cosine transform, we have

$$V = \sqrt{\frac{2}{\pi}}\int_0^\infty V_c\cos sx\,ds$$

$$= \sqrt{\frac{2}{\pi}} \int_0^\infty \sqrt{\frac{2}{\pi}} \frac{\mu}{s^2}\left(-e^{-ks^2t}+1\right)_c \cos sx\, ds$$

$$V = \frac{2\mu}{\pi} \int_0^\infty \frac{\cos sx}{s^2}\left(1-e^{-ks^2t}\right) ds$$

Example 14 *Using the fourier sine transform, solve the differential equation* $\frac{\partial U}{\partial t} = k\frac{\partial^2 U}{dx^2}$, *for x>0, t>0, under the boundary conditions* $U=U_0$ *when* $x=0, t>0$ *and the initial condition* $U=0$, *when* $t=0$, $x>0$.

Solution Taking the fourier sine transform on both the sides, we have

$$\sqrt{\frac{2}{\pi}} \int_0^\infty \frac{\partial U}{\partial t} \sin sx\, dx = k\sqrt{\frac{2}{\pi}} \int_0^\infty \frac{\partial^2 U}{\partial x^2} \sin sx\, dx$$

or, $$\frac{d}{dt}\sqrt{\frac{2}{\pi}} \int_0^\infty U \sin sx\, dx = -ks^2 U_s + ks\sqrt{\frac{2}{\pi}}U(0,t)$$

or, $$\frac{dU_s}{dt} + ks^2 U_s = ks\sqrt{\frac{2}{\pi}}U_0 \qquad [\because U(0,t) = U_0]$$

which is a linear differential equation of first order.

$\therefore$ I.F. $= e^{\int ks^2 dt} = e^{ks^2 t}$

Hence, its solution is

$$U_s e^{ks^2 t} = \sqrt{\frac{2}{\pi}} \int ks U_0 e^{ks^2 t} dt + A$$

$$\sqrt{\frac{2}{\pi}} \frac{U_0}{s} e^{ks^2 t} + c \qquad \ldots(i)$$

But when $t = 0, U_s = 0$ $\quad [\because U = 0 \text{ when } t = 0]$

$\therefore$ From Eq (i), we have

$$0 = \frac{U_0}{s}\sqrt{\frac{2}{\pi}} + A$$

$$\Rightarrow \quad A = -\frac{U_0}{s}\sqrt{\frac{2}{\pi}}$$

$\therefore$ From Eq. (i), $U_s e^{ks^2 t} = \frac{U_0}{s}\sqrt{\frac{2}{\pi}}\left(e^{ks^2 t} - 1\right)$

or, $$U_s = \frac{U_0}{s}\sqrt{\frac{2}{\pi}}\left(1-e^{-ks^2 t}\right)$$

Now, applying the inverse fourier transform, we have

$$U = \frac{2}{\pi}U_0 \int_0^\infty \frac{1-e^{-ks^2 t}}{s}\sin sx\, ds$$

or, $$U = \frac{2U_0}{\pi}\left[\int_0^\infty \frac{\sin sx}{s}ds - \int_0^\infty \frac{e^{-ks^2 t}}{s}\sin sx\, ds\right]$$

$$= \frac{2U_0}{\pi}\left[\frac{\pi}{2} - \int_0^\infty \frac{e^{-ks^2 t}}{s}\sin sx\, ds\right]$$

or, $$U(x,t) = U_0\left[1 - \frac{2}{\pi}\int_0^\infty \frac{e^{-ks^2 t}}{s}\sin sx\, ds\right]$$

Example 15 *Use the complex form of the fourier transform to show that* $V = \frac{1}{2\sqrt{\pi t}}\int_{-\infty}^{\infty} f(u)\,\text{exp.}\left\{\frac{-(x-u)}{4t}\right\}du$ *is the solution of the boundary value problem* $\frac{\partial V}{\partial t} = \frac{\partial^2 V}{\partial x^2}, -\infty < x < \infty, t > 0, V = f(x),$ *when* $t = 0$

Solution Taking the complex fourier transform on both the sides, we have

$$\frac{1}{\sqrt{2\pi}}\int_{-\infty}^{\infty}\frac{\partial V}{\partial t}e^{isx}dx = \frac{1}{\sqrt{2\pi}}\int_{-\infty}^{\infty}\frac{\partial^2 V}{\partial x^2}e^{isx}dx$$

or, $$\frac{d}{dt}V(s) = (-is)^2 V(s)$$

or, $$\frac{dV(s)}{dt} = -s^2 V(s)$$

whose solution is $V(s) = Ae^{-s^2 t}$

But when $t = 0, A = V(s)$

At $t = 0$, $V(s) = \frac{1}{\sqrt{2\pi}}\int_{-\infty}^{\infty} f(u)e^{isu}du \quad [\because V = f(x), \text{when } t = 0]$

$$\therefore\ V(s) = e^{-s^2 t}\frac{1}{\sqrt{2\pi}}\int_{-\infty}^{\infty} f(u)e^{isu}du$$

Taking the inverse fourier transform, we have

$$V(x,t)=\frac{1}{\sqrt{2\pi}}\int_{-\infty}^{\infty}\left[\frac{e^{-s^2t}}{\sqrt{2\pi}}\int_{-\infty}^{\infty}f(u)e^{isu}du\right]e^{-isx}ds$$

$$=\frac{1}{2\pi}\int_{-\infty}^{\infty}f(u)\left[\int_{-\infty}^{\infty}e^{-s^2t-is(x-u)}ds\right]du$$

$$=\frac{1}{2\pi}\int_{-\infty}^{\infty}f(u)\left[\int_{-\infty}^{\infty}e^{-t\left[s+\frac{i(x-u)}{2t}\right]^2-\frac{(x-u)^2}{4t}}ds\right]du$$

$$=\frac{1}{2\pi}\int_{-\infty}^{\infty}f(u)\left[\int_{-\infty}^{\infty}e^{-t\left[s+\frac{i(x-u)}{2t}\right]^2}ds\right]e^{\frac{-(x-u)^2}{4t}}du$$

Putting $\sqrt{t}\left[s+\dfrac{i(x-u)}{2t}\right]=y$

So that, $ds=\dfrac{dy}{\sqrt{t}}$

$$=\frac{1}{2\pi}\int_{-\infty}^{\infty}f(u)\left[\frac{1}{\sqrt{t}}\int_{-\infty}^{\infty}e^{-y^2}dy\right].e^{\frac{-(x-u)^2}{4t}}du$$

$$=\frac{1}{2\pi}\int_{-\infty}^{\infty}f(u)\frac{1}{\sqrt{t}}\sqrt{\pi}e^{\frac{-(x-u)^2}{4t}}du,$$

$$\left[\because \text{since} \int_{-\infty}^{\infty}e^{-y^2}dy=\sqrt{\pi}\right]$$

or, $$V=\frac{1}{2\sqrt{\pi t}}\int_{-\infty}^{\infty}f(u)\exp.\left\{\frac{-(x-u)^2}{4t}\right\}du$$

Example 16 *An infinitely long string having one end at x=0, is initially at rest along the x-axis. The end x=0 is given a transverse displacement f(t), t>0. Find the displacement of any point of the string at any time.*

Solution. Let U(x,t) be the transverse displacement of any point x of the string at any time t. Then we have to solve the wave equation

$$\frac{\partial^2U}{\partial t^2}=c^2\frac{\partial^2U}{\partial x^2},x>0,t>0 \qquad ...(i)$$

Subject to the condition $U(x,0)=0,U_t(x,0)=0,U(0,t)=f(t)$ and the displacement $U(x,t)$ is bounded.

The laplace transform of Eq. (i) defined as

$$L\{U(x,t)\} = \overline{U}(x,s) \text{ is}$$

$$S^2\overline{U} - SU(x,0) - \frac{\partial U(x,0)}{\partial t} = c^2 \frac{\partial^2 \overline{U}}{dx^2}$$

By using the first two conditions, we have

$$\frac{d^2\overline{U}}{dx^2} = \left(\frac{s}{c}\right)^2 \overline{U} \qquad \text{...(ii)}$$

Similarly, the laplace transforms of the third and fourth conditions

$$\overline{U}(0,s) = \overline{f}(s) \text{at x=0} \qquad \text{...(iii)}$$

and $\overline{U}(x,s)$ is bounded. ...(iv)

By solving Eq. (ii), we have

$$\overline{U}(x,s) = c_1 e^{\frac{sx}{c}} + c_2 e^{\frac{-sx}{c}}$$

From (iv) condition, we get $c_1 = 0$

From (iii) condition, we get $c_2 = \overline{f}(s)$

$$\therefore \quad \overline{U}(x,s) = \overline{f}(s) e^{\frac{-sx}{c}}$$

Using the complex inversion formula, we get

$$U = \frac{1}{2\pi i}\int_{\alpha-i\infty}^{\alpha+i\infty} e^{\left(t-\frac{x}{c}\right)s}\,\overline{f}(s)ds$$

$$= f\left(t - \frac{x}{c}\right)$$

Example 17 *Use finite fourier transform to solve:*

$$\frac{\partial V}{\partial t} = \frac{\partial^2 V}{\partial x^2}, V(0,t) = 0, V(4,t) = 0, V(x,0) = 2x,$$

where $0 < x < 4, t > 0$.

Solution Taking the finite fourier sine transform on both sides, we have

$$\int_0^4 \frac{\partial V}{\partial t} \sin\frac{\pi s x}{4} dx = \int_0^4 \frac{\partial^2 V}{\partial x^2} \sin\frac{\pi s x}{4} dx$$

or, $$\frac{d}{dt}V_s = \frac{-s^2\pi^2}{16}V_s + \frac{s\pi}{4}[V(0,t) - V(4,t)\cos s\pi]$$

or, $$\frac{dV_s}{dt} = \frac{-s^2\pi^2}{16}V_s$$

whose solution is $V_s = Ae^{-\frac{s^2\pi^2 t}{16}}$ ...(i)

Since, $V(x,0) = 2x, 0 < x < 4$

Taking finite fourier sine transform, we have

at $t = 0$, $V_s = \int_0^4 2x \sin\frac{\pi s x}{4} dx$

$$= \left[-2x.\frac{4}{\pi s}\cos\frac{\pi s x}{4} + 2.\frac{16}{s^2\pi^2}\sin\frac{s\pi x}{4}\right]_0^4$$

$$= \frac{-32\cos s\pi}{\pi s} = \frac{32}{\pi s}(-1)^{s+1} = A$$

$\therefore$ From Eq. (i)

$$V_s = \frac{32}{\pi s}(-1)^{s+1} e^{\frac{-s^2\pi^2 t}{16}}$$

Taking the inverse finite fourier sine transform, we have

$$V(x,t) = \frac{2}{4}\sum_{s=1}^{\infty}\frac{32(-1)^{s+1}}{\pi s}.e^{\frac{-\pi^2 s^2 t}{16}}.\sin\frac{\pi s x}{4}$$

$$= \frac{16}{\pi}\sum_{s=1}^{\infty}\frac{(-1)^{s+1}}{s}.e^{\frac{-\pi^2 s^2 t}{16}}.\sin\frac{\pi s x}{4}$$

which is the required solution.

Example 18 *Use the finite cosine transform to solve* $\frac{\partial V}{\partial t} = k\frac{\partial^2 V}{dx^2}$, *with the boundary conditions* $\frac{\partial V}{\partial x} = 0$, *when* $x = 0$ *and* $x = \pi, t > 0$ *and the initial condition* $V = f(x)$, *when* $t = 0, 0 < x < \pi$.

Solution. Taking the finite fourier cosine transform on both the sides, we have

$$\int_0^{\pi} \frac{\partial V}{\partial t} \cos sx\, dx = k \int_0^{\pi} \frac{\partial^2 V}{\partial x^2} \cos sx\, dx$$

or, $$\frac{d}{dt} V_c = k[-s^2 V_c - \{V.(0,t) - V(\pi,t)\cos \pi s\}]$$

or, $$\frac{dV_c}{dt} = -ks^2 V_c$$

whose solution is

$$V_c = Ae^{-ks^2 t} \quad \text{...(i)}$$

Now, taking finite fourier cosine transform of initial condition, we have at $t=0$,

$$V_c = \int_0^{\pi} f(x) \cos sx\, dx = A$$

$\therefore$ From Eq. (i)

$$V_c = e^{-ks^2 t} \int_0^{\pi} f(x) \cos sx\, dx$$

Taking the inverse finite fourier cosine transform, we have

$$V(x,t) = \frac{1}{\pi} V_c(0) + \frac{2}{\pi} \sum_{s=1}^{\infty} V_c(s) \cos sx$$

$$= \frac{1}{\pi} \int_0^{\pi} f(x) dx + \frac{2}{\pi} \sum_{s=1}^{\infty} \left[e^{-ks^2 t} \cos sx \int_0^{\pi} f(x) \cos sx\, dx \right]$$

Practice Problems

1. Find the complex transform of $f(x)$, if

$$f(x) = \begin{cases} e^{iwx} & , a < x < b \\ 0 & , x < a, x > b \end{cases}$$

$$\left[\text{Ans. } \frac{i}{\sqrt{2\pi}} \left[\frac{e^{-i(s+w)b} + e^{i(s+w)a}}{s+w} \right] \right]$$

2. Find fourier sine transform of $f(x)=\frac{1}{x}$.

$$\left[\text{Ans.} \sqrt{\frac{\pi}{2}}\right]$$

3. Find the fourier sine transform of $f(x)=\frac{1}{x(a^2+x^2)}$.

$$\left[\text{Ans.} \frac{\pi(1-e^{-sa})}{a^2\sqrt{2\pi}}\right]$$

4. Find the inverse fourier transform of $F(s)=e^{-|s|y}$.

$$\left[\text{Ans.} \frac{y\sqrt{2}}{\sqrt{2}(y^2+x^2)}\right]$$

5. Find the $f(x)$ if its cosine transform is $F_c(s)=\begin{cases}\frac{1}{\sqrt{2\pi}}\left(a-\frac{s}{2}\right) & ,s<2a \\ 0 & ,s\geq 2a\end{cases}$

$$\left[\text{Ans.} \frac{\sin^2 ax}{\pi s^2}\right]$$

6. Find the fourier sine and cosine transform of the function x^{m-1}.

$$\left[\text{Ans.} F_s(s)=\frac{\Gamma m}{s^m}\sqrt{\frac{2}{\pi}}\sin\frac{m\pi}{2},\ F_c(s)=\frac{\Gamma m}{s^m}\sqrt{\frac{2}{\pi}}\cos\frac{m\pi}{2}\right]$$

7. Solve for $f(x)$ the integral equation $\int_0^\infty f(x)\sin xt\, dx=\begin{cases}1, & 0\leq t<1 \\ 2, & 1\leq t<2 \\ 0, & t\geq 2\end{cases}$

$$\left[\text{Ans.} \frac{1}{\pi x}(2+2\cos x-4\cos 2x)\right]$$

8. Show that $\int_0^\infty \frac{\cos \mu x}{\mu^2+1}d\mu=\frac{\pi}{2}e^{-x}, x\geq 0$

9. Find the finite fourier sine and cosine transforms of $f(x)=x$.

$$\left[\text{Ans.} F_s(s)=\frac{\pi(-1)^{s+1}}{s},\ F_c(s)=\frac{(-1)^{s-1}}{s^2}\right]$$

10. Show that the finite sine transform of $\frac{x}{\pi}$ is $(-1)^{s+1}\frac{1}{s}$.

11. Find the finite fourier cosine transform of $f(x)$, if $f(x) = -\frac{\cos k(\pi - x)}{k \sin k\pi}$.

$$\left[\text{Ans.} \frac{1}{s^2 - k^2}\right]$$

Boundary Value Problems

12. Solve $\frac{\partial U}{\partial t} = \frac{\partial^2 U}{\partial x^2}$; $x > 0, t > 0$ subject to the conditions

(i) $U = 0$, when $x = 0, t > 0$

(ii) $U = \begin{cases} 1 & , 0 < x < 1 \\ 0 & , x \geq 1 \end{cases}$, when $t = 0$

(iii) $U(x,t)$ is bounded.

$$\left[\text{Ans.} U(x,t) = \frac{2}{\pi}\int_0^\infty \frac{1 - \cos s}{s} . e^{-s^2 t} . \sin sx \, ds\right]$$

13. Solve $\frac{\partial U}{\partial t} = 2\frac{\partial^2 U}{\partial x^2}$, if $U(0,t) = 0$, $U(x,0) = e^{-x}$, $x > 0, U(x,t)$ is bounded where $x > 0, t > 0$.

$$\left[\text{Ans.} U(x,t) = \frac{2}{\pi}\int_0^\infty \frac{s e^{-2s^2 t} \sin sx}{1 + s^2} ds\right]$$

14. If the flow of heat is linear so that the variation of temperature θ with z and y may be neglected and if it is assumed that no heat is generated in the medium, then solve the differential equation $\frac{\partial \theta}{\partial t} = k\frac{\partial^2 \theta}{\partial x^2}$, where $-\infty < x < \infty$ and $\theta = f(x)$ when $t = 0$, $f(x)$ being a given function of x.

$$\left[\text{Ans.} \theta(x,t) = \frac{1}{\sqrt{2\pi}}\int_{-\infty}^\infty F(s) e^{-ks^2 t - isx} ds\right]$$

15. Use finite fourier transform to solve $\frac{\partial V}{\partial t} = \frac{\partial^2 V}{\partial x^2}$, $0 < x < 6, t > 0$ and $V(0,t) = 0, V(6,t) = 0, V(x,0) = 2x$.

$$\left[\text{Ans.} V(x,t) = 6 + \frac{24}{\pi^2}\sum_{s=1}^{\infty}\left(\frac{\cos\pi s - 1}{s^2}\right)e^{\frac{-\pi^2 s^2 t}{36}}\cos\frac{\pi s x}{6}\right]$$

16. Find the solution of the wave equation $\frac{\partial^2 U}{\partial t^2} = 4\frac{\partial^2 U}{\partial x^2}$ subject to the conditions $U(0, t) = 0$, $U(\pi, t) = 0$, $U(x, 0) = (0.1)\sin x + (0.01)\sin 4x$ and $U_t(x, 0) = 0$ for $0 < x < \pi$, $t > 0$.

[**Ans.** $U(x, t) = (0.1)\cos 2t \sin x + (0.01)\cos 8t \sin 4x$]

17. Find the steady temperature $V(x, y)$ in a long square bar of sides π when on face is kept at constant and the other faces at zero temperature. Also, $V(x, y)$ is bounded.

$$\left[\text{Ans.} V(x,y) = \frac{4V_0}{\pi}\sum_{n=0}^{\infty}\frac{\sin h\,(2n+1)y \sin(2n+1)x}{(2n+1)\sin h\,(2n+1)\pi}\right]$$

18. Use a cosine transform to show that the steady temperature in the semi-infinite solid $y > 0$ when the temperature on the surface $y = 0$ is kept at unity over the strip$|x| < a$ and at zero outside the strip, is $\frac{1}{\pi}\left[\tan^{-1}\left(\frac{a+x}{y}\right) + \tan^{-1}\left(\frac{a-x}{y}\right)\right]$

The result $\int_0^{\infty} e^{-sx}x^{-1}\sin rx\,dx = \tan^{-1}\frac{r}{s},\ r > 0, s > 0$ may be used.

Chapter - 7

The Discrete Fourier Transform

The Discrete Fourier Transform

Fourier transform of discrete time signal $S(n)$ is called Discrete time Fourier Transform (DTFT) and it is represented by $S(w)$, which is a continuous function of frequency (w). This frequency domain representation of a signal is called Discrete Fourier Transform.

DTFT of a discrete time signal $S(n)$ is given by

$$S(w) = S(e^{jw}) = \sum_{n=-\infty}^{\infty} S(n)\, e^{-jwn}$$

Inverse DTFT is given by the following equation

$$S(n) = \frac{1}{2\pi} \int_{2\pi} S(e^{jw})\, e^{jwn}\, dw$$

DFT of a discrete time sequence $S(n)$ is related to DTFT of same sequence is given by

$$S(k) = \sum_{n=0}^{N-1} S(n)\, e^{\frac{-j\,2\pi\, kn}{N}} \quad , n = 0.1,2,\ldots\ldots, N-1$$

and Inverse DFT is defined as

$$S(n) = \frac{1}{N} \sum_{n=0}^{N-1} S(k)\, e^{\frac{j\,2\pi\, kn}{N}} \quad , n = 0.1,2,\ldots\ldots, N-1$$

Example 1 *A finite duration sequence of length L is given as*

$$S(n) = \begin{cases} 1 \ , 0 \le n \le L-1 \\ 0 \ , \quad \textit{otherwise} \end{cases}$$

Determine the N – point DFT of sequence for $N \ge L$.

Solution DFT of sequence $S(n)$ is

$$S(k) = \sum_{n=0}^{N-1} S(n)\, e^{\frac{-j2\pi kn}{N}} = \sum_{n=0}^{L-1} 1.e^{\frac{-j2\pi kn}{N}}$$

$$= e^{\frac{-j2\pi k.0}{N}} + e^{\frac{-j2\pi k.1}{N}} + \ldots + e^{\frac{-j2\pi k}{N}(L-1)}$$

$$= 1 + e^{\frac{-j2\pi k}{N}} + \ldots + e^{\frac{-j2\pi k}{N}(L-1)}$$

Which is a geometric progression form

$$\therefore \quad S(k) = \frac{1.\left[1 - e^{\frac{-j2\pi k}{N}(L-1+1)}\right]}{1 - e^{\frac{-j2\pi k}{N}}}$$

$$= \frac{1 - e^{\frac{-j2\pi kL}{N}}}{1 - e^{\frac{-j2\pi k}{N}}}, \; k = 0, 1, \ldots\ldots, N\text{-}1$$

This is the N – point DFT of the sequence $S(n)$.

Example 2 *Compute the 8-point DFT of the signal*

$S(n) = \{1, 1, 1, 1, 1, 1, 0, 0\}$

Also, sketch its magnitude and phase.

Solution DFT of a signal $S(n)$ is defined as

$$S(k) = \sum_{n=0}^{N-1} S(n)\, e^{\frac{-j2\pi kn}{N}}$$

Given, $N = 8$

$$\therefore \quad S(k) = \sum_{n=0}^{7} S(n)\, e^{\frac{-j\pi kn}{4}}$$

$$= S(0)\, e^{\frac{-j\pi k.0}{4}} + S(1)\, e^{\frac{-j\pi k}{4}} + S(2)\, e^{\frac{-j\pi k.2}{4}} + S(3)\, e^{\frac{-j\pi k.3}{4}} + S(4)\, e^{\frac{-j\pi k.4}{4}} + S(5)\, e^{\frac{-j\pi k.5}{4}} + S(6)\, e^{\frac{-j\pi k.6}{4}} + S(7)\, e^{\frac{-j\pi k.7}{4}}$$

$$= 1 + e^{\frac{-j\pi k}{4}} + e^{\frac{-j2\pi k}{4}} + e^{\frac{-j3\pi k}{4}} + e^{\frac{-j4\pi k}{4}} + e^{\frac{-j5\pi k}{4}}$$

$$\left(\because e^{-jk} = \cos x - j\sin x\right)$$

Now, k = 0, S (0) = 6

k = 1, S (1) = - 0.7071- j 1.7071

k = 2, S (2) = 1- j

k = 3, S (3) = 0.7071- j 0.2929

k = 4, S (4) = 0

k = 5, S (5) = 0.7071-j 0.2929

k = 6, S (6) = 1 + j

k = 7, S (7) = - 0.7071-j 1.7071

∴ DFT of $S(n)$ is

$$S(k) = \{ S(0), S(1), S(2), S(3), S(4), S(5), S(6), S(7)\}$$

$$= \{6, -0.7071-j\,1.7071, 1-j, 0.7071+j\,0.2929, 0, 0.7071-j\,0.2929, 1+j, -0.7071+j\,1.7071\}$$

$$\text{Amplitude}=\sqrt{(\text{Real part})^2+(\text{Imaginary part})^2}$$

$$|S(k)| = \{6, 1.8478, 1.4142, 0.7654, 0, 0.7654, 1.4142, 1.8478\}$$

$$\text{Phase} = \tan^{-1}\left(\frac{\text{Imaginary part}}{\text{Real part}}\right)$$

$$\angle S(k) = \{0,-1.9635,-0.785, 0.3927, 0,-0.3927, 0.785, 1.9635\}$$

The DFT as a Linear Transformation

DFT and IDFT can be computed by using matrices, Computation of N – point DFT requires $N \times N$ complex multiplication and $N(N-1)$ complex additions, N – point vector S_N of frequency sample is defined as

$$S_N = \begin{bmatrix} S(0) \\ S(1) \\ \vdots \\ S(N-1) \end{bmatrix}$$

N – Point vector S_N of signal sequence $S(n)$ is defined as

$$S_n = \begin{bmatrix} S(0) \\ S(1) \\ \vdots \\ S(n-1) \end{bmatrix}$$

$$\text{DFT } S(k) = \sum_{n=0}^{N-1} S(n)\, e^{\frac{-j\,2\pi\,kn}{N}} = \sum_{n=0}^{N-1} S(n)\, W_N^{kn},\ k = 0, 1 \ldots, N\text{-}1$$

$$\text{IDFT } S(n) = \frac{1}{N} \sum_{k=0}^{N-1} S(k)\, e^{\frac{j\,2\pi\,kn}{N}} = \frac{1}{N} \sum_{k=0}^{N-1} S(k)\, W_N^{kn},\ n = 0, 1 \ldots., N\text{-}1$$

Where $W_N = e^{\frac{-j\,2\pi}{N}}$ is an Nth root of unity.

$N-$ Point DFT in $N \times N$ matrix form as

$$S_N = W_N . S_N$$

Where W_N is the matrix of the linear transformation and it is a symmetric matrix.

$N-$ Point IDFT in matrix form as

$$S_N = W_N^{-1}\, S_N = \frac{1}{N}\, W_N^{*}\, S_N$$

Where W_N^{*} denotes the complex conjugate of the matrix W_N and W_N^{-1} is the inverse of W_N.

The relationship between W_N and W_N^{*} will be

$$W_N . W_N^{*} = N . I_N$$

Where I_N is an $N \times N$ identity matrix.

W_N is given by

$$W_N = \begin{bmatrix} 1 & 1 & 1 & \cdots\cdots & 1 \\ 1 & W_N & W_N^2 & \cdots\cdots & W_N^{N-1} \\ \vdots & W_N^2 & W_N^4 & \cdots\cdots & W_N^{2(N-1)} \\ \vdots & \vdots & \vdots & \cdots\cdots & \\ 1 & W_N^{N-1} & W_N^{2(N-1)} & \cdots\cdots & W_N^{(N-1)(N-1)} \end{bmatrix}$$

Example 3 *Compute theDFT of the four- point sequence* $S(n) = \{1, 2, 3, 4\}$, *using DFT transformation matrix. Also determine IDFT.*

Solution The first step is to determine the matrix W_4 because S(n) has 4-points.

$$W_4 = \begin{bmatrix} W_4^0 & W_4^0 & W_4^0 & W_4^0 \\ W_4^0 & W_4^1 & W_4^2 & W_4^3 \\ W_4^0 & W_4^2 & W_4^4 & W_4^6 \\ W_4^0 & W_4^3 & W_4^6 & W_4^9 \end{bmatrix} \quad \text{where } W_4^0 = \left[e^{\frac{-j\,2\pi}{4}} \right]^0 = 1$$

$$= \begin{bmatrix} 1 & 1 & 1 & 1 \\ 1 & W_4^1 & W_4^2 & W_4^3 \\ 1 & W_4^2 & W_4^0 & W_4^2 \\ 1 & W_4^3 & W_4^2 & W_4^1 \end{bmatrix} = \begin{bmatrix} 1 & 1 & 1 & 1 \\ 1 & -j & -1 & j \\ 1 & -1 & 1 & -1 \\ 1 & j & -1 & -j \end{bmatrix}$$

Then $S_4 = W_4 S_4$

$$= \begin{bmatrix} 1 & 1 & 1 & 1 \\ 1 & -j & -1 & j \\ 1 & -1 & 1 & -1 \\ 1 & j & -1 & -j \end{bmatrix} \times \begin{bmatrix} 1 \\ 2 \\ 3 \\ 4 \end{bmatrix}$$

$$= \begin{bmatrix} 1+2+3+4 \\ 1-2j-3+4j \\ 1-2+3-4 \\ 1+2j-3-4j \end{bmatrix} = \begin{bmatrix} 10 \\ -2+2j \\ -2 \\ -2-2j \end{bmatrix}$$

Inverse DFT of S_4 is determined by

$$S_4 = \frac{1}{4} W_4^* S_4$$

$$= \frac{1}{4} \begin{bmatrix} 1 & 1 & 1 & 1 \\ 1 & j & -1 & -j \\ 1 & -1 & 1 & -1 \\ 1 & -j & -1 & j \end{bmatrix} \times \begin{bmatrix} 10 \\ -2+2j \\ -2 \\ -2-2j \end{bmatrix}$$

$$= \frac{1}{4}\begin{bmatrix} 10 -2+2j-2-2-2j \\ 10-2j+2j^2+2+2j+2j^2 \\ 10+2-2j-2+2+2j \\ 10+2j-2j^2+2-2j-2j^2 \end{bmatrix} = \frac{1}{4}\begin{bmatrix} 4 \\ 8 \\ 12 \\ 16 \end{bmatrix} = \begin{bmatrix} 1 \\ 2 \\ 3 \\ 4 \end{bmatrix} \quad (\because j^2 = -1)$$

Remark $W_4^4 = W_4^{(0+4)} = W_4^0, W_4^1 = e^{\frac{-j2\pi}{4}} = -j$

$$W_4^6 = W_4^{(2+4)} = W_4^2, W_4^2 = \left(e^{\frac{-j2\pi}{4}}\right)^2 = -1$$

$$W_4^9 = W_4^{(1+2\times4)} = W_4^1, W_4^3 = \left(e^{\frac{-j2\pi}{4}}\right)^3 = j$$

Properties of the DFT

(i) Periodicity : If $S(n)$ is a discrete-time signal and $S(k)$ = is the N-point DFT of $S(n)$.

Then $S(n+N) = S(n)$, for all n

$S(k+N) = S(k)$, for all k

(ii) Linearity: If $S_1(n) \xleftrightarrow[N-\text{point}]{DFT} S_1(k)$

$$S_2(n) \xleftrightarrow[N-\text{point}]{DFT} S_2(k)$$

Then for any real-valued or complex-valued constants a_1 and a_2.

$$a_1S_1(n) + a_2S_2(n) \xleftrightarrow[N-\text{point}]{DFT} a_1S_1(k) + a_2S_2(k)$$

(iii) Symmetry Property: Let us assume that the N- point sequence $S(n)$ and its DFT are both complex-valued. Then the sequence can be expresses as

$$S(n) = S_R(n) + j\, S_1(n), 0 \le n \le N-1$$

$$S(k) = S_R(k) + j\, S_1(k), 0 \le k \le N-1$$

By substituting into the expression for the DFT, we have

$$S_R(k) = \sum_{n=0}^{N-1}\left[S_R(n)\cos\frac{2\pi kn}{N} - S_1(n)\sin\frac{2\pi kn}{N}\right] \quad \text{... (i)}$$

$$S_1(k) = -\sum_{n=0}^{N-1}\left[S_R(n)\sin\frac{2\pi kn}{N} - S_1(n)\cos\frac{2\pi kn}{N}\right] \quad \text{... (ii)}$$

and IDFT given by

$$S_R(n) = \frac{1}{N}\sum_{k=0}^{N-1}\left[S_R(k)\cos\frac{2\pi kn}{N} - S_1(k)\sin\frac{2\pi kn}{N}\right] \quad \text{... (iii)}$$

$$S_1(n) = \frac{1}{N}\sum_{k=0}^{N-1}\left[S_R(k)\sin\frac{2\pi kn}{N} + S_1(k)\cos\frac{2\pi kn}{N}\right] \quad \text{... (iv)}$$

Case I **If the sequence** $S(n)$ is real. Then

$$S(N-k) = S^*(k) = S(-k)$$

Consequently $|S(N-k)| = |S(k)|$

and $\angle S(N-k) = -\angle S(k)$

Case II If $S(n)$ is real and even. *i.e.,*

$$S(n) = S(N-n), 0 \le n \le N-1$$

Then (ii) yield $S_1(k) = 0$. Hence, the DFT reduces to

$$S(k) = \sum_{n=0}^{N-1} S(n) \cos \frac{2\pi kn}{N}, 0 \le k \le N-1$$

and IDFT reduces to

$$S(n) = \frac{1}{N} \sum_{k=0}^{N-1} S(k) \cos \frac{2\pi kn}{N}, 0 \le n \le N-1$$

Case III If $S(n)$ is real and odd. *i.e.,*

$$S(n) = -S(N-n), \; 0 \le n \le N-1$$

then (i) yield $S_R(k) = 0$. Hence

$$S(k) = -j \sum_{n=0}^{N-1} S(n) \sin \frac{2\pi kn}{N}, 0 \le k \le N-1$$

which is purely imaginary and odd.
The IDFT reduces to

$$S(n) = j \frac{1}{N} \sum_{k=0}^{N-1} S(k) \sin \frac{2\pi kn}{N}, 0 \le n \le N-1$$

Case IV If $S(n)$ is purely imaginary *i.e.,* $S(n) = jS_1(n)$.

Then (i) and (ii) reduces to

$$S_R(k) = \sum_{n=0}^{N-1} S_1(n) \sin \frac{2\pi kn}{N}$$

$$S_1(k) = \sum_{n=0}^{N-1} S_1(n) \cos \frac{2\pi kn}{N}$$

We see that $S_R(k)$ is odd and $S_1(k)$ is even

If $S_1(n)$ is odd, then $S_1(k) = 0$ and hence $S(k)$ is purely real. On the other hand, if $S_1(n)$ is even, then $S_R(k) = 0$ and hence $S(k)$ is purely imaginary.

(iv) Circular convolution: Circular convolution is defined as

$$S_3(m)=\sum_{n=0}^{N-1} S_1(n)S_2\left[(m-n)\right]_N, \textit{where } m=0,1,\ldots.., N-1.$$

where $S_2\left[(m-n)\right]_N$ is a circular shift of sequence $S_2(n)$ and $S_1(n)$ and $S_2(n)$ are two discrete-time sequences of finite length of *N*- point. Circular convolution is represented as

$$S_3(n)=S_1(n) \textcircled{N} S_2(n)=S_2(n)\textcircled{N}S_1(n)$$

where $\textcircled{N}$ denotes *N*-point circular convolution of two sequences.

Modulo *N*- operation: The circular shift of the sequence *S* (*n*) can be expressed as the index modulo *N*.

$$S_1(n)=S(n-k,\text{modulo } N)=S\left[(n-k)\right]_N$$

For example: If $k=2$ and $N=4$, then

for $n=0$, $\quad S_1(0)=S\left[(0-2)\right]_4=S(2)$

$n=1$, $\quad S_1(1)=S\left[(1-2)\right]_4=S\left[(-1)\right]_4=S(3)$

$n=2$, $\quad S_1(2)=S\left[(2-2)\right]_4=S\left[(0)\right]_4=S(0)$

$n=3$, $\quad S_1(3)=S\left[(3-2)\right]_4=S\left[(1)\right]_4=S(1)$

Example 4 *Determine the circular convolution of two sequences* $S_1(n)$ *and* $S_2(n)$ *defined*

as $\quad S_1(n)=S_2(n)=1,\ \textit{for } 0\le n\le N-1$

$\quad = 0$, otherwise

Solution DFT of these sequences will be

$$S_1(k)=S_2(k)$$

$$=\sum_{n=0}^{N-1} 1.e^{\frac{-(j\,2\pi kn)}{4}}$$

Above computation is applied for $N=4$

$$S_1(k)=S_2(k)=\sum_{n=0}^{3} e^{\frac{-(j\,2\pi kn)}{4}}$$

$$= 1 + e^{\frac{-jk\pi}{2}} + e^{-jk\pi} + e^{\frac{-jk3\pi}{2}}$$

when, $k = 0, S_1(0) = S_2(0) = 4$

$k = 1, S_1(1) = S_2(1) = 0$

$k = 2, S_1(2) = S_2(2) = 0$

$k = 3, S_1(3) = S_2(3) = 0$

This implies that $S_1(k) = S_2(k) = \begin{cases} N & , \text{for } k = 0 \\ 0 & , \text{otherwise} \end{cases}$

Now, the resultant of circular convolution of two sequences

$$S_1(n) = S_2(n) = \begin{cases} 1 & , 0 \le n \le N-1 \\ 0 & , \text{otherwise} \end{cases}$$

is given by

$$S_3(k) = S_1(k) \,.\, S_2(k)$$
$$= N \,.\, N = N^2, \text{ for } k = 0$$

and IDFT is

$$S_3(n) = \frac{1}{N} \sum_{k=0}^{N-1} S_3(k)\, e^{\frac{i2\pi kn}{N}}$$

$$= \frac{1}{N} . N^2 = N, \text{ for } 0 \le n \le N-1$$

Example 5 *Compute the circular convolution of the following two sequences:*

$$S_1(n) = \{2,1,2,1\},\ S_2(n) = \{1,2,3,4\}$$

Solution Circular convolution is defined as

$$S_3(m) = \sum_{n=0}^{N-1} S_1(n) S_2\left[(m-n)\right]_N,$$

where $m = 0,1,\ldots\ldots, N-1$.

Since the given sequence has 4-points *i.e.*, $N = 4$

The sequence $S_2\left[(m-n)\right]_4$ are graphed in a counter clockwise direction on a circle. Beginning with $m = 0$, we have

$$S_3(0) = \sum_{n=0}^{3} S_1(n)\ S_2\left[(-n)\right]_4$$

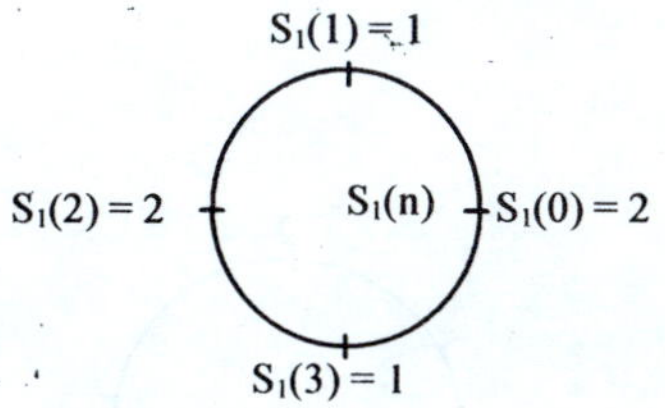

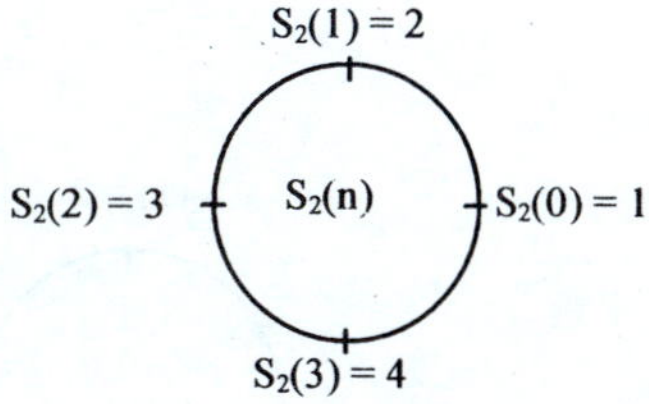

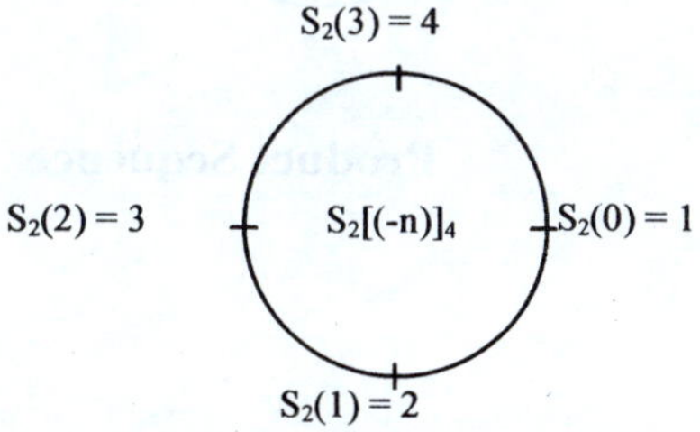

Folded Sequence of $S_2(n)$

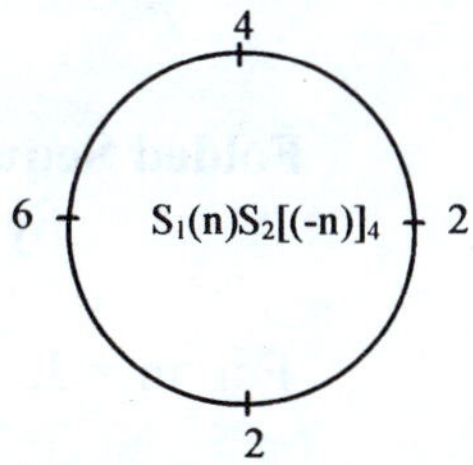

Product Sequence

$$\therefore \; S_3(0) = 2+4+6+2 = 14$$

For m=1

$$S_3(1) = \sum_{n=0}^{3} S_1(n)\, S_2\left[(1-n)\right]_4$$

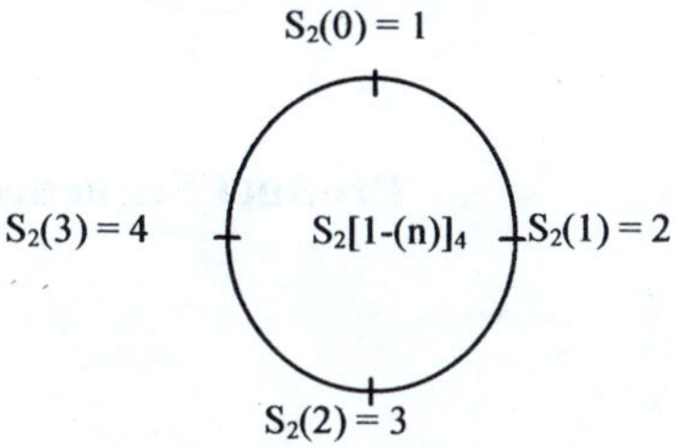

Folded Sequence rotated by one unit

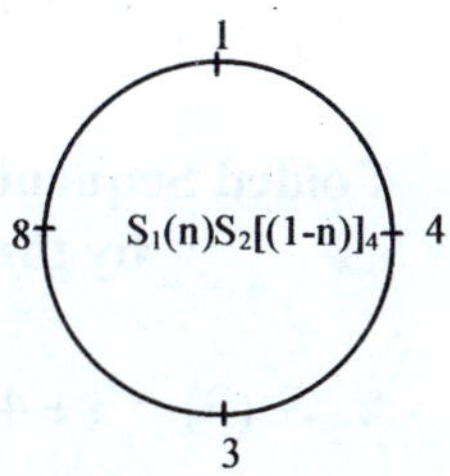

Product Sequence

$$\therefore \; S_3(1) = 1+8+3+4 = 16$$

For m=2

$$S_3(2) = \sum_{n=0}^{3} S_1(n)\, S_2\left[(2-n)\right]_4$$

$$\therefore \; S_3(2) = 2+2+4+6 = 14$$

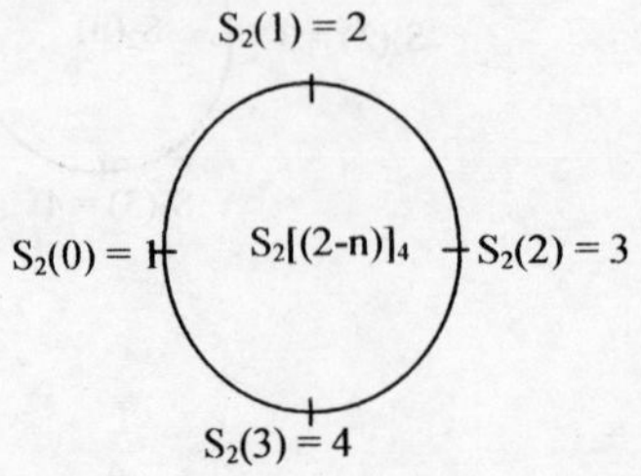

Folded Sequence rotated by two unit

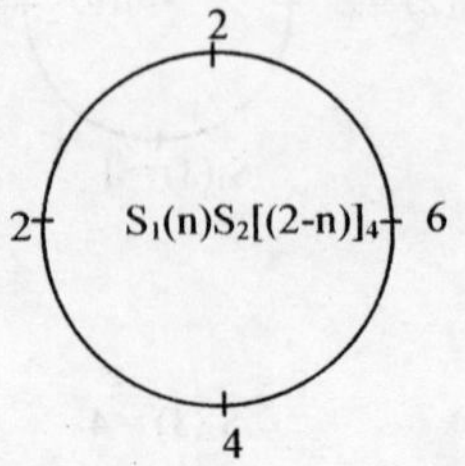

Product Sequence

For m = 3,

$$S_3(3) = \sum_{n=0}^{3} S_1(n)\, S_2\left[(3-n)\right]_4$$

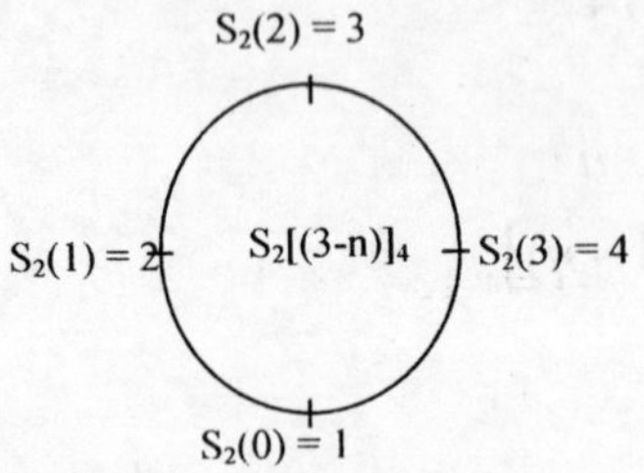

Folded Sequence rotated by three unit

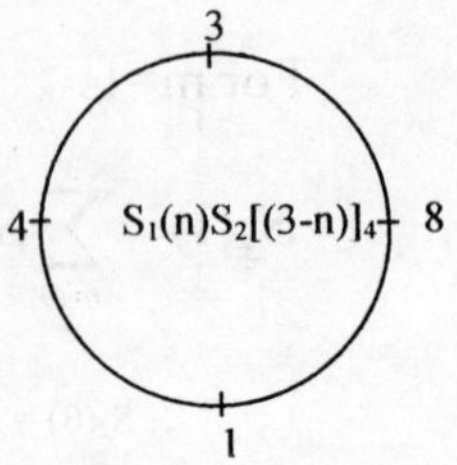

Product Sequence

$$\therefore\ S_3(3) = 3 + 4 + 1 + 8 = 16$$

Therefore, the circular convolution of two sequences $S_1(n)$ and $S_2(n)$ yields the sequence

$$S_3(n) = \{14, 16, 14, 16\}$$

Practice Problems

1. Determine the DFT of the sequence given as follows:

$$S(n)=\begin{cases}\frac{1}{3}, & 0\le n\le 2\\ 0, & \text{otherwise}\end{cases}$$

2. Deter mine the eight point DFT of the signal

$$S(n)=\{1, 1, 0, 1, 1, 0, 0, 0\}$$

and sketch its magnitude and phase.

3. Evaluate the N-point DFT of the sequence:

$$S(n)=\begin{cases}1, & n=3,7\\ 0, & n=1, 2, 4, 5, 6, 8, 9\end{cases}, \text{ Given } N=10$$

4. The first five points of the 8-point DFT of a real-value sequence are {0.25, 0.125-j0.3018, 0, 0.125-j *0.0518, 0*}. Determine the remaining three points.

5. Determine the circular convolution of the following sequences using the time-domain formula.

   (i) $S_1(n)=\{1, 2, 3, 1\}$ and $S_2(n)=\{4, 3, 2, 2\}$

   (ii) $S_1(n)=\{1, 2, 3, 1\}$ and $S_2(n)=\{1, 1, 1, 1\}$

6. Give the two 5-point sequence as

$$S_1(n)=\{0, 1, 2, 3, 4\} \text{ and } S_2(n)=\{0, 1, 0, 0, 0\}$$

Determine $S_1(n)=S_1(n) \textcircled{N} S_2(n)$ by DFT method.

7. Find the DFT of the sample sequence $x(n)=\{1,1,2,2,3,3\}$ and compute the amplitude and phase spectrum.

Chapter - 8

Wavelet and Haar Transform

Introduction

A wavelet is a "small wave", which has`its energy concentrated in time. It gives a tool widely used for many scientific purposes, but it is well suited only to the study of stationary signals where all frequencies have an infinite coherence time.

The fourier analysis brings only global information which is not sufficient to direct compact patterns. Gabor introduced a local fourier analysis, taking into account a sliding window, leading to a time frequency analysis. This method is only applicable to situations where the coherence time is independent of the frequency. Morlet introduced the wavelet transform in order to have a coherence time proportional to the period.

***Morlet's wavelet*:** The wavelet defined by morlet is,

$$\xi(w) = e^{-2\pi^2}(v - v_0)^2$$

It is a complex wavelet which can be decomposed in two parts, one for the real and the other for the imaginary part.

$$\xi_r(x) = \frac{1}{\sqrt{2\pi}} e^{\frac{-x^2}{2}} \cos(2\pi v_0 x)$$

and

$$\xi_i(x) = \frac{1}{\sqrt{2\pi}} e^{\frac{-x^2}{2}} \sin(2\pi v_0 x)$$

Where v_0 is a constant. The admissibilitv condition is verified only if $v_0 > 0.8$.

The graphs of real and imaginary parts shows the function as under :

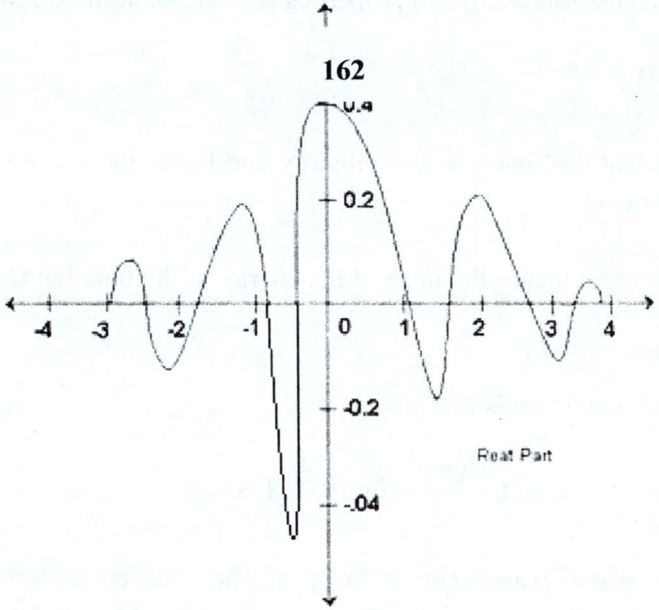

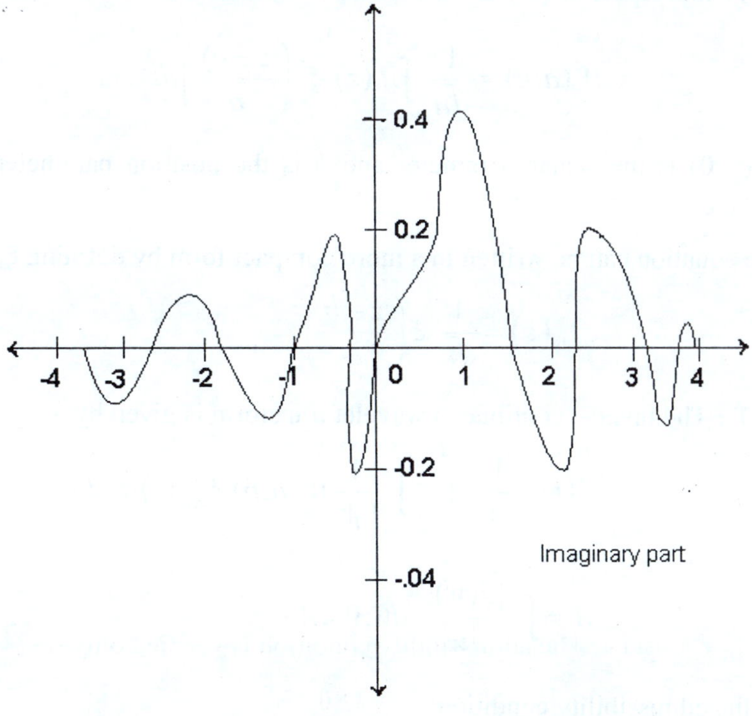

Mother wavelet : A real or complex-valued continuous function ξ (*t*) is called mother wavelet if it verified the following two properties and one admissible condition.

(i) $\int_{-\infty}^{\infty} \xi(t)dt = 0$

This means that the function is oscillatory and hence there is no need to use sine and cosine waves.

(ii) $\int_{-\infty}^{\infty} |\xi(t)|^2 \, dt < \infty$ it means the most of the energy of the function ξ(*t*) is confined to a finite duration.

(iii) Admissibility condition is given by

$$A = \int_{-\infty}^{\infty} \frac{|\xi(w)|^2}{|w|} dw : 0 < A < \infty$$

The continuous wave transform (CWT) : The Morlet-Grossmann defined the continuous wavelet transform of *f* (*z*), a square integrable function i.e., $f(z) \in L^2(R)$ with respect to ξ (*t*), a wavelet for 1 – D signal is

$$W(a,b) = \frac{1}{\sqrt{a}} \int_{-\infty}^{\infty} f(z).\xi^* \left(\frac{z-b)}{a}\right) dz$$

Where a (> 0) is the scalar parameter and *b* is the position parameter, $\xi^*(z)$ is the analyzing wavelet.

The above equation can be written in a more compact form by defining $\xi_{a,b}(z)$ as

$$\xi_{a,b}(z) = \frac{1}{\sqrt{a}} \xi\left(\frac{z-b}{a}\right)$$

Inverse CWT : The inverse continuous wavelet transform is given by

$$\xi(t) = \frac{1}{A} \int_{a=-\infty}^{\infty} \int_{b=-\infty}^{\infty} \frac{1}{|a|^2} W(a,b)\, \xi_{a,b}(t)\, da db$$

Where $$A = \int_{-\infty}^{\infty} \frac{|\xi(w)|^2}{|w|} dw; 0 < A < \infty$$

Which is the admissibility condition.

Properties of CWT : The CWT is characterized by the following three properties :

1. **Linerity :** If $f(t)$ and $g(t) \in L^2(R)$ and α and β are scalars,

   Then, $$W_\xi[\alpha f(t) + \beta g(t)] = aW_\xi[f(t)] + \beta W_\xi[g(t)]$$

2. **Translation :** The CWT of a translated function $f(z - u)$ is given by
$$W_\xi[f(z-u)] = W(a, b-u)$$

3. **Scaling :** The CWT of a scalar function is given by
$$W_\xi\left[\frac{1}{\sqrt{\alpha}} + f\left(\frac{z}{\alpha}\right)\right] = W\left[\frac{a}{\alpha}, \frac{b}{\alpha}\right]$$

Application of wavelet transform : Applications of wavelet transforms are

1. Data compression (image compression, video compression, audio compression).
2. Sound synthesis.
3. Computer and human vision.
4. Fingerprint compression.
5. Denoising noisy data.
6. Statistical analysis.

Haar Transform: The haar transform is the simplest useful energy compression process. In 1-D, the transform a 2-demant vector.

$$[x(1) \;\; x(2)]^T \text{ into } [y(1) \;\; y(2)]^T \text{ using}$$

$$\begin{bmatrix} y(1) \\ y(2) \end{bmatrix} = T \begin{bmatrix} x(1) \\ x(2) \end{bmatrix}$$

Where $$T = \frac{1}{\sqrt{2}}\begin{pmatrix} 1 & 1 \\ 1 & -1 \end{pmatrix}$$

Then $y(1)$ and $y(2)$ are simply the sum and difference of $x(1) + x(2)$, scaled by $\frac{1}{\sqrt{2}}$ to preserve energy. T is on orthogonal matrix because its rows are orthogonal to each other and they are normalized to unit magnitude. Therefore $T^1 = T^T$.

Hence we may recover X form Y using $\begin{bmatrix} x(1) \\ x(2) \end{bmatrix} = T^T \begin{bmatrix} y(1) \\ y(2) \end{bmatrix}$

In 2 – D, X and Y becomes 2 x 2 matrices. We may transform first the column of X, by premultyplying by T and then the rows of the result by post multiplying by T^T. Hence

$$Y = T\,X\,T^T$$

and to invert

$$X = T^T\,YT$$

If
$$X = \begin{bmatrix} a & b \\ c & d \end{bmatrix}$$

Then,
$$Y = \frac{1}{2}\begin{bmatrix} a+b+c+d & a-b+c-d \\ a+b-c-d & a-b-c+d \end{bmatrix}$$

Remark : The hear wavelet's mother wavelet function $\psi\,(t)$ can be described as

$$\psi(t) = \begin{cases} 1 & 0 \le t < \dfrac{1}{2} \\ -1 & \dfrac{1}{2} \le t < 1 \\ 0 & Otherwise \end{cases}$$

and its sealing function φ(t) can be described as

$$\phi(t) = \begin{cases} 1 & 0 \le t < 1 \\ 0 & otherwise \end{cases}$$

Chapter - 9

Theory of Probability

Introduction

The word "probable" refers to chance or likelihood of the happening of some event. The concept of probability is extremely important. It has found a very extensive application in the development of all physical sciences. We explain some definitions before discussing probability.

Random Experiment

An experiment is said to be a random experiment if it can be performed a large number of times under essentially unchanged conditions, admitted several possible outcomes.

If a performance is called an outcomes of case.

For example : Tossing a coin, throwing a dice.

Sample point

The outcome of the random experiment is called a sample point.

Sample Space

The set of all possible outcomes of a random experiment is called a sample space. A sample space is represented by S. The sample space is also known as 'Universal Set'.

Discrete Sample Space

A sample space is said to be discrete if it contains finite or infinite but countable sample points.

For example: If we throw dice as many times we required six number. Then sample space

$$S = \{1\}, \{0,1\} \{0,0,1\} \text{ etc.}$$

Continuous Sample Space

A sample space is said to be continuous if it contains uncountable sample points.

For example : All the points on a straight line

Event

The subsets of the sample space are said to be events.

For example : In a single throw of dice. The event is getting an even number is **E** = {2, 4, 6}

Sample Space is **S** = {1,2,3,4,5,6}

$\Rightarrow$ $\mathbf{E \subseteq S}$

Types of Events

(i) **Composite event :** The union of two events *X* and *Y* denoted by $X \cup B$ or $X + Y$ is called composite event.

(ii) **Compound event :** The intersection of two events *X* and *Y* denoted by $X \cap Y$ or *A-B* is called a compound event.

Mutually Exclusive Event

There is no common element between two events *X* and *Y*. Then these events are called mutually exclusive *i.e.* $X \cap Y = \phi$. It is also known as incompatible event.

For example : If we throw a dice, then

Sample space, S = {1,2,3,4,5,6}

Let, X = event of going a number less than 3

Y = event of getting a number more than 4

$\therefore$ X = {1,2}, Y = {5,6}

$\Rightarrow$ $X \cap Y = \phi$

Exhaustive Event

Two events are said to be exhaustive if their union is equal to the sample space. *i.e.*

$$S = X \cup Y$$

Complement Event

If X be an event. Then non – happening event $\bar{A}$ is said to be complement *i.e.* $\bar{A}$ contains all those elements which are not in *A*.

Independent Event

Two events X and Y are said to be independent if and only if

$$P(X \cup Y) = P(X) \,.\, P(Y)$$

For example : The toss of two coins or the drawing of two cards with replacement after each draw are independent.

Odd in Favour of an Event

If an event X can happen in a ways and fail in b ways, then odd in favour of an event is $a{:}b$ and probability is $P(X) = \dfrac{a}{a+b}$

Odd Against the Event

If an event X can happen in a ways and fail in b ways, then odd in against of an event is $b{:}a$ and probability is $P\left(\overline{X}\right) = \dfrac{b}{a+b}$

Probability of a Event

Let S be sample space in a random experiment and A be a subset of S then A is an event the probability of an event A is defined as

$$P(A) = \frac{\text{Number of favourable cases of A}}{\text{Total number of possible cases}}$$

$$\therefore \qquad P(A) = \frac{n(A)}{n(S)}$$

Addition Theorem of Probabilities

If $P_1, P_2, \ldots\ldots\ldots P_n$ be probabilities of n mutually exclusive events $A_1, A_2, \ldots\ldots\ldots\ldots A_n$, then the probability P, that any of these events will happen is given by.

$$P = P_1 + P_2 + \ldots\ldots\ldots + P_n$$

Proof : Let $A_1, A_2, \ldots\ldots\ldots A_n$ be the mutually exclusive events, Whose probabilities are $P_1, P_2, \ldots\ldots\ldots\ldots P_n$ respectively. Let N be the total number of trials, then any one of them can happen on

$$P_1N + P_2N + \ldots\ldots\ldots\ldots\ldots + P_nN \text{ occasions.}$$

Hence, probability P is given by

$$P = \frac{P_1N + P_2N + \ldots\ldots\ldots\ldots + P_nN}{N}$$

$$P(A_1 + A_2 + \ldots\ldots + A_n) = \frac{P_1 N}{N} + \frac{P_2 N}{N} + \ldots\ldots\ldots + \frac{P_n N}{N}$$

$$= P(A_1) + P(A_2) + \ldots\ldots\ldots + P(A_n)$$

$$\Rightarrow \quad P = P_1 + P_2 + \ldots\ldots\ldots\ldots + P_n$$

Particular Cases

(a) If A and B are defined on the sample space S. Then

$$n(A \cup B) = n(A) + n(B)$$

$$\therefore P(A \cup B) = \frac{n(A \cup B)}{n(S)}$$

$$= \frac{n(A) + n(B)}{n(S)}$$

$$= \frac{n(A)}{n(S)} + \frac{n(B)}{n(S)}$$

$$P(A \cup B) = P(A) + P(B)$$

(b) Since S and ϕ are mutually exclusive events and $S \cup \phi = S$.

$$\therefore \quad P(S \cup \phi) = P(S)$$

$$\Rightarrow \quad P(S) + P(\phi) = P(S)$$

$$\Rightarrow \quad P(\phi) = 0$$

(c) Let A and $\bar{A}$ are complement events. Then

$$\therefore \quad P(A \cup \bar{A}) = P(S)$$

$$\Rightarrow \quad P(A) + P(\bar{A}) = 1 \quad [P(S) = 1]$$

$$\Rightarrow \quad P(\bar{A}) = 1 - P(A)$$

(d) If two events A and B are not mutually exclusive. Then either A or B or both occur is defined as

$$P(A \cup B) = P(A) + P(B) - P(A \cap B)$$

Proof : Let $A \cup B$ as the union of two disjoint events $A \cap \bar{B}$ and B, then

$$P(A \cup B) = P[(A \cup \bar{B}) \cup B]$$

$$= P(A \cap \bar{B}) + P(B)$$

$$= P(A) - P(A \cap B) + P(B)$$

$$\Rightarrow \quad P(A \cup B) = P(A) + P(B) - P(A \cap B)$$

Conditional Probability

When an event A, depend upon the another event B. Then the probability of A is called the conditional probability and is denoted by $P\left(\frac{A}{B}\right)$ Thus,

$$P\left(\frac{A}{B}\right) = \frac{n(A \cup B)}{n(B)}$$

$$= \frac{\frac{n(A \cup B)}{n(S)}}{\frac{n(B)}{n(S)}}$$

$$\therefore \quad P\left(\frac{A}{B}\right) = \frac{P(A \cup B)}{P(B)}$$

Provided $P(B) \neq 0$.

Multiplication Theorem (Theorem of Compound Probability)

If A and B are two independent events, then the probability that both will happen is

$$P(A \cap B) = P(A) \cdot P(B)$$

Proof : Let the ways m be favourable to event A and n ways favourable to both the events A and B. Then the probability that the events A and B will happen simultaneously.

$$= \frac{n}{N}$$

$$= \frac{m}{N} \cdot \frac{n}{m} = P(A)$$

Now, $\frac{m}{N}$ be the probability that A will happen and $\frac{n}{m}$ be the conditional probability that B will happen when A already happen $= P\left(\frac{B}{A}\right)$

If $P(A \cap B)$ be the probability that A and B will occur simultaneously, then

$$P(A \cap B) = P(A).\ P\left(\frac{B}{A}\right)$$

Similarly, $P(A \cap B) = P(B).\ P\left(\frac{A}{B}\right)$

But, if *A* and *B* are independent events, then $P\left(\frac{B}{A}\right)$ is the same as *P(B)*.

Hence, $P(A \cap B) = P(A) \,.\, P(B)$

Illustrative Examples

Example 1 *What is the chance that a leap year, selected at random will contain 53 Sunday?*

Solution A leap year contains 366 days and therefore 52 weeks and 2 days. There are 52 Sunday in 52 weeks.

For the remaining 2 days, there may be

(i) Sunday and Monday
(ii) Monday and Tuesday
(iii) Tuesday and Wednesday
(iv) Wednesday and Thursday
(v) Thursday and Friday
(vi) Friday and Saturday
(vii) Saturday and Sunday

Now, for having 53 Sunday in the leap year, one of above 2 days must be Sunday. Thus, out of the 7 possibilities, 2 favour the events that one of the two days is a Sunday.

$\therefore$ Required probability $= \frac{2}{7}$

Example 2 *Form a pack of 52 cards, two cards are drawn at random. Find the probability that both cards are of spade.*

Solution Two cards can be drawn from a pack of 52 cards is

$$n(S) = {}^{52}C_2 = \frac{52 \times 51}{2 \times 1} = 1326$$

Now, 2 cards can be drawn out of 13 cards of spade.

$$n(A) = {}^{13}C_2 = \frac{13 \times 12}{2 \times 1} = 78$$

$\therefore$ Required probability $P(A) = \dfrac{n(A)}{n(S)} = \dfrac{78}{1326} = \dfrac{1}{17}$

Example 3 *Four persons are chosen at random from a group containing 3 men, 2 women and 4 children. Show that the chance that exactly two of them will be children is* $\dfrac{10}{21}$.

Solution Let A be the two of them are children.

Out of 9 persons 4 can be chosen in ${}^9C_4 = n(S)$

Since two of them are children *i.e.* ${}^4C_2$

Other two are to be selected out of 5 persons *i.e.* ${}^5C_2$

Thus, happening the event A are

$$n(A) = {}^4C_2 \times {}^5C_2$$

Hence, the required probability

$$P(A) = \frac{n(A)}{n(S)}$$

$$= \frac{{}^4C_2 \times {}^5C_2}{{}^9C_4} = \frac{10}{21}.$$

Example 4 *The chance of one event happening is the square of the chance of a second event, but the odds against the first are cube of the odd against the second. Find the chance of each.*

Solution Let A be the first event and B be the second event.

Let $$P(B) = \frac{1}{x} = \frac{1}{1+(x-1)}$$

$$P(A) = [P(B)]^2 = \frac{1}{x^2} = \frac{1}{1+(x^2-1)}$$

Odd against $$A = \frac{P(\bar{A})}{P(A)} = \frac{x^2-1}{1}$$

Odd against $$B = \frac{P(\bar{B})}{P(B)} = \frac{x-1}{1}$$

But, odd against A = (odd against B)^3

$\therefore \quad x^2-1 = (x-1)^3$

or $\quad x^3 - 4x^2 + 3x = 0$

$\Rightarrow \quad x(x-1)(x-3) = 0$

$\Rightarrow \quad x = 0, x = 1, x = 3$

Here, $x = 3$ satisfies the assumption.

Hence, $P(A) = \frac{1}{9}, P(B) = \frac{1}{3}$

Example 5 *A box containing 6 red, 4 white and 5 black balls. A person draw 4 balls from the box at random. Find the probability that among the balls drawn there is at least one ball of each colour.*

Solution Total number of balls = 6+4+5 =15

Let, A be at least one ball of each colour.

This event can happen in any form of three mutually exclusive events.

A_1 = 2 red, 1 white, 1 black

A_2 = 1 red, 2 white, 1 black

A_3 = 1 red, 1 white, 2 black

$$\therefore \quad P(A_1) = \frac{{}^6C_2 \times {}^4C_1 \times {}^5C_1}{{}^{15}C_4} = \frac{300}{1365}$$

$$P(A_2) = \frac{{}^6C_1 \times {}^4C_2 \times {}^5C_1}{{}^{15}C_4} = \frac{180}{1365}$$

$$P(A_3) = \frac{{}^6C_1 \times {}^4C_1 \times {}^5C_2}{{}^{15}C_4} = \frac{240}{1365}$$

$\therefore$ Required probability

$$P(A) = P(A_1) + P(A_2) + P(A_3)$$

$$= \frac{300}{1365} + \frac{180}{1365} + \frac{240}{1365} = \frac{144}{273}$$

Example 6 *X and Y are two independent events. The probability that both X and Y occur is 1/6 and the probability that neither of them occurs is 1/3. Find the probability of the occurrence of X.*

Solution Here, $P(X \cap Y) = \frac{1}{6}$, $P(\overline{X} \cap \overline{Y}) = \frac{1}{3}$

We know that,

$$P(X \cup Y) = 1 - P(\overline{X} \cap \overline{Y})$$

$$\Rightarrow \quad P(X) + P(Y) - P(X \cap Y) = 1 - P(\overline{X} \cap \overline{Y})$$

$$\Rightarrow \quad P(X) + P(Y) = 1 - P(\overline{X} \cap \overline{Y}) + P(X \cap Y)$$

$$= 1 - \frac{1}{3} + \frac{1}{6} = \frac{5}{6} \quad \text{...(i)}$$

Now, $[P(X) - P(Y)]^2 = [P(X) + P(Y)]^2 - 4\, P(X)\, P(Y)$

$$= \left(\frac{5}{6}\right)^2 - 4 \times \frac{1}{6} \quad [P(X \cap Y) = P(X) \,.\, P(Y)]$$

$$= \frac{1}{36}$$

$$\therefore \quad P(X) - P(Y) = \pm \frac{1}{6} \quad \text{...(ii)}$$

On adding Eqs. (i) and (ii), we get

$$2P(X) = \frac{5}{6} \pm \frac{1}{6}$$

$$\therefore \quad P(X) = \frac{1}{2} \text{ or } P(X) = \frac{1}{3}$$

Example 7 *Urn A contains 2 white and 5 black balls and urn B contains 3 white and 6 black balls. A ball is taken out at random from urn A and transferred to urn B. Then a ball is drawn at random from urn B. What is the probability that it is a black ball?*

Solution Let A_1 and A_2 be the events that a ball is taken out at random from urn A and transferred to urn B and E be the probability to draw the black ball. Then

$$E = (A_1 \cap E) \cup (A_2 \cap E)$$

$$\therefore \quad P(E) = P(A_1 \cap E) + P(A_2 \cap E)$$

$$= P(A_1)\, P\left(\frac{E}{A_1}\right) + P(A_2)\, P\left(\frac{E}{A_2}\right)$$

$$= \frac{2}{7} \times \frac{6}{9+1} + \frac{5}{7} \times \frac{6+1}{9+1}$$

$$= \frac{2}{7} \times \frac{6}{10} + \frac{5}{7} \times \frac{7}{10}$$

$$= \frac{12}{70} + \frac{35}{70} = \frac{47}{70}$$

Example 8 *In a college 25% students in Mathematics. 15% students in Physics and 10% students in Mathematics and Physics both are failed. A student is selected at random.*

(i) *If he is failed in Physics, then find the chance of his failure in Mathematics.*

(ii) *If he is failed in Mathematics, then find the chance of his failure in Physics.*

(iii) *Find the chance of his failure in Mathematics or Physics.*

Solution Let A and B are the events of failure in Mathematics and Physics respectively.

Let us assume that total number of students be 100. Therefore, $n(S)$ = 100. Since 25% students in Mathematics, then

$$n(A) = 25$$

and $$P(A) = \frac{25}{100} = \frac{1}{4}$$

Since 15% students in Physics, so $n(B) = 15$

and $$P(A) = \frac{15}{100} = \frac{3}{20}$$

The students failed in Mathematics and Physics both are 10%

So, $n(A \cap B) = 10$

and $P(A \cap B) = \frac{10}{100} = \frac{1}{10}$

(i) The chance of failure in Mathematics while he is failed in Physics is

$$P\left(\frac{A}{B}\right) = \frac{P(A \cap B)}{P(B)} = \frac{1/10}{3/20} = \frac{2}{3}$$

(ii) The chance of failure in Physics while he is failed in Mathematics is

$$P\left(\frac{B}{A}\right) = \frac{P(A \cap B)}{P(A)} = \frac{1/10}{1/4} = \frac{2}{5}$$

(iii) The chance of failure in Mathematics or Physics is

$$P(A \cup B) = P(A) + P(B) - P(A \cap B)$$

$$= \frac{1}{4} + \frac{3}{20} - \frac{1}{10}$$

$$= \frac{5+3-2}{20} = \frac{6}{20} = \frac{3}{10}$$

Example 9 *Three groups of children contain 3 girls and 1 boy, 2 girls and 2 boys. 1 girl and 3 boys. One child is selected at random from each group. Show that the chance that three selected consist of 1 girl and 2 boys is 13/32.*

Solution The selection of 1 girl and 2 boys can be made in such a manner:

	Group I,	Group II,	Group III	Probability
(i)	Boy,	boy,	girl	$\frac{1}{4} \times \frac{2}{4} \times \frac{1}{4} = \frac{1}{32}$
(ii)	Boy,	girl,	boy	$\frac{1}{4} \times \frac{2}{4} \times \frac{3}{4} = \frac{3}{32}$
(iii)	Girl,	boy,	boy	$\frac{3}{4} \times \frac{2}{4} \times \frac{3}{4} = \frac{9}{32}$

Since all these events are mutually exclusive.

$\therefore$ The required probability $= \frac{1}{32} + \frac{3}{32} + \frac{9}{32} = \frac{13}{32}$

Example 10 *Each of the three bags contains 5 white and 7 red balls. One ball is drawn from the first bag and is placed in second bag. Then one ball is drawn from the second bag and placed in the third bag. What is the probability of drawing a white ball from the third bag?*

Solution Let us assume that X be the white ball from the third bag which can happen mutually exclusive events an such a manner:

X_1 = white ball from first, white ball from second and white ball from third bag.

X_2 = white ball from first, red from second and white ball from third bag.

X_3 = red ball from first, white ball from second and white ball from third bag.

X_4 = red ball form first, red ball from second and white ball form third bag.

The probabilities are

$$P(X_1) = \frac{5}{12} \times \frac{6}{13} \times \frac{6}{13} = \frac{180}{2028}$$

$$P(X_2) = \frac{5}{12} \times \frac{7}{13} \times \frac{5}{13} = \frac{175}{2028}$$

$$P(X_3) = \frac{7}{12} \times \frac{5}{13} \times \frac{6}{13} = \frac{210}{2028}$$

$$P(X_4) = \frac{7}{12} \times \frac{8}{13} \times \frac{5}{13} = \frac{280}{2028}$$

$\therefore$ The required probability

$$P(X) = P(X_1) + P(X_2) + P(X_3) + P(X_4)$$

$$= \frac{180 + 175 + 210 + 280}{2028}$$

$$= \frac{845}{2028}$$

Example 11 *If four squares are chosen at random on a chess-board, find the chance that they should be in a diagonal line.*

Solution A chess-board is a square divided into 64 equal squares, we can choose 4 squares in ${}^{64}C_4$ ways. Then

$$n(S) = {}^{64}C_4$$

There are only four squares and hence the selection can be in ${}^4C_4$ ways and so on. Hence, the number of ways in a diagonal line can be chosen in ΔABD is

$${}^4C_4 + {}^5C_4 + {}^6C_4 + {}^7C_4 + {}^8C_4$$

Similarly, in ΔBCD, the squares can be chosen parallel to BD in an equal number of ways. Hence, the total number of ways in a diagonal line is

$${}^8C_4 + 2({}^4C_4 + {}^5C_4 + {}^6C_4 + {}^7C_4)$$

Since the line BD is common to both the triangles.

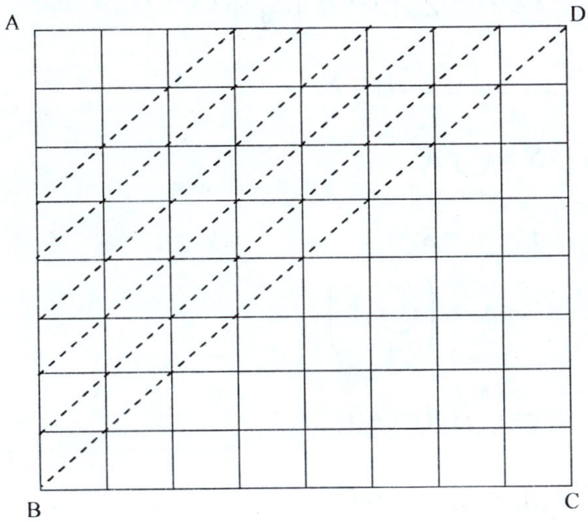

The same argument applies to squares parallel to AC an hence the total number of ways are

$$= 4({}^4C_4 + {}^5C_4 + {}^6C_4 + {}^7C_4) + 2.{}^8C_4 = 364$$

Hence, the required probability $= \dfrac{364}{{}^{64}C_4}$

$$= \frac{364 \times 4 \times 3 \times 2 \times 1}{64 \times 63 \times 62 \times 61}$$

$$= \frac{13}{22692}$$

Probability Compound Events Theorem

Let $A_1, A_2, \ldots\ldots\ldots A_n$ be the subsets of a sample space S, such that

(i) Each A_i, is a proper subset of S *i.e.*

$$A_i \subset S,\ \forall i = 1,2,\ldots\ldots\ldots\ldots\ldots n$$

(ii) $A_1 \cup A_2 \cup \ldots\ldots\ldots\ldots \cup A_n = S$

(iii) The subset A_i are pairwise disjoint, *i.e.*

$$A_i \cap A_j = \phi,\ \forall i,j = 1,2,\ldots\ldots\ldots\ldots n;\ i \neq j$$

If A is an event, then

$$P(A) = \sum_{i=1}^{n} P(A_i)P\left(\frac{A}{A_i}\right), P(A_i) \neq 0$$

Proof : Let $A_1 \cup A_2 \cup \ldots\ldots\ldots\ldots \cup A_n = S$

$$\Rightarrow \qquad S = \bigcup_{i=1}^{n} A_i$$

and therefore $A = A \cap S$

$$= \mathrm{A} \cap \left(\bigcup_{i=1}^{n} A_i\right)$$

$$= \bigcup_{i=1}^{n} (A \cap A_i)$$

Since, $A_i \cap A_j = \phi,\ i \neq j$

Then, $(A \cap A_i) \cap (A \cap A_j) = \phi, i \neq j$

Hence, $P(A) = \mathrm{P}\left[\bigcup_{i=1}^{n} (A \cap A_i)\right]$

By addition theorem,

$$P(A) = \sum_{i=1}^{n} P(A \cap A_i)$$

By product theorem,

$$P(A) = \sum_{i=1}^{n} (A_i)P\left(\frac{A}{A_i}\right)$$

Bay's Theorem

If the events $A_1, A_2, \dots\dots\dots A_n$ are pairwise independent and $A_1 \cup A_2 \cup \dots\dots\dots \cup A_n = S$ (a sample space), then for an event A, the probability is given by

$$P\left(\frac{A_i}{A}\right) = \frac{P(A_i)P\left(\frac{A}{A_i}\right)}{\sum_{i=1}^{n} P(A_i)P\left(\frac{A}{A_i}\right)}$$

Proof : By compound probability theorem we have

$$P(A \cap A_i) = P(A_i)\, P\left(\frac{A}{A_i}\right) = P(A)\, P\left(\frac{A_i}{A}\right)$$

$$\Rightarrow \quad P\left(\frac{A_i}{A}\right) = \frac{P(A_i)P\left(\frac{A}{A_i}\right)}{P(A)} \qquad \dots\text{(i)}$$

Since the event A can happen in the forms $A \cap A_1$, $A \cap A_2$, …………… $A \cap A_n$. Then from the theorem of total probability.

$$P(A) = P(A \cap A_1) + P(A \cap A_2) + \dots\dots\dots\dots + P(A \cap A_n)$$

$$= \sum_{i=1}^{n} P(A_i)P\left(\frac{A}{A_i}\right) \qquad \dots\text{(ii)}$$

$\therefore$ From Eqs. (i) and (ii), we have

$$P\left(\frac{A_i}{A}\right) = \frac{P(A_i)P\left(\frac{A}{A_i}\right)}{\sum_{i=1}^{n} P(A_i)P\left(\frac{A}{A_i}\right)}$$

Example 12 *There are 3 bags containing 2 white and 3 red balls, 3 white and 2 red balls and 4 white and 1 red balls respectively. There is equal probability of each bag being choosen. One ball is drawn from a bag chosen at random. What is the probability that a white ball is drawn?*

Solution Let A_i be the event and there are equal probability of each bag, then $P(A_i) = \frac{1}{3}$

B be the event that a white ball is drawn. Then

$$P(B) = P(A_1)P\left(\frac{B}{A_1}\right) + P(A_2)P\left(\frac{B}{A_2}\right) + P(A_3)P\left(\frac{B}{A_3}\right),$$

Now, $P(A_1) = P(A_2) = P(A_3) = \frac{1}{3}$

And $P\left(\frac{B}{A_1}\right) = \frac{2}{5}, P\left(\frac{B}{A_2}\right) = \frac{3}{5}, P\left(\frac{B}{A_3}\right) = \frac{4}{5}$

$\therefore$ $P(B) = \frac{2}{5} \times \frac{1}{3} + \frac{3}{5} \times \frac{1}{3} + \frac{4}{5} \times \frac{1}{3} = \frac{3}{5}$

The required probability of white ball is

$$P\left(\frac{A_1}{B}\right) = \frac{P(A_1)P\left(\frac{B}{A_1}\right)}{P(B)}$$

$$= \frac{\frac{1}{3} \times \frac{2}{5}}{\frac{3}{5}} = \frac{2}{9}$$

Example 13 *A bag contains 3 white and 4 red balls. Two balls are drawn at random and they are not replaced. What is the probability that first ball is white while second ball is red?*

Solution Let A_1, A_2 be the events and they can happen first ball white, first ball red and second ball red.

The event A can occur either A_1 or A_2. Such that:

$$A = (A \cap A_1) \cup (A \cap A_2)$$

$$P(A) = P(A \cap A_1) + P(A \cap A_2)$$

$$= P(A_1)\, P\left(\frac{A}{A_1}\right) + P(A_2)\, P\left(\frac{A}{A_2}\right)$$

$$= \frac{3}{7} \times \frac{4}{6} + \frac{4}{7} \times \frac{3}{6} = \frac{4}{7}$$

$$\therefore \text{The required probability, } P\left(\frac{A_1}{A}\right) = \frac{P(A \cap A_1)}{P(A)} = \frac{\frac{2}{7}}{\frac{4}{7}} = \frac{1}{2}.$$

Example 14 *There are three similar bags and they contains 2 white and 3 red balls, 3 white and 2 red balls, 4 white and 1 red balls respectively. The probability of selecting each bag is same. A bag is selected at random and a ball is drawn from it.*

(i) *Find the chance that a white ball is drawn.*

(ii) *If it is known that the ball is white, what is the probability that it comes from second bag?*

Solution Let A be the event of getting one white ball and A_1, A_2, A_3 be the events of first, second and third bag respectively.

The probability $P(A_1) = \frac{1}{3}, P(A_2) = \frac{1}{3}, P(A_3) = \frac{1}{3}$

The probability that a white ball is drawn from first bag i.e.

$$P\left(\frac{A}{A_1}\right) = \frac{2}{5}$$

Similarly. $P\left(\frac{A}{A_2}\right) = \frac{3}{5},\ P\left(\frac{A}{A_3}\right) = \frac{4}{5}$

(i) $$P(A) = \sum_{i=1}^{3} P(A_i) P\left(\frac{A}{A_i}\right)$$

$$= P(A_1)\ P\left(\frac{A}{A_1}\right) + P(A_2)\ P\left(\frac{A}{A_2}\right) + P(A_3)\ P\left(\frac{A}{A_3}\right)$$

$$= \frac{1}{3} \times \frac{2}{5} + \frac{1}{3} \times \frac{3}{5} + \frac{1}{3} \times \frac{4}{5} = \frac{3}{5}$$

(ii) $$P\left(\frac{A_2}{A}\right) = \frac{P(A_2) P\left(\frac{A}{A_2}\right)}{\sum_{i=1}^{3} P(A_i) P\left(\frac{A}{A_i}\right)}$$

$$= \frac{\frac{1}{3} \times \frac{3}{5}}{\frac{3}{5}} = \frac{1}{3}$$

Example 15 *In a bolt factory, machines A, B and C manufacture 25, 35 and 40 percent of the total of their output 5,4 and 2 percent are defective. A bolt is drawn at random and is found to be defective. What are the probabilities that it was manufactured by the machines A, B and C ?*

Solution Let the probability that the bolt was manufactured by machine A, B and C are

$$P(A) = \frac{25}{100} = \frac{1}{4}, P(B) = \frac{35}{100}, P(C) = \frac{40}{100}$$

The probability that a defective bolt manufactured by machine A, B and C are

$$= P\left(\frac{D}{A}\right) = \frac{5}{100}, P\left(\frac{D}{B}\right) = \frac{4}{100}, P\left(\frac{D}{C}\right) = \frac{2}{100}$$

∴ From Bay's theorem,

The probability that a defective bolt drawn at random is manufactured by machine A is

$$P\left(\frac{A}{D}\right) = \frac{P(A)\,P\left(\frac{D}{A}\right)}{P(A)\,P\left(\frac{D}{A}\right) + P(B)\,P\left(\frac{D}{B}\right) + P(C)\,P\left(\frac{D}{C}\right)}$$

$$= \frac{\frac{1}{4} \times \frac{5}{100}}{\frac{1}{4} \times \frac{5}{100} + \frac{35}{100} \times \frac{4}{100} + \frac{40}{100} \times \frac{2}{100}} = \frac{25}{69}$$

Similarly, $P\left(\frac{B}{D}\right) = \frac{28}{69}$ and $P\left(\frac{C}{D}\right) = \frac{16}{69}$

Discrete Random Variable

The variate which can take certain values depending on chance is called a random variable (or stochastic variate). If a random variable assumes only a finite or countable infinite number of values is called discrete random variable.

For example : $1, \frac{1}{2}, \frac{1}{3} \ldots\ldots\ldots \frac{1}{n}, \ldots\ldots\ldots\ldots\ldots$

Continuous Random Variable

If a random which can take any value in some interval is called a continuous random variable.

For example: $f(x) = \cos x, (0 \le x \le \pi)$

Probability Function of a Discrete Random Variable

Let X be the random variable where $x_1, x_2, x_3 \ldots\ldots\ldots\ldots\ldots x_n \in X$, then the probability that $X = x_i$ is denoted by $P(X = x_i)$ or $P(x_i)$ and is called the probability function of X. Here

$$P(x_i) \ge 0, \sum_{i=1}^{n} P(x_i) = 1$$

Mathematical Expectation

Let X be a discrete variable which may take n mutually exclusive values x_i, $i = 1,2 \ldots\ldots\ldots\ldots n$, with respective probabilities P_i, $i = 1,2 \ldots\ldots\ldots\ldots\ldots\ldots n$

The expectation of X is

$$E(X) = p_1x_1 + p_2x_2 + \ldots\ldots\ldots + p_nx_n$$

$$= \sum_{i=1}^{n} p_i x_i$$

Theorem of Expectation of the Sum of Two Discrete Random Variables

The expectation of a sum of a number of random variables is equal to the sum of their expectations, *i.e.*

$$E(x+y) = E(x) + E(y)$$

Proof: Let the random variables are x and y. Let x assume the values x_i $(i=1,2,\ldots\ldots\ldots m)$ and y the values y_j $(j = 1,2\ldots\ldots\ldots n)$ with probabilities P_i and P_j. The sum $(x + y)$ is a random variable which assumes mn values.

$$x_i + y_j,\ i = 1,2\ldots\ldots\ m, j = 1,2\ldots\ldots, n. \text{ with probability } P_{ij}.$$

Then

$$E(x+y) = \sum_{i=1}^{m}\sum_{j=1}^{n}(x_i + y_j)p_{ij}$$

$$= \sum_{i=1}^{m}\sum_{j=1}^{n} x_i p_{ij} + \sum_{i=1}^{m}\sum_{j=1}^{n} y_j p_{ij}$$

$$= \sum_{i=1}^{m} x_i \sum_{j=1}^{n} p_{ij} + \sum_{j=1}^{n} y_j \sum_{i=1}^{m} p_{ij}$$

Since $\sum_{j=1}^{n} p_{ij}$ represents the probability P_i of x so that $\sum_{j=1}^{n} p_{ij} = p_i.$

Similarly, $\sum_{i=1}^{m} p_{ij} = p_j$

Hence, $E(x+y) = \sum_{i=1}^{m} x_i p_i + \sum_{j=1}^{n} y_i p_j$

$$= E(x) + E(y)$$

Theorem of Expectation of the Product of two Discrete Random Variables

The expectation of the product of a number of two independent random variables is equal to the product of their expectation i.e.

$$E(xy) = E(x).\ E(y)$$

Proof: By using the above assumption,

$$E(xy) = \sum_{i=1}^{m} \sum_{j=1}^{n} x_i y_j p_i p_j$$

Since the variates are independent, then by the law of compound probabilities we have

$$p_{ij} = p_i\, p_j$$

Hence,

$$E(xy) = \sum_{i=1}^{m} \sum_{j=1}^{n} x_i y_j p_i p_j$$

$$= \sum_{i=1}^{m} x_i p_i \sum_{j=1}^{n} y_j p_j$$

$$= E(x).\ E(y)$$

Example 16 *In four tosses of a coin. Let a be the number of tails. Calculate the expected value of a. What can you say about the expectation of number of heads?*

Solution Let probability of tail = probability of head = $\frac{1}{2}$

$$\therefore \quad P(a=4)=\frac{^4C_4}{16}=\frac{1}{16}$$

$$P(a=3)=\frac{^4C_3}{16}=\frac{1}{4}$$

$$P(a=2)=\frac{^4C_2}{16}=\frac{3}{8}$$

$$P(a=1)=\frac{^4C_1}{16}=\frac{1}{4}$$

$$P(a=0)=\frac{^4C_0}{16}=\frac{1}{16}$$

Hence, the probability distribution of a is

$a = r$	0	1	2	3	4
$P(a = r)$	$\frac{1}{16}$	$\frac{1}{4}$	$\frac{3}{8}$	$\frac{1}{4}$	$\frac{1}{16}$

$\therefore$ Expectation $E(a) = \sum_{r=0}^{4} r.P(r)$

$$= 0\times\frac{1}{16}+1\times\frac{1}{4}+2\times\frac{3}{8}+3\times\frac{1}{4}+4\times\frac{1}{16}$$

$$= 2$$

Now, let number of heads = b, then

$$P(b)=\frac{^4C_b}{16},\ b = 0,1,2,3 \ldots\ldots\ldots.$$

$$\therefore \quad E(b) = E(a) = 2$$

Example 17 *The probability function of a discrete random variable is as follows.*

$X = x$	0	1	2	3	4	5	6	7
$P(x)$	0	k	$2k$	$2k$	$3k$	k^2	$2k^2$	$7k^2+k$

Find (i) k (ii) $P(X < 6)$ and $P(X \geq 6)$ (iii) $P(0 < X < 5)$ (iv) Distribution function X.

Solution (i) We know that $\sum_{x=0}^{7} P(x) = 1$

$\Rightarrow \quad 0 + k + 2k + 2k + 3k + k^2 + 2k^2 + 7k^2 + k = 1$

$\Rightarrow \quad 10k^2 + 9k - 1 = 0$

$\Rightarrow \quad (10k-1)(k+1) = 0$

$\Rightarrow \quad k = \frac{1}{10}, -1$

$\Rightarrow \quad k = \frac{1}{10}$ ($\because k = -1$ is not possible)

(ii) $P(X < 6) = P(X=0) + P(X=1) + P(X=2) + P(X=3) + P(X=4) + P(X=5)$

$= 0 + k + 2k + 2k + 3k + k^2$

$= 0 + \frac{1}{10} + 2\times\frac{1}{10} + 2\times\frac{1}{10} + 3\times\frac{1}{10} + \frac{1}{100}$

$= 0.81$

$P(X \geq 6) = P(X=6) + P(X=7)$

$= 2k^2 + 7k^2 + k = 9k^2 + k$

$= 9\times\frac{1}{100} + \frac{1}{10} = 0.19$

(iii) $P(0 < X < 5) = P(X = 1) + P(X = 2) + P(X = 3) + P(X = 4)$

$= k + 2k + 2k + 3k$

$= 8k$

$= 8\times\frac{1}{10} = 0.8$

(iv) Distribution function of X, $f(x) = P(X \leq x)$

$f(0) = 0, f(1) = \frac{1}{10}, f(2) = \frac{3}{10}, f(3) = \frac{5}{10}, f(4) = \frac{8}{10},$

$f(5) = \frac{81}{100}, f(6) = \frac{83}{100}, f(7) = 1.$

Example 18 *Four balls are drawn from a bag containing 5 red, 6 white and 7 black balls. Find the expectation of white balls drawn.*

Solution Let the white ball is X.

Total number of balls in a bag = 5 + 6 + 7 = 18

The number of white balls draw from a bag may be 0,1,2,3,4.

∴ Probability of white balls

$$P(X=r)=\frac{{}^{6}C_{r}\times{}^{12}C_{4-r}}{{}^{18}C_{4}}, r=0,1,2,3,4$$

We have, $P(X=0)=\dfrac{{}^{6}C_{0}\times{}^{12}C_{4}}{{}^{18}C_{4}}=\dfrac{495}{3060}$

$$P(X=1)=\frac{{}^{6}C_{1}\times{}^{12}C_{3}}{{}^{18}C_{4}}=\frac{1320}{3060}$$

$$P(X=2)=\frac{{}^{6}C_{2}\times{}^{12}C_{2}}{{}^{18}C_{4}}=\frac{990}{3060}$$

$$P(X=3)=\frac{{}^{6}C_{3}\times{}^{12}C_{1}}{{}^{18}C_{4}}=\frac{240}{3060}$$

$$P(X=4)=\frac{{}^{6}C_{4}\times{}^{12}C_{0}}{{}^{18}C_{4}}=\frac{15}{3060}$$

∴ Expectation of white balls

$$E(X)=0\times\frac{495}{3060}+1\times\frac{1320}{3060}+2\times\frac{990}{3060}+3\times\frac{240}{3060}+4\times\frac{15}{3060}$$

$$=\frac{4080}{3060}=\frac{4}{3}$$

Example 19 *A bag contains 2 white and 3 black balls. Four persons A, B, C and D in the order named each draw one ball and do not replace it. The person to draw a white ball receives Rs. 200. Determine their expectations.*

Solution A's chance of drawing a white ball $= \dfrac{2}{5}$

B get a chance if A fails and hence the probability of his getting the chance and succeeding $= \dfrac{3}{5} \times \dfrac{2}{4} = \dfrac{3}{10}$

Similarly, C's probability of getting the chance

$$= \frac{3}{5} \times \frac{2}{4} \times \frac{2}{3} = \frac{1}{5}$$

D's probability of getting a chance $= \dfrac{3}{5} \times \dfrac{2}{4} \times \dfrac{1}{3} \times 1 = \dfrac{1}{10}$

Their respective expectations are

$$A's = \text{Rs.} \frac{2}{5} \times 200 = \text{Rs. } 80$$

$$B's = \text{Rs.} \frac{3}{10} \times 200 = \text{Rs. } 60$$

$$C's = \text{Rs.} \frac{1}{5} \times 200 = \text{Rs. } 40$$

$$D's = \text{Rs.} \frac{1}{10} \times 200 = \text{Rs. } 20$$

Example 20 *A coin is tossed until the head appears. What is expectation of the number of tosses?*

Solution Let x denotes the number of tosses until the head appears. Their probabilities are as below:

Event – *H, TH, TTH*, ...

x – 1,2,3,..............

$$P(x) = \frac{1}{2}, \left(\frac{1}{2}\right)^2, \left(\frac{1}{2}\right)^3, \ldots\ldots\ldots$$

$$\therefore \quad E(x) = 1 \times \frac{1}{2} + 2 \times \left(\frac{1}{2}\right)^2 + 3 \times \left(\frac{1}{2}\right)^3 + \ldots\ldots\ldots$$

Multiply by $\frac{1}{2}$,we get

$$\frac{1}{2}E(x)=1\times\left(\frac{1}{2}\right)^2+2\times\left(\frac{1}{2}\right)^3+3\left(\frac{1}{2}\right)^4+..........$$

Subtracting, we get

$$\frac{1}{2}E(x)=\frac{1}{2}+\frac{1}{2^2}+\frac{1}{2^3}+.......... \quad \text{(geometric series)}$$

$$=\frac{\frac{1}{2}}{1-\frac{1}{2}}=1$$

$\therefore \quad E(x) = 2$

Example 21 *Balls are taken one by one out of an urn containing a white and b black balls until the first white ball is drawn. Show that the expectation of the number of black balls preceding the first white ball is* $\frac{b}{a+1}$.

Solution Let x_1 be the number of black balls before the first white ball is drawn and x_{a+1} be the number of black balls left after the white balls have been drawn.

Then

$$x_1+x_2+..............+x_{a+1}=b$$

Hence, $\quad E(x_1+x_2+..............+x_{a+1}) = E(b)$

$\Rightarrow \quad E(x_1)+E(x_2)+..................+E(x_{a+1})= E(b)$

The probability of every system of numbers x_1, x_2,.............., x_{a+1} are the same.

Thus, $\quad E(x_1) = E(x_2)=..................= E(x_{a+1}) = k(say)$

Therefore, $\quad (a + 1)\, k = b$

$$\Rightarrow \quad k=\frac{b}{a+1}$$

$$\therefore \quad E(x_1)=\frac{b}{a+1}$$

Example 22 *What is the expectation of the sum of points on n dice?*

Solution Let x_i, be the number of points on the ith dice, the sum of points on n dice is

$$x_1 + x_2 + \ldots\ldots + x_n$$

$$\therefore \quad E(x_1 + x_2 + \ldots\ldots + x_n) = E(x_1) + E(x_2) + \ldots\ldots + E(x_n)$$

For every single dice

$$E(x_i) = \frac{7}{2}, \; \text{I} = 1,2, \ldots\ldots\ldots\ldots\ldots n$$

$$\therefore \quad E(x_1 + x_2 + \ldots\ldots + x_n) = \frac{7}{2} + \frac{7}{2} + \ldots\ldots\ldots\ldots \; n \text{ terms}$$

$$= \frac{7}{2} n$$

Example 23 *Find the expectation of the product of points on n dice.*

Solution. Let the x_i be number of points on the ith dice, the product of points on n dice is $x_1.x_2. \ldots\ldots x_n$

$$\therefore \quad E(x_1 . x_2 . \ldots\ldots x_n) = E(x_1)\; E(x_2) \ldots\ldots E(x_n)$$

for every single dice $E(x_i) = \dfrac{7}{2}$, $i = 1,2,\ldots\ldots\ldots n,$

$$\therefore \quad E(x_1 . x_2 . \ldots\ldots x_n) = \frac{7}{2} . \frac{7}{2} . \ldots\ldots\ldots\ldots \; n \text{ terms}$$

$$= \left(\frac{7}{2}\right)^n$$

Practice Problems

1. Form a pack of 52 cards, two cards are drawn at random. Find the probability that both cards are of space.

[**Ans.** $\dfrac{1}{17}$]

2. The chance of one event happening is the square of the chance of a second event, but the odd against the first are the cube of the odd against the second. Find the chance of each.

[**Hint**. $P(A) = \{P(B)\}^2$, odd against A = (odd against B)^3, $P(A) = \frac{1}{9}$, $P(B) = \frac{1}{3}$]

3. Two cards are drawn at random from a pack of cards. Find the probability that both cards are of black colour or both are aces.

[**Ans.** $\frac{55}{221}$]

4. Probability that a boy will pass an examination is $\frac{3}{5}$ and that for a girl it is $\frac{2}{5}$. What is the probability that at least one of them passes examination?

[**Ans.** $\frac{19}{25}$]

5. What is the chance of throwing a total of 11 with two dice if the digit on first die is 5?

[**Ans.** $P\left(\frac{A}{B}\right) = \frac{n(A \cap B)}{n(B)} = \frac{1}{6}$]

6. A speak the truth in 60% and B in 75% of the cases. In what percentage of cases are they likely to contradict each other in stating the same fact?

[**Ans.** 45%]

7. One bag contains 4 white, 6 red and 15 black balls and a second bag contains 11 white, 5 red and 9 black balls. One ball from each bag is drawn. Find the probability of the following events.
   (i) both balls are white (ii) both balls are red
   (iii) both balls are black (iv) both balls are of the same colour.

$$\left[\text{Ans. } (i)\ \frac{44}{625}\ (ii)\ \frac{6}{125}\ (iii)\ \frac{27}{125}\ (iv)\ \frac{209}{625}\right]$$

8. A bay contains 3 black and 4 red balls. Two balls are drawn at random one at a time without replacement. What is the probability that the first ball selected in black if the second ball is known to be red?

[**Ans.** $\frac{1}{2}$]

9. There are three boxes containing 1 white, 2 red, 3 black balls; 2 white, 3 red 1 black balls; 3 white, 1 red, 2 black balls. A box is chosen at random and from it two balls are drawn at random. The two balls are one red and one white. What is the probability that they come form the (i) first box (ii) second box (iii) third box?

[**Hint**. Let A be the event of getting two balls and B_1, B_2, B_3 are boxes then for first box

$$P\left(\frac{B_1}{A}\right) = \frac{P(B_1)P\left(\frac{A}{B_1}\right)}{P(B_1)P\left(\frac{A}{B_1}\right) + P(B_2)P\left(\frac{A}{B_2}\right) + P(B_3)P\left(\frac{A}{B_3}\right)} = \frac{2}{11}$$

$$P\left(\frac{B_2}{A}\right) = \frac{6}{11}, P\left(\frac{B_3}{A}\right) = \frac{3}{11}]$$

10. The probability that a person can hit a target is $\frac{3}{5}$ and the probability that another person can hit the same target is $\frac{2}{5}$. But the first person can fire 8 shoots in the time the second person fire 10 shoots. They fire together. What is the probability that the second person shoots the target?

[**Ans.** $\frac{5}{11}$]

11. Thirteen cards are drawn simultaneously from a pack of 52 cards. If aces count 1, face card 10 and others according to denomination. Find the expectation of the total score on the 13 cards.

[**Ans.** 85]

12. A bag contains two one rupee coins and 3 twenty paise coins. A person is asked to draw two coins randomly. Fine the value of his expectations.

[**Hint**. $x_1 = 1+1 = 2$, when both are one rupee coin

$x_2 = 1 + \frac{20}{100}$, when both are 20 paise, $x_3 = \frac{20}{100} + \frac{20}{100}$, when both are 20 Paise

$\therefore\ E(X) = x_1 P_1 + x_2 P_2 + x_3 P_3 =$ Rs. 104]

13. A box contains a white and b black balls, c balls are drawn. Show that the expectation of the number of white balls drawn is $\frac{ca}{a+b}$.

14. In four tosses of a coin. Let x be the number of heads. Calculate the expected value of x^2.

[**Hint.** $E(x^2) = \sum_{x=0}^{4} x^2.P(x) = 5$]

15. X and Y throw one dice for a prize of Rs. 11 which is to be won by a player who first throws 6. If X has the first throw, what are their respective expectations?

[**Hint.** X's chance $= \frac{1}{6} + \left(\frac{5}{6}\right)^2 . \frac{1}{6} + \ldots\ldots\ldots\ldots = \frac{6}{11}$

Y's chance $= 1 - \frac{1}{6} = \frac{5}{11}$

$\therefore$ $E(X)$ = Rs. 6, $E(Y)$ = Rs. 5]

16. A person drawn 2 balls from a bay containing 3 white and 5 black balls. If he is to receive Rs. 10 for every white ball which he drawn and Rs. 1 for every black balls. What is his expectation?

[**Hint.** Expectation $= 2\times\frac{10}{28} + 11\times\frac{15}{28} + 20\times\frac{3}{28}$ = Rs. 8.75]

Chapter - 10

Theoretical Distribution

Theoretical Distribution

According to some rule, when the values of the variate in the universe are distributed, which can be derived mathematically, such distributions are called theoretical distribution.

The theoretical distributions are Binomial, the poisson and the normal distribution.

Binomial Distributions

Let the happening of an event, the probability of its success is p and probability of its failure be q in one trial. Let the number of trials are independent and the probability of success p and probability of failure q remain unchanged. Let us assume that the event be tried n time. The number of successes in n trials be r (say), than $P(x=r) = {}^nC_r p^r q^{n-r}$.

Which is called Binomial distribution or Bernoulli distribution.

If there be N number of sets, the number of sets with r successes is $N\,{}^nC_r p^r q^{n-r}$.

The number of sets corresponding the probabilities of 0,1,2,.....,r....,n are respectively.

then

$$Nq^n, N\,{}^nC_1 pq^{n-1}, N\,{}^nC_2 p^2 q^{n-2} ...., N\,{}^nC_r p^r q^{n-r}, ..., Np^n.$$

If there be N sets of each n trials, the frequencies of $0,1,2,\ldots,n$ successes are given by the terms of the expression

$$N[q^n + {}^nC_1 pq^{n-1} + {}^nC_2 p^2 q^{n-2} + ... + {}^nC_r p^r q^{n-r} + ... + p^n]$$

$\Rightarrow N(q+p)^n$, which is a Binomial expression

Constants of Binomial Distribution

Taking out origin at 0 success.

(i) First moment about origin (Mean)

$$\mu_1^1 = \sum_{r=0}^{n} r\,{}^nC_r p^r q^{n-r}\text{, where } q+p=1$$

$$= 0.q^n + 1.{}^nC_1 p^1 q^{n-1} + 2.{}^nC_2 p^2 q^{n-2} + \ldots + n.p^n$$

$$= npq^{n-1} + n(n-1)p^2q^{n-2} + \frac{1}{2}n(n-1)(n-2)p^3q^{n-3} + \ldots. + np^n$$

$$= np[q^{n-1} + (n-1)pq^{n-2} + \frac{1}{2}(n-1)(n-2)p^2q^{n-3} + \ldots + p^{n-1}]$$

$$= np[q^{n-1} + {}^{(n-1)}C_1 pq^{n-2} + {}^{(n-1)}C_2 p^2q^{n-3} + \ldots + p^{n-1}]$$

$$= np(q+p)^{n-1}$$

$= np$, since $q+p=1$

Hence, mean $= np$.

(ii) Second moment about origin:

$$\mu_2^1 = \sum_{r=0}^{n} r^2.{}^nC_r p^r q^{n-r}$$

$$= \sum_{r=o}^{n} [r(r-1)+r].n_{c_r} p^r q^{n-r}$$

$$= \sum_{r=0}^{n} r\,{}^nC_r p^r q^{n-r} + \sum_{r=0}^{n} r(r-1)\,{}^nC_r p^r q^{n-r}$$

$$= np + 0.q^n + 0.{}^nC_1 p^1 q^{n-1} + 2.1.{}^nC_2 p^2 q^{n-2} + 3.2\,{}^nC_3 p^3 q^{n-3} + \ldots + n(n-1)p^n$$

$$= np + 2.\frac{n(n-1)}{2}p^2q^{n-2} + 3.2.\frac{n(n-1)(n-2)}{3.2}.p^3q^{n-3} + \ldots. + n(n-1)p^n$$

$$= np + n(n-1)p^2q^{n-2} + n(n-1)(n-2)p^3q^{n-3} + \ldots + n(n-1)p^n$$

$$= np + n(n-1)p^2[q^{n-2} + (n-2)pq^{n-3} + \ldots. + p^{n-2}]$$

$$= np + n(n-1)p^2(q+p)^{n-2}$$

$$np + n(n-1)p^2, \text{since } q + p = 1$$

(iii) Third moment about origin

$$\mu_3^1 = \sum_{r=0}^{n} r^{3.n}C_r p^r q^{n-r}$$

$$= \sum_{r=0}^{n} [r(r-1)(r-2) + 3r(r-1) + r].^nC_r p^r q^{n-r}$$

$$[\because r^3 = r(r-1)(r-2) + 3r(r-1) + r]$$

$$= n(n-1)(n-2)p^3(q+p)^{n-3} + 3n(n-1)p^2(q+p)^{n-2} + np(q+p)^{n-1}$$

$$= n(n-1)(n-2)p^3 + 3n(n-1)p^2 + np$$

(iv) Fourth moment about origin

$$\mu_4^1 = \sum_{r=0}^{n} r^{4.\,n}c_r p^r q^{n-r}$$

$$= \sum_{r=0}^{n} [r(r-1)(r-2)(r-3) + 6r(r-1)(r-2] + 7r(r-1) + r].^nC_r p^r q^{n-r}$$

$$= n(n-1)(n-2)(n-3)p^4 + 6n(n-1)(n-2)p^3 + 7n(n-1)p^2 + np$$

(v) First moment about mean:

$$\mu_1 = 0$$

(vi) Second moment about mean:

$$\mu_2 = \mu_2^1 - (\mu_1^1)^2$$

$$= n(n-1)p^2 + np - (np)^2$$

$$= n^2p^2 - np^2 + np - n^2p^2$$

$$= np - np^2$$

$$= np(1-p)$$

$=npq$, *since* $q+p=1$

$\therefore$ Variance, $\sigma^2 = npq$

Standard deviation, $\sigma = \sqrt{npq}$

(vii) Third moment about mean:

$$\mu_3 = \mu_3^1 - 3\mu_1^1\mu_2^1 + 2(\mu_1^1)^3$$

$$= n(n-1)(n-2)p^3 + 3n(n-1)p^2 + np - 3np[np + n(n-1)p^2] + 2(np)^3$$

$$= np[(n-1)(n-2)p^2 + 3(n-1)p + 1 - 3np - 3n(n-1)p^2 + 2n^2p^2]$$

$$= np[(n^2 - 3n + 2)p^2 + 3np - 3p + 1 - 3np - 3n^2p^2 + 3np^2 + 2n^2p^2]$$

$$= np[n^2p^2 - 3np^2 + 2p^2 - 3p + 1 - n^2p^2 + 3np^2]$$

$$= np[2p^2 - 3p + 1]$$

$$= np(1-p)(1-2p)$$

$$= npq(1\text{-}2p)$$

$$= npq(q\text{-}p)$$

(viii) Fourth moment about mean

$$\mu_4 = \mu_4^1 - 4\mu_3^1\mu_1^1 + 6\mu_2^1(\mu_1^1)^2 - 3(\mu_1^1)^4$$

$$= n(n-1)(n-2)(n-3)p^4 + 6n(n-1)(n-2)p^3 + 7n(n-1)p^2$$
$$+np - 4[n(n-1)(n-2)p^3 + 3n(n-1)p^2 + np]np$$
$$+6[n(n-1)p^2 + np]n^2p^2 - 3n^4p^4$$

$$= p^4[n(n-1)(n-2)(n-3) - 4n^2(n-1)(n-2) + 6n^3(n-1) - 3n^4]$$
$$+p^3[6n(n-1)(n-2) - 12n^2(n-1) + 6n^3] + p^2[7n(n-1) - 4n^2] + np$$

$$= p^4(3n^2 - 6n) + p^3(12n - 6n^2) + p^2(3n^2 - 7n) + np$$

$$= 3n^2(p^4 - 2p^3 + p^2) - n(6p^4 - 12p^3 + 7p^2) + np$$

$$= 3n^2p^2(p^2 - 2p + 1) - n(6p^4 - 12p^3 + 6p^2) - np^2 + np$$

$$= 3n^2p^2(1-p)^2 - 6np^2(p^2 - 2p + 1) + np(1-p)$$

$$= 3n^2p^2q^2 - 6np^2q^2 + npq$$

$$= 3n^2p^2q^2 + npq(1 - 6pq)$$

(ix) Karl pearson's coefficients

$$\beta_1 = \frac{\mu_3^2}{\mu_2^3} = \frac{[npq(q-p)]^2}{(nnq)^3}$$

$$= \frac{(q-p)^2}{npq} = \frac{(1-2p)^2}{npq}$$

$$\beta_2 = \frac{\mu_4}{\mu_2^2} = \frac{3n^2p^2q^2 + npq(1-6pq)}{(npq)^2} = 3 + \frac{1-6pq}{npq}$$

$$\gamma_1 = \sqrt{\beta_1} = \frac{1-2p}{\sqrt{npq}}$$

$$\gamma_2 = \beta_2 - 3 = \frac{1-6pq}{npq}$$

Recurrence Formula

The recurrence relation for the binomial distribution is given by

$$\mu_{r+1} = pq\left(nr\mu_{r-1} + \frac{d\mu_r}{dp}\right)$$

where μ_r id the rth moment about the mean.

Proof: We have

$$\mu_r = \sum_{x=0}^{n}(x-np)^r.\,{}^nC_x p^x q^{n-x}$$

differentiating with respect to *p*, we get

$$\frac{d\mu_r}{dp} = \sum_{x=0}^{n} r(x-np)^{r-1}(-n)\,{}^nC_x p^x q^{n-x} + \sum_{x=0}^{n}(x-np)^r\,{}^nC_x.xp^{x-1}q^{n-x}$$

$$+\sum_{x=0}^{n}(x-np)^r\,{}^nC_x p^x.(n-x)q^{n-x-1}(-1), \quad \text{by putting } q = 1\text{-p}$$

$$= -nr\mu_{r-1} + \sum_{x=0}^{n}(x-np)^r.\,{}^nc_x p^x q^{n-x}\left\{\frac{x}{p} - \frac{(n-x)}{q}\right\}$$

$$= -nr\mu_{r-1} + \frac{1}{pq}\sum_{x=0}^{n}(x-np)^r\,{}^nC_x p^x q^{n-x}(qx - np + xp)$$

$$= -nr\mu_{r-1} + \frac{1}{pq}\sum_{x=0}^{n}(x-np)^r\,{}^nC_x p^x q^{n-x}(x - np)$$

$$pq\frac{d\mu_r}{dp} = -nrpq\mu_{r-1} + \sum_{x=0}^{n}(x-np)^{r+1}.{}^nC_x p^x q^{n-x}$$

$$\therefore \quad pq\left(\frac{d\mu_r}{dp} + nr\mu_{r-1}\right) = \mu_{r+1}$$

This recurrence formula is due to 'Renovsky'.

Moment Generating Function

The relative frequency of x success is $p(x) = {}^nC_x p^x q^{n-x}, x = 0,1,2,...,n.$

Then m.g.f. about origin

$$M_0(t) = \sum_{x=0}^{n} {}^nC_x p^x q^{n-x} e^{tx}$$

$$= \sum_{x=0}^{n} {}^nC_x (pe^t)^x q^{n-x}$$

$$= (q + pe^t)^n$$

The m.g.f. about mean np is

$$M_{np}(t) = e^{-npt}(q + pe^t)^n$$

$$= [qe^{-pt} + pe^{(t-p)t}]^n$$

$$= [qe^{-pt} + pe^{qt}]^n$$

Now, $\quad M_x(t) = (q+pe^t)^n$

$$= \left[q + p\left(1 + t + \frac{t^2}{2!} + \frac{t^3}{3!} + ...\right)\right]^n$$

$$= \left[(q+p) + pt + \frac{pt^2}{2!} + \frac{pt^3}{3!} + ...\right]^n$$

$$= \left[1 + pt + \frac{pt^2}{2!} + \frac{pt^3}{3!} + ...\right]^n$$

$$= 1 + {}^nC_1\left(pt + \frac{pt^2}{2!} + \frac{pt^3}{3!} + ...\right) + {}^nC_2\left(pt + \frac{pt^2}{2!} + \frac{pt^3}{3!} + ...\right)^2 + ...$$

$$= 1 + n\left(pt + \frac{pt^2}{2!} + \frac{pt^3}{3!} + \ldots\right) + \frac{n(n-1)}{2}\left(pt + \frac{pt^2}{2!} + \frac{pt^3}{3!} + \ldots\right)^2 + \ldots$$

we get,

$$\mu_1^{\prime} = \text{Coefficient of } t = np$$

$$\mu_2^{\prime} = \text{Coeff. of } \frac{t^2}{2!} = np + \frac{n(n-1)}{2}.2p^2$$

$$= np + (n^2 - n)p^2$$

$$= np + n^2p^2 - np^2$$

$$= np(1-p) + n^2p^2$$

Variance, $\mu_2 = \mu_2^{\prime} - (\mu_1^{\prime})^2$

$$= npq + n^2p^2 - n^2p^2$$

$$= npq$$

Standard deviation, $\sigma = \sqrt{npq}$

Cumulant Generating Function

It can be expressed as the logarithm of the m.g.f. $M_x(t)$ and is denoted by $K(t)$.

$$K(t) = \log M_x(t)$$

$$= \log(q + pe^t)^n$$

$$= n\log(q + pe^t)$$

$$= n\log\left[1 + pt + \frac{pt^2}{2!} + \frac{pt^3}{3!} + \ldots\right]$$

$$= n\left[\left(pt + \frac{pt^2}{2!} + \frac{pt^3}{3!} + \ldots\right) - \frac{1}{2}\left(pt + \frac{pt^2}{2!} + \frac{pt^3}{3!} + \ldots\right)^2 + \ldots\right]$$

The m.g.f. about the mean np is given by

$$K_{np}(t) = \log M_{np}(t)$$

$$= \log\left(qe^{-pt} + pe^{qt}\right)^n$$

$$= n\log\left(qe^{-pt} + pe^{qt}\right).$$

$$= n\log\left[q\left(1 - pt + \frac{p^2t^2}{2!} - \frac{p^3t^3}{3!} + \ldots\right) + p\left(1 + qt + \frac{q^2t^2}{2!} + \frac{q^3t^3}{3!} + \ldots..\right)\right]$$

$$= n\log\left[1 + \frac{pqt^2}{2!} + pq\left(q^2 - p^2\right)\frac{t^3}{3!} + pq\left(q^3 + p^3\right)\frac{t^4}{4!} + \ldots\right]$$

Expanding RHS and equating coefficients of $\frac{t^2}{2!}, \frac{t^3}{3!}, \ldots,$ we get

$$K_2 = npq, K_3 = npq\left(q^2 - p^2\right) = npq(q - p)$$

$$K_4 = npq\left(q^3 + p^3\right) - 3np^2q^2$$

$$= npq\left[\left(q^3 + p^3\right) - 3pq\right]$$

$$= npq\left[(q + p)^3 - 3pq(q + p) - 3pq\right]$$

$$= npq[1 - 6pq]$$

Illustrative Examples

Example 1 *In four throws, with a pair of dice, what is the chance of throwing doublets twice at least?*

Solution Here, only six doublets, which are (1,1), (2,2), (3,3), (4,4), (5,5), (6,6).

Hence, the chance of throwing doublet in one throw $(p) = \frac{6}{6^2} = \frac{1}{6}$

The chance of not throwing a doublet $(q) = 1 - \frac{1}{6} = \frac{5}{6}$

$\therefore$ The chance of atleast two doublets in 4 throws

$$= {}^4C_2\left(\frac{1}{6}\right)^2\left(\frac{5}{6}\right)^2 + {}^4C_3\left(\frac{1}{6}\right)^3\left(\frac{5}{6}\right) + {}^4C_4\left(\frac{1}{6}\right)^4$$

$$= \frac{19}{144}$$

Example 2 *In tossing 10 coins simultaneously, what is the probability of (i) having exactly 5 heads and (ii) having at least 7 heads.*

Solution Let probability of head = p and probability of tail = q

Let $p=\frac{1}{2}, q=\frac{1}{2}, n=10$ (no. of coins)

(i) $p(x=5)={}^nC_x\, p^x\, q^{n-x}$

$$= {}^{10}C_5\left(\frac{1}{2}\right)^5\left(\frac{1}{2}\right)^{10-5}$$

$$= {}^{10}C_5\left(\frac{1}{2}\right)^{10}$$

$$= \frac{10!}{5!5!}\times\frac{1}{1024}=\frac{63}{256}$$

(ii) p (x = at least 7 heads)

$$= {}^{10}C_7\left(\frac{1}{2}\right)^7\left(\frac{1}{2}\right)^{10-7} + {}^{10}C_8\left(\frac{1}{2}\right)^8\left(\frac{1}{2}\right)^{10-8} + {}^{10}C_9\left(\frac{1}{2}\right)^9\left(\frac{1}{2}\right)^{10-9} + {}^{10}C_{10}\left(\right.$$

$$= {}^{10}C_7\left(\frac{1}{2}\right)^{10} + {}^{10}C_8\left(\frac{1}{2}\right)^{10} + {}^{10}C_9\left(\frac{1}{2}\right)^{10} + {}^{10}C_{10}\left(\frac{1}{2}\right)^{10}$$

$$= \frac{1}{2^{10}}\left({}^{10}C_7+{}^{10}C_8+{}^{10}C_9+{}^{10}C_{10}\right)$$

$$= \frac{1}{1024}(120+45+10+1)=\frac{176}{1024}=\frac{11}{64}$$

Example 3 A perfect cubical die is thrown a large number of times in a set of 8. The occurrence of 5 or 6 is called a success. In what proportion of the sets you expect 3 successes?

Solution Here, probability of success (p) $=\frac{2}{6}=\frac{1}{3}$

and $q=\frac{2}{3}$, no. of sets n = 8, successes r = 3

The binomial distribution is $N(q+p)^n$.

The number of set in which 3 successes are

$$N.{}^{8}C_{3}\left(\frac{1}{3}\right)^{3}\left(\frac{2}{3}\right)^{8-3}$$

$$= N.\frac{8\times7\times6}{3\times2\times1}\times\frac{1}{27}\times\frac{32}{243}$$

$$= 0.2731\,N$$

$$\therefore \text{ Percentage} = 0.2731\,N\times\frac{100}{N}$$

$$= 27.31\%$$

Example 4 *If 10% of the bolts produced by a machine are defective, determine the probability that out of 10 bolts chosen at random (i) 1 (ii) none (iii) at most 2 bolts wills be defective.*

Solution The probability of defective bolts $p = 10\% = 0.1$

The probability of non-defective bolts $q = 1 - 0.1 = 0.9$

Given $n = 10$ bolts

By binomial distribution $P(x) = {}^{n}C_{x}p^{x}q^{n-x}$

(i) The probability of 1 bolt defective is

$$P(1) = {}^{10}C_{1}(0.1)^{1}(0.9)^{10-1}$$

$$= 0.3874$$

(ii) The probability of non-defective bolts

$$P(0) = {}^{10}C_{0}(0.1)^{0}(0.9)^{10} = 0.3486$$

(iii) The chance of at most 2 defective bolts

$$= P(0) + P(1) + P(2)$$

$$= 0.3486 + 0.3874 + {}^{10}C_{2}(0.1)^{2}(0.9)^{10-2}$$

$$= 0.3486 + 0.3874 + 0.1937$$

$$= 0.9297$$

Example 5 Criticise the following statement: "For binomial distribution means is 5 and standard deviation is 3".

Solution We know that in a binomial distribution

Mean $= np = 5$

Standard deviation $= \sqrt{npq} = 3$

$$\Rightarrow \qquad npq = 9$$

$$\therefore \qquad \frac{npq}{np} = \frac{9}{5}$$

$$q = 1.8$$

But q can never be greater than 1.

Hence, the given statement is false.

Example 6. In a binomial distribution, the mean and variance are 4 and $\frac{4}{3}$ respectively.

Find the probability of (i) 2 successes, (ii) more than 2 successes, (iii) 3 or more than three successes.

Solution The binomial distribution $P(r) = {}^nC_r p^r q^{n-r}$

Here, mean $= np = 4$

variance $= npq = \frac{4}{3}$

Then, $\frac{npq}{np} = \frac{4/3}{4} \Rightarrow q = \frac{1}{3}\, and\ p = \frac{2}{3}, n = 6$

(i) The probability of two success are

$$p(r=2) = {}^6C_2\left(\frac{2}{3}\right)^2\left(\frac{1}{3}\right)^{6-2} = \frac{20}{243}$$

(ii) The probability of more than 2 successes

$$P(r>2) = 1 - P(r \le 2)$$

$$= 1 - \sum_{r=0}^{2} {}^6C_r\left(\frac{2}{3}\right)^r\left(\frac{1}{3}\right)^{6-r}$$

$$= 1 - \frac{73}{1029} = \frac{956}{1029}$$

(iii) The probability of 3 or more than 3 successes

$$P(r \geq 3) = P(r > 2)$$

$$\frac{956}{1029}$$

Example 7 If the probability of hitting a target is 10% and 10 shots are fired independently, what is the probability that the target will be hit atleast once?

Solution The probability of hitting a target is 10% $\Rightarrow p = 0.1$

and $q = 1 - p = 0.9, n = 10$

The probability that the target will hit at least once

$= 1 - P(0)$

$= 1 - {}^{10}C_0 (0.1)^0 (0.9)^{10-0}$

$= 0.6513$

Example 8 In a bombing action, there is 50% change that any bomb will strike the target. Two direct hits are needed to destroy the target completely. How many bombs are required to be dropped to give a 99% chance or better completely destroying the target?

Solution Let the number of bombs be n.

The probability of 50% chance for strike is $p = \frac{50}{100} = \frac{1}{2}, q = \frac{1}{2}$

By binomial distribution $\left(\frac{1}{2} + \frac{1}{2}\right)^n = \frac{1}{2^n}[1 + {}^nC_1 + {}^nC_2 + \ldots + {}^nC_n]$

For at least two hits, $\frac{1}{2^n}[{}^nC_2 + {}^nC_3 + \ldots + {}^nC_n] \geq 0.99$

or $\quad 1 - (1+n)\frac{1}{2^n} \geq 0.99$

or $\quad (1+n)\frac{1}{2^n} \leq 0.01$

or $\quad 100 + 100n - 2^n \leq 0$

or $\quad 2^n \geq 100 + 100n$

By trial method, we have $n = 11$

Hence, the required number of bombs is 11.

Example 9 *Determine the binomial distribution for which the mean is 4 and variance is 3 and find its mode.*

Solution The binomial distribution is

$$P(r) = {}^nC_r p^r q^{n-r}; r = 0,1,2,...,n$$

Then, mean = np = 4 and variance = npq = 3

$$\therefore \quad \frac{npq}{np} = \frac{3}{4} \Rightarrow q = \frac{3}{4}, p = \frac{1}{4}, n = 16$$

$$\text{Hence,} \quad P(r) = {}^{16}C_r \left(\frac{1}{4}\right)^r \left(\frac{3}{4}\right)^{16-r} ; r = 0,1,2,...,16$$

Now, mode = $np + p$

$$= 4 + \frac{1}{4} = 4.25, \text{ which is not integer.}$$

Therefore, for one mode = integer part of $np + p = 4$.

Example 10 *Prove that the difference of mean and mode is not greater than unity in a binomial distribution.*

Solution Let Binomial distribution is $P(r) = {}^nC_r p^r q^{n-r}; r = 0,1,2,...,n$

Then, mean = np

There are three possibilities, if mode are

(i) If np is a positive integer then mean and mode are equal

$\therefore$ | mean – mode | = 0<1

(ii) If $np + p$ is a positive integer then mode is $np + p$

$\therefore$ | mean – mode | = | $np - np - p$| = p<1

(iii) If $np + p$ is not a positive integer then mode (say r) is given by

$$np + p - 1 \le r \le np + p$$

$$\Rightarrow \quad p - 1 \le r - np \le p$$

$$\Rightarrow \quad -p \leq r - np \leq p$$

$$\Rightarrow \quad -p \leq \text{mode} - \text{mean} \leq p$$

$$\Rightarrow \quad |\text{ mode} - \text{mean }| \leq 1$$

Example 11 *A variable takes values 0,1,2,..., n. with frequencies proportional to the binomial coefficients.*

$$1, \binom{n}{1}, \binom{n}{2}, \ldots\ldots, \binom{n}{n}$$

Find the mean and variance of the distribution and show that the variance is half of the mean.

Solution The total frequency is

$$1 + {}^nC_1 + {}^nC_2 + \ldots + {}^nC_n = (1+1)^n = 2^n$$

are respectively

$$\frac{1}{2^n}, \frac{{}^nC_1}{2^n}, \ldots, \frac{{}^nC_n}{2^n}.$$

This is binomial distribution $(q+p)^n = \left(\frac{1}{2} + \frac{1}{2}\right)^n$

In which $p = \frac{1}{2}, q = \frac{1}{2}$

Now, mean $= np = \frac{n}{2}$ ...(i)

Variance $= npq = n \times \frac{1}{2} \times \frac{1}{2} = \frac{n}{4}$ ...(ii)

Comparing = Eqs. (i) and (ii), we get

Variance $= \frac{1}{2}$(mean)

Example 12 *Calculate the value of p, if the ratio of the probability of an event happening exactly r times in n trials to the probability of event happening exactly n-r times in n trials is independent of n (0<p<1).*

Solution The probability of the event happening *r* times in *n* trials

$$= {}^nC_r p^r (1-p)^{n-r}$$
$$(\because q = 1-p)$$

Similarly, the probability of the event happening *n-r* times in *n* trials

$$= {}^nC_{n-r} p^{n-r} (1-p)^r$$

$$\therefore \text{ The ratio } = \frac{{}^nC_r p^r (1-p)^{n-r}}{{}^nC_{n-r} p^{n-r} (1-p)^r}$$

$$= \left(\frac{1-p}{p}\right)^{n-2r}$$

It can be seen that the only value for $\left(\frac{1-p}{p}\right)$ which would make this ratio independent of *n* is unity.

Hence, $\frac{1-p}{p} = 1$

or $p = \frac{1}{2}$

Example 13 *If a coin is tossed n times, where n is a large even number. S how that the probability of exactly* $\frac{1}{2}n - r$ *heads and* $\frac{1}{2}n + r$ *tails is*

$$\left(\frac{2}{nr}\right)^{\frac{1}{2}} e^{\frac{-2r^2}{n}}$$

Solution The probability of head is $\frac{1}{2}$ and a tail is also $\frac{1}{2}$ in a single throw of a coin.

The probability of *n* trials in binomial distribution is $\left(\frac{1}{2} + \frac{1}{2}\right)^n$

Since, *n* is very large, we may assume $n = 2k$, where *k* is very large. The probability of $\left(\frac{1}{2}n - r\right)$ heads and $\left(\frac{1}{2}n + r\right)$ tails in *n* trial is given by

$$f_r = {}^nC_{\frac{n}{2}-r} \left(\frac{1}{2}\right)^{\frac{n}{2}-r} \left(\frac{1}{2}\right)^{\frac{n}{2}+r}$$

$$= {}^nC_{\frac{n}{2}-r}\left(\frac{1}{2}\right)^n$$

$$= {}^{2k}C_{k-r}\left(\frac{1}{2}\right)^{2k} \qquad \text{...(i)}$$

Also, the probability of $\frac{1}{2}n$ heads and $\frac{1}{2}n$ tails in n trials is

$$f_0 = {}^{2k}C_k\left(\frac{1}{2}\right)^{2k} \qquad \text{.....(ii)}$$

Now, $\dfrac{f_r}{f_0} = \dfrac{{}^{2k}C_{k-r}}{{}^{2k}C_k} = \dfrac{k!k!}{(k+r)!(k-r)!}$

According to stirling's theorem, we have

$$k! = \sqrt{2\pi}k^{k+\frac{1}{2}}e^{-k}$$

$$\therefore \qquad \frac{f_r}{f_0} = \frac{\sqrt{2\pi}k^{k+\frac{1}{2}}e^{-k}.\sqrt{2\pi}k^{k+\frac{1}{2}}e^{-k}}{\sqrt{2\pi}(k+r)^{k+r+\frac{1}{2}}e^{-(k+r)}.\sqrt{2\pi}(k-r)^{k-r+\frac{1}{2}}e^{-(k-r)}}$$

$$= \frac{1}{\left(1+\frac{r}{k}\right)^{k+r+\frac{1}{2}}\left(1-\frac{r}{k}\right)^{k-r+\frac{1}{2}}}$$

or $\qquad \log\dfrac{f_r}{f_0} = -\left(k+r+\dfrac{1}{2}\right)\log\left(1+\dfrac{r}{k}\right) - \left(k-r+\dfrac{1}{2}\right)\log\left(1-\dfrac{r}{k}\right)$

$$= -\left(k+r+\frac{1}{2}\right)\left(\frac{r}{k}-\frac{r^2}{2k^2}+...\right)+\left(k-r+\frac{1}{2}\right)\left(\frac{r}{k}+\frac{r^2}{2k^2}+...\right)$$

$= \dfrac{-r^2}{k} + ...$ as k is a large number.

$$\frac{f_r}{f_0} = e^{\frac{-r^2}{k}}$$

or $$\frac{f_r}{f_o} = e^{\frac{-2r^2}{n}}$$

or $$f_r = f_0 \, e^{\frac{-2r^2}{n}} \qquad \text{...(iii)}$$

Now, $$f_o = \frac{(2k)!}{k!k!}\left(\frac{1}{2}\right)^{2k}$$

$$= \frac{\sqrt{2\pi} e^{-2k} (2k)^{2k+\frac{1}{2}}}{\left(\sqrt{2\pi} e^{-k} k^{k+\frac{1}{2}}\right)^2}\left(\frac{1}{2}\right)^{2k}$$

$$= \left(\frac{1}{\pi k}\right)^{\frac{1}{2}} = \left(\frac{2}{\pi n}\right)^{\frac{1}{2}} \qquad \text{.....(iv)}$$

From Eqs. (iii) and (iv), we get

$$f_r = \left(\frac{2}{\pi n}\right)^{\frac{1}{2}} e^{\frac{-2r^2}{n}}$$

Example 14 In litters of 4 mice the number of litters which contained 0, 1, 2, 3, 4 female were noted. The figures are given in the table below :

No of female mice	0	1	2	3	4	Total
No of litters	8	32	34	24	5	103

If the chance of obtaining a female in a single trial is assumed constant, estimate this constant of unknown probability. Find also expected frequencies.

Solution The mean of observation

$$m = \frac{0\times 8 + 1\times 32 + 2\times 34 + 3\times 24 + 4\times 5}{103}$$

$$= \frac{192}{103} = 1.864$$

since, $n = 4$, mean $= np = m$

$$\therefore \qquad p = \frac{m}{n} = \frac{1.864}{4} = 0.466$$

$$q = 1 - p = 1 - 0.466 = 0.534$$

Therefore, expected frequencies are the respective terms of the binomial distribution of

$$N(q+p)^n$$

$$=103\ (0.534 + 0.466)^4$$

Theoretical frequency distribution table are

r	f	$p(r)={}^4C_r(0.534)^{4-r}(0.466)^r$	$f_e = 103p(r)$
0	8	${}^4C_0(0.543)^4 = 0.0813$	8.375
1	32	${}^4C_1(0.534)^3(0.466) = 0.283$	29.23
2	34	${}^4C_2(0.534)^2(0.466)^2 = 0.3715$	38.26
3	24	${}^4C_3(0.534)(0.466)^3 = 0.2161$	22.26
4	5	${}^4C_4(0.466)^4 = 0.0471$	4.857
Total	103		102.98

Practice Problems

1. Find the binomial distribution whose mean is 5 and variance is $\frac{10}{3}$.

[**Ans.** ${}^{15}C_r\left(\frac{1}{3}\right)^r\left(\frac{2}{3}\right)^{15-r}$]

2. A coin is tossed six times. What is the probability of obtaining (i) 4 heads (ii) 5 heads (iii) 6 heads (iv) getting 4 or more heads.

[**Ans.** (i) $\frac{15}{64}$ (ii) $\frac{3}{32}$ (iii) $\frac{1}{64}$ (iii) $\frac{11}{32}$]

3. If 10% bulbs produced by a factory are defective, find the probability that out of 5 bulbs chooses at random.

(i) None will be defective (ii) one will be defective (iii) at least two be defective.

[Ans. (i) 0.59049 = 0.6065 (ii) 0.32805 = 0.3032 (iii) 0.08146 = 0.9855]

4. Assumingthat half the population are consumers of chocolate, so that the chance of an individual being a consumer is $\frac{1}{2}$ and assuming that 100 investigators each take 10 individuals to see whether they are consumers, how many investigators would you expect to report that three people or less were consumers?

[Ans. 17 (nearly)]

5. Show that if *np* be a whole number, the mean of the binomial distribution coincides with the greatest term.

6. Let *r* be a binomially distributed random variable with mean 10 and variance 5.Show that

(i) $p(r>6) = \left(\frac{1}{2}\right)^{20} \sum_{r=7}^{20} {}^{20}c_r$ (ii) $p\,(3<r<12) = \left(\frac{1}{2}\right)^{20.} \sum_{r=4}^{11} {}^{20}c_r$

7. Determine the probability of getting 9 exactly twice in 3 throws with a pair of fair dice.

[Ans. $\frac{8}{243}$]

8. Out of 800 families with 5 children each, how many would you expect to have (i) 3 boys (ii) 5 girls (iii) either 2 or boys

[Ans. (i) 250 (ii) 25 (iii) 500]

9. The following data are the number of seeds germinating out of 10 on damp filter for 80 sets of seeds. Fit a binomial distribution to these data.

x	0	1	2	3	4	5	6	7	8	9	10	Total
f	6	20	28	12	8	6	0	0	0	0	0	80

[Ans. $80(0.7825 + 0.2175)^{10}$, 6.9, 19.1, 24.0, 17.8, 8.6, 2.9, 0.7, 0.1, 0, 0,0]

10. Seven coins are tossed and the number of heads are noted. The experiment is repeated 128 times and following distribution is obtained

No. of heads	0	1	2	3	4	5	6	7	Total
Frequency	7	6	19	35	30	23	7	1	128

Fit the binomial distribution considering the coin unbiased.

$$\left[\text{Ans.}128\left(\frac{1}{2}+\frac{1}{2}\right)^7\right]$$

Poisson Dustribution

The poisson distribution is generally derived as the limiting form of the binomial distribution $(q+p)^n$ when $n \to \infty$ and $p \to 0$ so that np is a finite quantity (say m).

The probability of r successes in the binomial distribution

$$p(r) = {}^nC_r p^r q^{n-r}$$

$$= \frac{n!}{r!(n-r)!} p^r q^{n-r}$$

Since, mean $m = np \Rightarrow p = \dfrac{m}{n}$

$$\therefore \qquad p(r) = \frac{n!}{r!(n-r)!}\left(\frac{m}{n}\right)^r \left(1-\frac{m}{n}\right)^{n-r}$$

$$= \frac{m^r}{r!}\left(1-\frac{m}{n}\right)^n \frac{n!}{(n-r)!\,n^r\left(1-\dfrac{m}{n}\right)^r}$$

Taking limits as $n \to \infty$, so that

$$\lim_{n\to\infty}\left(1-\frac{m}{n}\right)^n = e^{-m} \text{ and } \lim_{n\to\infty}\left(1-\frac{m}{n}\right)^r = 1$$

Then, Probability of r successes is

$$p(r) = \frac{m^r}{r!}e^{-m}\lim_{n\to\infty}\frac{n!}{(n-r)!\,n^r}$$

By using stirling's theorem,

$$n! = \sqrt{2\pi}\, n^{n+\frac{1}{2}} e^{-n}$$

Then, $$p(r) = \frac{m^r}{r!}e^{-m}\lim_{n\to\infty}\frac{\sqrt{2\pi}\,.n^{n+1/2}.e^{-n}}{\sqrt{2\pi}\,(n-r)^{n-r+\frac{1}{2}}.n^r.e^{-(n-r)}}$$

$$= \frac{m^r}{r!} e^{-m} \lim_{n\to\infty} \frac{1}{\left(1-\frac{r}{n}\right)^n \left(1-\frac{r}{n}\right)^{-r+\frac{1}{2}}} \cdot \frac{1}{}$$

$$= \frac{m^r e^{-m}}{r! e^r} \cdot \frac{1}{e^{-r}}$$

$$p(r) = \frac{m^r e^{-m}}{r!}$$

Hence, the probabilities of 0, 1, 2, …, r successes are given by

$$e^{-m}, me^{-m}, \frac{m^2 e^{-m}}{2!}, \ldots, \frac{m^r e^{-m}}{r!}$$

Constants of the Poisson Distribution

(i) First moment about origin (Mean)

$$\mu_1^1 = \sum_{r=0}^{\infty} r.\frac{m^r e^{-m}}{r!}$$

$$= \sum_{r=0}^{\infty} \frac{m^r e^{-m}}{(r-1)!}$$

$$= e^{-m}\left(m + \frac{m^2}{1!} + \frac{m^3}{2!} + \ldots\ldots\right)$$

$$= me^{-m}\left(1 + \frac{m}{1!} + \frac{m^2}{2!} + \ldots\ldots\right)$$

$$= me^{-m}.e^m$$

$$= m$$

(ii) Second moment about origin

$$\mu_2^1 = \sum_{r=0}^{\infty} r^2.\frac{m^r e^{-m}}{r!}$$

$$= \sum_{r=0}^{\infty} [r + r(r-1)] \frac{m^r e^{-m}}{r!}$$

$$= \sum_{r=0}^{\infty} \frac{m^r e^{-m}}{(r-1)!} + \sum_{r=0}^{\infty} \frac{m^r e^{-m}}{(r-2)!}$$

$$= m + e^{-m}\left[m^2 + \frac{m^3}{1!} + \frac{m^4}{2!} + ...\right]$$

$$= m + e^{-m}m^2\left[1 + \frac{m}{1!} + \frac{m^2}{2!} + ...\right]$$

$$= m + m^2\, e^{-m}.e^m$$

$$= m + m^2$$

(iii) Third moment about origin

$$\mu_3^1 = \sum_{r=0}^{\infty} r^3.\frac{m^r e^{-m}}{r!}$$

$$= e^{-m} \sum_{r=0}^{\infty} [r(r-1)(r-2) + 3r(r-1) + r]\frac{m^r}{r!}$$

$$= e^{-m}\, m^3 \sum_{r=3}^{\infty} \frac{m^{r-3}}{(r-3)!} + 3e^{-m}\, m^2 \sum_{r=2}^{\infty} \frac{m^{r-2}}{(r-2)!} + me^{-m} \sum_{r=1}^{\infty} \frac{m^{r-1}}{(r-1)!}$$

$$= m^3 + 3m^2 + m$$

(iv) Fourth moment about origin

$$\mu_4^1 = \sum_{r=0}^{\infty} r^4.\frac{m^r e^{-m}}{r!}$$

$$= e^{-m} \sum_{r=0}^{\infty} [r(r-1)(r-2)(r-3) + 6r(r-1)(r-2) + 7r(r-1) + r]\frac{m^r}{r!}$$

$$= e^{-m}\, m^4 \sum_{r=4}^{\infty} \frac{m^{r-4}}{(r-4)!} + 6e^{-m}\, m^3 \sum_{r=3}^{\infty} \frac{m^{r-3}}{(r-3)!} + 7e^{-m}\, m^2 \sum_{r=2}^{\infty} \frac{m^{r-2}}{(r-2)!}$$

$$+ e^{-m}.m\sum_{r=1}^{\infty} \frac{m^{r-1}}{(r-1)!}$$

$$= m^4 + 6m^3 + 7m^2 + m$$

(v) First moment about mean

$$\mu_1 = 0$$

(vi) Second moment about mean (variance)

$$\mu_2 = \mu_2^1 - (\mu_1^1)^2$$

$$= m^2 + m - m^2$$

$$= m$$

$$\text{standard deviation} = \sqrt{\mu_2} = \sqrt{m}$$

(vii) Third moment about mean

$$\mu_3 = \mu_3^1 - 3\mu_2^1\,\mu_1^1 + 2(\mu_1^1)^3$$

$$= (m^3 + 3m^2 + m) - 3(m^2 + m)\,m + 2m^3$$

$$= m$$

(viii) Fourth moment about mean

$$\mu_4 = \mu_4^1 - 4\mu_3^1\,\mu_1^1 + 6\mu_2^1\,(\mu_1^1)^2 - 3(\mu_1^1)^4$$

$$= m^4 + 6m^3 + 7m^2 + m - 4(m^3 + 3m^2 + m)\,m + 6\,(m^2 + m)\,m^2 - 3m^4$$

$$= 3m^2 + m$$

(ix) Karl pearson's Coefficients-

$$\beta_1 = \frac{\mu_3^2}{\mu_2^3} = \frac{m^2}{m^3} = \frac{1}{m}$$

$$\beta_2 = \frac{\mu_4}{\mu_2^2} = \frac{3m^2 + m}{m^2} = 3 + \frac{1}{m}$$

$$\gamma_1 = \sqrt{\beta_1} = \frac{1}{\sqrt{m}}$$

$$\gamma_2 = \beta_2 - 3 = \frac{1}{m}$$

Recurrence Formula for Poisson Distribution

For poisson distribution $P(r) = \dfrac{e^{-m}m^r}{r!}$

and $$P(r+1) = \frac{e^{-m} m^{r+1}}{(r+1)!}$$

$$\therefore \quad \frac{P(r+1)}{P(r)} = \frac{e^{-m} m^{r+1}}{(r+1)!} \cdot \frac{r!}{e^{-m} . m^r}$$

$$= \frac{m.r!}{(r+1).r!}$$

$$= \frac{m}{r+1}$$

$$P(r+1) = \frac{m}{r+1} P(r);\ r = 0,1,2,\ldots\ldots\ldots$$

IIlustrative Examples

Example 1 *If the variance of the poisson distribution is 2, find the probabilities for r = 1,2,3,4 from the recurrence relation of the poisson distribution. Also, find $P(x \geq 4)$.*

Solution Poisson distribution is $P(r) = \frac{m^r e^{-m}}{r!}$

The recurrence relation is $P(r+1) = \frac{m}{r+1} p(r)$

$$P(r+1) = \frac{m}{r+1} \cdot \frac{m^r e^{-m}}{r!}$$

$$P(r+1) = \frac{m^{r+1} . e^{-m}}{(r+1).r!}$$

Given, variance = m = 2

Putting r = 1,2,3,4, in Eq. (i), we get

For $r = 1$, $P(x = 2) = \frac{2^2 e^{-2}}{2.1!} = 2e^{-2} = \frac{2}{e^2}$

For $r = 2$ $P(x = 3) = \frac{2^3 e^{-2}}{3.2!} = \frac{4}{3e^2}$

For $r = 3$ $P(x = 4) = \frac{2^4 e^{-2}}{4.3!} = \frac{2}{3e^2}$

For $r = 4\ P(x = 5) = \dfrac{2^5 e^{-2}}{5.4!} = \dfrac{4}{15e^2}$

For $r \geq 4P(x \geq 4) = \dfrac{4}{15e^2}$

Example 2 *Six coins are tossed 6400 times. Using poisson distribution, find the approximate probability of getting six heads x times and 2 times.*

Solution The probability of getting a head = The probability of getting a tail for each coin.

$\therefore$ The probability of getting 6 head with 6 coins $= \left(\dfrac{1}{2}\right)^6 = \dfrac{1}{64} = p$

$$\therefore \qquad m = np = 6400 \times \frac{1}{64} = 100$$

Then, by poisson distribution

$$P(X = r) = \frac{e^{-m} m^r}{r!}$$

$$= \frac{e^{-100}(100)^r}{r!}$$

for x times $P(X = x) = \dfrac{e^{-100}(100)^x}{x!}$

For 2 times $P(X = 2) = \dfrac{e^{-100}(100)^2}{2!}$

$$= 5000e^{-100}$$

Example 3 *Find the probability that at most 5 defective fuses will be found in a box of 200 fuses if experience show that 2% of such fuses are defective. ($e^{-4} = 0.0183$)*

Solution Here, $m = np = 200 \times \dfrac{2}{100} = 4$

Hence, $P(r \leq 5) = \displaystyle\sum_{r=0}^{5} \frac{e^{-4} 4^r}{r!}$

$$= e^{-4}\left[1+4+\frac{4^2}{2!}+\frac{4^3}{3!}+\frac{4^4}{4!}+\frac{4^5}{5!}\right]$$

$$= 0.0183 \times 51.41 = 0.9408$$

Example 4 *The probability that an injection manufactured by a company will be defective is* $\frac{1}{10}$ *. If 12 such injection are manufactured, find the probability that (i) Exactly two will be defective (ii) At least two will be defective (iii) None will be defective* $(e^{-1.2} = 0.3012)$

Solution The probability of defective injection $p = \frac{1}{10}$, $n = 12$

Since, $m = np = 12 \times \frac{1}{10} = 1.2$

(i) Exactly two will be defective

$$P(X = 2) = \frac{e^{-1.2}(1.2)^2}{2!} = 0.216$$

(ii) At least two will be defective

$$P(\text{at least two}) = P(2) + P(3) + \ldots\ldots\ldots + P(12)$$

$$= 1-[P(0) + P(1)]$$

$$= 1-\left[\frac{e^{-1.2}(1.2)^0}{0!}+\frac{e^{-1.2}(1.2)^1}{1!}\right]$$

$$= 1-e^{-1.2}\,[1+1.2]$$

$$= 1- 0.3012 \times 2.2 = 0.3373$$

(iii) None will be defective

$$P(X = 0) = \frac{e^{-1.2}(1.2)^0}{0!}$$

$$= e^{-1.2} = 0.3012$$

Example 5 *A poisson distribution has a double mode at x = 4 and x = 5. Find the probability that x will have either of these values P(x=4) = P(x=5)*

Solution Let the poisson distribution be $P(x = r) = \frac{e^{-m}m^r}{r!}$; r = 0,1,2,3, ………

$$P(x=4)=\frac{e^{-m}m^4}{4!}$$

and $\quad P(x=5)=\frac{e^{-m}m^5}{5!}$

Here, $\quad P(x=4)=P(x=5)$

$$\Rightarrow \frac{e^{-m}m^4}{4!}=\frac{e^{-m}m^5}{5!}\Rightarrow m=5$$

Now, $e^{-5}=0.006738$

The required probability $= P(x=4)+P(x=5)=2P(x=4)$

$$=2.\frac{e^{-5}5^4}{4!}$$

$$=\frac{2\times 0.006738\times 5^4}{4!}=0.035$$

Example 6 *A car-hire firm has two cars, which it hires out day by day. The number of demands for a car on each day is distributed as poisson distribution with mean 1.5. Calculate the proportion of days on which neither car is used and the proportion of days on which some demand is refused. ($e^{-1.5}=0.2231$).*

Solution The proportion of days when no car will be required is

$$\frac{e^{-m}m^0}{0!}=0.2231$$

The probability when no car, one car, two car will be required is

$$=\frac{e^{-1.5}(1.5)^0}{0!}+\frac{e^{-1.5}(1.5)^1}{1!}+\frac{e^{-1.5}(1.5)^2}{2!}$$

$$=e^{-1.5}(1+1.5+1.125)$$

$$=0.2231\times 3.625$$

$$=0.8087$$

$\therefore$ The proportion of days on which some demand is refused

$$=1-0.8087$$

$$=0.1913$$

Example 7 In a certain factory turning razor blades, there is a small chance, $\frac{1}{500}$ for any blade to be defective. The blades are in packets of 10. Using poisson distribution to calculate the approximate number of packets containing no defective, one defective and two defective blades respectively in a consignment of 10,000 packets. ($e^{-0.02} = 0.9802$).

Solution Given, N = 10,000, n = 10, $p = \frac{1}{500}$

$$\therefore m = np = 10 \times \frac{1}{500} = 0.02$$

The frequencies of 0,1,2,.......... defective blades are given by

$$N\frac{e^{-m}m^r}{r!}$$

$$\Rightarrow \quad 10000 \times \frac{e^{-0.02}(0.02)^r}{r!}$$

(i) Number of packet with no defective blades = $10000 \times e^{-0.02}$

$= 10000 \times 0.9802$

$= 9802$

(ii) Number of packet with one defective blades

$$= \frac{10000 \times e^{-0.02} \times 0.02}{1!} = 196$$

(iii) Number of packet with two defective blades

$$= \frac{10000 \times e^{-0.02} \times (0.02)^2}{2!} = 1.96 = 2 \text{ (nearly)}$$

Example 8 *Show that in a poisson distribution with unity mean, the mean deviation about the mean is $\frac{2}{e}$ times the standard deviation.*

Solution The poisson distribution is

$$P(X = r) = \frac{e^{-m}m^r}{r!}; r = 0,1,2.........$$

In the case, $m = 1$

The probabilities are $\frac{e^{-1}}{0!}, \frac{e^{-1}}{1!}, \frac{e^{-1}}{2!}$............

The mean deviation about mean is

$$\text{M.D.} = \sum_{x=0}^{\infty} p_i[x_i - m]$$

$$= \sum_{x=0}^{\infty} \frac{e^{-1}|x-1|}{x!}$$

$$= e^{-1}\left[1 + \frac{1}{2!} + \frac{2}{3!} + \frac{3}{4!} + ...\right]$$

$$= \frac{1}{e}\left[1 + \frac{2-1}{2!} + \frac{3-1}{3!} + \frac{4-1}{4!} + ...\right]$$

$$= \frac{1}{2}\left\{\left[1 + 1 + \frac{1}{2!} + \frac{1}{3!} + ...\right] - \left[\frac{1}{2!} + \frac{1}{3!} + ...\right]\right\}$$

$$= \frac{1}{e}\{e - (e-2)\}$$

$$= \frac{1}{e} \times 2 = \frac{2}{e}$$

Example 9 *If X and Y be independent poisson variates, show that the conditional distribution of X, given X+Y= n, is binomial.*

Solution Let X and Y be independent poisson variate with mean m_1 and m_2.

$$P(X = r : X + Y = n) = \frac{P(X = r, X + Y = n)}{P(X + Y = n)}$$

$$= \frac{P(X = r).P(Y = n - r)}{P(X + Y = n)}$$

$$= \frac{\dfrac{e^{-m_1}.m_1^r}{r!} \cdot \dfrac{e^{-m_2}m_2^{n-r}}{(n-r)!}}{\dfrac{e^{-(m_1+m_2)}(m_1+m_2)^n}{n!}}$$

$$= \frac{n!}{r!(n-r!)} \cdot \frac{m_1^r m_2^{n-r}}{(m_1+m_2)^n}$$

$$= {}^nC_r p^r q^{n-r}, \text{where } p = \frac{m_1}{m_1+m_2}, q = \frac{m_2}{m_1+m_2}$$

Hence, the conditional distribution of X, is binomial.

Example 10 *Find the mean and S.D. for the table of deaths of women over 85 years old recorded in a three year period.*

No. of deaths	0	1	2	3	4	5	6	7
No. of days	364	376	218	89	33	13	2	1

Find the expected number of days with one death recorded for the poisson series fitted to the data.

Solution The frequency distribution is

x	f	fx	fx^2
0	364	0	0
1	376	376	376
2	218	436	872
3	89	267	801
4	33	132	528
5	13	65	325
6	2	12	72
7	1	7	49
Total	1096	1295	3023

Gven, $N = 1096$

$$\therefore \text{Mean, m} = \frac{\sum fx}{N} = \frac{1295}{1096} = 1.18$$

$$\text{and standard deviation,} \sigma = \sqrt{\frac{\sum fx^2}{N} - \left(\frac{\sum fx}{N}\right)^2}$$

$$= \sqrt{\frac{3023}{1096} - (1.18)^2}$$

$$= \sqrt{2.758 - 1.392}$$

$$= \sqrt{1.366}$$

$$= 1.17 \text{ (approx.)}$$

Now, $e^{-1.18} = 1 - 1.18 + \dfrac{(1.18)^2}{2!} - \dfrac{(1.18)^3}{3!} + \dfrac{(1.18)^4}{4!} - \dfrac{(1.18)^5}{5!} + \dots\dots\dots$

$$= 1 - 1.18 + 0.6962 - 0.2738 + 0.0808 - 0.019$$

$$= 0.31$$

$\therefore$ For one death, frequency $= \dfrac{Ne^{-m}.m^1}{1!}$

$$= 1096 \times e^{-1.18} \times 1.18$$

$$= 1096 \times 0.31 \times 1.18$$

$$= 400.91$$

Practice Problems

1. If the probability of a bad reaction from a certain injection is 0.001, determine the chance that out of 2000 individuals more than two will get a bad reaction. ($e^{-2} = 0.1353$).

[Ans. 0.3235]

2. Using poisson distribution. Find the probability that the ace of spades will be drawn from a pack of well shuffled cards at least once is 104 consecutive trials. ($e^{-2} = 0.136$)

[Ans. 0.864]

3. A telephone switch board handles 600 calls on the average during a rush hour. The board can make a maximum of 20 connections per minute. Using poisson distribution to estimate the probability that the board will be over taxed during any given minute. ($e^{-1} = 0.00004539$)

[Ans. 1-0.00004539 $\sum_{r=0}^{20} \dfrac{10^r}{r!}$]

4. Let λ and μ_r denote the mean and rth moment about mean of a poisson distribution. Obtain the recurrence formula

$$\mu_{r+1} = \lambda r \mu_{r-1} + \lambda \frac{d\mu_r}{d\lambda}$$

5. The letters were received in an office on each of 100 days. Assuming the following data to form a random sample from a poisson distribution. Fit the distribution and calculate the expected frequencies. ($e^{-4} = 0.0183$)

No.of letters	0	1	2	3	4	5	6	7	8	9	10
Frequencies	1	4	15	22	21	20	8	6	2	0	1

[**Ans.** Expected frequencies = $\dfrac{100 \times e^{-4} \times 4^r}{r!}$, 1.83, 7.32, 14.64, 9.22, 19.52, 15.62, 10.41, 5.95, 2.975, 1.322, 5.59]

6. Fit poisson distribution to the following and calculate theoretical frequencies.

Deaths	0	1	2	3	4
Frequency	122	60	15	2	1

[**Ans.** $\dfrac{200 \times 0.61 \times (0.5)^r}{r!}$, 122, 61, 15, 2, 0]

7. A manufacturer of pins known that 5% of his product is defective. If he sells pins in boxes of 100 and guarantees that no more than 10 pins will be defective. What is the approximate probability that a box will fail to meet the guarantees quality.

[**Ans.** $1 - \sum_{r=0}^{10} \dfrac{e^{-5} . 5^r}{r!}$]

8. A poisson distribution has a mode $P(x = 1) = P(x = 2)$, then find $P(x = 4)$ and $P(x \leq 4)$.

[**Ans.** $P(x = 4) = \dfrac{2}{3} e^{-2}$, $P(x \leq 4) = 7e^{-2}$]

9. A poisson distribution has a double mode at x = 3 and x = 4, what is the probability that x will have one or the other of these two values.

[**Ans.** $\dfrac{64}{3} e^{-4}$]

Normal Distribution

The normal distribution was introduced by the mathematician De-Moivre in 1733 as the limiting form of the binomial distribution.

Its density function is given by

$$f(x) = \frac{1}{\sigma\sqrt{2\pi}} e^{-\frac{1}{2}\left(\frac{x-m}{\sigma}\right)^2}$$

Where μ and σ are the parameters of the distribution. *f(x)* is non negative and $\int_{-\infty}^{\infty} f(x)dx = 1$. It is also called "Gaussian Distribution".

Proof :
$$\int_{-\infty}^{\infty} f(x)dx = \frac{1}{\sigma\sqrt{2\pi}} \int_{-\infty}^{\infty} e^{-1/2\left(\frac{x-m}{\sigma}\right)^2} dx$$

By putting $\frac{x-m}{\sigma} = z$ and $dx = \sigma\, dz$, then

$$= \frac{1}{\sqrt{2\pi}} \int_{-\infty}^{\infty} e^{-\frac{1}{2}.z^2} dz$$

$$= \frac{2}{\sqrt{2\pi}} \int_{0}^{\infty} e^{-\frac{1}{2}.z^2} dz$$

Again, putting $\frac{1}{2} z^2 = t$ and $zdz = dt \Rightarrow dz = \frac{dt}{\sqrt{2t}}$ then

$$= \sqrt{\frac{2}{\pi}} \int_{0}^{\infty} e^{-t} \frac{dt}{\sqrt{2t}}$$

$$= \frac{1}{\sqrt{\pi}} \int_{0}^{\infty} e^{-t}.t^{-\frac{1}{2}} dt$$

$$= \frac{1}{\sqrt{\pi}} \int_{0}^{\infty} e^{-t} t^{\frac{1}{2}-1} dt$$

$$= \frac{1}{\sqrt{\pi}} \Gamma\frac{1}{2} \qquad \left(\because \Gamma n = \int_{0}^{\infty} e^{-t}.t^{n-1} dt\right)$$

$$= \frac{\sqrt{\pi}}{\sqrt{\pi}} = 1 \qquad \left(\because \Gamma\frac{1}{2} = \sqrt{\pi}\right)$$

Derivation of Normal Distribution as a limiting case of Binomial Distribution when p = q

Let the binomial distribution be $N(p+q)^n$. If $p = q = \dfrac{1}{2}$.

The frequency of r successes is $f(r) = N\,{}^nC_r p^r q^{n-r}$

If n is an even integer (say $2k$), k being an integer.

At $n \to \infty$ the frequencies of r and $r+1$ successes are

$$f(r) = N.{}^{2k}C_r\left(\frac{1}{2}\right)^{2k}, f(r+1) = N.{}^{2k}C_{r+1}\left(\frac{1}{2}\right)^{2k}$$

$$\therefore \quad \frac{f(r+1)}{f(r)} = \frac{{}^{2k}C_{r+1}}{{}^{2k}C_r}$$

$$= \frac{(2k)!}{(r+1)!(2k-r-1)!} \Big/ \frac{(2k)!}{r!(2k-r)!}$$

$$= \frac{r!(2k-r)!}{(r+1)!(2k-r-1)!}$$

$$= \frac{2k-r}{r+1} \qquad \ldots\text{(i)}$$

The frequency of r successes will be greater than $r+1$, then

$$f(r) > f(r+1)$$

$$\Rightarrow \quad \frac{f(r+1)}{f(r)} < 1$$

$$\Rightarrow \quad 2k - r < r + 1$$

$$\Rightarrow \quad r > k - \frac{1}{2} \qquad \text{[from Eq. (i)]}\ldots\text{(ii)}$$

Let frequency is maximum (say y_0), and $r = k$, then

$$y_0 = N.{}^{2k}C_k\left(\frac{1}{2}\right)^{2k}$$

$$= N.\frac{(2k)!}{k!k!}\left(\frac{1}{2}\right)^{2k}$$

Let y_x be the frequency of $k+x$ successes, then

$$y_x = N.\,^{2k}C_{k+x}\left(\frac{1}{2}\right)^{2k}$$

$$= N.\frac{(2k)!}{(k+x)!(k-x)!}\left(\frac{1}{2}\right)^{2k}$$

$$\therefore \qquad \frac{y_x}{y_0} = \frac{k!k!}{(k+x)!(k-x)!}$$

$$= \frac{k(k-1)(k-2).....(k-x+1)}{(k+x)(k+x-1).....(k+1)}$$

$$= \frac{\left(1-\frac{1}{k}\right)\left(1-\frac{2}{k}\right).......\left(1-\frac{x-1}{k}\right)}{\left(1+\frac{1}{k}\right)\left(1+\frac{2}{k}\right)..........\left(1+\frac{x}{k}\right)}$$

Taking log on both sides, we have

$$\log\frac{y_x}{y_0} = \left[\log\left(1-\frac{1}{k}\right)+\log\left(1-\frac{2}{k}\right)+......+\log\left(1-\frac{x-1}{k}\right)\right]$$

$$-\left[\log\left(1+\frac{1}{k}\right)+\log\left(1+\frac{2}{k}\right)+.......+\log\left(1+\frac{x}{k}\right)\right]$$

Expand logarithm series for each term and neglecting higher terms of $\frac{x}{k}$, we get

$$\log\frac{y_x}{y_0} = -\frac{1}{k}\ \{1+2+3+.....+(x-1)\} - \frac{1}{k}\{1+2+3+......+(x-1)+x\}$$

$$= -\frac{2}{k}\{1+2+3+.....+(x-1)\} - \frac{x}{k}$$

$$= -\frac{2}{k}\,\frac{(x-1)(x-1+1)}{2} - \frac{x}{k}$$

$$= -\frac{x}{k}(x-1) - \frac{x}{k}$$

$$= -\frac{x^2}{k} + \frac{x}{k} - \frac{x}{k}$$

$$= -\frac{x^2}{k}$$

$$\therefore \qquad \frac{y_x}{y_0} = e^{\frac{-x^2}{k}}$$

$$\Rightarrow \qquad y_x = y_0 e^{\frac{-x^2}{k}} \qquad \text{...(iii)}$$

In the binomial distribution,

$$\sigma^2 = npq$$

$$= n \times \frac{1}{2} \times \frac{1}{2} = \frac{n}{4} = \frac{2k}{4} = \frac{k}{2}$$

$$\therefore \qquad k = 2\sigma^2 \qquad \text{...(iv)}$$

$\therefore$ From Eqs. (iii) and (iv), we get

$$y_x = y_0 e^{\frac{-x^2}{2\sigma^2}}$$

Derivation of Normal Distribution as a Limiting Case of Binomial Distribution when $p \neq q$ Let the binomial distribution be $N(p+q)^n$. If the frequency of r successes is $f(r) = N.{}^nC_r p^r q^{n-r}$ and frequency of $(r+1)$ successes is $f(r+1) = N.\ {}^nC_{r+1} p^{r+1} q^{n-r-1}$

$$\therefore \qquad \frac{f(r+1)}{f(r)} = \frac{{}^nC_{r+1}}{{}^nC_r} \cdot \frac{p}{q}$$

$$= \frac{n-r}{r+1} \cdot \frac{p}{q}$$

The frequency of r successes will be greater than $r+1$, then

$$f(r) > f(r+1)$$

$$\Rightarrow \qquad \frac{f(r+1)}{f(r)} < 1$$

$$\Rightarrow \quad \frac{n-r}{r+1}.\frac{p}{q} < 1$$

$$\Rightarrow \quad (n-r)p < (r+1)q$$

$$\Rightarrow \quad r(p+q) > np - q$$

$$\Rightarrow \quad r > np - q$$

Let us take that np is a whole number.

If y_0 be the maximum frequency, then

$$y_0 = f(np) = N.\frac{n!}{np!np!}p^{np}.q^{nq}$$

and $$y_x = f(np+x) = \frac{n!}{(np+x)!(nq-x)!}p^{np+x}.q^{nq-x}$$

$$\therefore \quad \frac{y_x}{y_0} = \frac{np!np!}{(np+x)!(nq-x)!}\left(\frac{p}{q}\right)^x$$

Using stirling formula, n! = $\sqrt{2\pi}\ e^{-n}\ n^{n+\frac{1}{2}}$, we get

$$\frac{y_x}{y_0} = \frac{\sqrt{2\pi}e^{-np}(np)^{np+1/2}.\sqrt{2\pi}e^{-np}(np)^{np+1/2}}{\sqrt{2\pi}e^{-(np+x)}(np+x)^{np+x+1/2}.\sqrt{2\pi}e^{-(nq-x)}(nq-x)^{nq-x-1/2}}\left(\frac{np}{nq}\right)^x$$

$$= \frac{1}{\left(1+\frac{x}{np}\right)^{np+x+1/2}\left(1-\frac{x}{nq}\right)^{nq-x+1/2}}$$

taking log on both sides, we get

$$\log\frac{y_x}{y_0} = -\left(np+x+\frac{1}{2}\right)\log\left(1+\frac{x}{np}\right) - \left(nq-x+\frac{1}{2}\right)\log\left(1-\frac{x}{nq}\right)$$

$$= -\left(np+x+\frac{1}{2}\right)\left(\frac{x}{np}-\frac{x^2}{2n^2p^2}+\frac{x^3}{3n^3p^3}-\ldots\ldots\right)+\left(nq-x+\frac{1}{2}\right)\left(\frac{x}{nq}+\frac{x^2}{2n^2q^2}+\frac{x^3}{3n^3q^3}+\ldots\ldots\right)$$

$$= x\left(-1-\frac{1}{2np}+1+\frac{1}{2nq}\right)+x^2\left(\frac{1}{2np}-\frac{1}{np}+\frac{1}{4n^2p^2}+\frac{1}{2nq}-\frac{1}{nq}+\frac{1}{4n^2q^2}\right)$$

$$+x^3\left(\frac{1}{2n^2p^2}-\frac{1}{3n^2p^2}-\frac{1}{6n^3p^3}+\frac{1}{3n^3q^3}+\frac{1}{6n^3q^3}-\frac{1}{2n^2q^2}\right)$$

$$=\frac{p-q}{2npq}x+\frac{p^2+q^2}{4n^2p^2q^2}x^2-\frac{1}{2npq}x^2+\text{....... higher terms}$$

Neglecting the terns of order of $\frac{1}{n^2}$ and higher orders, we get

$$\log\frac{y_x}{y_0}=-\frac{q-p}{2npq}x-\frac{1}{2npq}x^2$$

Since, $P < 1, q < 1$ then $q - p$ is very small in comparison to n and therefore the first term can be neglected. Hence,

$$\log\frac{y_x}{y_0}=-\frac{x^2}{2npq}$$

Since, $\sigma^2 = npq$, therefore

$$\log\frac{y_x}{y_0}=\frac{-x^2}{2\sigma^2}$$

$$\therefore \qquad y_x = y_0 e^{\frac{-x^2}{2\sigma^2}}$$

Normal Curve A curve $y_x = y_0 e^{-\frac{1}{2}\left(\frac{x-m}{\sigma}\right)^2}$ is known as normal curve. If mean is at origin, then

$$y_x = y_0 e^{\frac{-x^2}{2\sigma^2}}$$

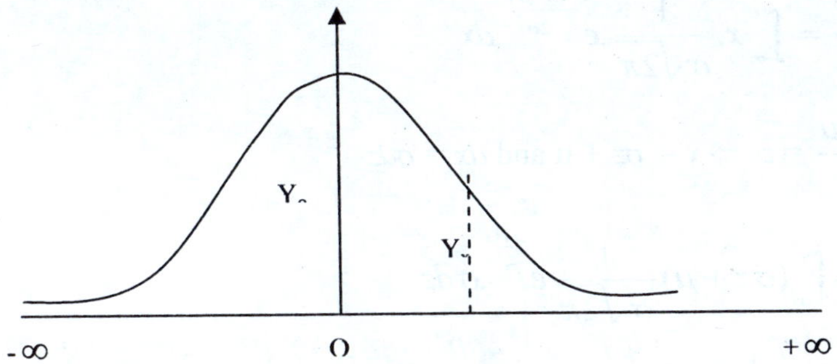

Properties of the Normal Distribution

(1) The curve is symmetrical about $x = m$ and lies about the x-axis. The median, mode coincide with mean.

(2) If the equation of the normal curve is taken as the origin at the mean, the value of y_0 is calculated by making the total area under the curve to be unity.

(3) The maximum value of $y_0 = \dfrac{1}{\sigma\sqrt{2\pi}}$

(4) The point of inflexion of the normal curve, where $\dfrac{d^2y}{dx^2} = 0$ and $\dfrac{d^3y}{dx^3} \neq 0$ given by $x = \pm\sigma$ where the origin is taken at the mean.

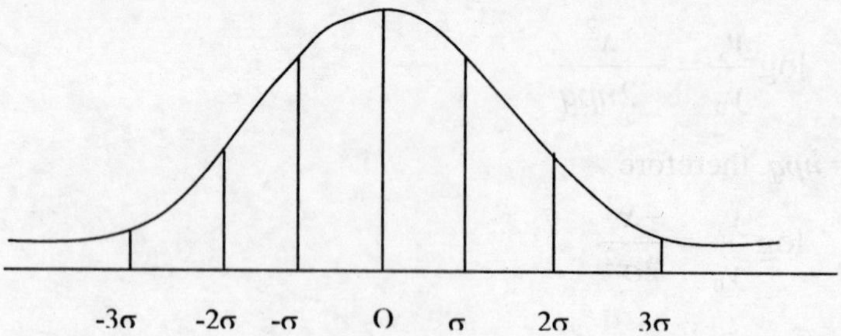

Constants of the Normal Distribution

Mean $E(X) = \int_{-\infty}^{\infty} xf(x)dx$

$$= \int_{-\infty}^{\infty} x.\frac{1}{\sigma\sqrt{2\pi}}.e^{-\frac{(x-\mu)^2}{2\sigma^2}}dx$$

Put $\dfrac{x-\mu}{\sigma} = z \Rightarrow x = \sigma z + \mu$ and $dx = \sigma dz$

$$= \int_{-\infty}^{\infty} (\sigma z + \mu)\frac{1}{\sigma\sqrt{2\pi}}e^{\frac{-z^2}{2}}.\sigma dz$$

$$= \frac{1}{\sqrt{2\pi}}\left[\int_{-\infty}^{\infty} \sigma z e^{\frac{-z^2}{2}}dz + \mu\int_{-\infty}^{\infty} e^{\frac{-z^2}{2}}dz\right]$$

Since, the integrand is odd in first term and even in second term, then

$$= \frac{1}{\sqrt{2\pi}}\left[0 + \mu\int_{-\infty}^{\infty} e^{\frac{-z^2}{2}}\,dz\right]$$

$$= \frac{2\mu}{\sqrt{2\pi}}\int_0^{\infty} e^{\frac{-z^2}{2}}\,dz$$

$$= \frac{\sqrt{2}}{\sqrt{\pi}}\mu\int_0^{\infty} e^{-t}\frac{dt}{\sqrt{2t}} \qquad \text{Put } \frac{z^2}{2} = t \Rightarrow zdz = dt$$

$$= \frac{\mu}{\sqrt{\pi}}\int_0^{\infty} e^{-t}t^{-\frac{1}{2}}dt$$

$$= \frac{\mu}{\sqrt{\pi}}\int_0^{\infty} e^{-t}t^{\frac{1}{2}-1}\,dt$$

$$= \frac{\mu}{\sqrt{\pi}}\Gamma\frac{1}{2} = \frac{\mu}{\sqrt{\pi}}\sqrt{\pi} = \mu$$

Variance

$\text{var}\,(X) = E\,(X - \mu)^2$

$$= \frac{1}{\sigma\sqrt{2\pi}}\int_{-\infty}^{\infty}(x-\mu)^2 e^{-\frac{(x-\mu)^2}{2\sigma^2}}\,dx$$

$$= \frac{1}{\sigma\sqrt{2\pi}}\int_{-\infty}^{\infty}\sigma^2 z^2 e^{-\frac{z^2}{2}}.\sigma dz \qquad \text{Put } \frac{x-\mu}{\sigma} = z$$

$$= \frac{2\sigma^2}{\sqrt{2\pi}}\int_0^{\infty} z^2 e^{-\frac{z^2}{2}}\,dz.$$

$$= \frac{2\sigma^2}{\sqrt{2\pi}}\int_0^{\infty} 2t.e^{-t}\frac{dt}{\sqrt{2t}} \qquad \text{Put } \frac{z^2}{2} = t$$

$$= \frac{2\sigma^2}{\sqrt{\pi}}\int_0^{\infty} e^{-t}.t^{1/2}dt$$

$$= \frac{2\sigma^2}{\sqrt{\pi}}\int_0^{\infty} e^{-t}.t^{\frac{3}{2}-1}\,dt$$

$$= \frac{2\sigma^2}{\sqrt{\pi}}\Gamma\frac{3}{2} = \frac{2\sigma^2}{\sqrt{\pi}}.\frac{1}{2}\sqrt{\pi} = \sigma^2$$

Median

Consider the equation is

$$\frac{1}{\sigma\sqrt{2\pi}}\int_{-\infty}^{m} e^{-\frac{(x-\mu)^2}{2\sigma^2}}\,dx = \frac{1}{2}$$

Put $\frac{x-\mu}{\sigma} = z$, then

$$\frac{1}{\sqrt{2\pi}}\int_{-\infty}^{\frac{m-\mu}{\sigma}} e^{-\frac{z^2}{2}}\,dz = \frac{1}{2} \quad \text{...(i)}$$

But by symmetry of the function

$$\frac{1}{\sqrt{2\pi}}\int_{-\infty}^{0} e^{-\frac{z^2}{2}}\,dx = \frac{1}{\sqrt{2\pi}}\int_{0}^{\infty} e^{-\frac{z^2}{2}}\,dz = \frac{1}{2} \quad \text{...(ii)}$$

Comparing Eqs. (i) and (ii) we get

$$\frac{m-\mu}{\sigma} = 0$$

$$\Rightarrow \quad m = \mu$$

Mode Let $f(x) = \frac{1}{\sigma\sqrt{2\pi}}.e^{-\frac{(x-\mu)^2}{2\sigma^2}}$

To determine the mode, we have

$$\frac{df(x)}{dx} = \frac{1}{\sigma\sqrt{2\pi}}e^{-\frac{(x-\mu)^2}{2\sigma^2}}\left(-\frac{x-\mu}{\sigma^2}\right)$$

Now, $\frac{df(x)}{dx} = 0 \Rightarrow \frac{x-\mu}{\sigma^2} = 0 \Rightarrow x = \mu$

and $$\frac{d^2 f(x)}{dx^2} = -\frac{1}{\sigma^2\sqrt{2\pi}}e^{-\frac{(x-\mu)^2}{2\sigma^2}} + \frac{1}{\sigma\sqrt{2\pi}}e^{-\frac{(x-\mu)^2}{2\sigma^2}}\left(\frac{x-\mu}{\sigma}\right)^2$$

at $x = \mu$,

$$\frac{d^2 f(x)}{dx^2} = -\frac{1}{\sigma^2\sqrt{2\pi}}e^{-\frac{(x-\mu)^2}{2\sigma^2}}, \text{which is negative}$$

Thus, $x = \mu$ is a mode.

Remark : Mean, median and mode coincide at the origin.

Quartiles

The first quartile Q_1 is

$$\frac{1}{\sigma\sqrt{2\pi}}\int_{-\infty}^{Q_1} e^{-\frac{(x-\mu)^2}{2\sigma^2}}\,dx = \frac{1}{4}$$

By solving this equation with the help of "Normal Probability Integral Table", we get $Q_1 = \mu - 0.6745\,\sigma$.

The third quartile Q_3 is

$$\frac{1}{\sigma\sqrt{2\pi}}\int_{-\infty}^{Q_3} e^{-\frac{(x-\mu)^2}{2\sigma^2}}\,dx = \frac{3}{4}$$

or $\quad Q_3 = \mu + 0.6745\,\sigma$

Mean Deviation About Mean

$$E(|x-\mu|) = \frac{1}{\sigma\sqrt{2\pi}}\int_{-\infty}^{\infty}|x-\mu|\,e^{-\frac{(x-\mu)^2}{2\sigma^2}}\,dx$$

$$= \frac{\sigma}{\sqrt{2\pi}}\int_{-\infty}^{\infty}|z|\,e^{-\frac{1}{2}z^2}\,dz \qquad \textbf{Put } \frac{x-\mu}{\sigma} = z$$

$$= \frac{\sigma}{\sqrt{2\pi}}\left[\int_{-\infty}^{0}(-z)e^{-\frac{1}{2}z^2}\,dz + \int_{-\infty}^{\infty} z e^{-\frac{1}{2}z^2}\,dz\right]$$

$$= \frac{2\sigma}{\sqrt{2\pi}}\int_{0}^{\infty} z e^{-\frac{1}{2}z^2}\,dz$$

$$= \frac{2\sigma}{\sqrt{2\pi}}\int_{0}^{\infty} e^{-t}\,dt \qquad \textbf{Put } \frac{z^2}{2} = t$$

$$= \frac{2\sigma}{\sqrt{2\pi}}[-e^{-t}]_0^{\infty}$$

$$= \frac{2\sigma}{\sqrt{2\pi}} = \sigma\sqrt{\frac{2}{\pi}}$$

$$= 0.7979\sigma$$

$$= \frac{4}{5}\sigma \text{ (approx.)}$$

Quartile Deviation (Probable Error)

$$\text{Q.D.} = \frac{Q_3 - Q_1}{2}$$

$$= \frac{(\mu + 0.6745\sigma) - (\mu - 0.6745\sigma)}{2}$$

$$= 0.6745\ \sigma = \frac{2}{3}\sigma \text{ (approx.)}$$

Point of inflexion

Let the function

$$f(x) = \frac{1}{\sigma\sqrt{2\pi}} e^{-\frac{(x-\mu)^2}{2\sigma^2}}$$

Now, $$\frac{df(x)}{dx} = \frac{1}{\sigma\sqrt{2\pi}} e^{-\frac{(x-\mu)^2}{2\sigma^2}} \left(-\frac{x-\mu}{\sigma^2}\right)$$

and $$\frac{d^2 f(x)}{dx^2} = -\frac{1}{\sigma^3\sqrt{2\pi}} e^{-\frac{(x-\mu)^2}{2\sigma^2}} + \frac{(x-\mu)^2}{\sigma^5\sqrt{2\pi}} e^{-\frac{(x-\mu)^2}{2\sigma^2}}$$

Now, $$\frac{d^2 f(x)}{dx^2} = 0$$

$$\Rightarrow \quad \frac{1}{\sigma^3\sqrt{2\pi}} e^{-\frac{(x-\mu)^2}{2\sigma^2}} \left[-1 + \frac{(x-\mu)^2}{\sigma^2}\right] = 0$$

$$\Rightarrow \quad \frac{(x-\mu)^2}{\sigma^2} = 1 \quad \Rightarrow x - \mu = \pm\sigma$$

$$\Rightarrow \quad x = \mu \pm \sigma$$

Again, $$\frac{d^3 f(x)}{dx^3} = \frac{1}{\sigma^3\sqrt{2\pi}} e^{-\frac{(x-\mu)^2}{2\sigma^2}} \left(-\frac{x-\mu}{\sigma^2}\right) + \frac{(x-\mu)^2}{\sigma^5\sqrt{2\pi}} e^{-\frac{(x-\mu)^2}{2\sigma^2}} \left(-\frac{x-\mu}{\sigma^2}\right)$$

$$+ \frac{1}{\sigma^5\sqrt{2\pi}} e^{-\frac{(x-\mu)^2}{2\sigma^2}} 2(x-\mu)$$

Putting $x - \mu = \pm\sigma$, we get

$$\frac{d^3 f(x)}{dx^3} = \frac{1}{\sigma^3\sqrt{2\pi}} e^{-\frac{1}{2}}\left(\mp\frac{1}{\sigma}\right) \mp \frac{1}{\sigma^4\sqrt{2\pi}} e^{-\frac{1}{2}} \pm \frac{1}{\sigma^4\sqrt{2\pi}} e^{-\frac{1}{2}}$$

$$= \pm\frac{2}{\sigma^4\sqrt{2\pi}} e^{-\frac{1}{2}} \neq 0$$

$\therefore$ The point of inflexion are $x = \mu \pm \sigma$.

The Odd Moments About Mean are Zero

Proof: $\mu_{2n+1} = \int_{-\infty}^{\infty} (x-\mu)^{2n+1} \frac{1}{\sigma\sqrt{2\pi}} e^{-\frac{(x-\mu)^2}{2\sigma^2}} dx$

$$= \frac{\sigma^{2n+1}}{\sqrt{2\pi}} \int_{-\infty}^{\infty} z^{2n+1} e^{\frac{-z^2}{2}} dz \qquad \text{Put } \frac{x-\mu}{\sigma} = z$$

= 0 (because integrand is an odd function)

Thus, all odd moments are zero *i.e.*

Put $n = 0,1,2,\ldots\ldots\ldots\ldots$ $\mu_1 = \mu_3 = \mu_5 = \ldots\ldots\ldots = \mu_{2n+1} = 0$

The even moments about mean

$$\mu_{2n} = \sigma^2 (2n-1)\, \mu_{2n-2}$$

or

$$\mu_{2n} = (2n-1)(2n-3)\ldots.3.1.\sigma^{2n}$$

Proof: $\mu_{2n} = \frac{1}{\sigma\sqrt{2\pi}} \int_{-\infty}^{\infty} (x-\mu)^{2n} e^{-\frac{(x-\mu)^2}{2\sigma^2}} dx$

$$= \frac{\sigma^{2n}}{\sqrt{2\pi}} \int_{-\infty}^{\infty} z^{2n} e^{-\frac{z^2}{2}} dz \qquad \text{Put } \frac{x-\mu}{\sigma} = z$$

$$= \frac{2\sigma^{2n}}{\sqrt{2\pi}} \int_{0}^{\infty} z^{2n} e^{-\frac{z^2}{2}} dz$$

$$= \frac{2\sigma^{2n}}{\sqrt{2\pi}} \int_{0}^{\infty} (2t)^{2n} e^{-t} . \frac{dt}{\sqrt{2t}} \qquad \text{Put } \frac{z^2}{2} = t$$

$$= \frac{2\sigma^{2n}}{\sqrt{2\pi}} \int_{0}^{\infty} 2^{n-\frac{1}{2}} t^{n-\frac{1}{2}} e^{-t} dt$$

$$= \frac{2^n \sigma^{2n}}{\sqrt{\pi}} \int_{0}^{\infty} e^{-t} t^{\left(n+\frac{1}{2}-1\right)} dt$$

$$= \frac{2^n \sigma^{2n}}{\sqrt{\pi}} \Gamma\left(n+\frac{1}{2}\right) \qquad \ldots\text{(i)}$$

Similarly, $\mu_{2n-2} = \frac{2^{n-1}\sigma^{2n-2}}{\sqrt{\pi}}\Gamma\left(n-\frac{1}{2}\right)$ ...(ii)

Dividing Eqs. (i) and (ii), we get

$$\frac{\mu_{2n}}{\mu_{2n-2}} = 2\sigma^2 \frac{\Gamma\left(n+\frac{1}{2}\right)}{\Gamma\left(n-\frac{1}{2}\right)}$$

$$= 2\sigma^2\Gamma\left(n-\frac{1}{2}\right) = \sigma^2 (2n-1)$$

$$\mu_{2n} = \sigma^2 (2n-1)\mu_{2n-2} \quad \text{...(iii)}$$

Putting $n = n$-1, n-2,...... 2,1 we have

$$\mu_{2n-2} = \sigma^2 (2n-3)\mu_{2n-4}$$

$$\mu_{2n-4} = \sigma^2 (2n-5)\mu_{2n-6}$$

.

.

.

.

$$\mu_4 = \sigma^2.3.\mu_2$$

$$\mu_2 = \sigma^2.1.\mu_0 = \sigma^2$$

Substituting these values one by one in Eq. (iii), we get

$$\mu_{2n} = (2n-1)(2n-3).....3.1.\sigma^{2n} \quad \text{...(iv)}$$

In particular n = 1,2......

$$\mu_2 = \sigma^2,\ \mu_4 = 3\sigma^4,\ ....$$

Beta and Gamma Coefficients

$$\beta_1 = \frac{\mu_3^2}{\mu_2^3} = 0, \qquad \beta_2 = \frac{\mu_4}{\mu_2^2} = \frac{3\sigma^4}{\sigma^4} = 3$$

$$\gamma_1 = \sqrt{\beta_1} = 0, \qquad \gamma_2 = \beta_2 - 3 = 0$$

The odd Moments About Origin are Vanish

Proof : $$\mu^1_{2p+1} = \int_{-\infty}^{\infty} \frac{1}{\sigma\sqrt{2\pi}} e^{-\frac{x^2}{2\sigma^2}} x^{2p+1} dx$$

$$= \frac{1}{\sigma\sqrt{2\pi}} \int_{-\infty}^{\infty} x^{2p+1} e^{-\frac{x^2}{2\sigma^2}} dx$$

= 0 (because integrand is an odd function)

$\therefore$ $$\mu^1_1 = \mu^1_3 = ... = 0$$

The Even Moments About Origin is Given by

$$\mu^1_{2p} = \frac{2^p \sigma^{2p}}{\sqrt{\pi}} \Gamma\left(P + \frac{1}{2}\right)$$

or $$\mu^1_{2p} = (2p-1)(2p-3)...3.1.\sigma^{2p}$$

Proof : $$\mu^1_{2p} = \frac{1}{\sigma\sqrt{2\pi}} \int_{-\infty}^{\infty} x^{2p} e^{-\frac{x^2}{2\sigma^2}} dx$$

$$= \frac{2}{\sigma\sqrt{2\pi}} \int_{0}^{\infty} x^{2p} e^{-\frac{x^2}{2\sigma^2}} dx$$

Putting $$\frac{x^2}{2\sigma^2} = t \;\; \Rightarrow dx = \frac{\sigma dt}{\sqrt{2t}}$$

We get, $$\mu^1_{2p} = \frac{2}{\sigma\sqrt{2\pi}} \int_{0}^{\infty} e^{-t}.2^p \sigma^{2p} t^p.\frac{\sigma}{\sqrt{2t}} dt$$

$$= \frac{2^p \sigma^{2p}}{\sqrt{\pi}} \int_{0}^{\infty} e^{-t} t^{\left(p+\frac{1}{2}\right)-1} dt$$

$$= \frac{2^p \sigma^{2p}}{\sqrt{\pi}} \Gamma\left(p + \frac{1}{2}\right) \qquad ...(i)$$

Putting $p = p\text{-}1$ in Eq.(i), we get

$$\mu^1_{2p-2} = \frac{2^{p-1} \sigma^{2p-2}}{\sqrt{\pi}} \Gamma\left(p - \frac{1}{2}\right) \qquad ...(ii)$$

Dividing Eqs.(i) and (ii), we get

$$\frac{\mu^{1}_{2p}}{\mu^{1}_{2p-2}} = 2\sigma^2 \frac{\Gamma\left(p+\frac{1}{2}\right)}{\Gamma\left(p-\frac{1}{2}\right)}$$

$$\mu^{1}_{2p} = 2\sigma^2\left(p-\frac{1}{2}\right)\mu^{1}_{2p-2} \qquad \text{...(iii)}$$

Putting $p = p\text{-}1, p\text{-}2, \ldots\ldots\ldots 2,1$ we get

$$\mu^{1}_{2p-2} = \sigma^2 (2p\text{-}3)\, \mu^{1}_{2p-4}$$

$$\mu^{1}_{2p-4} = \sigma^2 (2p\text{-}5)\, \mu^{1}_{2p-6}$$

.

.

.

.

$$\mu^{1}_{4} = \sigma^2.3.\ \mu^{1}_{2}$$

$$\mu^{1}_{2} = \sigma^2.1.\ \mu^{1}_{0} = \sigma^2$$

Substituting these values in Eq. (iii), we get

$$\mu^{1}_{2p} = (2p\text{-}1)(2p\text{-}3).....3.1.\sigma^{2p} \qquad \text{...(iv)}$$

Moment Generating Function

(i) About origin

$$M_x(t) = \int_{-\infty}^{\infty} e^{tx}.\frac{1}{\sigma\sqrt{2\pi}} e^{-\frac{(x-\mu)^2}{2\sigma^2}} dx$$

$$= \frac{1}{\sigma\sqrt{2\pi}} \int_{-\infty}^{\infty} e^{-t(\mu+\sigma z)} e^{-\frac{z^2}{2}} .\sigma dz \qquad \text{Put} \frac{x-\mu}{\sigma} = z$$

$$= \frac{e^{\mu t}}{\sqrt{2\pi}} \int_{-\infty}^{\infty} e^{\sigma t z} e^{-\frac{z^2}{2}} dz$$

$$= \frac{e^{\mu t}}{\sqrt{2\pi}} \int_{-\infty}^{\infty} e^{-\frac{1}{2}(z^2 - 2\sigma t z + t^2\sigma^2) + \frac{1}{2}t^2\sigma^2} dz$$

$$= \frac{e^{\mu t + \frac{1}{2}t^2\sigma^2}}{\sqrt{2\pi}} \int_{-\infty}^{\infty} e^{-\frac{1}{2}(z - t\sigma)^2} dz$$

$$= \frac{2e^{\mu t + \frac{1}{2}t^2\sigma^2}}{\sqrt{2\pi}} \int_{0}^{\infty} e^{-r} . \frac{dr}{\sqrt{2r}} \qquad \text{Put } \frac{1}{2}(z - t\sigma)^2 = r$$

$$= \frac{e^{\mu t + \frac{1}{2}t^2\sigma^2}}{\sqrt{\pi}} \int_{0}^{\infty} e^{-r} r^{\frac{1}{2}-1} dr \qquad \Rightarrow (z - t\sigma)dz = dr$$

$$= \frac{e^{\mu t + \frac{1}{2}t^2\sigma^2}}{\sqrt{\pi}} \Gamma\frac{1}{2} \qquad \Rightarrow dz = \frac{dr}{\sqrt{2r}}$$

$$= e^{\mu t + \frac{1}{2}t^2\sigma^2} \qquad \left(\because \Gamma\frac{1}{2} = \sqrt{\pi}\right)$$

(ii) About mean

$$M_{x-\mu}(t) = \frac{1}{\sigma\sqrt{2\pi}} \int_{-\infty}^{\infty} e^{t(x-\mu)} e^{-\frac{(x-\mu)^2}{2\sigma^2}} dx$$

$$= \frac{1}{\sqrt{2\pi}} \int_{-\infty}^{\infty} e^{t\sigma z} e^{-\frac{z^2}{2}} dz \qquad \text{Put } \frac{x-\mu}{\sigma} = z$$

$$= \frac{2}{\sqrt{2\pi}} \int_{0}^{\infty} e^{-\frac{1}{2}(z^2 - 2t\sigma z + t^2\sigma^2) + \frac{1}{2}\sigma^2 t^2} dz$$

$$= \frac{2}{\sqrt{2\pi}} e^{\frac{1}{2}\sigma^2 t^2} \int_{0}^{\infty} e^{-\frac{1}{2}(z - t\sigma)^2} dz$$

$$= \frac{2}{\sqrt{2\pi}} e^{\frac{1}{2}\sigma^2 t^2} \int_{0}^{\infty} e^{-\frac{1}{2}r^2} dr \qquad \text{Put } z\text{-}t\sigma = r$$

$$= e^{\frac{1}{2}\sigma^2 t^2}, \textit{Since} \frac{1}{\sqrt{2\pi}} \int_{-\infty}^{\infty} e^{-\frac{1}{2}r^2} dr = 1$$

Illustrative Examples

Example 1. *In a normal distribution whose mean is 2 and S.D. is 3. Find a value of the variate such that the probability of the interval from the mean to the value is 0.4115. Find another value such that the probability for the interval from x = 3.5 to that value is 0.2307.*

Solution. Let the function be $f(x) = \frac{1}{\sigma\sqrt{2\pi}} e^{-\frac{(x-\mu)^2}{2\sigma^2}}$

Let the required value be x, then

$$P(2 < X < x) = \frac{1}{\sigma\sqrt{2\pi}} \int_2^x e^{-\frac{(x-\mu)^2}{2\sigma^2}} dx$$

Given, $\mu = 2$, $\sigma = 3$

Substituting, $t = \frac{x-\mu}{\sigma} = \frac{x-2}{3}$ we get

$$P(2 < X < x) = \frac{1}{\sigma\sqrt{2\pi}} \int_2 e^{-\frac{1}{2}t^2} dt = 0.4115$$

is found to be $t = 1.35$ (from the table)

Hence, $t = \frac{x-2}{3} \Rightarrow x = 3t + 2$

$$= 3 \times 1.35 + 2$$

$$x = 6.05$$

Again, from $x = 3.5$, we have $t = \frac{3.5-2}{3} = 0.5$

Hence, the area under normal curve between $t = 0.5$ to t_1 is 0.2307.

From the table, area between $t = 0$ to $t = 0.5$ is 0.1915.

Hence, area between $t = 0$ to $t = t_1$ is

$$0.1915 + 0.2307 = 0.4222$$

The value of $t_1 = 1.42$ (form the table)

Hence, $\frac{x-2}{3} = 1.42 \Rightarrow x = 3 \times 1.42 + 2$

$$x = 6.26$$

Example 2. *Assume the mean height of soldiers to be 68.22 inches with a variance of 10.8 inches square. How many soldiers in a regiment of 10,000 would you expect to be over 6 feet tall, given that the area under the standard normal curve between x = 0 and x = 0.35 is 0.1368 and, between x = 0 and x = 1.15 is 0.3746.*

Solution. Here, m = 68.22 inches, $\sigma^2 = 10.8 \Rightarrow \sigma = 3.28$ inches

$x = 6$ feet = 72 inches $\quad \because (1' = 12'')$

$$\text{Now, } t = \frac{x-m}{\sigma} = \frac{72-68.22}{3.28} = 1.15$$

The area between x = 0 and x = 1.15 is 0.3746.

Thus, the area between t = 1.15 and t = ∞ is 0.5 - 0.3746 = 0.1254

Hence, the expected number of soldiers above 6 feet = 0.1254 × 10000

= 1254

Example 3. *If skulls are classified as A,B, C according as the length, breadth index is under 75, between 75 and 80 or over 80, find approximately (assuming that the distribution is normal) the mean and S.D. of a series in which A are 58%. B are 38% and C are 4% being given that if*

$$f(t) = \frac{1}{\sqrt{2\pi}} \int_0^t \exp.(-t^2)dt,$$

then f (0.2) = 0.08 and f (1.75) = 0.46

Solution. Let the mean be *m* and S.D. be σ .

As for *A*, the area under 75 is 0.58.

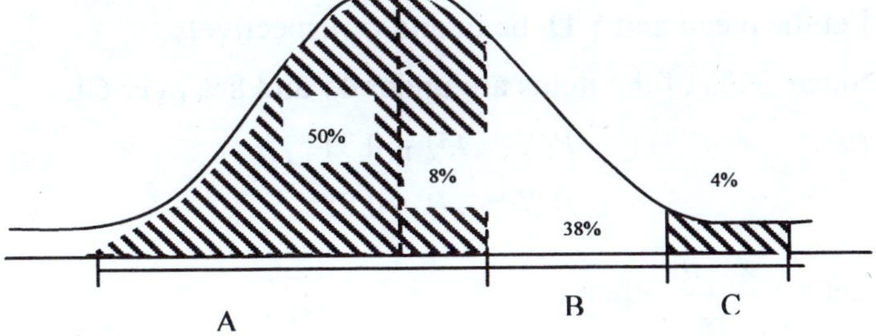

The area between the mean and 75 is 0.08.

Hence, $$\frac{75-m}{\sigma}=0.2 \Rightarrow m+0.2\,\sigma=75 \qquad \text{...(i)}$$

Similarly, the area above 80 is 0.4.

The area between mean and 80 is 0.46.

Hence, $$\frac{80-m}{\sigma}=1.75 \Rightarrow m+1.75\,\sigma=80 \qquad \text{...(ii)}$$

Solving Eqs. (i) and (ii), we get

$$m=74.4, \sigma=3.2$$

Example 4. *Prove that for the normal distribution the quartile deviation, the mean deviation and the standard deviation are approximately in the ratio 10 : 12 : 15.*

Solution. Here, quartile deviation $=\frac{2}{3}\sigma$

Mean deviation $=\frac{4}{5}\sigma$

Standard deviation $=\sigma$

$$\therefore \qquad \text{Q.D. : M.D. : S.D.} = \frac{2}{3}\sigma : \frac{4}{5}\sigma : \sigma$$

$$= 10\sigma : 12\sigma : 15\sigma$$

$$= 10 : 12 : 15$$

Example 5. *In a distribution which is exactly normal, 31% of the items are under 45 and 8% over 64. Find the mean and S.D. of the distribution.*

Solution. Let the mean and S.D. be m and σ respectively.

Since, 31% of the items are under 45 and 8% over 64.

i.e. $$P(X<45)=0.31$$

$$P(X>64)=0.08$$

Let $t=\frac{x-m}{\sigma}$, then

$$P(x<45)=p\,(m+t\sigma<45)$$

$$=p\left(t<\frac{45-m}{\sigma}\right)=0.31$$

From the table, $\frac{45-m}{\sigma}$ = -0.5 (approx.) ...(i)

And $P(X > 64) = P(m + t\sigma > 64)$

$$= P\left(t < \frac{64-m}{\sigma}\right) = 0.08$$

From the table, $\frac{64-m}{\sigma} = 1.4$ ...(ii)

From, Eqs. (i) and (ii), $45 - m = -0.5\sigma$

and $64 - m = 1.4\sigma$

Solving these two equations, we get $\sigma = 10$, $m = 50$.

Example 6. *If two normal universes have the same total frequency but the S.D. of one is k times that of the others. Show that the maximum frequency of the first is $\frac{1}{k}$ that of the other.*

Solution. If the total frequency is N and σ and $k\sigma$ be the standard deviation of the two universes. Then equations of the normal curves are

$$y = y_0 e^{-\frac{x^2}{2\sigma^2}} \quad \text{and} \quad y = y_0^1 e^{-\frac{x^2}{2k^2\sigma^2}}$$

If the total frequencies are same, then

$$\int_{-\infty}^{\infty} y_0 e^{-\frac{x^2}{2\sigma^2}} dx = \int_{-\infty}^{\infty} y_0^1 e^{-\frac{x^2}{2k^2\sigma^2}} dx$$

In the L.H.S. Put $\frac{x}{\sqrt{2}\sigma} = t \Rightarrow dx = \sqrt{2}\,\sigma dt$

$$\text{L.H.S.} = y_0 \int_{-\infty}^{\infty} e^{-t^2} \sqrt{2}\sigma dt = \sqrt{2}\sigma y_0 \frac{1}{2}\sqrt{\pi} \quad \text{...(i)}$$

In the R.H.S. Put $\frac{x}{\sqrt{2}k\sigma} = t \Rightarrow dx = \sqrt{2}\,k\sigma\, dt$

$$\text{R.H.S.} = y_0^1 \int_{-\infty}^{\infty} e^{-t^2} \sqrt{2}k\sigma dt = \sqrt{2}k\sigma y_0^1 \frac{1}{2}\sqrt{\pi} \quad \text{...(ii)}$$

Equating (1) and (2), we get

$$y_0 = k y_0^1$$

Hence, $$\frac{y_0^1}{y_0} = \frac{1}{k}$$

Example 7. *The marks obtained in statistics in a certain examination are found to be normally distributed. If 12.5% of the candidates obtains 60% or more marks. 39% obtain less than 30 marks, find the mean number of marks obtained by the candidates given*

$\frac{x}{\sigma}$	0.27	0.28	0.29	1.14	1.15	1.16
A	0.6064	0.6102	0.6104	0.8727	0.8749	0.8770

Where $A = \frac{1}{\sqrt{2\pi}} \int_{-\infty} e^{-\frac{1}{2}t^2} dt, t = \frac{x}{\sigma}$, *x being the deviation from the mean.*

Solution. Since the area of the standard normal curve to the right of $t = 1.15$ is 0.1251 and the proportion of the candidates, obtaining more that 60 marks, is 0.125 we have

$$\frac{60 - m}{\sigma} = 1.15 \qquad \text{...(i)}$$

Now, the proportion of candidates getting less than 30 marks, is 0.39. The area between -∞ to 30 + 2 (m - 30) of the standard normal curve is 0.39 + 2(0.5-0.39) = 0.61

The value of $\frac{x}{\sigma}$ for $A = 0.61$ comes out to be 0.279.

Hence,
$$\frac{x}{\sigma} = \frac{m - 30}{\sigma} = 0.279 \qquad \text{...(ii)}$$

Solving Eqs. (i) and (ii), we get

$$\frac{60 - m}{m - 30} = \frac{1.15}{0.279}$$

$$\Rightarrow \qquad m = 35.8$$

Example 8. *A certain examination was taken by 1000 students who are to be classified into sub groups, A, B, C, D, E according to ability, range of ability to be equal in the sub groups. On the assumption that the trait measured is normally distributed, how many students should be placed in each group, given that f(0.6) = 0.225, f(1.8) = 0.463, f(3.0)= 0.499.*

Where $f(x) = \frac{1}{\sqrt{2\pi}} \int_0^a \exp.\left(-\frac{x^2}{2}\right) dx.$

Solution. Here, $t = \dfrac{x-m}{\sigma}$, the range of ability has been divided into five equal 6 sub groups ranging from m ± 3σ. So that each sub size 6 groups is of interval = $\dfrac{6\sigma}{5} = 1.2\sigma$.

The number of students between m and $m + 0.6\sigma = 0.225 \times 1000 = 225$.

The number of students between $m - 0.6\sigma$ and $m + 0.6\sigma = 2 \times 225 = 450$.

Similarly, the number of students between $m + 0.6\sigma$ and $m + 1.8\sigma$.

$$= (0.463 - 0.225) \times 1000 = 238$$

The number of student between $m + 1.8\sigma$ and $m + 3\sigma$

$$= (0.499 - 0.463) \times 1000 = 36$$

Since the normal distribution is symmetrical about the mean. We have the number of student in the various classes are as:-

	Sub group	No. of students
A.	m-3 σ to m - 1.8 σ	36
B.	m-1.8 σ to m - 0.6 σ	238
C.	m-0.6 σ to m + 0.6 σ	450
D.	m+0.6 σ to m + 1.8σ	238
E.	m+1.8 σ to m + 3σ	36
	Total	998

The total number of student came out to be 998 in place of 1000. The remaining two students can be allotted the two extreme classes since at the end, the intervals are open.

Fitting of Normal Distribution: For a given set of values, a normal distribution can be fitted by two methods.

(i) Method of ordinates (ii) Method of areas

(i) Method of Ordinates Find the mid-value of each class interval and also the mean and S.D. of the given data for each mid-value. Find $t = \dfrac{x-m}{\sigma}$ and from the tables find the corresponding ordinates for each t.

Finally we get the theoretical frequencies by multiplying these values of the ordinates by $\frac{N.i}{\sigma}$ where N is the total frequency and i be the length of the corresponding class interval.

Example 9. *Fit a normal distribution to the following data:*

Interval mid-point	100	95	90	85	80	75	70	65	60	55	50	45
Frequencies	0	1	3	2	7	12	10	9	5	3	2	0

Solution.

Class(x)	f	$d=\frac{x-75}{5}$	fd^2	fd^2	$t=\frac{x-m}{\sigma}$
100	0	5	0	0	2.88
95	1	4	4	16	2.38
90	3	3	9	27	1.88
85	2	2	4	8	1.38
80	7	1	7	7	0.88
75	12	0	0	0	0.38
70	10	-1	-10	10	-0.12
65	9	-2	-18	36	-0.62
60	5	-3	-15	45	-1.12
55	3	-4	-12	48	-1.62
50	2	-5	-10	50	-2.12
45	0	-6	0	0	-2.62
Total	54 = N		-41	247	

Let assumed mean $a = 75, i = 5$

$$\text{Mean } m = a + \frac{\sum fd}{\sum f} \times i$$

$$= 75 - \frac{41}{54} \times 5$$

$$= 75 - 3.79 = 71.2$$

$$\sigma^2 = \left[\frac{\sum fd^2}{\sum f} - \left(\frac{\sum fd}{\sum f}\right)^2\right] \times i^2$$

$$= \left[\frac{247}{54} - \left(-\frac{41}{54}\right)^2\right] \times (5)^2$$

$$= [4.57 - 0.576] \times 25$$

$$= 3.994 \times 25 = 99.85$$

$$= 100 (\text{approx.})$$

$$\therefore \ \sigma = 10$$

The value of ordinates of the normal curve

$$\phi(t) = \frac{1}{\sqrt{2\pi}} e^{-\frac{t^2}{2}}$$

For the above values of t from the table are respectively $[\phi(-t) = \phi(t)]$:-

$\phi(t)$	Theoretical frequency $27 \times \phi(t)$
0.0063	0.17
0.0235	0.63
0.0681	1.83
0.1539	4.15
0.2709	7.31
0.3712	10.0
0.3961	10.7
0.3292	8.88
0.2131	5.75
0.1074	2.89
0.0422	1.14
0.0129	0.34

Since, $i = 5$ $\therefore$ $\frac{\sigma}{i} = \frac{10}{5} = 2$

Hence, the values of the ordinates are multiplied by

$$\frac{N.i}{\sigma} = \frac{54}{2} = 27.$$

(ii) Method of area- Find the mean and S.D. of the given data. Calculate $t = \frac{x-m}{\sigma}$ and read off the area to the left of t from the table of area under the standard normal curve. Convert these area lying between intervals. Multiplying the areas between class interval by N, the total frequency. We obtain theoretical frequencies corresponding to the class intervals.

Example 10. *Fit a normal distribution to the following data :-*

Class	15-20	20-25	25-30	30-35
Frequency	8	13	19	10

Solution.

Class	Mid value x	f	$d = \frac{x-27.5}{5}$	fd	fd^2
15-20	17.5	8	-2	-16	32
20-25	22.5	13	-1	-13	13
25-30	27.5	19	0	0	0
30-35	32.5	10	1	10	10
TOTAL	-	$N = 50$	-	-19	55

Let assumed mean, $a = 27.5, i = 5$

$$\text{Mean, m} = a + \frac{\sum fd}{\sum f} \times i$$

$$= 27.5 - \frac{19}{50} \times 5 = 25.6$$

$$\sigma^2 = \left[\frac{\Sigma fd^2}{\Sigma f} - \left(\frac{\Sigma fd}{\Sigma f}\right)^2\right] \times i^2$$

$$= \left[\frac{55}{50} - \left(\frac{-19}{50}\right)^2\right] \times 25$$

$$= (1.10 - 0.144) \times 25 = 23.89$$

$$\sigma = 4.89$$

x	$t = \frac{x-m}{\sigma}$	Area to the left of t	Area in the center p	Expected frequencies NP
$-\infty$	$-\infty$	0		
20	−1.14	0.1271	0.1271	6.4
25	−0.12	0.4522	0.3251	16.2
30	0.90	0.8159	0.3637	18.2
$+\infty$	$+\infty$	1.0000	0.1841	9.2
Total	-	-	10000	50.0

Practice Problems

1. Find the ordinate at X=2 of the normal curve with mean 3 and standard deviation 1.6.

   [Ans. 0.205]

2. A factory turns out an article by mass production methods. From past experience it appears that 20 articles on an average are rejected out of every batch of 100. Find the variance of the number of rejects in a batch. What is the probability that the number of rejects in a batch exceed 30?

   [Ans. t = 2.5, p = 0.0062]

3. In a intelligence test administrated to 1000 children, the average score is 42 and S.D. is 24. Find (i) the number of children whose score exceeds 60 (ii) the number of children with score lying between 20 and 40. It is given that, if

$$f(x)=\frac{1}{\sqrt{2\pi}}\int_0^x e^{-\frac{x^2}{2}}\,dx., \text{ then}$$

$$f(0.75)=0.2734, f(0.91)=0.3184, f(0.08)=0.0319.$$

[**Ans.** $p(x>60)=0.2266.$ No. of children $=226.6$

for $x=20, t=-0.91,$ for $x=40, t=-0.08$

$p(20<X<40)=0.2865,$ No. of children $=286.5$]

4. If x is normal variate with mean 8 and standard deviation 4, find

(i) $p(X\le 5)(ii)\,p(5\le X\le 10)$

[**Ans.** $(i)\,p(X\le 5)=0.2266(ii)\,p(5\le X\le 10)=0.4649$]

5. A minimum height is to be prescribed to government services such that 60% of the young men will have a fair chance of coming upto the standard. The heights of young men are normally distributed with mean 60.6 inches and standard deviation 2.55 inches. Determine the minimum specification.

[**Ans.** minimum height is 59.95 inches.]

6. One thousand candidates in an examination were grouped into 3 classes A,B,C in descending order of merit. The number in the first two classes were 50 and 350 respectively. The heighest and lowest marks in classes B were 60 and 50 respectively. Assuming the distribution to the normal prove that the average marks is approximately 48.2 and standard deviation is approximately 7.1. The following data may be used the area U is measured from the mean zero to any ordinate x.

$\frac{x}{\sigma}$	0.2	0.3	0.4	1.5	1.6	1.7
U	0.079	0.118	0.155	0.433	0.445	0.455

[**Ans.** $m=48.18, \sigma=7.16.$

Hint. Frequency of

$$A=\frac{60}{1000}=0.05, B=\frac{350}{1000}=0.35, C=\frac{1000-(50+350)}{1000}=0.6.]$$

7. Two hundred and fifty five metal rods were cut roughly 6 inches over size. Finally the lengths of the over size amount we measured exactly and grouped with 1 inch

intervals, there being in all 12 groups $\frac{1}{2}"-1\frac{1}{2}",1\frac{1}{2}"-2\frac{1}{2}",....,11\frac{1}{2}"-12\frac{1}{2}"$. The frequency distribution for 255 lengths was as follows-

Length (inches)	1	2	3	4	5	6	7	8	9	10	11	12
Frequency	2	10	19	25	40	44	41	28	25	15	5	1

Fit a normal curve to this data.

[**Ans.** 166

8. Fit a normal distribution to the following data

Class	42.5-47.5	47.5-52.5	52.5-57.5	57.5-62.5	62.5-67.5	67.5-72.5	72.5-77.5	77.5-82.5
Frequency	0	2	3	5	9	10	12	7

Class	82.5-87.5	87.5-92.5	92.5-97.5	97.5-102.5
Frequency	2	3	1	0

[**Ans.** $m = 71.2, \sigma = 10$ (nearly)

t = -2.62,-2.12,-1.62,-1.12,-0.62,-0.12,0.38,0.88,1.38,1.88,2.38,2.88

Frequencies are 0.3,1.1,2.9,5.8,8.9,10.5,10.0,7.3,4.2,1.8,0.6,0.2]

Chapter - 11

Sampling Distribution (Large Samples)

Introduction

The purpose of sampling is to get information about the population from the sample. The importance of the theory of sampling lies in the fact that for a large population or universe, it is neither practical nor necessary to collect data for each and every number of the population. The population measures are mean, standard deviation etc., are called parameters while the measures obtained from the samples are called statistics.

Types of Sampling

Mainly three types of sampling are as follows:

(i) **Random sampling:** The most important method of drawing a sample is that of random sampling, which means that each number of population has the same change of being selected from the sample.

e.g. drawing a lottery system, throw of a die etc.

(ii) **Purposive sampling:** In this method personal element has a great chance of entering into the selection of the sample.

(iii) **Mixed sampling:** The random sampling and purposive sampling are mixed together is known as mixed sampling.

Sampling of Attributes

The sampling of attributes may be considered as the drawing samples from a population containing *A* and not *A*. The presence of the attribute *A* may be called a success and its absence a failure.

For example: The sample consisted of 100 events, 60 of which were successes and 40 failures.

Simple Sampling

Simple sample is a random sampling in which each event has the same chance of success and in which the chance of success of different events are independent of whether previous trials have been made or not.

For example: From a pack of playing cards, the chance of getting a queen is not affected by what was obtained on the previous trials and remains the same *i.e.*, $\frac{4}{52}$ chance.

Mean and standard Deviation in Simple Sampling of Attributes

Let we take *N* samples of size *n*, If each individuals has a chance *p* for success and $q = 1 - p$ a change of failure. The frequencies of samples with 0, 1, 2,……., *n* successes are in terms of binomial distribution of.

$$N\,(q+p)^n,\ where\ p+q=1$$

This means that the probabilities of 0,1,2,….. *n* times are $q^n, {}^nC_1q^{n-1}p, {}^nC_2q^{n-2}p^2, \ldots\ldots, p^n$. The binomial probability distribution thus determined is called the sampling distribution of the number of successes. The mean value of this distribution is *np*. The standard deviation is $\sqrt{nqp}$ and the variance is *nqp*.

If proportion of successes are given.

The mean proportion of successes will be $\frac{np}{p} = p$ and the standard deviation (SD) of the proportion of successes $= \frac{\sqrt{nqp}}{n} = \sqrt{\frac{pq}{n}}$

Tests of Significance for Large Samples

For a large values of n, the binomial distribution is very closed to the normal distribution . For normal distribution, at 99.7% of its members lie within a range $\pm 3\sigma i.e., \pm 3\sqrt{npq}$ on either side of the mean np. Again at only 5% of its members of a normal distribution lie outside the range mean $\pm 2\sigma$. Thus, if the number of successes in a large sample of size n differs from the expected value np by more than $3\sqrt{npq}$, we call this difference highly significant. Sometimes difference of more than $2\sqrt{npq}$ maybe called significant whereas a difference of less than $2\sqrt{npq}$ is called insignificant.

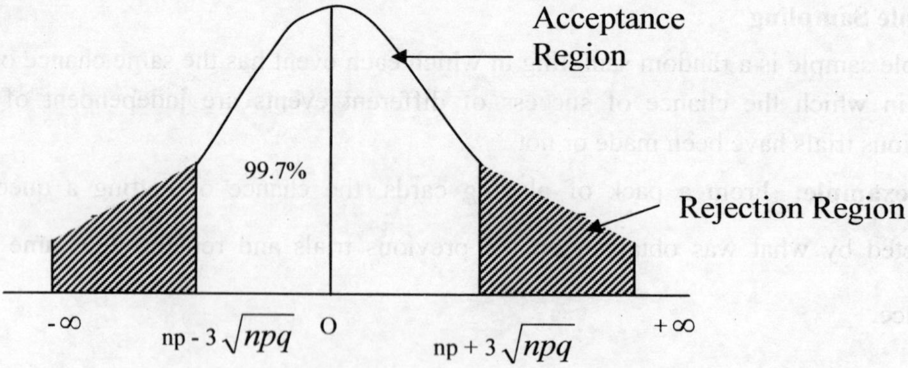

Standard Error

The Standard deviation of simple sampling is called standard error. It is denoted by SE. The standard error is given due to the fact that the expected value as the true value and divergence from it as error of estimation due to fluctuations of sampling.

Probable Error

The probable error is 0.67449 times the standard error. If p denotes the probable error, then p= 0.67449 σ.

Test of Significance in Case of Attributes

The rule for a test of significance is as follows:

1. Assume null hypothesis H_o and alternative hypothesis H_1.

2. Define a test statistic

$$Z = \frac{x - np}{\sqrt{npq}} \sim N(0,1), \text{ where x is the number of successes.}$$

3. Decide the level of significance (∝)

(a) for $\propto = 5\%$, we reject H_0, if $|z| > 1.96$

(b) for $\propto = 1\%$, we reject H_0, if $|z| > 2.58$

(c) for $\propto = 0.27\%$, we reject H_0, if $|z| > 3$

(d) for $\propto = 5\%$, we accept H_0, if $|z| < 1.96$

Illustrative Examples

Example 1 Out of 200 individuals 40 percent show a certain trait and that the number expected on a certain theory is 50 percent, find whether the number observed differs significantly from expectation?

Solution. Given,

$$P = 50\% = \frac{50}{100} = 0.5 = \frac{1}{2}$$

And $q = 1\text{-}p = 1\text{-}\frac{1}{2} = \frac{1}{2}$

$n = 200$

$$\text{The standard error} = \sqrt{\frac{pq}{n}} = \sqrt{\frac{1}{2}\times\frac{1}{2}\times\frac{1}{200}} = 0.035$$

$$\text{Now, } z = \frac{0.5-0.4}{0.035} = 2.8$$

Hence, the difference is significant.

Example 2 A certain cubical die was thrown 9000 times and a 5 or 6 was obtained 3240 times. On assumption of the random throwing do the data indicate an unbiased die?

Solution The chance of throwing a 5 or 6 with one die = $p = \frac{2}{6} = \frac{1}{3}$

and hence $q = 1 - p = 1\text{-}\ \frac{1}{3} = \frac{2}{3}$

$\therefore$ The expected number of successes = np

$$= 9000 \times \frac{1}{3} = 3000$$

The standard error of number of successes = $\sqrt{npq}$

$$= \sqrt{9000 \times \frac{1}{3} \times \frac{2}{3}}$$

$$= 44.72$$

Hence, $z = \frac{x - np}{\sqrt{npq}}$

$$= \frac{3240 - 3000}{44.72} = 5.37$$

Since, $|z| > 3$

Hence, we reject H_0 at 0.27% level of significance.

i.e., The difference x ~ np is highly significant..

We therefore conclude that the die is biased one.

Example 3 In breeding certain stocks, 408 hairy and 126 glabrous plants were obtained. If the expectation is one-fourth glabrous. Is the divergence significant, or might it have occurred as a fluctuation of sampling?

Solution The total number of plants = 408 + 126 = 534

Expected number of glabrous plants $= \frac{1}{4} \times 534 = 133.5$

Observed number of glabrous plants = 126

Difference between expected and observed number of glabrous plants = 133.5 – 126 = 7.5

Hence, $P = \frac{1}{4}$, $q = 1 - \frac{1}{4} = \frac{1}{3}$ and $n = 534$

The standard error of the number of glabrous plants= $\sqrt{npq}$

$$= \sqrt{534 \times \frac{1}{4} \times \frac{3}{4}} = 10 \text{(approx.)}$$

$$\therefore \quad z = \frac{7.5}{10} = 0.75$$

which shows that the difference is very in significant and might have occurred as a fluctuation of sampling.

Example 4 In a locality of 18000 families a sample of 840 families was selected. Of these 840 families, 206 families were found to have a monthly income of Rs 50 or less. It is desired to estimate how many out of the 18000 families have a monthly income of Rs 50 or less. Within what limits would you place your estimate?

Solution Here, $p = \dfrac{206}{840} = 0.245 \quad q = 1 - 0.245 = 0.755$

$\therefore$ The standard error of the proportion of families having a monthly income of Rs 50 or less

$$= \sqrt{\frac{pq}{n}}$$

$$= \sqrt{\frac{0.245 \times 0.755}{840}}$$

$$= 0.015$$

$$= 1.5 \text{ percent}$$

Hence, the required limits are $p \pm 3\sqrt{\dfrac{pq}{n}}$

$$P = 24.5\%, \sqrt{\frac{pq}{n}} = 1.5\%$$

$\therefore$ Limits are $24.5\% \pm (3 \times 1.5)\%$

$$= 20\% \text{ and } 29\%$$

$\therefore$ The limits for number of families out of 18000 families are

$$\frac{20}{100} \times 18000 \text{ and } \frac{29}{100} \times 18000$$

i.e., 3600 and 5220 families.

Example 5 A random sample of 500 pineapples was taken from a large consignment and 65 were found to be bad. Estimate the population of bad pineapples in the consignment, as well as the standard error of the estimate. Deduce that the percentages of bad pineapples in the consignment almost certainly lies between 8.5 and 17.5.

Solution Here, $p = \frac{65}{500} = \frac{13}{100}$ and $q = 1 - p = 1 - \frac{13}{100} = \frac{87}{100}$

$n = 500$

∴ The standard error of the proportion of bad pineapples in the consignment are

$$= \sqrt{\frac{pq}{n}} = \sqrt{\frac{13}{100} \times \frac{87}{100} \times \frac{1}{500}}$$

$= 0.015 = 1.5$ percent

∴ The limits of the bad pine apples in the consignment are $p \pm 3\sqrt{\frac{pq}{n}}$

$$\frac{13}{100} \times 100 \pm (3 \times 1.5)\% = (13 \pm 3 \times 1.5) \text{ percent}$$

$= 8.5\%$ and 17.5%

Example 6 Certain crosses of pea gave 5321 yellow and 1804 green seeds. The expectation is 25 percent of green on a mendelian hypothesis. Can the divergence from the expected value have arisen from the fluctuation of simple sampling only?

Solution The total number of Pea seeds $= 5321 + 1804$

$= 7125$

The expected value of green seed = 25% of the total pea seeds

$$= \frac{1}{4} \times 7125$$

$= 1781$

Observed value = 1804 green seeds

∴ The difference between the expected value and observed value

$= 1804\text{-}1781$

$= 23$

Also, the standard error of green seeds

$= \sqrt{npq}$

$= \sqrt{7125 \times \frac{1}{4} \times \frac{3}{4}} = 36.6$

Hence, $\quad Z = \frac{23}{36.6} = 0.6$

Since the divergence of the observed value from the expected value is only 0.6 times the standard error, we may very well say that this divergence is due to fluctuation of simple sampling.

Practice Problems

1. In some dice-throwing experiment, Weldon threw dice 49152 times and of these 25145 yielded 4 or 5 or 6. Is this consistent with the hypothesis that the dice were unbiased?

   [**Ans.** The data is not consistent with the hypothesis i.e., die were biased, $|z|$=5.13]

2. Show that for a random sample of size 100 drawn with replacement, the standard error osample proportion cannot exceed 0.05.

   [Hint. taking $p = q = \frac{1}{2}$,n=100]

3. A sample of 900 days is taken from meteorological records of a certain district and 100 of them are found to be foggy. What are the probable limits to the percentage of foggy days in the district?

   [**Ans.** SE of proportion = 1.05%, Limits are 14.26% and 7.96%]

4. Experience has shown that 20 percent of a manufactured product is top quality. In one day's production of 400 articles, there were only 50 of the top quality. Does this contradict our hypothesis of 20 percent? Given that, if x is normal with mean m and variance σ^2, then $P\{|x-m| \geq 1.96\sigma\}$=0.05.

   [**Ans.** SE = 0.02,Z = 3.75, The difference is significant and the hypothesis of

20 percent is contradicted].

5. A coin is tossed 400 times and it turns up head 216 times. Discuss whether the coin may be unbiased one.

[**Ans.** SE= 10, The result is fluctuation of sampling *i.e.*, the coin may be taken as unbiased one.]

6. Of 10,000 babies born in a state 5200 are male children. Taking this to be a random samples of the births in the state. Show that it throws considerable doubt on the hypothesis that the sexes are born in equal proportions.

[**Ans.** $|z|=8$]

Comparison of Two Large Samples

Let the two large samples n_1 and n_2 from different population give proportions of A's as p_1 and p_2 respectively. There are two cases arise.

(i) If the two populations are similar as regards the proportions of A's, can the difference between the two proportion have arisen merely as a fluctuation of simple sampling. If p_0 be the estimate, then it is given by

$$p_0 = \frac{p_1 n_1 + p_2 n_2}{n_1 + n_2}$$

Let E_1 and E_2 be the standard error in two samples, then

$$E_1^2 = \frac{p_0 q_0}{n_1} \text{ and } E_2^2 = \frac{p_0 q_0}{n_2}$$

If E be the standard error in the proportion of successes in the population, then

$$E^2 = E_1^2 + E_2^2$$

$$= \frac{p_0 q_0}{n_1} + \frac{p_0 q_0}{n_2}$$

$$= p_0 q_0 \left(\frac{1}{n_1} + \frac{1}{n_2}\right)$$

If $p_1 \sim p_2 < 3E$, it may be due to fluctuations of sampling, but if $p_1 \sim p_2 > 3E$, it may be taken a real difference in the population proportions.

(ii) If the proportion of A's are not the same in the proportions from which the samples are drawn and if p_1 and p_2 are the true values of the proportions, the standard error E of the difference is given by

$$E^2 = \frac{p_1 q_1}{n_1} + \frac{p_2 q_2}{n_2}$$

Thus, if $p_1 \sim p_2 < 3E$, it may be taken due fluctuation of sampling.

Illustrative Examples

Example 7 In a random sample of 500 men from a particular district of M.P., 300 are found to be smokers. In one of 1000 men from another district,550 are smokers. Do the data indicate that the two districts are significantly different with respect to the prevalence of smoking among men?

Solution. Given, $$p_1 = \frac{300}{500} = \frac{3}{5}, p_2 = \frac{550}{1000} = \frac{11}{20}$$

$$n_1 = 500, n_2 = 1000$$

$$p_0 = \frac{p_1 n_1 + p_2 n_2}{n_1 + n_2}$$

$$= \frac{300 + 550}{500 + 1000} = \frac{850}{1500} = \frac{17}{30}$$

and $$q_0 = 1 - p_0 = 1 - \frac{17}{30} = \frac{13}{30}$$

$\therefore$ $$E^2 = p_0 q_0 \left(\frac{1}{n_1} + \frac{1}{n_2} \right)$$

$$= \frac{17}{30} \times \frac{13}{30} \left(\frac{1}{500} + \frac{1}{1000} \right)$$

$$= \frac{17}{30} \times \frac{13}{30} \times \frac{3}{1000}$$

$$= \frac{663}{900000} = 0.00074$$

$$E = 0.0271$$

and $$p_1 - p_2 = \frac{3}{5} - \frac{11}{20} = \frac{1}{20} = 0.05$$

Here, $p_1 - p_2 < 3E$, hence the difference is not significant i.e., the districts are different with respect to prevalence of smoking among men.

Example 8 In two large populations there are 30 and 25 percent respectively of fair-haired people. Is this difference likely to be hidden in samples of 1200 and 900 respectively two populations?

Solution Here, $$p_1 = \frac{30}{100} = 0.30, p_2 = \frac{25}{100} = 0.25$$

and $$q_1 = 1 - p_1 = 1 - 0.30 = 0.70, q_2 = 1 - 0.25 = 0.75$$

$$n_1 = 1200, n_2 = 900$$

Now, $$p_1 - p_2 = 0.30 - 0.25 = 0.05$$

and $$E^2 = \frac{p_1 q_1}{n_1} + \frac{p_2 q_2}{n_2}$$

$$= \frac{0.30 \times 0.70}{1200} + \frac{0.25 \times 0.75}{900}$$

$$= 0.000175 + 0.000208$$

$$= 0.000383$$

$$E = 0.0195$$

We notice that $p_1 - p_2 < 3E$. Hence, it is unlikely that real difference will be hidden.

Example 9 956 children were born in a city x in one year of which 52.5 percent were male, while 1406 children were born in cities x and y both out of which proportion of male was 0.496. Is the difference in the proportion of male children in two cities significant?

Solution Here, n_1=956, n_2 =1406-956=450

$$p_1 = 52.5\% = \frac{52.5}{100} = 0.525, p_0 = 0.496$$

$$q_0 = 1 - p_0 = 1 - 0.496 = 0.504$$

Now, $p_0 = \dfrac{p_1 n_1 + p_2 n_2}{n_1 + n_2}$

$$0.496 = \frac{0.525 \times 956 + 450 p_2}{1406}$$

$$0.496 \times 1406 = 501.9 + 450\ p_2$$

$$450\ p_2 = 697.38 - 501.9$$

$$= 195.5$$

$$p_2 = \frac{195.5}{450}$$

$$= 0.434$$

We have, $p_1 - p_2 = 0.525\text{-}0.434 = 0.091$

Also, $E^2 = p_0 q_0 \left(\dfrac{1}{n_1} + \dfrac{1}{n_2} \right)$

$$0.496 \times 0.504 \left(\frac{1}{956} + \frac{1}{450} \right) = 0.00081$$

$\therefore$ E = *0.0286*

We notice that $p_1 - p_2 > 3E$. Hence, the difference is significant. Thus, the difference in the proportion of male children in two cities is significant is rejected.

Example 10 If for one- half of n events, the chance of success is p and chance of failure is q, whilst for the other half, the change of success is q and the chance of failure is p. Show that the standard deviation of the number of success is the same as if the chance of successes were p in all the cases i.e., $\sqrt{npq}$,but the mean of the number of successes is $\dfrac{n}{2}$ and not np.

Solution Let the number of successes in first and second half be a and b respectively.

$$E(a) = \frac{np}{2}, E(b) = \frac{nq}{2}$$

$$\text{var}(a)=\frac{npq}{2},\ \text{var}(b)=\frac{npq}{2}$$

$$\therefore\quad E(a+b)=E(a)+E(b)$$

$$=\frac{n}{2}(p+q)=\frac{n}{2}\qquad(\therefore p+q=1)$$

which is not equal to np.

and $$\text{var}(a+b)=\text{var}(a)+\text{var}(b)$$

$$=\frac{npq}{2}+\frac{npq}{2}$$

$$=npq$$

$$\Rightarrow\quad \sigma=\sqrt{npq}$$

Example 11 A machine puts out 16 imperfect articles in a sample of 500. After machine overhauled, puts out 3 imperfect articles in a batch of 100. Has the machine been improved?

Solution Here,

$$p_1=\frac{16}{500}=0.032,\ p_2=\frac{3}{100}=0.03$$

$$n_1=500, n_2=100$$

On the condition that the machine has not improved i.e., $p_1 = p_2$,

$$\therefore\quad p_0=\frac{p_1n_1+p_2n_2}{n_1+n_2}$$

$$=\frac{16+3}{500+100}=\frac{19}{600}$$

$$q_0=1-p_0$$

$$1-\frac{19}{600}=\frac{581}{600}$$

$$\therefore\quad E^2=p_0q_0\left(\frac{1}{n_1}+\frac{1}{n_2}\right)$$

$$= \frac{19}{600} \times \frac{581}{600}\left(\frac{1}{500} + \frac{1}{100}\right)$$

$$= \frac{19}{600} \times \frac{581}{600} \times \frac{6}{500}$$

$$= 0.000367$$

$$E = 0.0192$$

Now, $\quad p_1 - p_2 = 0.032 - 0.03 = 0.002$

which is less than $3E$. Thus, the difference is not significant. We conclude that the machine has not been improved is correct.

Practice Problems

1. In a simple sample of 600 men from a certain large city, 400 are found to be smokers. In one of 900 from another large city, 450 are smokers. Do the data indicate that cities are significantly different with respect to prevalence of smoking among men?

[**Ans.** Since $p_1 - p_2 (= 0.166) > 3E(= 0.078)$. Thus, it is highly significant.

Hence, the

2. Population are similar is almost certainly wrong].

1912 children were born in a city x in one year out of which 52.5 percent were male, while 2812 children were born in city x and y both out of which proportion of male was 0.496. Is the difference in the proportion of male children in two cities significant?

[**Ans.** $p_1 - p_2 (= 0.091) > 3E(= 0.057)$, The difference is significant *i.e.*, the difference in the proportion of male children in two cities significant].

3. In a large city x, 20 percent of a random sample of 900 school boys had a certain slight physical defect. In another large city y, 18.5 percent of a random sample of 1600 school boys had the same defect. Is the difference between the proportion significant?

[**Ans.** $p_1 - p_2(=0.815) < 3E(=0.12)$. The difference between the proportion is not significant.]

4. In two large cities, there are 35 percent and 30 percent of fair haired people. Is the difference likely to be revealed by samples of 1500 and 1000 respectively from the two cities?

[**Ans.** It is unlikely that the difference will be hidden.]

5. The following table gives the proportion of dark coloured people in the two cities.

City	Total Population Observed	Percentage of People
Bhopal	250	42
Gwalior	450	48

Can the difference observed in the percentage of dark coloured people be due to the fluctuation of sampling?

[**Ans.** $p_1 - p_2(=6\%) < 3E(=11.7)$, Hence the divergence can be taken only for the fluctuation of sampling.]

Sampling Distribution

If a large number of sample is taken from a universe and some function like mean, median, standard deviation etc., is calculated for each sample, they will form a frequency distribution. Such a distribution is said to be sampling distribution.

Let the sampling distribution of a statistic be $y = f(x)$, where x is a variate and y is the corresponding frequency. The total frequency of all the samples, which give a value of x greater than a given value x_0 will be given by the area to the right of the ordinate at x_0 will be given by the area to the right of the ordinate at x_0 divided by the total area of the curve. If p denotes this probability, then

$$p = \int_{x_0}^{\infty} f(x)dx \Big/ \int_{-\infty}^{\infty} f(x)dx$$

Similarly, the chance that a sample would give a value of x lying between x_1 and x_2 is given by

$$p = \int_{x_1}^{x_2} f(x)dx \Big/ \int_{-\infty}^{\infty} f(x)dx$$

If the total area under the curve is unity, then probability is given by

$$x_1 = \int_{x_0}^{\infty} f(x)dx \text{ and } \int_{x_1}^{x_2} f(x)dx \text{ as } \int_{-\infty}^{\infty} f(x)dx = 1$$

Standard Error of Sampling Distribution of Means

Let $x_1, x_2, \ldots\ldots\ldots, x_n$ are random samples. Then

$$\text{Sample mean is } \overline{x} = \frac{\sum x}{n}$$

$$\therefore \quad \text{var}(\overline{x}) = \frac{1}{n^2}[\text{var}(x_1 + x_2 + \ldots\ldots + x_n)]$$

$$= \frac{1}{n_2}[\text{var}(x_1) + \text{var}(x_2) + \ldots.. + \text{var}(x_n)]$$

$$= \frac{1}{n_2}[\sigma^2 + \sigma^2 + \ldots. + \sigma^2, n \text{ times}]$$

$(\because \text{var}(x_1) = \text{var}(x_2) = \ldots\sigma^2)$

$$= \frac{1}{n^2} n\sigma^2$$

$$= \frac{\sigma^2}{n}$$

$$\therefore \text{ Standard error of mean} = \frac{\sigma}{\sqrt{n}}$$

Sampling Distribution of Mean

Let us take the size of sample be n are drawn without replacement from a finite universe N. If $\overline{x}$ and $\overline{\sigma}$ be the mean and standard deviation of the sampling distribution and m and σ be the mean and standard deviation of the universe. Then

$$\overline{x} = m, \overline{\sigma} = \frac{\sigma}{\sqrt{n}}\sqrt{\frac{N-n}{N-1}}$$

If the universe is finite with replacement, then

$$\overline{x} = m, \overline{\sigma} = \frac{\sigma}{\sqrt{n}}$$

Distribution of the Difference Between Two Sample Means

Let we take two population with mean m_x and m_y and standard deviation σ_x and σ_y respectively. Let N be the pairs of random sample that are formed from each of two populations each having n_1 variates from the first and n_2 variates from the second population. Let the means of first populations be $\bar{x}_1, \bar{x}_2, \ldots, \bar{x}_n$ and $\bar{y}_1, \bar{y}_2, \ldots \bar{y}_n$ from the second.

Then the difference of the means are $\bar{x}_1 - \bar{y}_1, \bar{x}_2 - \bar{y}_2, \ldots, \bar{x}_n - \bar{y}_n$

Then

$$E(\bar{x}-\bar{y}) = \frac{1}{n}[(\bar{x}_1-\bar{y}_1)+(\bar{x}_2-\bar{y}_2)+\ldots+(\bar{x}_n-\bar{y}_n)]$$

$$= \frac{1}{n}(\bar{x}_1+\bar{x}_2+\ldots+\bar{x}_n) - \frac{1}{n}(\bar{y}_1+\bar{y}_2+\ldots+\bar{y}_n)$$

$$= E(\bar{x}) - E(\bar{y})$$

$$= m_x - m_y$$

Also $\quad \text{Var}(\bar{x}-\bar{y}) = \text{var}(\bar{x}) + \text{var}(\bar{y})$

$$= \frac{\sigma_x^2}{n_1} + \frac{\sigma_y^2}{n_2}$$

As the two samples are independent and $\sigma_{\bar{x}} = \frac{\sigma_x}{\sqrt{n_1}}, \sigma_{\bar{y}} = \frac{\sigma_y}{\sqrt{n_2}}$.

Hence, the standard error of the difference between the means of samples from two independent population is $\sqrt{\sigma_{\bar{x}}^2 + \sigma_{\bar{y}}^2}$, where $\sigma_{\bar{x}}$ and $\sigma_{\bar{y}}$ are the SE of the means two samples.

Test of Significance for Mean

Let us take a random variable x of size n with mean μ and standard deviation σ.

Then, $$z = \frac{\bar{x}-\mu}{\sigma/\sqrt{n}}$$

If z calculated is greater than 3, we reject the null hypothesis $H_0 : \mu = \mu_0,$

Where, $$z = \frac{\bar{x} - \mu_0}{\sigma / \sqrt{n}}.$$

We observe that, if $|z| > 1.96$, the difference is significant on 5% level of significance and if $|z| > 2.58$, the difference is significant on 1% level.

Test of Significance of the Means of Two Large Samples

Let the random sample of size n_1 and mean $\bar{x}$ and the other of size n_2 and mean $\bar{y}$. The variance of the difference of the means of two samples is $\sigma^2\left(\frac{1}{n_1} + \frac{1}{n_2}\right)$, where $\frac{\sigma}{\sqrt{n_1}}$ and $\frac{\sigma}{\sqrt{n_2}}$ are the standard deviations of the means of the sample. Hence,

$$Z = \frac{\bar{x} - \bar{y}}{\sigma\sqrt{\frac{1}{n_1} + \frac{1}{n_2}}}$$

If $|z| > 1.96$, the difference is significant at 5% level of significance

If $|z| > 3$, it is highly significant.

If the samples to be taken from two populations with mean μ_1 and μ_2 and standard deviations σ_1 and σ_2. Than

$$z = \frac{(\bar{x} - \bar{y}) - (\mu_1 - \mu_2)}{\sqrt{\frac{\sigma_1^2}{n_1} + \frac{\sigma_2^2}{n_2}}}$$

If $\mu_1 = \mu_2$, we have

$$z = \frac{\bar{x} - \bar{y}}{\sqrt{\frac{\sigma_1^2}{n_1} + \frac{\sigma_2^2}{n_2}}}$$

Fiducial or Confidence Limits

Let a random sample of size n with mean μ and standard deviation σ (unknow). Let $\bar{x}$ be the mean of the sample, then standard error of $\bar{x}$ is $\frac{(\bar{x} - \bar{\mu})}{\sigma}\sqrt{n}$.

If $\bar{x}$ is not significant at 5% level of significance, this relative deviation must be less than that which is exceeded with a probability of 5%.

Such a relative deviation is 1.96. Hence,

$$\left|\frac{(\bar{x}-\mu)\sqrt{n}}{\sigma}\right| < 1.96$$

i.e., If
$$p\left\{\bar{x}-\frac{1.96\sigma}{\sqrt{n}} < \mu < \bar{x}+\frac{1.96\sigma}{\sqrt{n}}\right\} = 0.95$$

The values $\bar{x} \pm \frac{1.96\sigma}{\sqrt{n}}$ are called 95% fiducial limit or confidence limit.

Some Standard error of Other Parameters

When the parent population is normal, some standard error of parameters are as follows :

S. No.	Parameters	Standard errors
(i)	Mean	$\frac{\sigma}{\sqrt{n}}$
(ii)	Median	$\frac{1.25331\sigma}{\sqrt{n}}$
(iii)	Quartiles	$\frac{1.36263\sigma}{\sqrt{n}}$
(iv)	Standard deviation	$\frac{\sigma}{\sqrt{2n}}$
(v)	Semi inter-quartile range	$\frac{0.7867\sigma}{\sqrt{n}}$
(vi)	Variance	$\sigma^2\sqrt{\frac{2}{n}}$
(vii)	Coefficient of correlation	$\frac{(1-r^2)}{\sqrt{n}}$

(viii)	μ_3	$\sigma^3\sqrt{\frac{96}{n}}$
(ix)	μ_4	$\sigma^4\sqrt{\frac{96}{n}}$
(x)	Coefficient of variation(ν)	$\frac{v}{\sqrt{2n}}$ (approx.)

Illustrative Examples

Example 12 *A sample of 400 items is taken from a normal distribution whose mean is 4 and variance is 4. If the sample mean is 4.45 .Can the sample be regarded as a truly random sample?*

Solution Given, mean of the population μ =4 and variance $\sigma^2 = 4$

Standard deviation $\sigma = 2$, sample mean $\overline{x}$ =4.45, n=400

Hence,
$$z = \frac{\overline{x}-\mu}{\sigma/\sqrt{n}} = \frac{4.45-4}{2/\sqrt{400}}$$
$$= \frac{0.45\times 20}{2} = 4.5$$

Hence, the deviation of the mean of the sample is 4.5 standard error.

$\therefore$ It is highly significant.

Therefore, the sample cannot be regarded as a random sample.

Example 13 *Suppose that the standard deviation of stature in mean is 2.48 inches. One hundred male students in a large university are measured and their average height is found to be 68.52 inches. Determine the 98% confidence limits for the mean height of the university male students. Given that at 2% level of probability the relative deviation is 2.32.*

Solution We know that

$$Z = \frac{(\overline{x}-\mu)\sqrt{n}}{\sigma} \sim N(0,1)$$

$$\therefore \quad p[|z| \le 2.32] = 0.98$$

or $$p\left[\left|\frac{(\overline{x}-\mu)\sqrt{n}}{\sigma}\right| \le 2.32\right] = 0.98$$

or $$p\left[\overline{x}-\frac{2.32\sigma}{\sqrt{n}} \le \mu \le \overline{x}+\frac{2.32\sigma}{\sqrt{n}}\right] = 0.98$$

Where $\overline{x}\pm\frac{2.32\sigma}{\sqrt{n}}$ are the required confidence limits for the mean height of the universe.

Now, given $\overline{x}=68.52, \sigma=2.48, n=100$

$\therefore$ Limits are $\overline{x}\pm\frac{2.32\sigma}{\sqrt{n}}$

or $68.52\pm\frac{2.32\times 2.48}{\sqrt{100}}$

or 68.52 ± 0.57536

or 68.52 ± 0.575

or 67.945 and 69.095

Example 14 *The mean height of 9339 children of age 5 years is 41.26 inches, the standard deviation is 2.238 inches. Find the odd against the possibility that the mean of a random sample of 100 is greater than 41.70.*

Solution Given,

$$\overline{x}=41.70, \mu=41.26, \sigma=2.238, n=100$$

$$z=\frac{(\overline{x}-\mu)\sqrt{n}}{\sigma}$$

$$=\frac{(41.70-41.26)\sqrt{100}}{2.238}$$

$$=\frac{4.4}{2.238}=1.96$$

from the table, we see that for z = 1.96, the area of the left is 0.9750 and so the area to the right is 1-0.9750=0.025

Hence, the probability that the mean of random sample of 100 is greater than 41.70 is 0.025.

Example 15 *If the mean breaking strength of copper wire is 575 lbs with a standard deviation of 8.3 lbs. How large a sample must be used in order that there be one chance in 100 that the mean breaking strength of the sample is less than 572 lbs. Given that at 99% the value of z is 2.33.*

Solution Given, $\bar{x} = 572, \mu = 575, \sigma = 8.3$

$$\therefore \quad |z| = \left|\frac{(\bar{x}-\mu)}{\sigma}\sqrt{n}\right|$$

$$= \left|\frac{(572-575)}{8.3}\sqrt{n}\right|$$

$$|z| = \frac{3}{8.3}\sqrt{n}$$

Now, the probability that $\bar{x}$ <572 is $\frac{1}{100}$=0.01.

So, at 99%, $|z| = 2.33$

Therefore, $2.33 = \frac{3}{8.3}\sqrt{n}$

or $$\sqrt{n} = \frac{2.33 \times 8.3}{3} = \frac{19.339}{3} = 6.446$$

n = 41.55

n = 42 nearly.

Example 16 *A research worker wishes to estimate mean of a population by using sufficiently large sample. The probability is 95% that sample mean will not differ from the true mean by more than 25% of the standard deviation. How large a sample should be taken?*

Solution We know that,

$$z = \frac{\bar{x}-\mu}{\sigma/\sqrt{n}} \sim N(0,1)$$

The area under the curve between the values −1.96 and 1.96 of the standard normal variate is 0.95. Hence, we have

$$\left|\frac{\bar{x}-\mu}{\sigma/\sqrt{n}}\right| \le 1.96$$

or $$|\bar{x}-\mu| \le 1.96\frac{\sigma}{\sqrt{n}} \quad \text{...(i)}$$

Again,

$$|\bar{x}-\mu| \le 25\% \text{ of } \sigma$$

or $$|\bar{x}-\mu| \le \frac{\sigma}{4} \quad \text{...(ii)}$$

$\therefore$ From Eqs. (i) and (ii), we have

$$1.96\frac{\sigma}{\sqrt{n}} \le \frac{\sigma}{4}$$

or $$\sqrt{n} \ge 1.96 \times 4$$

$$n \ge (7.84)^2$$

$$n \ge 61.4656$$

$$n = 62 \textit{nearly}.$$

Example 17 *The mean of simple of 1000 and 2000 are 67.5 and 68.0 inches respectively. Can the sample be regarded as drawn from the same population of standard deviation 2.5 inches?*

Solution Given,

$$\bar{x} = 67.5, \bar{y} = 68.0, n_1 = 1000,$$

$$n_2 = 2000, \sigma = 2.5$$

We have,

$$z = \frac{\bar{x}-\bar{y}}{\sigma\sqrt{\frac{1}{n_1}+\frac{1}{n_2}}}$$

$$= \frac{67.5-68.0}{2.5\sqrt{\left(\frac{1}{1000}+\frac{1}{2000}\right)}}$$

$$|z| = \frac{0.5}{2.5 \times 0.0387} 5.1$$

Hence, the difference between the means is more than three times the standard error and so it is statistically significant. The samples cannot be regarded from the same population of standard deviation 2.5 inches.

Example 18 *A random sample of 200 villages was taken from Bhopal district and the average population per village was found to be 485 with a standard deviation of 50. Another random sample of 200 villages from the district gave an average population of 510 per village with a standard deviation of 40. Is the difference between the average of the two sample statistically significant?*

Solution Given,

$$\bar{x} = 485, \sigma_1 = 50, n_1 = 200, \bar{y} = 510, \sigma_2 = 40, n_2 = 200$$

So that, $$z = \frac{\bar{x} - \bar{y}}{\sqrt{\frac{\sigma_1^2}{n_1} + \frac{\sigma_2^2}{n_2}}}$$

$$= \frac{485 - 510}{\sqrt{\frac{(50)^2}{200} + \frac{(40)^2}{200}}} = \frac{-25}{4.53}$$

$$|z| = 5.5 > 3$$

Hence, the difference between the means of the sample is highly significant and could not have arisen from causes due to fluctuations of sampling.

Example 19 *A sample of heights of 6400 soldiers has a mean of 67.85 inches and a standard deviation of 2.56 inches while a simple sample of height of 1600 sailors has a mean of 68.55 inches and a standard deviation of 2.52 inches. Do the data indicate that the sailors are on the average taller than soldiers?*

Solution Given,

$$\bar{x} = 67.85, \sigma_1 = 2.56, n_1 = 6400,$$

$$\bar{y} = 68.55, \sigma_2 = 2.52, n_2 = 1600$$

The standard error of the difference of the mean height

$$E = \sqrt{\frac{\sigma_1^2}{n_2} + \frac{\sigma_2^2}{n_2}}$$

$$= \sqrt{\frac{(2.56)^2}{6400} + \frac{(2.52)^2}{1600}}$$

$$= \sqrt{0.001024 + 0.003969}$$

$$= 0.07 \text{ nearly.}$$

The difference between the mean $= \overline{x} - \overline{y}$

$$= 67.85\text{-}68.55 = 0.7 \text{ inches}$$

which is nearly 10 times the standard error of the difference of the means and this is highly significant. Hence, the data indicates that the sailors with greater mean height are on the on average taller than the soldiers.

Practice Problems

1. A sample of 400 male students is found to have a mean height of 67.47 inches. Can it be reasonably regarded as sample from a large population with mean height 67.39 inches and standard deviation 1.30 inches?

   [**Ans.** z = 1.23, which is significant]

2. An industry desires to make a survey of the mean weekly wages of 10,000 of its workers. The mean weekly wages are rupees thirty with standard deviation of rupees 2.5. Since study of all workers is impossible, a representative sample of 400 workers is selected; by how much would the results differ from the above sample?

   [**Ans.** E = Rs. 0.125, The mean weekly wages lie between Rs 29.725 and Rs 30.375]

3. An unbiased coin is thrown n times. It is desired that the relative frequency of the appearance of heads should lie between 0.49 and 0.51. Find the smallest value of n that will ensure this result with 90% confidence.

   [**Hint.** at 90%, $z = 1.645$, $E = \dfrac{1}{2\sqrt{n}}$

   Hence, $0.5\text{-}1.645 \times \dfrac{1}{2\sqrt{n}} = 0.49$ and $0.5\text{+}1.645 \times \dfrac{1}{2\sqrt{n}} = 0.51$

   $\therefore \qquad n = 68$ nearly.]

4. A sample of 1600 members found to have a mean 5.7, could it be reasonably regarded as a simple sample from a large population whose mean is 4.5 and standard deviation is 2.8.

   [**Ans.** $z > 3$, the sample cannot regarded as a simple sample.]

5. Random samples of 500 and 400 have means 11.5 and 10.9 respectively. Can the samples be regarded as drawn from the population of the standard deviation 5.

[**Ans.** Difference of means = 0.6; standard error = 0.335, the difference is greater than SE. Hence, it is not significant,.]

6. A potential buyer of light bulbs bought 50 bulbs of each of two brands. Upon testing these bulbs he found that brand A had a mean life of 1282 hours with a SD of 80 hours, whereas brand B had a mean life of 1208 hours with a SD of 94 hours. Can the buyer be quite certain that the two brands differ in quality?

[**Ans.** Difference of mean=74 hours, SE = 17.46 hours. The difference is significant, the hypothesis not be accepted.]

7. Intelligence test on two groups boys and girls, give the following results. Examine if the difference is significant.

Boys	Mean = 81	SD = 12 no. = 81
Girls	Mean = 84	SD = 10 no. = 121

[**Ans.** Deviation from expected value cannot be due to fluctuation of sampling, $|z| = 3.16$.]

8. In a wheat variety test conducted over a wide area, the mean difference between two variates was found to be 5.5 bushels to the acre. The standard error of this difference was 1.4 bushels per acre and was determined from 100 pairs o plots. Set up the confidence limits at the 5% probability level for the mean difference in yield between the two variates.

[**Ans.** $(5.5 \pm 1.96 \times 1.4)$ bushels *i.e.*, 2.756 and 8.244 bushels.

Chapter - 12

Theory of Estimation

Introduction

The purpose of performing all scientific experiment is to get information about the unknown parameter, is known as estimation. The general theory connected with estimation is known as the theory of estimator. The estimator is a rule of determining the unknown population parameter such as mean, variance etc.

Consider a random sample $x_1, x_2, \ldots . x_n$ from the universe with probability density function $f(x,\theta)$, where θ is an unknown parameter. In order to find the estimate in terms of the sample values. There are infinite many ways of choosing an estimator and the problem is to choose the best estimator. The distribution of the estimator should be concentrated as closely as the true parameter value of θ.

The estimator can be done in either of the two ways.

(i)Point estimation (ii) Interval estimation.

Point Estimation

A single value is calculated as an estimate from an unknown population parameter, is called point estimation.

For example: The distance between two points are 40.15cm because the distance measured in single point.

Interval Estimation

An estimate of a population parameter given by two values between which the parameter may be considered to lie called an interval estimation. The interval (θ_1, θ_2) is called the confidence interval for a parameter θ based on random sample is $p(\theta_1 \leq \theta \leq \theta_2) = 1 - a$, then θ_1 is called the lower confidence limit, θ_2 the upper confidence limit and $(1-a)$ is the confidence coefficient.

For example: If the distance is 30.23 ± 4.26cm i.e., the distance lies between 34.49 and 25.97 [The interval is (25.97, 34.49).]

Properties of best estimator

An estimator is said to be best estimator if it is

(i) Unbiased (ii) consistent (iii) efficient and (iv) sufficient.

Unbiased Estimator

If the mean of the sample population of a statistic and the corresponding population parameter both are equal, the statistic is called unbiased estimator of the parameter, otherwise a biased estimator.

An estimator $\hat{\theta}$ of the parameter θ is said to be unbiased, if

$$E(\hat{\theta}) = \theta$$

If $E(\hat{\theta}) > \theta$, the estimator is positively biased, and

If $E(\hat{\theta}) < \theta$, the estimator is negatively biased.

Unbiasedness is an important property of good estimators.

It is said asymptotically unbiased if we replace $\hat{\theta}$ as t_n, **then**

$$\lim_{n\to\infty} E(t_n) = \theta$$

Illustrative Examples

Examples 1 **If t is an unbiased estimator of** θ, show that t^2 and $\sqrt{t}$ are biased estimator of

θ^2 and $\sqrt{\theta}$ respectively.

Solution We know that

$$\text{var}(t) = E(t-\theta)^2$$

$$= E\{t^2 - 2t\theta + \theta^2\}$$

$$= E(t)^2 - 2\theta E(t) + \theta^2$$

$$= E(t)^2 - 2\theta^2 + \theta^2 \qquad \text{(: Given E(t)=θ)}$$

$$= E(t)^2 - \theta^2$$

but var (t) $\neq 0$, hence $E(t)^2 - \theta^2 \neq 0$

or $$E(t)^2 \neq \theta^2$$

Hence, t^2 is biased estimator of θ^2.

Also, var $(\sqrt{t}\ ) = E\left[\sqrt{t} - E(\sqrt{t})\right]^2$

$$= E\left[t - 2\sqrt{t}\,E(\sqrt{t}) + \left\{E(\sqrt{t})\right\}^2\right]$$

$$= E(t) - 2E(\sqrt{t})E(\sqrt{t}) + \left\{E(\sqrt{t})\right\}^2$$

$$= \theta - 2\left\{E(\sqrt{t})\right\}^2 + \left\{E(\sqrt{t})\right\}^2 \qquad (\because \text{Given } E(t) = \theta)$$

$$= \theta - \left\{E(\sqrt{t})\right\}^2$$

But var $((\sqrt{t}) \neq 0)$

Hence, $$\theta - \left\{E(\sqrt{t})\right\}^2 \neq 0$$

or $$\left\{E(\sqrt{t})\right\}^2 \neq \theta$$

or $$E(\sqrt{t}) \neq \sqrt{\theta}$$

Hence, $\sqrt{t}$ is biased estimator of $\sqrt{\theta}$.

Example 2 *Prove that for a random sample of size n viz $x_1, x_2, \ldots.. x_n$ taken from a finite population.*

$$S^2 = \frac{1}{n}\sum_{i=1}^{n}(x_i - \bar{x})^2$$

Is not unbiased estimation of the parameter σ^2.Further, show that it is asymptotically unbiased estimator of the parameter σ^2.

Solution Given, E(S^2)=E $\left[\frac{1}{n}\sum_{i=1}^{n}(x_i - \bar{x})^2\right]$

$$=\mathbf{E}\left[\frac{1}{n}\sum_{i=1}^{n}\{(x_i-\mu)-(\bar{x}-\mu)\}^2\right]$$

$$=\mathrm{E}\left[\frac{1}{n}\sum_{i=1}^{n}(x_i-\mu)^2-\frac{2}{n}\sum_{i=1}^{n}(x_i-\mu)(\bar{x}-\mu)+\frac{1}{n}\sum_{i=1}^{n}(\bar{x}-\mu)^2\right]$$

$$=\mathbf{E}\left[\frac{1}{n}\sum_{i=1}^{n}(x_i-\mu)^2-2(\bar{x}-\mu)\frac{1}{n}\sum_{i=1}^{n}(x_i-\mu)+\frac{1}{n}\sum_{i=1}^{n}(\bar{x}-\mu)^2\right]$$

$$=\mathbf{E}\left[\frac{1}{n}\sum_{i=1}^{n}(x_i-\mu)^2-2(\bar{x}-\mu)(\bar{x}-\mu)+(\bar{x}-\mu)^2\right]$$

$$=\frac{1}{n}\sum_{i=1}^{n}(x_i-\mu)^2-2E(\bar{x}-\mu)^2+E(\bar{x}-\mu)^2$$

$$\left[\because \frac{1}{n}\sum_{i=1}^{n}(x_i-\mu)=\bar{x}-\mu \text{ and } \bar{x}-\mu \text{ is a constant}\right]$$

$$=\sigma^2-E(\bar{x}-\mu)^2$$

$$=\sigma^2-\frac{\sigma^2}{n}$$

$$=\sigma^2\left(1-\frac{1}{n}\right)$$

$$E(s^2)=\sigma^2\left(\frac{n-1}{n}\right)$$

i.e., $E(s^2)\neq\sigma^2$

Hence, S^2 is not an unbiased estimator of the parameter σ^2.

But $\lim\limits_{x\to\infty}E(s^2)=\lim\limits_{x\to\infty}\left[\frac{n-1}{n}\sigma^2\right]$

$$=\lim_{x\to\infty}\left(1-\frac{1}{n}\right)\sigma^2=\sigma^2$$

Hence, S^2 is asymptotically unbiased estimator of the parameter σ^2.

Example 3 Let $x_1, x_2, ...x_n$ be random sample drawn from a normal distribution with mean μ and variance σ^2. Show that the sample mean $\bar{x} = \frac{1}{n}\sum_{i=1}^{n} x_i$ is an unbiased estimate of population mean μ.

Solution Let $\bar{x} = \frac{1}{n}\sum_{i=1}^{n} x_i$

$$\mathbf{E}(\bar{x}) = E\left[\frac{1}{n}\sum_{i=1}^{n} x_i\right]$$

$$= \frac{1}{n} E\left[\sum_{i=1}^{n} x_i\right]$$

$$= \frac{1}{n} E\left[x_1 + x_2 + ... + x_n\right]$$

$$= \frac{1}{n}\left[E(x_1) + E(x_2) + ... + E(x_n)\right]$$

Now, given $E(x_i) = \mu$

$$\therefore \quad E(\bar{x}) = \frac{1}{n}\{\mu + \mu + ... + \mu\} = \frac{n\mu}{n} = \mu$$

Hence, sample mean $\bar{x}$ is an unbiased estimate of population mean μ.

Example 4 If $x_1, x_2, ...x_n$ is a random sample from a normal distribution of mean $\bar{x}$ and standard deviation 1. Show that $t = \frac{1}{n}\sum_{i=1}^{n} x_i^2$ is an unbiased estimator of $\bar{x}^2 + 1$

Solution Given that $E(x_i) = x, V(x_i) = 1$

Now, $E(x_i^2) = V(x_i) + \{E(x_i)\}^2$

$$\therefore \quad E(x_i^2) = 1 + \bar{x}^2$$

or $$E(t) = E\left[\frac{1}{n}\sum_{i=1}^{n} x_i^2\right]$$

$$= \frac{1}{n}\sum_{i=1}^{n} E(x_i^2)$$

$$= \frac{1}{n}\sum_{i=1}^{n} (1+\overline{x}^2)$$

$$= 1+\overline{x}^2$$

Hence, t is an unbiased estimator of $1+\overline{x}^2$.

Example 5 $x_1, x_2, ...x_n$ are random observations on a Bernoulli variate x taking the value 1 with probability θ and value θ with probability (1-θ). Show that $\frac{t(t-1)}{n(n-1)}$ is an unbiased estimate of θ^2, where $t = \sum_{i=1}^{n} x_i$.

Solution $E(x) = 1.\theta + 0.(1-\theta) = \theta$

And $E(x^2) = 1^2.\theta + 0^2.(1-\theta) = \theta$

Now, $V(x) = E(x^2) - \{E(x)\}^2$

$$= \theta - \theta^2$$

$$= \theta(1-\theta)$$

$\therefore$ $E(t) = E(x_1 + x_2 + ... + x_n)$

$$= E(x_1) + E(x_2) + ... + E(x_n)$$

$$= \theta + \theta + ... + \theta$$

$$= n\theta$$

$$\text{var}(t) = \text{var}(x_1 + x_2 + ... + x_n)$$

$$= \text{var}(x_1) + \text{var}(x_2) + ... + \text{var}(x_n)$$

$$= n\theta(1-\theta)$$

$$E(t^2) = \text{var}(t) + \{E(t)\}^2$$

$$= n\theta(1-\theta) + n^2\theta^2$$

$$= n\theta(1-\theta+n\theta)$$

$$\therefore \quad E\left[\frac{t(t-1)}{n(n-1)}\right]=\frac{1}{n(n-1)}\left\{E(t^2)-E(t)\right\}$$

$$=\frac{1}{n(n-1)}\left\{n\theta(1-\theta+n\theta)-n\theta\right\}$$

$$=\frac{1}{n(n-1)}n\theta(1-\theta+n\theta-1)$$

$$=\frac{1}{n(n-1)}n\theta^2(n-1)$$

$$=\theta^2$$

Hence, $\dfrac{t(t-1)}{n(n-1)}$ is an unbiased estimate of θ^2.

Consistent Estimator

If t_n is an estimator of the parameter θ of the population, it is said to be a consistent estimator if the probability that t_n assumes a value which differs from θ by more than an arbitrary constant ε approaches zero as n$\rightarrow\infty$ i.e.,

$$P\left\{\left|\theta-t_n\right|\right\}>\varepsilon\rightarrow 0 \text{ as n}\rightarrow\infty$$

In other words, t_n is a sample statistic computed from a sample size n and θ is the population to be estimated then t_n will be consistent estimator of θ, if for an arbitrary positive number ε, Such that

$$P\left\{\left|t_n-\theta\right|\right\}\le\varepsilon\rightarrow 1 \text{ as } n\rightarrow\infty$$

Illustrative Examples

Example 6 Show that if t is a consistent estimator of θ, then t^2 is a consistent estimator of θ^2.

Solution Given $P\left\{\left|t-\theta\right|\right\}<\varepsilon\rightarrow 1 \text{ as } n\rightarrow\infty$

or $P\left\{-\epsilon<(t-\theta)<\epsilon\right\}\rightarrow 1$

or $P\{\theta-\in<t<\theta+\in\}\to 1$

or $P\{(\theta-\in)^2<t^2<(\theta+\in)^2\}\to 1$

or $P\{\theta^2+\in^2-2\theta\in<t^2<\theta^2+\in^2+2\theta\in\}\to 1$

or $P\{\in^2-2\theta\in<t^2-\theta^2<\in^2+2\theta\in\}\to 1$

or $P\{-\alpha<t^2-\theta^2<\alpha\}\to 1$

$(\because \in \text{ is small than } \alpha \text{ is small})$

or $P\{|t^2-\theta^2|\}<\alpha\to 1$ as $n\to\infty$

Hence, t^2 is a consistent estimator of θ^2

Example 7 Show that in a random sampling from a normal population, the sample mean is a consistent estimator of the universe mean.

Solution Let the random sample be $x_1, x_2, ...x_n$ which mean $\bar{x}$ and let μ be the mean and σ^2 be the variance of the normal population. If z is a normal variate, then

$$z=\frac{(\bar{x}-\mu)\sqrt{n}}{\sigma}$$

Now, $$P\{|\bar{x}-\mu|<\in\}=P\left\{\frac{\sigma}{\sqrt{n}}|z|<\in\right\}$$

$$=P\left\{|z|<\frac{\in\sqrt{n}}{\sigma}\right\}$$

$$=P\left\{-\frac{\in\sqrt{n}}{\sigma}<z<\frac{\in\sqrt{n}}{\sigma}\right\}$$

$$=\int_{-\frac{\in\sqrt{n}}{\sigma}}^{\frac{\in\sqrt{n}}{\sigma}}\frac{1}{\sqrt{2\pi}}e^{-\frac{1}{2}z^2}dz$$

$$\lim_{n\to\infty}P\{|\bar{x}-\mu|<\in\}=\lim_{n\to\infty}\int_{-\frac{\in\sqrt{n}}{\sigma}}^{\frac{\in\sqrt{n}}{\sigma}}\frac{1}{\sqrt{2\pi}}e^{-\frac{1}{2}z^2}dz$$

$$=\frac{1}{\sqrt{2\pi}}\int_{-\infty}^{\infty}e^{-\frac{1}{2}z^2}dz=1$$

Hence, $\bar{x}$ is a consistent estimator of μ.

Example 8 *Show that the sample median is a consistent estimator for universe mean of a normal distribution.*

Solution Let m_d be the sample median of the random sample $x_1, x_2, ...x_n$ from the normal population with mean μ, then the normal variate.

$$z = \frac{(m_d - \mu)\sqrt{2n}}{\sigma\sqrt{\pi}}$$

$$\therefore P\{|m_d - \mu| < \in\} = P\left\{|z| < \frac{\in\sqrt{2n}}{\sigma\sqrt{\pi}}\right\}$$

$$= \int_{\frac{-\in\sqrt{n}}{\sigma\sqrt{\pi}}}^{\frac{\in\sqrt{n}}{\sigma\sqrt{\pi}}} \frac{1}{\sqrt{2\pi}} e^{-\frac{1}{2}z^2} dz$$

$$= \int_{-\infty}^{\infty} \frac{1}{\sqrt{2\pi}} e^{-\frac{1}{2}z^2} dz, \text{ when } n \to \infty$$

$$= 1$$

Hence, the sample median m_d is a consistent estimator for the universe mean μ.

Example 9 Show that the sampling variance is a consistent estimatior for the population variance of a normal distribution.

Solution Let S^2 be the sample variance and σ^2 be the population variance of the sample $x_1, x_2, ...x_n$.

If z be normal variate, then

$$Z = \frac{(s^2 - \sigma^2)}{\sigma^2\sqrt{2/n}}$$

Now, $$P\{|s^2 - \sigma^2| < \in\} = P\left\{\sigma^2\sqrt{\frac{2}{n}}|z| < \in\right\}$$

$$= P\left\{|z| < \frac{\in}{\sigma^2}\sqrt{\frac{n}{2}}\right\}$$

$$= \int_{-\frac{\epsilon}{\sigma^2}\sqrt{\frac{n}{2}}}^{\frac{\epsilon}{\sigma^2}\sqrt{\frac{n}{2}}} \frac{1}{\sqrt{2\pi}} e^{-\frac{1}{2}z^2} dz$$

$$= \int_{-\infty}^{\infty} \frac{1}{\sqrt{2\pi}} e^{-\frac{1}{2}z^2} dz, when \ \ n \to \infty$$

$$= 1$$

Hence, the sample variance S^2 is a consistent estimator of population variance σ^2 of the normal distribution.

Efficient Estimator

In the sampling distribution, if two unbiased. Statistics have the same mean, the statistic with the smaller variance is said to be an efficient estimator of the mean while the other variance is called an inefficient estimator.

Suppose two consistent estimator t_n and t'_n from the same sample and of the same parameter θ. The sampling distribution of t_n has a variance σ^2 and t_n'has a variance σ'^2 respectively.

If $\sigma^2 < \sigma'^2$, then t_n is called more efficient.

In other words, if var (t_n)< var (t'_n), for size n, then t_n will be more efficient.

If the variance of a consistent estimator t_n is less than that of all other consistent estimator for all n, then t_n is said to be the most efficient.

The efficiency 'e' of an estimator t_n'is

$$e = \frac{\text{var}(t_n)}{\text{var}(t_n')}$$

A statistic with an efficiency of 1 is said to be most efficient.

IIIustrative Examples

Example 10 *Show that, in estimating for the mean for random sample from a normal distribution. Sample mean is more efficient than a sample median. Compare their relative efficiencies.*

Solution **Var(mean)** $=\dfrac{\sigma^2}{n}$**, for any n**

and Var (median) $=\dfrac{\pi\sigma^2}{2n}$, for large n

Here, var (mean)<var(median)

Thus, the mean is more efficient than the median.

If e be the efficiency of sample median, then

$$e=\frac{\text{var}(mean)}{\text{var}(median)}$$

$$=\frac{\sigma^2/n}{\pi\sigma^2/2n}=\frac{2}{\pi}$$

Example 11 *Show that for a sample of size n from normal population with mean μ and variance σ^2, that statistic $t=\frac{1}{n+1}\sum_{i=1}^{n}x_i$ is the most efficient for estimating μ though it is biased.*

Solution Given $t=\dfrac{1}{n+1}\sum_{i=1}^{n}x_i$

$$E(t)=\frac{1}{n+1}E\left[\sum_{i=1}^{n}x_i\right]$$

$$=\frac{1}{n+1}E(n\bar{x})$$

$$=\frac{n}{n+1}E(\bar{x})$$

$$=\frac{n}{n+1}\mu$$

Hence, t is a biased estimator of μ.

Now, var (t) = var $\left(\dfrac{n}{n+1}\bar{x}\right)$

$$=E\left[\frac{n}{n+1}\bar{x}-E\left(\frac{n}{n+1}\bar{x}\right)\right]^2$$

$$=\left(\frac{n}{n+1}\right)^2 E\left[\bar{x}-E(\bar{x})\right]^2$$

$$=\left(\frac{n}{n+1}\right)^2 \operatorname{var}(\bar{x})$$

$$\therefore \quad =\frac{\operatorname{var}(t)}{\operatorname{var}(\bar{x})}=\left(\frac{n}{n+1}\right)^2<1$$

$$\therefore \quad \operatorname{var}(t)<\operatorname{var}(\bar{x})$$

Hence, t is the most efficient estimator of μ.

Example 12 If $t_3=\frac{1}{2}(t_1+t_2)$, where t_1 and t_2 are both the most efficient with variance σ, then prove that

$$\operatorname{var}(t_3)=\frac{1}{2}\sigma(1+r)$$

Where r is the coefficient of correlation between t_1 and t_2.

Solution Consider

$$\operatorname{var}(a_1x_1+a_2x_2)=a_1^2\operatorname{var}(x_1)+a_2^2\operatorname{var}(x_2)+2a_1a_2\operatorname{cov}(x_1,x_2)$$

$$\therefore \quad \operatorname{var}(t_3)=\operatorname{var}\left\{\frac{1}{2}(t_1+t_2)\right\}$$

$$=\frac{1}{4}\operatorname{var}(t_1)+\frac{1}{4}\operatorname{var}(t_2)+2.\frac{1}{2}.\frac{1}{2}\operatorname{cov}(t_1,t_2)$$

$$=\frac{1}{4}\sigma+\frac{1}{4}\sigma+\frac{1}{2}r\sqrt{\operatorname{var}(t_1).\operatorname{var}(t_2)}$$

$$=\frac{1}{2}\sigma+\frac{1}{2}r\sigma$$

$$=\frac{1}{2}\sigma(1+r)$$

Sufficient Estimator

As estimator t_n contains all the information in the sample regarding the population parameter θ, then t_n is called sufficient estimator.

Let f(x, θ) be the frequency function with θ be the parameter and $x_1, x_2, ..., x_n$, be the random variables of a sample size n. Then the function is given by

$$L(x_1, x_2, ..., x_n, \theta) = f(x_1, \theta), f(x_2, \theta), ...f(x_n, \theta)$$

Where L is known as likelihood function

If $L(x_1, x_2, ..., x_n, \theta) = L_1(t_n, \theta) L_2(x_1, x_2, ..., x_n)$

Where the second function on RHS does not contain θ, then t_n is said to be sufficient for θ.

The statistic t_n is said to be sufficient, if with respect to any other statistic t'_n, then probability is

$$P(t_n, t_n') = P_1(t_n, \theta), P_2(t_n' / t_n)$$

Illustrative Examples

Example 13 Show that if σ^2 is known, in random sampling normal distribution, $\bar{x}$ is sufficient estimator for μ, but if μ is known $S^2 = \frac{1}{n}\sum (x_i - \bar{x})^2$ is not a sufficient estimator for σ^2.

Solution For the normal distribution

$$f(x, \mu) = \frac{1}{\sigma\sqrt{2\pi}} e^{-\frac{1}{2}\cdot\frac{(x-\mu)^2}{\sigma^2}}$$

$$L(x_1, x_2, ..., x_n, \mu) = \frac{1}{\sigma\sqrt{2\pi}} e^{-\frac{1}{2}\cdot\frac{(x_1-\mu)^2}{\sigma^2}} \cdot \frac{1}{\sigma\sqrt{2\pi}} e^{-\frac{1}{2}\cdot\frac{(x_2-\mu)^2}{\sigma^2}}$$

$$... \frac{1}{\sigma\sqrt{2\pi}} e^{-\frac{1}{2}\cdot\frac{(x_n-\mu)^2}{\sigma^2}}$$

$$= \left(\frac{1}{\sigma\sqrt{2\pi}}\right)^n e^{-\frac{1}{2\sigma^2}\sum_{i=1}^{n}(x_i-\mu)^2}$$

$$=\left(\frac{1}{\sigma\sqrt{2\pi}}\right)^n e^{-\frac{1}{2\sigma^2}\sum_{i=1}^{n}\left\{(x_i-\bar{x})+(\bar{x}-\mu)\right\}^2}$$

$$=\left(\frac{1}{\sigma\sqrt{2\pi}}\right)^n e^{-\frac{1}{2\sigma^2}\sum_{i=1}^{n}\left\{(x_i-\bar{x})^2+(\bar{x}-\mu)^2+2(\bar{x}-\mu)(x_i-\bar{x})\right\}}$$

$$=\left(\frac{1}{\sigma\sqrt{2\pi}}\right)^n e^{-\frac{1}{2\sigma^2}\sum_{i=1}^{n}\left\{(x_i-\bar{x})^2+(\bar{x}-\mu)^2\right\}}\left[\because \sum_{i=1}^{n}(x_i-\bar{x})=0\right]$$

$$=\left(\frac{1}{\sigma\sqrt{2\pi}}\right)^n e^{-\frac{1}{2\sigma^2}\left\{\sum_{i=1}^{n}(x_i-\bar{x})^2+\sum_{i=1}^{n}(\bar{x}-\mu)^2\right\}}$$

$$=\left(\frac{1}{\sigma\sqrt{2\pi}}\right)^n e^{-\frac{1}{2\sigma^2}\sum_{i=1}^{n}(x_i-\bar{x})^2}\ e^{-\frac{1}{2\sigma^2}\sum_{i=1}^{n}(\bar{x}-\mu)^2}$$

$$=\left(\frac{1}{\sigma\sqrt{2\pi}}\right)^n e^{-\frac{ns^2}{2\sigma^2}}\ e^{-\frac{n}{2\sigma^2}(\bar{x}-\mu)^2}$$

Thus, the function is broken into two factors, the second factor contains $\bar{x}$ (estimator) and μ (parameter). The first factor does not contain μ. Thus, if σ^2 is known, $\bar{x}$ is sufficient for estimating μ.

However, since the first factors contains S^2 and σ^2 while the second factor also contains σ^2. If μ is known. Therefore, S^2 is not a sufficient estimator for σ^2.

Example 14 *Show that in estimating for the parameter m in the poisson distribution* $\frac{e^{-m}.m^x}{x!}$, $\bar{x}$ *is sufficient estimator for m.*

Solution For the poisson distribution

$$f(x,m)=\sum_{i=0}^{\infty}\frac{e^{-m}.m^x}{x!}$$

$$L(x_1, x_2, ..., x_n, m) = \frac{e^{-m} m^{x_1}}{x_1 !} \cdot \frac{e^{-m} m^{x_2}}{x_2 !} .... \frac{e^{-m} m^{x_n}}{x_n !}$$

$$= \frac{(e^{-m}.e^{-m} ....ntimes)(m^{x_1}.m^{x_2}....m^{x_n})}{x_1 ! x_2 !....x_n !}$$

$$= \frac{e^{-nm} m^{n\bar{x}}}{x_1 ! x_2 !....x_n !}$$

$$= L_1(x, m).L_2(x_1, x_2, ..., x_n)$$

Thus, the function is broken into two factors. The first factors contains estimator $\bar{x}$ and parameter m while the second factor contains only $x_1, x_2, ..., x_n$.

Hence, $\bar{x}$ is a sufficient estimate for m.

Example 15 *Let t_1, t_2 be two independent observations from a poisson distribution with mean θ. Prove that $t_1 + 2t_2$ is not sufficient for θ.*

Solution Let $t = t_1 + 2t_2$

Let us take $t_1 + 2t_2$=2(in particular)

$\Rightarrow \quad t_1 = 0, t_2 = 1 \, and \, t_1 = 2, t_2 = 0$

Fro the poisson distribution

$$P(x = r) = \frac{e^{-\theta} . \theta^r}{r!}, r = 0, 1, 2, ...\infty$$

Consider, $P(t_1 = 0, t_2 = 1 : t_1 + 2t_2 = 2)$

$$= \frac{P(t_1 = 0, t_2 = 1)}{P(t_1 + 2t_2 - 2)}$$

$$= \frac{P(t_1 = 0, t_2 = 1)}{P(t_1 = 0, t_2 = 1) + P(t_1 = 2, t_2 = 0)}$$

$$= \frac{e^{-\theta}.\theta e^{-\theta}}{e^{-\theta}.\theta e^{-\theta} + \frac{\theta^2}{2} e^{-\theta}.e^{-\theta}}$$

$$= \frac{1}{1 + \frac{\theta}{2}}$$

Which is not independent of θ.

Hence, $t_1 + 2t_2$ is not sufficient for θ.

Maximum Likelihood Parameter

In the frequency function f(x, θ), a maximum likelihood estimator t_n of the parameter θ is an estimator that maximize the likelihood function $L(x_1, x_2, ...x_n, \theta)$ as a function of θ.

Thus, the maximum likelihood estimator of θ is t_n, if it exists, then

$$f(x_1, x_2, ...x_n, t_n) > f(x_1, x_2, ...x_n, t'_n)$$

where t'_n is any other possible value of θ.

For finding maximum likelihood estimator for the parameter θ, we must maximize L with respect to θ, a variable, as such

$$\frac{dL}{d\theta} = 0, where \frac{d^2L}{d\theta^2} < 0$$

Since, L is positive, then replace L by log L, we have

$$\frac{d}{d\theta}\log L = 0 and \frac{d^2}{d\theta^2}\log L < 0$$

which is more convenient from.

Properties of Maximum Likelihood Estimators

The important properties are

(i) Maximum likelihood estimators are consistent.

(ii) Maximum likelihood estimators are most consistent.

(iii) The distribution of maximum likelihood estimator tends to normality for large samples with variance.

(iv) Maximum likelihood estimators are sufficient, if sufficient estimator exist.

(v) Maximum likelihood estimators are not necessarily unbiased.

Illustrative Examples

Example 16 *Show that the maximum likelihood estimator of the parameter θ of the distribution*

$$dP = \frac{e^{-x/\theta} \; .x^{p-1}}{\theta^p \Gamma P}; 0 \le x \le \infty, P > 1, \theta > 0$$

where P is known is $\frac{\bar{x}}{P}$.

Solution Let $L(x_1, x_2, ..., x_n, \theta) = \sum_{i=1}^{n} \frac{e^{-x_i/\theta} \; .x_i^{\;p-1}}{\theta^p \Gamma P}$

$$= \frac{e^{-\frac{1}{\theta}\sum_{i=1}^{n} X_i} .(x_1, x_2, ..., x_n)^{p-1}}{\theta^{pn} .(\Gamma P)^n}$$

$$\log L = \frac{-n\bar{x}}{\theta} + (P-1)\sum_{i=1}^{n} \log x_i - np\log\theta - n\log\Gamma P$$

$$\frac{\partial}{\partial\theta}\log L = \frac{n\bar{x}}{\theta^2} - \frac{np}{\theta} = 0$$

Now, $$\frac{1}{\text{var}(t)} = -nE\left\{\frac{\partial^2}{\partial\theta^2}\log f(x,\theta)\right\}$$

$$= -nE\left\{\frac{\partial^2}{\partial\theta^2}\log\frac{e^{-x/\theta} \; .x^{p-1}}{\theta^p \Gamma P}\right\}$$

$$= nE\left\{\frac{2x}{\theta^3} - \frac{P}{\theta^2}\right\}$$

$$= n\int_0^\infty \left(\frac{2x}{\theta^3} - \frac{P}{\theta^2}\right)\frac{e^{-x/\theta} \; .x^{p-1}}{\theta^p \Gamma P}dx$$

$$= \frac{2n}{\theta^3}\int_0^\infty \frac{x.e^{-x/\theta} \; x^{p-1}}{\theta^p \Gamma P}dx - \frac{nP}{\theta^2}\int_0^\infty \frac{e^{-x/\theta} \; x^{p-1}}{\theta^p \Gamma P}dx$$

Put $\frac{x}{\theta} = u, x = u\theta, dx = \theta du$

$$\therefore \frac{1}{\text{var}(t)} = \frac{2n}{\theta^3}\int_0^\infty \frac{\theta u e^{-u}(u\theta)^{p-1}}{\theta^p \Gamma P}\theta du - \frac{nP}{\theta^2}\int_0^\infty \frac{e^{-u}(u\theta)^{p-1}}{\theta^p \Gamma P}\theta du$$

$$= \frac{2n}{\theta^2}\cdot\frac{1}{\Gamma P}\int_0^\infty u^p e^{-u} du - \frac{np}{\theta^2 \Gamma P}\int_0^\infty u^{p-1}e^{-u} du$$

$$= \frac{2n}{\theta^2}\frac{1}{\Gamma P}\Gamma(P+1) - \frac{nP}{\theta^2}\cdot\frac{1}{\Gamma P}\Gamma P$$

$$= \frac{2nP}{\theta^2} - \frac{nP}{\theta^2}$$

$$= \frac{nP}{\theta^2}$$

$$\therefore \quad \text{Var(t)} = \frac{\theta^2}{nP}$$

$$\therefore \quad \text{var}\left(\frac{\bar{x}}{p}\right) = \frac{\theta^2}{nP} \qquad \left(\text{Given } t = \frac{\bar{x}}{p}\right)$$

Example17 *Show that for the binomial population with density function* $f(x,p) = {}^n c_x p^x q^{n-x}; x = 1,2,\ldots,n$

The maximum likelihood estimator for P is $\frac{x}{n}$ and variance is $\frac{Pq}{n}$.

Solution Given f(x,p) is the probability out of n trials. It is itself the likelihood function.

$$\therefore \quad L = {}^n c_x p^x q^{n-x}; x = 1,2,\ldots,n$$

$$\log L = \log {}^n c_x + x\log p + (n-x)\log q$$

$$= \log {}^n c_x + x\log p + (n-x)\log(1-p) \qquad (\because p+q=1)$$

$$\therefore \quad \frac{\partial}{\partial P}\log L = \frac{x}{p} - \frac{n-x}{1-p} = 0 \Rightarrow p = \frac{x}{n}$$

$$\frac{\partial^2}{\partial P^2}\log L = -\frac{x}{p^2} - \frac{n-x}{(1-p)^2}$$

Now, $$\frac{1}{\text{var}(p)} = -nE\left[\frac{\partial^2}{\partial P^2}\log L\right]$$

$$\text{var}(p) = -\frac{1}{E\left(\dfrac{\partial^2}{\partial p^2}\log L\right)}$$

$$= \frac{-1}{E\left(\dfrac{-x}{P^2} - \dfrac{n-x}{(1-p)^2}\right)}$$

$$= \frac{1}{\dfrac{1}{P^2}E(x) + \dfrac{n}{(1-P)^2} - \dfrac{1}{(1-P)^2}E(x)}$$

Since, E(x)=n p and 1-p=q then

$$\text{var}(p) = \frac{1}{\dfrac{n}{p} + \dfrac{n}{q^2} - \dfrac{np}{q^2}}$$

$$= \frac{1}{\dfrac{n(q^2 + p - p^2)}{pq^2}}$$

$$= \frac{pq^2}{nq} = \frac{pq}{n}$$

Example 18 *Show that the maximum likelihood estimator for θ based on n observations for the frequency function:*

$$f(x,\theta) = (1+\theta)x^{\theta}; 0 \le x \le 1, \theta > 0$$

$$is -1 - \frac{n}{\log(x_1, x_2, ..., x_n)}$$

Solution Given that $f(x,\theta) = (1+\theta)x^{\theta}; 0 \le x \le 1; \theta > 0,$ where $x_1, x_2, ..., x_n$ are the n sample observations.

Now, $$L=\sum_{i=1}^{n} f(x_i,\theta)$$

$$=\sum_{i=1}^{n}(1+\theta)x_i^{\theta}$$

$$L=(1+\theta)^n\sum_{i=1}^{n} x_i^{\theta}$$

$\therefore$ $$\log L = n\log(1+\theta)+\theta\sum_{i=1}^{n}\log x_i$$

$$\frac{\partial}{\partial\theta}\log L=\frac{n}{1+\theta}+\sum_{i=1}^{n}\log x_i=0$$

$$\theta=-\frac{n+\sum_{i=1}^{n}\log x_i}{\sum_{i=1}^{n}\log x_i}$$

$$=-\frac{n}{\sum_{i=1}^{n}\log x_i}-1$$

$$=-1-\frac{n}{\log(x_1,x_2,....,x_n)}$$

Method of Moments

Let f(x,θ) is a probability density function of a distribution with unknown parameter θ. Then rth moment about origin be

$$\mu_r'=\int_{-\infty}^{\infty} x^r f(x,\theta_1,\theta_2,...,\theta_k)dx, r=1,2,....,k$$

The solution of these k equation will give the required estimators of the parameter $\theta_1,\theta_2,...,\theta_k$.

Properties of the moment method

(i) These estimators are easy to understand.
(ii) These are not unbiased in general.
(iii) These are not efficient in general.
(iv) They are efficient when $n\to\infty$.
(v) These estimators obtained by this method are consistent.

Illustrative Examples

Example 19 *Estimate the parameter p in the sampling from binomial distribution* f(x,n,p)=$^{n}c_{x}p^{x}q^{n-x}$; $x = 0,1,2,....,n$. *By the method of moments.*

Solution We known that, for a binomial distribution,

The moment about origin be

$$\mu_1' = np$$

or $$p = \frac{\mu_1'}{n}$$

i.e., $$p = \frac{\bar{x}}{n}, \text{ where } \bar{x} \text{ is the sample mean.}$$

Example 20 *Estimate the parameter □ in sampling from a poisson distribution by the method of moments. Suppose the five observations 18,19,20,22 and 26 are taken.*

Solution For a poisson distribution

$$f(x,\theta) = \frac{e^{-\theta}.\lambda^{x}}{x!}; x = 0,1,2,....$$

First moment about origin μ_1'=θ ...(i)

Let $x_1, x_2,...., x_n$ be the random sample of size n from the distribution.

Then the first sample moment about origin is

$$m_1' = \frac{1}{n}\sum_{i=1}^{n} x_i = \bar{x} \qquad ...(ii)$$

From Eqs. (i) and (ii), we have

$$\mu_1' = m_1' \Rightarrow \theta = \bar{x}$$

i.e., The moment estimate of θ is sample mean x

Also, $$\theta = \frac{18+19+20+22+26}{5}$$

$$= \frac{105}{5} = 21$$

Practice Problems

1. Show that the sample mean $\bar{x}$ is an unbiased estimator of $\frac{1}{\theta}$ for distribution.

   $f(x,\theta)=\theta(1-\theta)^{x-1}; x=1,2,...;0<\theta<1.$

2. If $x_1, x_2,....., x_n$ be random sample of size n from a normal distribution with mean zero and variance σ^2. Show that $\frac{1}{n}\sum_{i=1}^{n} x^2{}_i$ is an unbiased statistic for σ^2. .

3. $x^2{}_1, x^2{}_2,....., x^2{}_r$ are r sample variance based on sample of size $n_1, n_2,....., n_r$ respectively. If t be the statistic given by $t=\frac{n_1x_1{}^2+n_2x_2{}^2+.......+n_rx_r{}^2}{a}$

   For estimating σ^2 and be unbiased. Find the value of a.

   $\left[Ans.a=n_1+n_2+.......+n_r-r\right].$

4. Show that for random sampling from Cauchy population with density function,

   $f(x,\mu)=\frac{1}{\pi}.\frac{1}{1-(x-\mu)^2}, -\infty\le x\le\infty.$

   The sample mean is not a consistent estimator for the population mean.

5. For the poisson parameter, show that $\frac{1}{\bar{x}}$ is a consistent estimator of $\frac{1}{\theta}$.

6. Show that an estimate t_n for which $E(t_n)=\theta_n\to 0$ and var $(t_n)\to 0$ as $n\to\infty$ is consistent.

7. If t the most efficient estimator of θ and t' another estimator with efficiency e, show that $\text{var}(t-t')=\left(\frac{1}{e}-1\right)\text{var}(t)$

   $\left[H\text{int}.\text{var}(t-t')=\text{var}(t)+\text{var}(t')-2\text{cov}(t,t'), \text{cov}(t,t')=e\sqrt{\text{var}(t).\text{var}(t')}\right]$

8. In normal samples, given that

   Var (mean deviation)= $\frac{2}{\pi}\frac{n-1}{n^2}\sigma^2\left(\frac{\pi}{2}+\sqrt{n(n-2)}-n+\sin^{-1}\frac{1}{n-1}\right)$

   var (standard deviation) = $\frac{\sigma^2}{n}$.

   Compare their relative efficiencies.

   (**Ans.** Relative efficiency =0.876, var (mean deviation)is more efficient.)

9. If T is the most efficient estimator and T' is an unbiased estimator with efficiency e. obtain the efficiency of $lT + (1-l)T'$, where l is constant.

10. Show that t_n is a sufficient estimator for θ, then $\frac{\partial}{\partial\theta}\log L$ is a function of t_n and θ only.

11. Let t_1, t_2 be two independent observations from a poisson distribution with mean θ prove that $t_1 + t_2$ is sufficient for θ.

12. Show that for the population, $dP = \frac{e^{-x/\theta}x^{p-1}}{\theta^p \Gamma P}dx, 0 \le x \le \infty, P > 0,$

Where P is known, x/P is a sufficient estimator for θ.

13. Find the maximum likelihood estimate for the parameter θ in the frequency function $f(x,\theta) = \theta e^{-\theta x}$.

(Ans. $\frac{1}{x}$)

14. Given the frequency function,

$$f(x,\theta) = \frac{e^{-x^2/2\theta^2}}{\theta\sqrt{2\pi}}, \infty < x < \infty$$

Find the maximum likelihood estimator for θ.

$$\left[Ans.\theta = \sqrt{\frac{x_1^2 + x_2^2 + \ldots + x_n^2}{n}}\right]$$

15. Given the frequency function, $f(x,\theta) = \frac{e^{-\theta}.\theta^x}{x!}$ where x can assume only non-negative integer values, given the six observations 6,11,4,8,7,and 6. Find the maximum likelihood estimate for θ.

($Ans.\theta = n\bar{x}$, observation value =42)

16. Given the frequency function, $f(x,\theta) = \frac{1}{\sigma\sqrt{2\pi}}e^{-\frac{1}{2}\frac{(x-\theta)^2}{\sigma^2}}$

Find the maximum likelihood estimator of θ.

(Ans. $\bar{x}$)

17. Let $x_1, x_2, \ldots, x_k$ be a random sample of size k from a binomial distribution f(x,p)=$^nc_x p^x q^{n-x}; x = 0,1,2,\ldots,n; p+q=1$

Show that the moment estimate of p is $\frac{\bar{x}}{n} = \frac{1}{kn}\sum_{i=1}^{k} x_i$

18. If x has the probability density function $f(x) = \frac{1}{\sigma\sqrt{2\pi}}.e^{-\frac{1}{2}\frac{(x-\mu)^2}{\sigma^2}}$

Obtain the estimates of parameters μ and σ^2 by the method of moments.

$$\left[Ans. \mu = \bar{x},\ \sigma^2 = \frac{1}{n}\sum_{i=1}^{n}(x_i - \bar{x})^2 \right]$$

Chapter - 13

Theory of Testing a Hypothesis

A statistical Hypothesis

A statistical hypothesis is a tentative statement about distribution of one or more random variables on the basis of sample information. By some testing we shall be decide whether a parameter is acceptable or to reject in the sample. This process is said to be a testing of hypothesis. There are two types of hypothesis.

(i) Null Hypothesis

(ii) Composite Hypothesis

Null Hypothesis

According to R.A. Fisher "A null hypothesis is a hypothesis which is tested for possible rejection under the assumption that it is true."

Null hypothesis is denoted by H_0.

Composite Hypothesis

The hypothesis, which is differ from a null hypothesis H_0 and is accepted when H_0 is rejected, is called an alternative or composite hypothesis. If statistical hypothesis determines the population completely, it is called simple hypothesis otherwise. Composite hypothesis

For example: In the experiment of tossing a coin, The probability of getting head $p=\frac{1}{2}$ is a simple hypothesis, whereas $p>\frac{1}{2}$ is a composite hypothesis.

Critical Region and Acceptance Region

If the sample point falls in the region of rejection, H_0 is rejected. The region is called critical region or region of rejection. It is denoted by W_c.

The rest of the sample space is called the acceptance region.

If $x_1, x_2, ...x_n$ be the sample space, W_c be the critical region and the whole region is *W*, then

$$W_a=W-W_c$$

Where W_a *is called the acceptance region.*

Types of Errors

When a null hypothesis H_0 is tested against an alternative hypothesis H_1, the following two types of errors may arise.

1. When a null hypothesis H_0 is rejected, although it is true, it is known as type I error 'or' error of first kind.
2. When a null hypothesis H_0 is accepted, although it is false. It is the type II error or error of second kind. The probabilities of committing the type I and type II errors are called sizes of errors and is denoted by α and β respectively. Thus, $\alpha = P$ (reject H_0 /when H_0 is true) $= P$ (reject H_0 /H_0) $=P$ ($W\in W_c/H_0$) and $\beta = P$ (accept H_0 /when H_0 is false) $= P$ (accept H_0/H_1) $= P(W\in W_a/H_0)$

Level of Significance

The power function level below which we reject the hypothesis is called the level of significance. The level of significance in testing of hypothesis are 5% and 1%, where 5%indicates that there are about 5 chance in 100 trials and 1% indicated that there is about one chance in 100 trials.

Power Function of a Test

The power function $P(\theta)$ of a test of hypothesis H_0 against an alternative of hypothesis H_1 is the function of the parameter where the probability that the sample point will fall in the critical region of the test where θ is the real parameter.

$$\begin{aligned} P(\theta) &= P \text{ (rejection } H_0/H_1 \text{ is true)} \\ &= P\,(W\in W_a/H_1) \\ &= 1- P\,(W\in W_c/H_1) \end{aligned}$$

$= 1 - P\,(\text{accept } H_0/H_1)$

$= 1 - \beta\,(\theta)$

Power of a Test

The value of the power function at a definite value of the parameter is known as the power of the test. Therefore, if we are to test null hypothesis.

$H_0 : \theta = \theta_0$ against alternative hypothesis

$H_1 : \theta = \theta_1$, then the power of the test is

$$1\text{-}\beta = P\,(W \in W_c/H_1)$$

$$P\,(\text{rejection } H_0/H_1 \text{ is true})$$

Best Critical Region

In testing the null hypothesis $H_0 : \theta = \theta_0$ against the alternative hypothesis $H_1 : \theta = \theta_1$ the critical region is best if the type II error is minimum or $(1\text{-}\beta)$ is maximum when compared to every other possible critical region of size α.

Thus, a test defined by the best critical region is called most powerful test.

Procedure for Testing a Hypothesis

the procedure for testing a hypothesis are

1. To set up the null hypothesis H_0 to be tested along with H_1, the alternative hypothesis.
2. Specification of the test statistic and its sampling distribution.
3. On the basis of distribution, to decide critical region and acceptance region.
4. To take a random sample and calculate the test statistic.

Illustrative Examples

Example 1 *Given the frequency function* $f(x,\theta) = \begin{cases} \dfrac{1}{\theta}; & 0 \le x \le \theta \\ 0; & \text{else where} \end{cases}$

If you are testing the hypothesis $H_0 : \theta = 1$ against $H_1 : \theta = 2$ by means of single observed value of x, what would be the size of the type I and type II *error, if you choose the interval (a)* $0.5 \le x$, *(b)* $1 \le x \le 1.5$ *as the critical regions?*

Solution (a) α = Probability that H_0 falls in the critical region when H_0 is true.

$$\alpha = P\{x \ge 0.5 \; when \; \theta = 1\}$$

$$= \int_{0.5}^{1} \frac{1}{\theta} dx$$

$$= \int_{0.5}^{1} dx = [x]_{0.5}^{1} = 1 - 0.5 = 0.5$$

= size of type I error

Also β = probability that the observed value falls in the region of acceptance when H_1 is true.

$$= P\{x \le 0.5 \; when \; \theta = 2\}$$

$$= \int_{0}^{0.5} \frac{1}{\theta} dx = \int_{0}^{0.5} \frac{1}{2} dx$$

$$= \frac{1}{2}[x]_{0}^{0.5} = \frac{0.5}{2} = 0.25$$

Type II error = 0.25

(b) $$\alpha = P\{1 \le x \le 1.5 \; when \; \theta = 1\}$$

$$= \int_{1}^{1.5} \frac{1}{\theta} dx = \int_{1}^{1.5} 0 \; dx = 0$$

Since, $f(x,\theta) = 0$, above $x = 1$

Also, β = P {x lies outside 1 and 1.5 when θ=2}

$$= 1 - \int_{1}^{1.5} \frac{1}{\theta} dx$$

$$= 1 - \frac{1}{2} \int_{1}^{1.5} dx$$

$$= 1 - \frac{1}{2}[x]_{1}^{1.5}$$

$$= 1 - \frac{1}{2}[1.5 - 1] = 1 - \frac{1}{2}(0.5) = \frac{3}{4}$$

Example 2 *A random variable follows the distribution.*

$$f(x,\theta) = \frac{1}{\theta} e^{-x/\theta}, 0 \le x \le \infty, \theta > 0.$$

The null hypothesis $H_0 : \theta = 5$ is rejected and alternative hypothesis $H_1 : \theta = 10$ is accepted. If an observation selected at random takes the value 15 or more than. Find the critical region, acceptance region and size of two types of errors.

Solution The frequency function is given by

$$f(x,\theta) = \frac{1}{\theta} e^{-x/\theta}, 0 \le x \le \infty, \theta > 0.$$

Null hypothesis $H_0 : \theta = 5$

Alternative hypothesis $H_1 : \theta = 10$

Critical region $W_C : x \ge 15$

Acceptance region $W_a : x < 15$

Type I error $\alpha = P$ {reject H_0 when H_0 is true}

$= P$ {x is in critical region when H_0 is true}

$$= P\left\{x \in {W_c}/{H_0}\right\}$$

$$= P\{x \ge 15 \text{ when } \theta = 5\}$$

$$= \int_{15}^{\infty} \frac{1}{\theta} e^{-x/\theta} dx$$

$$= \int_{15}^{\infty} \frac{1}{5} e^{-x/5} dx$$

$$= \frac{1}{5}[-5e^{-x/5}]_{15}^{\infty}$$

$$= -[e^{-\infty} - e^{-3}] \qquad [\because e^{-\infty} = 0]$$

$$= e^{-3}$$

Type II error $\beta = P$ {accept H_0 when H_1 is true}

$= P$ {x is in acceptance region when H_1 is true}

$$= P\left\{x \in W_a / H_1\right\}$$

$$= P\{x < 15 \text{ when } \theta = 10\}$$

$$= \int_0^{15} \frac{1}{\theta} e^{-x/\theta} dx$$

$$= \int_0^{15} \frac{1}{10} e^{-x/10} dx$$

$$= \frac{1}{10}[-10e^{-x/10}]_0^{15}$$

$$= -[e^{-15/10} - e^0]$$

$$= -[e^{-3/2} - 1]$$

$$= 1 - e^{-3/2}$$

Example 3 *Given a frequency distribution*

$$f(x,\theta) = \begin{cases} (1+\theta)x^\theta; & 0 \le x \le 1, \theta > 0 \\ 0; & \text{else where} \end{cases}$$

Let the hypothesis H_0 : $\theta = 2$ be tested by taking a single observation on x and using the interval 0.1 as a critical region. Calculate

(i) *The size of type I error*

(ii) *The probability of judgement that H_0 is false if the true value of θ is 3.*

(iii) *Power function.*

Solution Given that

Null hypothesis $H_0 : \theta = 2$

Critical region $W_c : x < 0.1$

Sample space $W: 0 \le x \le 1$

Acceptance region $W_a: x \ge 0.1$

(i) $\alpha = P$ {reject H_0 when H_0 is true}

$= P\{x \in W_c | H_0\}$

$= P\{x < 0.1 \text{ when } \theta = 2\}$

$$= \int_0^{0.1} (1+\theta)\, x^\theta\, dx$$

$$= \int_0^{0.1} 3x^2\, dx$$

$$= 3\left[\frac{x^3}{3}\right]_0^{0.1}$$

$= (0.1)^3 = 0.001$

(ii) P (judgement that H_0 is false when $\theta = 3$)

$= P\{\text{reject } H_0 \text{ when } \theta = 3\}$

$= P\{x \in W_c \text{ when } \theta = 3\}$

$= P\{x < 0.1 \text{ when } \theta = 3\}$

$$= \int_0^{0.1} (1+\theta)\, x^\theta\, dx$$

$$= 4\int_0^{0.1} x^3 dx$$

$$= \left[x^4\right]_0^{0.1} = (0.1)^4 = 0.0001$$

(iii) Power function = P {reject H_0 when alternative hypothesis H_1 is given}

$= P\{x \in W_c / \theta\}$

$= P\{x < 0.1/\theta\}$

$$= \int_0^{0.1} (1+\theta) x^\theta d\theta$$

$$= \left[\frac{(1+\theta)x^{\theta+1}}{(\theta+1)} \right]_0^{0.1}$$

$$= (0.1)^{\theta+1}$$

Example 4 *Given a binomial distribution*

$$f(x,p) = \begin{cases} {}^nC_x p^x q^{n-x}; & x = 0,1,2,3,4 \\ 0 \quad ; & elsewhere \end{cases}$$

It is desired to test $H_0 : P = P_0 = \frac{1}{3}$ *against* $H_1 : P = P_1 = \frac{1}{2}$ *by agreeing to accept* H_0, *if* $x \leq 2$ *in four trials and to reject otherwise.*

What are the probabilities of committing

(a) *type I error,* (b) *type II error*

Solution (a) The probability of type of I error

$$\alpha = P \,\{\text{reject } H_0 \text{ when } H_0 \text{ is true}\}$$

$$= P\{x \geq 2 \text{ when } P = P_0 = \frac{1}{3}\}$$

$$= \sum_{x=3}^{4} {}^4C_x \left(\frac{1}{3}\right)^x \left(\frac{2}{3}\right)^{4-x} \qquad [\because q = 1-p]$$

$$= {}^4C_3 \left(\frac{1}{3}\right)^3 \left(\frac{2}{3}\right)^{4-3} + {}^4C_4 \left(\frac{1}{3}\right)^4 \left(\frac{2}{3}\right)^{4-4}$$

$$= 4 \times \frac{1}{27} \times \frac{2}{3} + 1 \times \frac{1}{81}$$

$$= \frac{8}{81} + \frac{1}{81} = \frac{9}{81} = \frac{1}{9}$$

Hence, the hypothesis $H_0 : P = \frac{1}{3}$ is being tested at the level of significance $\alpha = \frac{1}{9}$.

(b) The probability of type II error

$$\beta = P \,\{\text{accept } H_0 \text{ when } H_1 \text{ is true}\}$$

$$= P \,\{ x \leq 2 \text{ when } P = P_0 = \frac{1}{2} \}$$

$$= \sum_{x=0}^{2} {}^4C_x \left(\frac{1}{2}\right)^x \left(\frac{1}{2}\right)^{4-x}$$

$$= {}^4C_0 \left(\frac{1}{2}\right)^0 \left(\frac{1}{2}\right)^{4-0} + {}^4C_1 \left(\frac{1}{2}\right)^1 \left(\frac{1}{2}\right)^{4-1} + {}^4C_2 \left(\frac{1}{2}\right)^2 \left(\frac{1}{2}\right)^{4-2}$$

$$= {}^4C_0 \left(\frac{1}{2}\right)^4 + {}^4C_1 \left(\frac{1}{2}\right)^4 + {}^4C_2 \left(\frac{1}{2}\right)^4$$

$$= \left(\frac{1}{2}\right)^4 \left[{}^4C_0 + {}^4C_1 + {}^4C_2\right]$$

$$= \frac{1}{2^4}[1+4+6] = \frac{11}{2^4} = \frac{11}{16}$$

Example 5 *Suppose P be the probability that a coin will fall head in a single toss in order to test H_0: $P = P_0 = \frac{1}{2}$ against $H_1 : P = P_1 = \frac{3}{4}$ The coin is tossed 5 times and H_0 is rejected if more than 3 heads are obtained. Find the probability of type I error and power function.*

Solution Let number of heads in 5 tosses of a coin = 0,1,2,3,4,5.

Let P is the probability that a coin will fall head in a single toss and q be the probability that a coin will fall tail.

$$\therefore \quad P(x=5) = {}^5C_x P^x q^{5-x} \quad , x = 0,1,2,3,4,5.$$

Now,

Sample space $W : x = 0,1,2,3,4,5.$

Critical region $W_c : x = 4,5$ *i.e.*, $x \geq 4$

Acceptance region $W_a : x = 0,1,2,3$ *i.e.*, $x \leq 3$

$\therefore$ Type I error

$$\alpha = P\,\{\text{reject } H_0 \text{ when } H_0 \text{ is true}\}$$

$$= P\{x \in W_c \text{ when } H_0 \text{ is true}\}$$

$$= P\,\{\, x \geq 4 \text{ when } P = P_0 = \frac{1}{2}\,\}$$

$$= \sum_{x=4}^{5} {}^5C_x \left(\frac{1}{2}\right)^x \left(\frac{1}{2}\right)^{5-x}$$

$$= {}^5C_4\left(\frac{1}{2}\right)^4\left(\frac{1}{2}\right)^{5-4} + {}^5C_5\left(\frac{1}{2}\right)^5\left(\frac{1}{2}\right)^{5-5}$$

$$= {}^5C_4\left(\frac{1}{2}\right)^5 + {}^5C_5\left(\frac{1}{2}\right)^5$$

$$= \left(\frac{1}{2}\right)^5\left[{}^5C_4 + {}^5C_5\right]$$

$$= \frac{1}{2^5}[5+1] = \frac{6}{32} = \frac{3}{16}$$

Power function

$$P(\theta) = 1 - \beta$$

$= P$ {reject H_0 when H_1 is true}

$= P\{x \in W_c$ when H_1 is true$\}$

$= P\ \{\ x \geq 4$ when $P = \frac{3}{4}\ \}$

$$= \sum_{x=4}^{5} {}^5C_x\left(\frac{3}{4}\right)^x\left(\frac{1}{4}\right)^{5-x}$$

$$= {}^5C_4\left(\frac{3}{4}\right)^4\left(\frac{1}{4}\right) + {}^5C_5\left(\frac{3}{4}\right)^5$$

$$= 5\times\frac{81}{1024} + \frac{243}{1024}$$

$$= \frac{648}{1024} = \frac{81}{128}$$

Example 6 *It is desired to test a hypothesis* $H_0 : P = P_0 = \frac{1}{2}$ *against* $H_1 : P = P_1 = \frac{3}{4}$ *on the basis of tossing a coin once, where P is the probability of getting a head in a single trial and agreeing to accept H_0, if a tail appears and to accept H_1 otherwise. Find the values of α and β.*

Solution The probability of type I error

$\alpha = P$ {reject H_0 when H_0 is true}

$= P\{$head appears$/P = P_0\}$

$= [P]$ when $P = P_0 = \frac{1}{2} = \frac{1}{2}$

The probability of type II error

$\beta = P\{\text{reject } H_0 \text{ when } H_1 \text{ is true}\}$

$= P\{\text{tail appears}/P = P_1\}$

$= [1\text{-}P] \text{ when } P = P_0 = \dfrac{3}{4}$

$= 1 - \dfrac{3}{4}$

$= \dfrac{1}{4}$

Example 7 *An urn contains either 3 red and 4 black of 4 red and 3 black balls. Three balls are drawn at random from the urn and on the basis of their colours, the inner object of the urn is decided. The null hypothesis H_0 is that the urn contains 3 red and 4 black balls. If H_0 is accepted in all the case other than the drawing of all the red balls from the urn (i.e. H_0 is rejected in the case if all the 3 red balls appears form the urn), then find the values of α and β.*

Solution Given, Null hypothesis H_0 : 3 red and 4 black balls.

Alternative hypothesis H_1 : 4 red and 3 black balls.

Let x be the number of red balls amongst the 3 drawn balls.

$\therefore$ Sample space $W : x = 0,1,2,3$

Critical region $W_c : x = 3$

Acceptance region $W_a : x = 0,1,2$

Now, $\alpha = P\{\text{reject } H_0 \text{ when } H_0 \text{ is true}\}$

$= P\{x \in W_c \mid H_0 \text{ is true}\}$

$= P\{x = 3 \text{ when 3 red and 4 black balls contains}\}$

$= \dfrac{^3C_3}{^7C_3} = \dfrac{1}{35}$

$\beta = P\{\text{accept } H_0 \text{ when } H_1 \text{ is true}\}$

$= P\{x = 0,1,2 \text{ when 4 red and 3 black balls contains}\}$

$= 1\text{-}P\{x = 3 \text{ when 4 red and 3 black balls contains}\}$

$= 1\text{-}\dfrac{^4C_3}{^7C_3}$

$= 1 - \dfrac{4}{35} = \dfrac{31}{35}$

Example 8 *An urn contains 4 balls of which 1 or 2 are white and the rest are red. To test the hypothesis H_0: one ball in white, balls are drawn one after another until a white ball appears. Suggest a good critical region for this test and find the values of α and β.*

Solution. Let x be the number of red balls drawn until a white ball appears.

Null hypothesis H_0: one ball is white.

Alternative hypothesis H_1 : two balls are white.

The first possible test may be accept H_0, if $x = 3$ and reject otherwise.

Now, $\alpha = P$ {reject H_0 when H_0 is true}

$= P\{x \neq 3$ when one ball is white$\}$

$= P\{ x = 0$ or 1 or 2/one ball is white$\}$

$$= \frac{1}{4} + \frac{3}{4}.\frac{1}{3} + \frac{3}{4}.\frac{2}{3}.\frac{1}{2} = \frac{3}{4}$$

$\beta = P$ {accept H_0 when H_1 is true}

$= P\{ x = 3$ when 2 balls are white$\}$

$= 0$

The second possible test may be to accept H_0 if $x = 2$ or 3 and reject otherwise.

$\alpha = P$ {reject H_0 when H_0 is true}

$= P\{ x = 0$ or 1 when one ball is white$\}$

$$= \frac{1}{4} + \frac{3}{4}.\frac{1}{3} = \frac{1}{2}$$

$\beta = P$ {accept H_0 when H_1 is true}

$= P\{ x = 2$ or 3 when two balls are white$\}$

$$= \frac{2}{4} \times \frac{1}{3}.\frac{2}{2} + 0 = \frac{1}{6}$$

Practice Problems

1. If $x \geq 1$ is the critical region for testing the null hypothesis $H_0: \theta = \theta_0 = 2$ against the alternative hypothesis $H_1 : \theta = \theta_1 = 1$, on the basis of a single observation from the population.

$$f(x,\theta)=\theta e^{-\theta x}, 0 \le x \le \infty.$$

Obtain the sizes of type I and type II errors. Also, find the power function of the test.

[Ans. e^{-2}, $1-e^{-1}$, e^{-1}]

2. Let x be a random variable having a binomial distribution.

$$f(x,P)={}^nC_x P^x q^{n-x}, \sum x_1 \le 1 \text{ or } \sum x_1 \ge 4$$

with probability of success P on the basis of a random sample size n it is proposed to reject the null hypothesis $H_0 : P = \frac{1}{2}$

(a) Find the level of significance of the test for $n = 5$.

(b) Find the power of the test at $P = \frac{2}{3}$

[Ans. (a) $\frac{3}{8}$, (b) $\frac{4}{8}$]

3. Given the frequency distribution

$$f(x,\theta)=\begin{cases}\frac{1}{\theta} & ; \quad 0 \le x \le \theta \\ 0 & ; \quad elsewhere\end{cases}$$

And that you are testing the null hypothesis $H_0 : \theta = \theta_0 = 3$ against $H_1 : \theta = \theta_1 = 2$ by means of a single observed value of x, what would be the sizes of the type I and type II errors, if

(i) critical region is $x < 1$.

(ii) critical region is the interval $1 < x < 2$.

[Ans. (a) $\alpha = \frac{1}{3}$, $\beta = \frac{1}{2}$, (b) $\alpha = \frac{1}{3}$. $\beta = \frac{1}{2}$]

4. A binomial distribution $f(x,P)=\begin{cases}{}^nC_x p^x q^{n-x}; & x=0,1,2,3 \\ 0; & elsewhere\end{cases}$

If is desired to test $H_0 : P = P_0 = \frac{1}{4}$ against $H_1 : P = P_1 = \frac{1}{2}$ by agreeing to accept H_0, if $x \le 1$ in 3 trials and to accept H_1 otherwise. What is the probability of committing (i) type I error (ii) type II error?

[Ans. (i) 5/32 (ii) 1/2]

5. A random variable has the following distribution.

$$f(x,\theta)=\frac{1}{\theta}e^{-x/\theta},0\le x<\infty,\theta>0.$$

If randomly selected observations assumes the value 20 or more than 20, then accept the alternative hypothesis $H_1 : \theta = 15$ and reject null hypothesis $H_0 : \theta = 5$. Find the critical interval and the size of both types of errors.

[**Ans.** $W_c = 20 \le x < \infty$ and $W_a = 0 \le x < 20$, $\alpha = e^{-4}$, $\beta = 1 - e^{-\frac{4}{3}}$]

6. Let r be the number of successes obtained in 5 minutes during a poisson distribution. It is desired to test null hypothesis $H_0 : m = 2$ against alternative hypothesis $H_1 : m = 3$, where m is the mean for the 5 minutes and $r \ge 4$ is used for the critical region. Compute the value of α and β.

[**Hint.** Poisson distribution $f(x,m) = \frac{e^{-m}.m^r}{r!}$, $r = 0,1,2\ldots\ldots\ldots,\infty$, $\alpha = 0.1431$, $\beta =$ 0.6474]

7. For testing a null hypothesis $H_0 : \theta = \theta_0 = \frac{1}{2}$ against alternative hypothesis H_1 : $\theta = \theta_1 = \frac{2}{3}$ on the basis of tossing a coin thrice and it is decided to reject H_0., if 3 heads appear and to accept H_1. Here θ, is the probability of a head in a single trial. Find the value of α and β.

[**Ans.** $\alpha = \frac{1}{8}, \beta = \frac{19}{27}$]

Recurred relation

Consider a function to compute on n input suggests splitting the inputs into k distinct subset, $1 \le k \le n$. yielding k sub problems. When it splits the input into two sub problems of the some kind as the original problem. This splitting is typical of many of the problems we examine here. Let P is the problem. Small p is a Boolean – valued function that determines whether the input size is small enough that the answer can be computed without splitting. The problem P is divided into smaller sub problems. These sub problems P_1, P_2, ... P_k are solved by recursive applications. Combine is a function that determines. The solution to P using the solutions to the k sub problems.If the size of p is n and the size of the k sub problems are n_i, n_2, ... n_k, repectively, then the computing time is described by the recurrence relation.

$$T(n) = \begin{cases} \xi(n) & ; \quad n \text{ small} \\ T(n_1) + T(n_2) + \ldots\ldots\ldots + T(n_k) + f(n); & \text{otherwise} \end{cases} \quad \ldots\text{(i)}$$

Where T(n) is the time on any input of size n and ξ (n) is the time to compute the answer directly for small inputs. The function f(n) is the time for dividing P and combining the solutions to sub problem.

The complexity is given by recurrences of the form

$$T(n) = \begin{cases} T(1) & ; \quad n = 1 \\ aT\left(\frac{n}{b}\right) + f(n); & n > 1 \end{cases} \quad \ldots\text{(ii)}$$

Where a and b are known constants. We assume that T(1) is known and n is a power of b *i.e.* $n = b^k$.

One of the method for solving any such recurrence relation is known as substitution method.

Example 1 Consider the case in which a = 2 , b = 2, T(1) = 2 and f(n) = n We have

$$T(n) = \left(\frac{n}{b}\right) + f(n)$$

$$= 2T\left(\frac{n}{b}\right) + n$$

$$= 2\left[2T\left(\frac{n}{4}\right) + \frac{n}{2}\right] + n$$

$$= 4T\left(\frac{n}{4}\right) + 2n = 4\left[2T\left(\frac{n}{8}\right) + \frac{n}{4}\right] + 2n$$

$$= 8T\left(\frac{n}{8}\right) + 3n$$

.

.

.

In general, we observe that $T(n) = 2^i\, T\left(\frac{n}{2i}\right) + in$, for any $\log_2 n \geq i \geq 1$.

In particular, then $T(n) = 2^{\log_2 n} T\left(n/_{2^{\log_2 n}} \right) + n\log_2 n$ corresponding to the choice of $i = \log_2 n$

Thus, $T(n) = nT(1) + n\log_2(n) = n\log_2 n + 2n$

Remark: Beginning with the recurrence relation (2) and using the substituting method, it can be shown that

$$T(n) = n^{\log_b a}\, [T(1) + u(n)]$$

Where $u(n) = \sum_{j=1}^{k} h(b^j)$ and $h(n) = f(n)/_{n^{\log_b a}}$

Let us consider some examples using by the table.

h(n)	u(n)
$0(n^r)$, r < 0	0(1)
~ $((\log n)^i)$, i ≥ 0	~ $((\log n)^{i+1}/(i+1))$
$\Omega\ (n^r)$, r > 0	~ h(n)

Example 2 Compute the recurrence when n is a power of 2.

$$T(n) = \begin{cases} T(1) & ,n=1 \\ T\left(n/_2\right)+C & ,n>1 \end{cases}$$

Solution comparing with equation (2), we observe that

$a = 1,\ b = 2,\ f(n) = c$

$\log_b(a) = 0$ and $h(n) = f(n)/n^{\log_b a}$

$\Rightarrow$ $c = c\,(\log n)^0 = \sim((\log n)^0)$

form the table, we have

$$u(n) = \sim \log n, \text{ so } T(n) = n^{\log_b a}[C + \sim(\log n)] = \sim(\log n)$$

Example 3 Compute the recurrence $T(n) = 7T\ (n/2) + 18\ n^2$, $n \geq 2$ and a power of 2.

Solution Here, $a = 7$, $b=2$, $f(n) = 18\ n^2$

So that $\log_b a = \log_2 7 \approx 2.81$

And hence $h(n) = 18n^2/n^{\log_2 7}$

$= 18n^{2}\text{-}\log_2 7$

$= 0(n^r)$

where $r = 2 - log_2 7 < 0$

Thus $u(n) = 0(1)$

Hence, $T(n) = n^{log_2 7} [T(1) + 0(1)]$

$= \sim n^{log_2 7}$ as $T(1)$ assumed to be constant

Example 4 Compute the recurrence $T(n) = 9T(n/3) + 4n^6$, $x \geq 3$ and a power of 3.

Solution Here, $a = 9$, $b = 3$, $f(n) = 4n^6$

So that $log_b a = log_3 9 = 2$

And hence $h(n) = f(x) / n^{log_b a}$

$= 4n^6 / n^{log_3 9}$

$= 4n^6 / n^2 = 4n^4$

$= \Omega(n^4)$

and $u(n) = \sim (h(n)) = \sim (n^4)$ (form the table)

Hence, $T(n) = n^2 [T(1) + \sim (n^4)]$

$= \sim (n^6)$, as $T(1)$ is assumed to be constant.

Practice Problems

1. State the recurrence relation for the following choice of a, b and f(n). (C being a constant).

   (i) $a = 5, b = 4, f(n) = cn^2$

   (ii) $a = 1, b = 2, f(n) = cn$

   (iii) $a = 28, b = 3, f(n) = cn^3$

2. Solve the recurrence relation by using the substitution method

   (i) $$T(n) = \begin{cases} 1 & , n \leq 4 \\ T\left(\sqrt{n}\right) + C & , n > 4 \end{cases}$$

   (ii) $$T(n) = \begin{cases} 1 & , n \leq 4 \\ 2T(\sqrt{n}) + \log n & , n > 4 \end{cases}$$

Chapter - 14

Markov Analysis

Introduction

Markov process is derived from Russian mathematician Andrei. A Markov in 1905. This type of probabilistic models known as stochastic process, in which the current state of a process depends on all of its previous state.

Markov process is used in examining and predicting the behaviour of consumers in terms of their brand loyalty and their switching patterns to other brands. It is usually constructed in terms of transition probabilities. It is also used to study the stock market price movements.

Stochastic Process

A stochastic process, is a family of random variables $\{x_\beta : \beta \in A\}$, where β is the index parameter assumes values in a certain range A. The range A is often called the index set and the values assumed by the process are called the states. In most practical cases x_β is the observation at time β and A is the time- range. Thus, we may define, A stochastic process is a collection of random variables $\{X(t) : t \in T\}$, where for each t, the index parameter, in the index set T, $X(t)$ state of the process at time t is a random variables.

Markov Process

A stochastic process $\{X(t): t \in T\}$ is called a Markov process if corresponding to every finite collection of integers $t_0 < t_1 < \dots\dots\dots < t_n$ in T, the time scale then the joint probability distribution of $X(t_0), X(t_1) \dots\dots X(t_n)$,is defined in such a way that the conditional probability.

$$P\{X(t_n) = x_n \mid X(t_{n-1}) = x_{n-1}, \dots\dots\dots X(t_0) = x_0\}$$

is identical with the conditional probability

$$P\{X(t_n) = x_n \mid X(t_{n-1}) = x_{n-1}\}$$

In other words, A stochastic process is called a Markov process if the future development of the process depends on its parts only through its present and is otherwise independent of the past history of the process.

Classification of Markov Processes

Depending on the nature of state space and parameter space, Markov Process are classified as under

(i) Discrete (time) Parameter Markov Processes or continuous (time) Parameter Markov Processes

(ii) Discrete state space Markov processes or continuous state space Markov process.

Transition Probability

The movement of probability from one state to another or remaining in the state during a single time period is called a transition probability. Mathematically.

$$P_{x_{n-1},x_n} = P\{X(t_n) = x_n | X(t_{n-1}) = x_{n-1}\}$$

The symbol n is the number of steps. This transition probability of the process is now in x_n state at time t_n, provided that it was in x_{n-1} state at time t_{n-1}. Some times this probability is known as one step transition probability.

Transition Probability Matrix: The square matrix with transition probabilities P_{ij} is called transition probability matrix and it is denoted by P. Thus,

$$P = \begin{pmatrix} P_{11} & P_{12} & \cdots\cdots & P_{1m} \\ P_{21} & P_{22} & \cdots\cdots & P_{2m} \\ \cdots\cdots & \cdots\cdots & \cdots\cdots & \cdots\cdots \\ P_{m1} & P_{m2} & \cdots\cdots & P_{mm} \end{pmatrix}$$

The sum of the elements in each row of the matrix is unity and each element is non-negative *i.e.,*

$$Pij \geq 0, \text{ for each i, } j$$

and $$\sum_{j=1}^{m} P_{ij} = 1, \text{ for each } i$$

n-Step Transition Probabilities

Let $P_{ij}^{(n)}$ be the probability that the process in state E_i at r^{th} trial will be in the state E_j after n trials. Then $P_{ij}^{(n)}$ are called transition probabilities of order n or n – step ransition probabilities matrix and is denoted by

$$P^{(n)} = \begin{pmatrix} P_{11}^{(n)} & P_{12}^{(n)} \ldots\ldots & P_{1m}^{(n)} \\ P_{21}^{(n)} & P_{22}^{(n)} \ldots\ldots & P_{2m}^{(n)} \\ \ldots\ldots\ldots & \ldots\ldots\ldots\ldots & \ldots\ldots\ldots \\ P_{m1}^{(n)} & P_{m2}^{(n)} \ldots\ldots & P_{mm}^{(n)} \end{pmatrix}$$

where $\quad P_{ij}^{(n)} \geq 0$ and $= \sum_{j=1}^{m} P_{ij} = 1$, for each i

Diagrammatic Representation of Transition Probabilities

The transition probabilities can be represented by two types of diagrams:

1. Transition diagram
2. Probability tree diagram

1. *Transition Diagram*

Such diagrams show the transition probabilities or shifts that can occur in any situation. Such a diagram is shown in Fig. I along with matrix of transition probabilities. The arrows from each state indicate the possible states to which a system can move from the given state. The matrix of transition probabilities which corresponds to above diagram is below.

$$P = \begin{matrix} & \begin{matrix} 1 & 2 & 3 \end{matrix} \\ \begin{matrix} 1 \\ 2 \\ 3 \end{matrix} & \begin{pmatrix} 1/2 & 1/2 & 0 \\ 0 & 1/2 & 1/2 \\ 1/2 & 1/4 & 1/4 \end{pmatrix} \end{matrix}$$

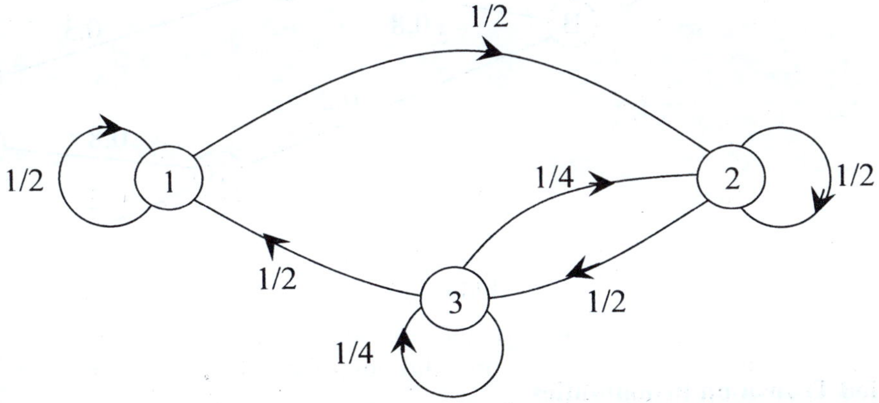

Fig.1

A zero element in the transition matrix indicates that the transition is not possible.

2. *Probability Tree Diagram*

This diagram emphasizes the probabilities and their movement from one step to another step, along with all possible branches or paths that may connected the outcomes over a period of time.

Example: Let probabilities of shifting from *A*to *A* itself and *B* as well as from state *B* to *A* and *B* itself are represented as elements of a transition matrix as follows

$$P = \begin{matrix} & A & B \\ A & 0.7 & 0.3 \\ B & 0.8 & 0.2 \end{matrix}$$

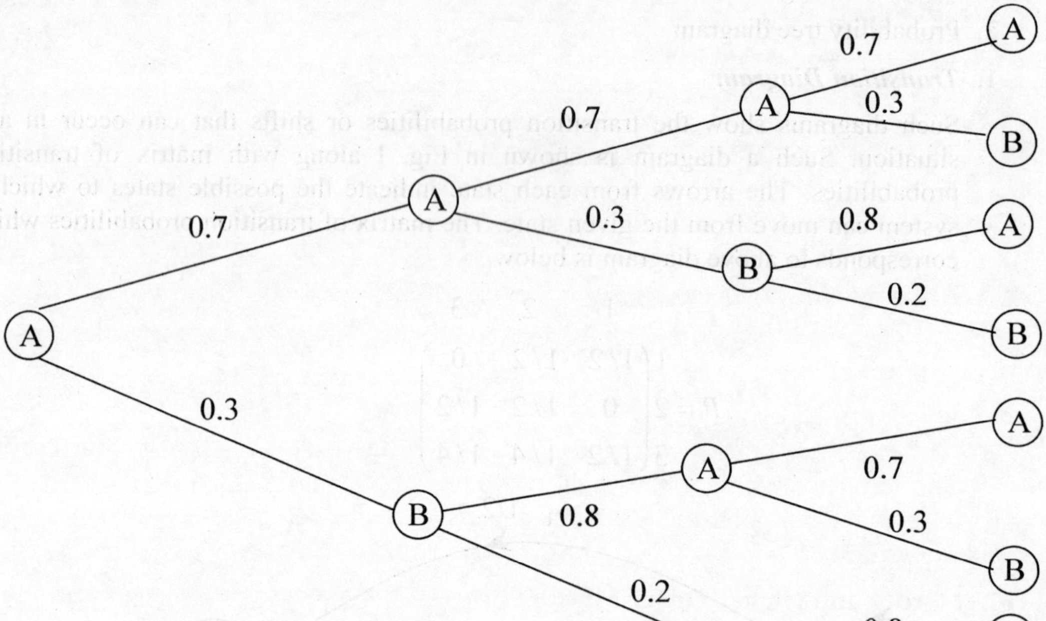

Fig 2

Multi-Period Transition Probabilities

A vector of state probability is a row vector $V_0 = [P_{11} \;\; P_{12} \;\ldots\ldots\ldots P_{1m}]$ in which the entries are non-negative in the interval [0,1] and add upto 1.

Using Probability vector we can determine the state probabilities at a future time.

If V_0 be the state probabilities at time $n = 0$ and P is the transition matrix then V_1 vector of state probabilities after 1 step (at time n=1) can be written in terms of row matrix as

$$V_1 = V_0 \times P$$

Similarly, V_2 vector of state probability after 2 step (at time n=2) can be obtained by

$$\begin{aligned} V_2 &= V_1 \times P \\ &= (V_0 \times P) \times P \\ &= V_0 \times P^2 \end{aligned}$$

After n step, V_n vector of state probabilities can be obtained by

$$\begin{aligned} V_n &= V_{n-1} \times P \\ &= V_0 \times P^n \end{aligned}$$

Step of Constructing Matrix of Transition Probabilities

Step 1: Determine retention probabilities: We divide the numbers of customers retained for the period under review by the number customers at the beginning of the period.

Step 2: Determine gains and loss probabilities

(i) For those customers who switch brands, show gains and losses among the brands for completing the matrix of transition.

(ii) To convert the customers switching of brands so that all gains and losses take the form of transition probabilities, divide the number of customers that each entity has gained or lost by the original number of customers it served.

Step 3: Develop matrix of transition probabilities: In a matrix of transition probabilities, retentions are shown in as values on the main diagonal. The row in the matrix show the retention and loss of customers while the columns show the retention and gain of customers.

First Order and Higher Order Markov Process

The first order Markov Process is based on the following three assumptions

1. The set of possible outcomes in finite.
2. The probability of the next outcome depends only on the immediately preceding outcome.
3. The transition probabilities are constant overtime.

The second order Markov Process assumes that the probability of the next outcome may be depend on the two previous outcomes. Similarly, a third order Markov Process assumes that the probability of the next outcome can be calculated by obtaining and taking account of the outcomes of the past three outcomes.

Example 1 *Two manufacturers A and B are competing with each other in a restricted market. Over the year, A's customers have exhibited a high degree of loyalty as measured by the fact that customers using A's product 80% of time. Also former customers purchasing the product from B have switched back to A's 60% of time. (a) Construct and interpret the state transition matrix in terms of (i) retention and loss (ii) retention and gain.*

(b) Calculate the probability of a customer purchasing A's product at the end of the second period.

Solution (a) The transition probabilities can be arranged in matrix form as shown below. Here, the probability of a customer's purchase at the next step n =1 (next purchase) depends upon the product which a customer is having now, *i.e.* at step n= 0 (present purchase). Each probability in the following matrix must, therefore, be a conditional probability for going from one state to another. Next purchase (n=1)

$$\begin{array}{c} \text{Present} \\ P = \text{Purchase} \\ (n=0) \end{array} \begin{array}{c} \\ A \\ \\ B \end{array} \overset{\quad A \qquad\quad B}{\begin{bmatrix} 0.80 & 0.20 \\ & \\ 0.60 & 0.40 \end{bmatrix}} \quad \text{Retention and Gain} \downarrow$$

— Retention and Loss →

Mathematically conditional probabilities in the matrix can be stated as

(i) $P\,(A_0 / A_1) = P_{11} = 0.80$, probability that customers now using *A*'s product at n=0 will again purchase *A*'s product at n=1 is 0.80. This means retention to *A*'s product.

(ii) $P\,(B_0 / A_1) = P_{21} = 0.60$, probability that customers now using *B*'s product at n=0 will purchase *A*'s product at n=1 is 0.60. This implies loss to *B*'s product.

(iii) $P\,(A_0 / B_1) = P_{12} = 0.20$, probability that customers now using *A*'s product at n=0 will purchase *B*'s product at n=1 is 0.20. This implies loss to *A*'s product.

(iv) $P\,(B_0 / B_1) = P_{22} = 0.40$, probability that customers now using *B*'s product at n=0 will again purchase *B*'s product at n=1 is 0.40. This means retention to *B*'s product.

(b) The transition probability can be represented by two types of diagrams.

(i) Transition diagram→

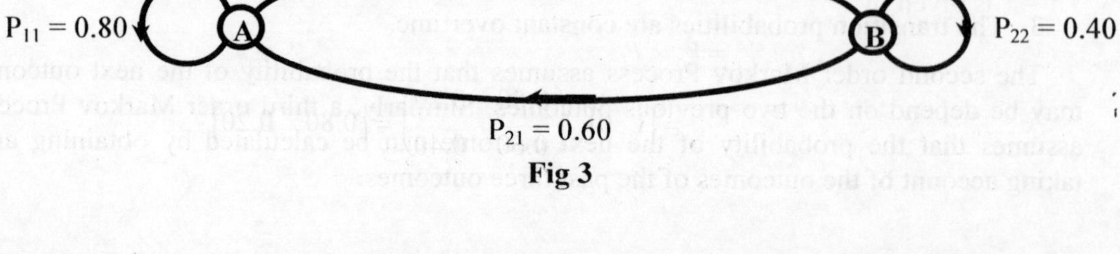

Fig 3

Nodes indicate states and arrows represent the transition probabilities between states.

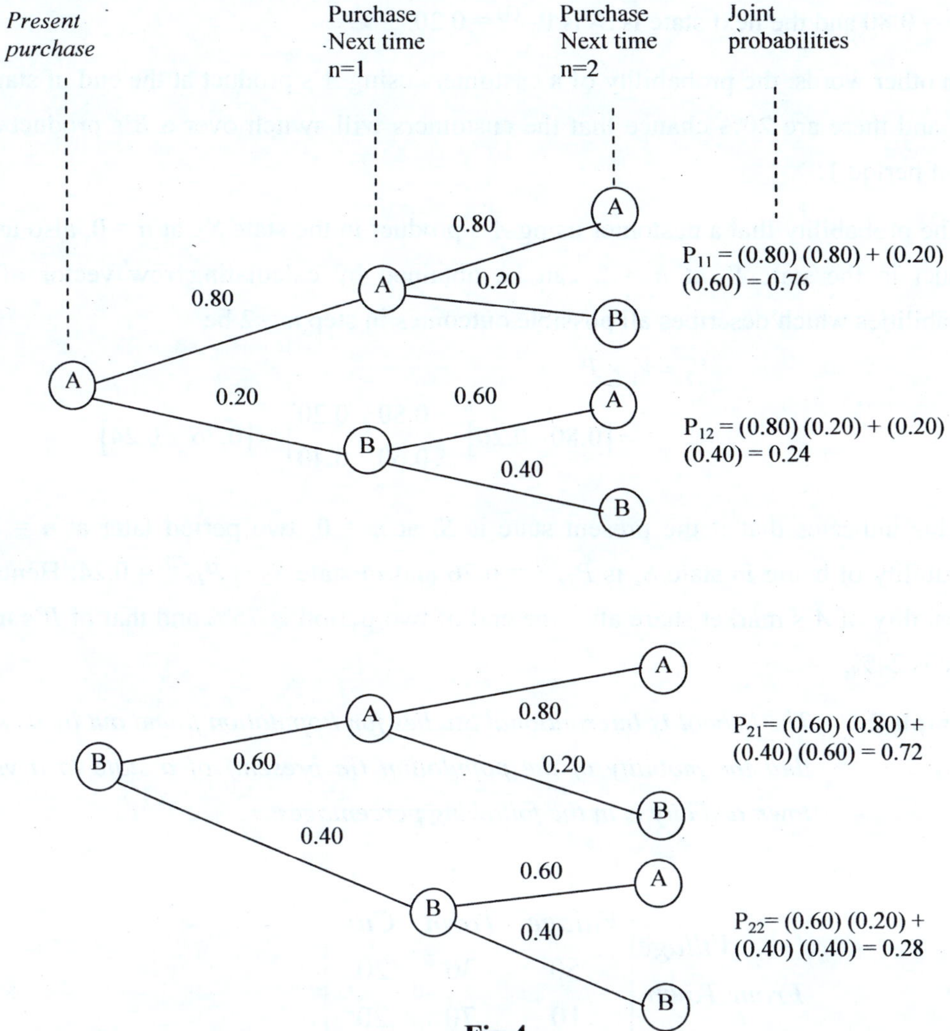

Fig 4

Probabilities computations: If we start with a customer's purchase of *A*'s product in the state at $n = 0$, then $P_{11} = 1$, $P_{12} = 0$, So, $V_o = [1,0]$

After the first transition, the row vector V_1 which describes all possible outcomes at $n = 1$ is given by

$$V_1 = V_0 \times P$$

$$= [1 \quad 0] \begin{pmatrix} 0.80 & 0.20 \\ 0.60 & 0.40 \end{pmatrix} = [0.80 \quad 0.20]$$

This implies that if the present state S_1, the probability that the next state is also S_1 is $P_{11}^{(1)} = 0.80$ and the next state is S_2 is $P_{12}^{(1)} = 0.20$.

In other words, the probability of a customers using *A*'s product at the end of state 1 is 80% and there are 20% chance that the customers will switch over a *B*'s product at the end of period 1.

The probability that a customer using *A*'s product in the state S_2, at $n = 0$, also use *A*'s product in the state S_2 at $n = 2$ can be obtained by calculating row vector of state probabilities which describes all possible outcomes in step $n = 2$ be

$$V_2 = V_1 \times P$$

$$= [0.80 \quad 0.20] \begin{pmatrix} 0.80 & 0.20 \\ 0.60 & 0.40 \end{pmatrix} = [0.76 \quad 0.24]$$

This indicates that if the present state is S_1 at $n = 0$, two period later at $n = 2$, the probability of being in state S_1 is $P_{11}^{(2)} = 0.76$ and in state S_2 is $P_{12}^{(2)} = 0.24$. Hence, the probability of *A*'s market share after the end of two period is 76% and that of *B*'s market share is 24%.

Example 2 *The School is International Studies for Population found out by its survey that the mobility of the population (in present) of a state to a village, town and city is in the following percentages.*

		To		
		Village	*Town*	*City*
	Village	50	30	20
From	*Town*	10	70	20
	City	10	40	50

What will be the proportion of population in village, town and city after two years, given that the present population has proportions of 0.7, 0.2 and 0.1 in the village, town and city respectively?

Solution Calculate for first year → The expected proportion of population in a village, town and city after the first year is obtained by multiplying the transition matrix by the matrix of the present population proportion of the village, town and city:

$$[0.7 \quad 0.2 \quad 0.1]\begin{pmatrix} 0.50 & 0.30 & 0.20 \\ 0.10 & 0.70 & 0.20 \\ 0.10 & 0.40 & 0.50 \end{pmatrix} = [0.38 \quad 0.39 \quad 0.23]$$

Present proportion Transition matrix Proportion after first year.

Calculation for second year $\rightarrow$ the expected proportion of population for the second year can be obtained in the same manner.

$$[0.38 \quad 0.39 \quad 0.23]\begin{pmatrix} 0.50 & 0.30 & 0.20 \\ 0.10 & 0.70 & 0.20 \\ 0.10 & 0.40 & 0.50 \end{pmatrix} = [0.252 \quad 0.479 \quad 0.269]$$

Proportion after first year Transition matrix Proportion after second year

Example 3 The number of units of an item that are withdrawn from inventory on a day-to-day basis of a Markov chain process in which requirements for tomorrow depend on today's requirements. A one day transition matrix is given below.

Number of units withdraw from inventory.

Tomorrow

$$\begin{array}{cc} & \begin{array}{ccc} 5 & 10 & 12 \end{array} \\ \begin{array}{r} 5 \\ \textit{Today}\,10 \\ 12 \end{array} & \begin{pmatrix} 0.6 & 0.4 & 0.0 \\ 0.3 & 0.3 & 0.4 \\ 0.1 & 0.3 & 0.6 \end{pmatrix} \end{array}$$

(a) Construct a tree diagram showing inventory requirements on two consecutive days.

(b) Develop a two-day transition matrix.

(c) Comment on how a two-day transition matrix might be helpful to a manager who is responsible for inventory management.

Solution

(a)

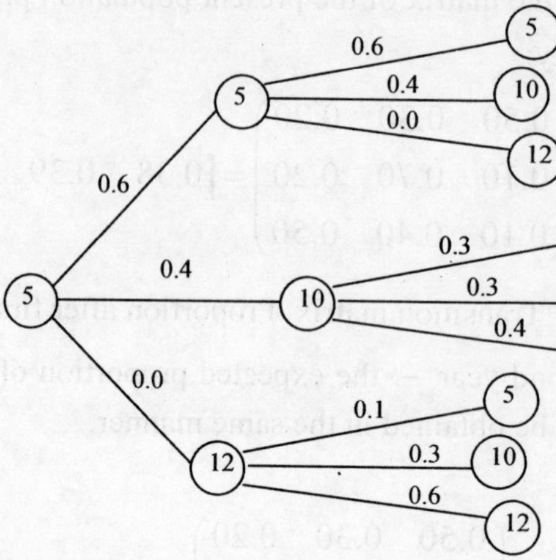

$P_{11} = (0.6)(0.6) + (0.4)(0.3) + (0.0)(0.1) = 0.48$

$P_{12} = (0.6)(0.4) + (0.4)(0.3) + (0.0)(0.3) = 0.36$

$P_{13} = (0.6)(0.0) + (0.4)(0.4) + (0.0)(0.6) = 0.16$

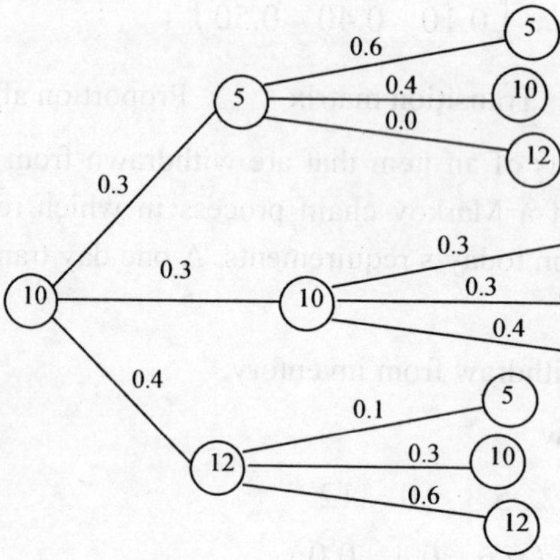

$P_{21} = (0.3)(0.6) + (0.3)(0.3) + (0.4)(0.1) = 0.31$

$P_{22} = (0.3)(0.4) + (0.3)(0.3) + (0.4)(0.3) = 0.33$

$P_{23} = (0.3)(0.0) + (0.3)(0.4) + (0.4)(0.6) = 0.36$

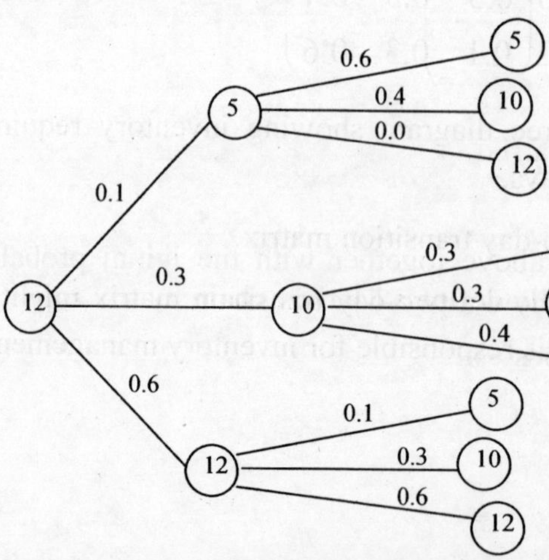

$P_{31} = (0.1)(0.6) + (0.3)(0.3) + (0.6)(0.1) = 0.21$

$P_{32} = (0.1)(0.4) + (0.3)(0.3) + (0.6)(0.3) = 0.31$

$P_{33} = (0.1)(0.0) + (0.3)(0.4) + (0.6)(0.6) = 0.48$

(b) Let the transition matrix be denoted by P. Then a two-day transition matrix is given by.

$$P^2 = \begin{pmatrix} 0.6 & 0.4 & 0.0 \\ 0.3 & 0.3 & 0.4 \\ 0.1 & 0.3 & 0.6 \end{pmatrix} \begin{pmatrix} 0.6 & 0.4 & 0.0 \\ 0.3 & 0.3 & 0.4 \\ 0.1 & 0.3 & 0.6 \end{pmatrix} = \begin{matrix} \\ 5 \\ 10 \\ 12 \end{matrix} \begin{matrix} 5 & 10 & 12 \\ \left(\begin{matrix} 0.48 \\ 0.31 \\ 0.21 \end{matrix}\right. & \begin{matrix} 0.36 \\ 0.33 \\ 0.31 \end{matrix} & \left.\begin{matrix} 0.16 \\ 0.36 \\ 0.48 \end{matrix}\right) \end{matrix}$$

(c) Suppose that each morning a manager must place an order for inventory replenishment. As a result of delivery time requirements, an order placed today arrives two day later.

The two-day transition matrix can be used for guiding ordering decision.

For example, if today the manager experiences a demand for five units, then 2 days later, the probability of requiring five units is 0.48, that of requiring ten units 0.36 and twelve units 0.16.

Markov Chain

Let $P_j^{(0)}$ (j =0,1,2,.............) be the absolute probability such that the system is in the state E_j at time t_0, where E_j denote the exhaustive and mutually exclusive outcomes of a system at any time. We now define

$$P_{ij} = P\{X(t_n) = j | X(t_{n-1}) = i\}$$

As the one-step transition probability of going from state i at time t_{n-1} to state j at time t_n . The probabilities from state E_i to state E_j are expressed in the matrix from as under

$$P = \begin{pmatrix} P_{00} & P_{01} & P_{02} \ldots\ldots \\ P_{10} & P_{11} & P_{12} \ldots\ldots \\ P_{20} & P_{21} & P_{22} \ldots\ldots \\ \ldots\ldots & & \end{pmatrix} \quad P_{ij} \geq 0 \text{ for all } i, j \text{ and } \sum P_{ij} = 1, \text{ for all } i.$$

This matrix is known as stochastic matrix.

Definition

The transition matrix P is defined above together with the initial probabilities $\{P_j^{(0)}\}$ associated with the states E_j completely define a Markov chain.

The Markov chains are of two types

(i) ergodic

(ii) regular

An ergodic Markov chain is to pass from one state to another in a finite number of steps, regardless of present state.

A regular Markov chain is defined as a chain having a transition matrix P such that for some hour of P it has only non-zero positive probability values.

Example 4 Determine if the following transition matrix is ergodic Markov Chain.

$$\text{Present States}\quad \begin{matrix} & \textit{Future States} \\ & \begin{matrix} 1 & 2 & 3 & 4 \end{matrix} \\ \begin{matrix} 1 \\ 2 \\ 3 \\ 4 \end{matrix} & \begin{bmatrix} 1/3 & 1/3 & 0 & 1/3 \\ 0 & 1/2 & 1/4 & 1/4 \\ 1/4 & 0 & 1/2 & 1/4 \\ 0 & 0 & 1/3 & 2/3 \end{bmatrix} \end{matrix}$$

Solution We observe that from state 1, it is possible to go directly to every other state except state 3. For state 3, it is possible to go from state 1 to state 2 to state 3. Therefore it is possible to go form state 1 to any other state. Similarly, from state 2, it is possible to go to state 3 or state 4, then from state 3 to state 1, or from state 4 to state 3 to state 1. Also, from state 3 it is possible to go directly to state 1. Finally, from state 4, it is possible to go to state 3, then from state 3 to state 1.

Hence, above transition matrix is an ergodic Markov chain.

Example 5 *Test the following transition matrix to see if the Markov chain is regular and ergodic where a is some positive P_{ij} value.*

$$P = \begin{matrix} & \begin{matrix} 1 & 2 & 3 & 4 \end{matrix} \\ \begin{matrix} 1 \\ 2 \\ 3 \\ 4 \end{matrix} & \begin{bmatrix} 0 & a & a & 0 \\ a & 0 & 0 & a \\ a & 0 & 0 & a \\ 0 & a & a & 0 \end{bmatrix} \end{matrix}$$

Solution We compute

$$P^2 = \begin{matrix} & \begin{matrix} 1 & 2 & 3 & 4 \end{matrix} \\ \begin{matrix} 1 \\ 2 \\ 3 \\ 4 \end{matrix} & \begin{bmatrix} a & 0 & 0 & a \\ 0 & a & a & 0 \\ 0 & a & a & 0 \\ a & 0 & 0 & a \end{bmatrix} \end{matrix} \quad P^4 = \begin{matrix} & \begin{matrix} 1 & 2 & 3 & 4 \end{matrix} \\ \begin{matrix} 1 \\ 2 \\ 3 \\ 4 \end{matrix} & \begin{bmatrix} a & 0 & 0 & a \\ 0 & a & a & 0 \\ 0 & a & a & 0 \\ a & 0 & 0 & a \end{bmatrix} \end{matrix} \quad P^8 = \begin{matrix} & \begin{matrix} 1 & 2 & 3 & 4 \end{matrix} \\ \begin{matrix} 1 \\ 2 \\ 3 \\ 4 \end{matrix} & \begin{bmatrix} a & 0 & 0 & a \\ 0 & a & a & 0 \\ 0 & a & a & 0 \\ a & 0 & 0 & a \end{bmatrix} \end{matrix}$$

Here, we observe that P raised to an even-number power gives the result as above. While P raised to an odd-number power will gibe the original matrix. Since, all the elements are not non-zero positive element, the above matrix is not regular. But, it is an ergodic since it is possible to go from state 1 to state 2 or state 3, from state 2 to state 1 or state 4, from state 2 to state 1, from state 3 to state 1, from state 4 to state 2 or state 1.

Steady State (Equilibrium) Condition

Steady state or equilibrium probabilities are the long term average probabilities for being in each state. After some step, state probabilities may become constant and will unchanged. At that point a steady condition reached to the system. Thus, in steady state condition, if P is a transition matrix for a Markov process and V is the probability vector, then we have $V = VP$.

Then, we refer to V as a steady state vector.

Method For Determining Steady State Condition

The following steps are involved in solving steady state condition.

Step 1: Construct a state transition matrix.

Step 2: The purpose of Markov analysis is to predict the future market value. This is given by using following equation.

$$\begin{bmatrix} \text{expected market} \\ \text{shares of period 2} \end{bmatrix} = \begin{bmatrix} \text{market shares} \\ \text{of period 1} \end{bmatrix} \text{[Transition matrix]}$$

$$\begin{bmatrix} \text{expected market} \\ \text{shares of period 3} \end{bmatrix} = \begin{bmatrix} \text{market shares} \\ \text{of period 2} \end{bmatrix} \text{[Transition matrix]}$$

$$\begin{bmatrix} \text{expected market} \\ \text{shares of period } n \end{bmatrix} = \begin{bmatrix} \text{market shares} \\ \text{of period } n-1 \end{bmatrix} \text{[Transition matrix]}$$

Step 3: To find a steady state condition for a Markov process with transition matrix P, we solve the system of simultaneous equation by $x+y+z+\dots\dots\dots = 1$

and	$[x\ y\ z\ \dots] =$	[x y z….]	P
	Period n	Period (n-1)	Transition matrix.
	Share	Share	

where x, y, z ………………are unknown as there are states in the markov system.

Example 6 A manufacturing company has a certain piece of equipment that is inspected at the end of each day and classified as just overhauled, good, fair or inoperative. If the item is inoperative it is overhauled, a procedure that takes one day. Let us denote the four classifications as states 1,2,3, and 4 respectively. Assume that the working condition of the equipment follows a Markov chain with the following transition matrix.

$$P = \text{Today}\begin{array}{c} \\ 1 \\ 2 \\ 3 \\ 4 \end{array}\overset{\textit{Tomorrow}}{\begin{array}{cccc} 1 & 2 & 3 & 4 \\ \end{array}}\begin{bmatrix} 0 & 3/4 & 1/4 & 0 \\ 0 & 1/2 & 1/2 & 0 \\ 0 & 0 & 1/2 & 1/2 \\ 1 & 0 & 0 & 0 \end{bmatrix}$$

If it costs Rs. 125 to overhaul a machine (including lost time) on the average and Rs. 75 as production lost if a machine is found inoperative. Using the steady-state probabilities, compute the expected per day cost of maintenance.

Solution. Let P_1, P_2, P_3 and P_4 be state probabilities representing the proportion of times that the machine will be in states 1,2,3 and 4 respectively. The steady state condition may be expressed as

$V=V.P$

$$[P_1 \ P_2 \ P_3 \ P_4] = [P_1 \ P_2 \ P_3 \ P_4]\begin{bmatrix} 0 & 3/4 & 1/4 & 0 \\ 0 & 1/2 & 1/2 & 0 \\ 0 & 0 & 1/2 & 1/2 \\ 1 & 0 & 0 & 0 \end{bmatrix}$$

We get,

$$P_1 = P_4, \quad P_2 = \frac{3}{4}P_1 + \frac{1}{2}P_2, \quad P_3 = \frac{1}{4}P_1 + \frac{1}{2}P_2 + \frac{1}{2}P_3$$

$$P_4 = \frac{1}{2}P_3, \quad P_1 + P_2 + P_3 + P_4 = 1$$

Rearranging all the equations, we have

$P_1 = P_4$

$$\frac{3}{4}P_1 - \frac{1}{2}P_2 = 0$$

$$\frac{1}{4}P_1 + \frac{1}{2}P_2 - \frac{1}{2}P_3 = 0$$

$$P_4 = \frac{1}{2}P_3$$

$$P_1 + P_2 + P_3 + P_4 = 1$$

Solving these equations, we get the required steady state probabilities are

$$P_1 = \frac{2}{11}, P_2 = \frac{3}{11}, P_3 = \frac{4}{11}, P_4 = \frac{2}{11}$$

Thus, on the average, 2 out of every 11 days the machine will be overhauled, 3 out of every 11 days it will be in good conditions, 4 out of every 11 days it will be in fair condition and 2 out of every 11 days it will be found inoperative at the end of the day.

Hence, the average cost per day of maintenance will be

$$\frac{2}{11}\times 125 + \frac{2}{11}\times 75 = Rs\ 36.36$$

Example 7 Suppose these are two market products of brands A and B respectively: Let each of these two brands have exactly 50% of the total market in same period and let market be of a fixed size. The transition matrix is given below.

$$\text{To} \qquad \begin{matrix} & A & B \end{matrix}$$

$$From \begin{matrix} A \\ B \end{matrix}\begin{pmatrix} 0.9 & 0.1 \\ 0.5 & 0.5 \end{pmatrix}$$

If the initial market share break down is 50% for each brand, then determine their market shares in the steady-state.

Solution The given transition matrix be $\begin{pmatrix} 0.9 & 0.1 \\ 0.5 & 0.5 \end{pmatrix}$

Given initial row vector for brands A and B are 50% each *i.e.*

$$V_0 = [0.50 \quad 0.50]$$

Determine the expected market shares for the next period $n = 1$ is

$$\begin{bmatrix} extected\ market \\ shars\ of\ Period\ 2 \end{bmatrix} = [marketshare\ of\ period\ 1][Transition\ matix]$$

i.e., $V_1 = V_0 \times P$

$$= [0.50 \quad 0.50]\begin{pmatrix} 0.9 & 0.1 \\ 0.5 & 0.5 \end{pmatrix}$$

$$= [0.7 \quad 0.3]$$

Again, determine the expected market share for the next period $n = 2$ is

$$V_2 = V_1 \times P$$

$$= [0.7 \quad 0.3]\begin{pmatrix} 0.9 & 0.1 \\ 0.5 & 0.5 \end{pmatrix}$$

$$= [0.78 \quad 0.22]$$

In similar way, we get $V_3 = V_2 \times P$

$$= [0.78 \quad 0.22]\begin{pmatrix} 0.9 & 0.1 \\ 0.5 & 0.5 \end{pmatrix} = [0.812 \quad 0.188]$$

$$V_4 = V_3 \times P$$

$$= [0.182 \quad 0.188]\begin{pmatrix} 0.9 & 0.1 \\ 0.5 & 0.5 \end{pmatrix} = [0.8248 \quad 0.1752]$$

$$V_5 = V_4 \times P$$

$$= [0.8248 \quad 0.1752]\begin{pmatrix} 0.9 & 0.1 \\ 0.5 & 0.5 \end{pmatrix} = [0.82992 \quad 0.17008]$$

$$V_6 = V_5 \times P$$

$$= [0.82992 \quad 0.17008]\begin{pmatrix} 0.9 & 0.1 \\ 0.5 & 0.5 \end{pmatrix} = [0.83 \quad 0.17]$$

Thus, we observe that starting with [0.50 0.50] of the market shares, after 6 periods the resulting market shares are about [0.83 0.17] *i.e.*, 83% and 17% respectively.

Thus, the steady-state market shares of A and B will be $\frac{5}{6}$ and $\frac{1}{6}$ respectively of the total market share.

Now, steady-state position can be obtained by the matrix equation.

$$[x \quad y]\begin{pmatrix} 0.9 & 0.1 \\ 0.5 & 0.5 \end{pmatrix} = [x \quad y]$$

$$0.9x + 0.5y = x$$

$$x + 0.5y = y$$

and $\quad x + y = 1$

Solving, we get $x = \dfrac{5}{6}, y = \dfrac{1}{6}$

Example 8 On January 1 (this year), bakery A had 40% of its local market share while the other two bakeries B and C had 40% and 20% respectively of the market share. Based upon a study **by a marketing research firm, the following facts were compiled. Bakery** A retains 90% of its customers while gaining 5% of competitor B's customers and 10% of C's customers. Bakery B retains 85% of its customers while gaining 5% of A's customers and 7% of C's customers. Bakery C retains 83% of its customers and gain 5% of A's customers and 10% of B's customers. What will each firm's share be on January 1, next year, and what will each firm's market share be at equilibrium?

Solution Using the above given data, the state transition matrix can be expressed as follows

$$P = \text{Bakery} \begin{matrix} & A & B & C \\ A & \\ B & \\ C & \end{matrix}\begin{pmatrix} 0.9 & 0.05 & 0.05 \\ 0.05 & 0.85 & 0.10 \\ 0.10 & 0.07 & 0.83 \end{pmatrix} \text{Retention and gain} \downarrow$$

Bakery (columns A, B, C)

—— Retention and loss →

At initial state, the market share of bakeries are 40, 40 and 20 percent on January 1 next year *i.e.* $V_0 = [0.40 \quad 0.40 \quad 0.20]$

The expected market share for bakeries *A, B* and *C* on January 1 next year are

$$\begin{bmatrix} \text{market share on} \\ \text{January 1 next} \\ \text{year} \end{bmatrix} = \begin{bmatrix} \text{market share on} \\ \text{January 1 this} \\ \text{year} \end{bmatrix} \times [\text{transition matrix}]$$

$$= [0.40 \ \ 0.40 \ \ 0.20] \begin{pmatrix} 0.90 & 0.05 & 0.05 \\ 0.05 & 0.85 & 0.10 \\ 0.10 & 0.07 & 0.83 \end{pmatrix}$$

$$= [0.400 \ \ 0.374 \ \ 0.226]$$

This means that market shares of bakeries A, B and C on January 1 next year will be 40% , 37.4% and 22.6% respectively.

Determination of equilibrium condition $\rightarrow$ The steady state conditions are

$$[x \ \ y \ \ z] = [x \ \ y \ \ z] \times P$$

$$= [x \ \ y \ \ z] \begin{pmatrix} 0.90 & 0.05 & 0.05 \\ 0.05 & 0.85 & 0.10 \\ 0.10 & 0.07 & 0.83 \end{pmatrix}$$

Solving above matrix equation, we have

$$x = 0.90x + 0.05y + 0.10z$$

$$y = 0.05x + 0.85y + 0.07z$$

$$z = 0.05x + 0.10y + 0.83z$$

and $\quad x + y + z = 1$

Solving these equations, we get

$$x = 0.43, y = 0.28, z = 0.29$$

Hence, the solution for equilibrium market shares are

$$A = 43\%, B = 28\%, C = 29\% \text{ of the total market.}$$

Characteristics of a Markov Chain

The following properties for a Markov chain at finite states are

1. There are finite number of states.
2. If current state is known, the conditional probability of the next state is independent of the state prior to the present state. i.e., The transiton probabilities depend only on the current state of the process.
3. The transition probabilities of moving to alternative states in the next time period, given a state in the current time period must sum to unity.
4. In a particular state, the probability for long-run will be constant over the time.

Practice Problems

1. A salesman's territory consists of three cities X, Y and Z. He never sells in the same city on successive days. If the sells in city X, then the next day he sells in city Y. However, if the sells in either Y or Z, then the next day he is twice as likely to sell in city X as in the other city. In the long run, how often does he sell in each of the cities?

[**Ans.** $X = 40\%, Y = 45\%, Z = 15\%$]

2. There are two white marble in urn A and 4 red marbles in urn B. At each step of process a marble is selected from each urn and the two marbles selected are interchanged. Let X_n be the number of red marbles in urn A after n interchanged.
   (i) Find the transition matrix.
   (ii) What is the probability that there are 2 red marbles in urn A after 3 steps?
   (iii)In the long run, what is the probability that there are 2 red marbles in urn A ?

$$\left[\text{Ans.(i) } P = \begin{pmatrix} 0 & 1 & 0 \\ 1/8 & 1/2 & 3/8 \\ 0 & 1/2 & 1/2 \end{pmatrix} \text{(ii)} \frac{3}{8} \text{ (iii)} \frac{2}{5}\right]$$

3. A grocer stocks his store with three types of powder P, Q and R. When brand P is sold out the probability is 0.7 that he stocks up with brand P again. When he sells out brand Q. The probability is 0.8 that he will stock up again with brand Q. Finally when he sells out brand R, the probability is 0.6 that he will stock up with brand R again. When he switches to another powder he does so with equal probability for the remaining two brands. Find the transition matrix. In the long run how does he stocks up with powder?

[**Ans.** P = 0.3077, Q = 0.4615, R = 0.2308]

4. A professor tries not to be late for class too often. If he is late one day, he is 90% sure to be on time next time. If he is on time then the next time there is a 30% chance of his being late. In the long run, how often is he late for class?

[**Ans.** 0.25 0.75]

5. The conclusion of an analysis of voting trends in a certain state is that the voting patterns of successive generations are described by the following matrix P.

$$P = \begin{array}{c} \\ Congress \\ BJP \\ Others \end{array} \begin{array}{c} \begin{array}{ccc} Congress & BJP & Others \end{array} \\ \begin{bmatrix} 80\% & 15\% & 5\% \\ 20\% & 70\% & 10\% \\ 60\% & 30\% & 10\% \end{bmatrix} \end{array}$$

Among the Congressmen of one generation, 80% of their next generation are Congressmen, 15% are BJP men, 5% are others and so on. Construct a stochastic matrix from P that defines a Markov chain model of these voting patterns. If there are 2.5 million who believe in Congress, 1.5 million in BJP and 0.25 million in other parties at a certain period. What is the distribution likely to be in the next generation?

[**Ans.** [0.5765 0.3529 0.0706]]

6. Honey Inc. had 35% of the total market for its cosmetics, while the two other manufactures of cosmetics Lace Inc. and Shalon Inc. have 40% and 25% shares respectively in the local market, as on 1st April of this year. A study by a market research firm has disclosed the following. Honey Inc. retains 86% of its customers, while it gains 4% and 6% of the customers from its two competitors. Lace and shalon respectively. Lace Inc. retain 90% of its customers, and gain 8% and 9% of customers respectively from Honey and shalon. Shalon retains 85% of its customers and gain 6% and 6% of customers from Lace and Honey respectively. What will be the share of each firm on April 1 next year? What will be the market share of each firm of equilibrium?

   [**Ans.** Honey Inc. 33.2%, Lace Inc. 41.05%, Shalon Inc. 25.75%. Market share of each firm at equilibrium will be 25%, 46% and 29%]

7. A market survey is made on three brands of break fast food P, Q and R. Every time the customer purchase a new package, he may buy the same brand or switch to another brand. The following estimates are obtained, expressed as decimal fraction. Brand just purchased

$$\text{Present brand} \quad \begin{array}{c} \\ P \\ Q \\ R \end{array}\!\!\begin{array}{c} \begin{array}{ccc} P & Q & R \end{array} \\ \begin{bmatrix} 0.7 & 0.2 & 0.1 \\ 0.3 & 0.5 & 0.2 \\ 0.3 & 0.3 & 0.4 \end{bmatrix} \end{array}$$

   At this time it is estimated that 30% of the people buy brand P, 20% brand Q and 50% brand R. What will be the distribution of customers be two time periods later and at equilibrium?

   [**Ans.** (i) [0.468 0.320 0.212] (ii) [0.5 0.3125 0.1875]]

8. A psychologist makes the following assumptions, concerning the behaviour of mice subject to a particular feeding schedule. For any particular trial 80% of the mice that went right in the previous experiment will go right in this trial and 60% of those mice that went left in the previous experiment will go right in this trial. If 50% went right in the first trial, what would he predict for (i) the second trial (ii) the third trial (iii) the thousand trial?

   [**Ans.** (i) [0.7 0.3] (ii) [0.74 0.26] (iii) [0.75 0.25]]

9. Consider the following transition matrix of brand switching

$$\begin{array}{c} \\ X \\ Y \\ Z \end{array}\begin{array}{c} \quad X \qquad Y \qquad Z \\ \begin{bmatrix} 0.50 & 0.15 & 0.35 \\ 0.20 & 0.55 & 0.25 \\ 0.25 & 0.30 & 0.45 \end{bmatrix} \end{array}$$

The manufacturer of brand Z and considering a marketing strategy to attract brand Y customers. It is estimated that this strategy will

(i) increase probability of customer's switch from Y to Z by 0.20

(ii) decrease probability of customer's switch from Z to Y by 0.01 and

(iii) decrease probability of customer's switch from X to Z by 0.25.

Should the new strategy be used?

10. A house wife buys three kinds of cereals: X, Y and Z: she never buys the same cereal in successive weeks. If she buys cereal X, the next week she buys Y. However, if she buys either Y or Z, then the next week she is three times as likely to buy X as the other brands. In the long run, how often does she buy each of the three brands?

$$\left[\text{Ans.} \begin{array}{c} \\ X \\ Y \\ Z \end{array}\begin{array}{c} \quad X \qquad Y \qquad Z \\ \begin{pmatrix} 0 & 1 & 0 \\ 3/4 & 0 & 1/4 \\ 3/4 & 1/4 & 0 \end{pmatrix} \end{array}, [0.428 \;\; 0.457 \;\; 0.114] \right]$$

Chapter - 15

Queuing Theory

Introduction

In every day life, it is seen that a queue form at bus stand, Cinema ticket window, doctor's clinic, bank counters and so on. Queue is called the waiting line, in order to maintain a proper discipline. Here the arriving people are called the customers and the person by whom he is serviced is called a server.

Queuing theory can be applied to a wide variety of operational situations where imperfect matching between the customers and service facilities is caused by one's inability to predict accurately the arrival and service times of customers.

Queuing theory is concerned with the statistical description of the behaviour of queues with finding, e.g., the probability distribution of the number in the queue from which the mean and variance of queue length and the probability distribution of waiting time for a customer or the distribution of a. server's busy periods can be found.

Important Definition in Queuing Theory

We define various terms, which are used in our queue model

1. **Queue length :** The number of customers waiting in the line at any time is defined as queue length.
2. **Average length of queue :** Average length of queue is defined by the number of customers in the queue per unit time.
3. **Waiting time :** Waiting time is the time upto which a unit has to wait in the queue before it is taken into service.
4. **Service time :** The time taken by the server for servicing of a unit is called service time.

5. **Busy period :** The time during which a server remains busy in servicing is called busy period. Therefore, the busy period is the time between the start of service of the first unit to the end of service of the last unit in the queue.

6. **Idle period :** The idle period of a server is the time during which it remains free because there is no customer present in the system. Therefore, the idle period of a server begins when all the customers in the queue are served and it continues upto the time of arrival of the customer.

7. **Mean arrival rate :** The mean arrival rate in a waiting line situation is defined as the expected number of arrivals occurring in a time interval of length unity.

8. **Mean service rate :** The mean service rate is defined as the expected number of services completed in a time interval of length unity.

Queuing System

A queuing system can be completely described by

(a) Input source (arrival pattern)

(b) Queuing process

(c) Queue discipline

(d) Service mechanism (Service pattern)

(e) Customer's behaviour

(a) Input source : The arrival pattern is the way in the inter arrival time, which is the time period between two succesive arrival, where the customer arrive and join the system. The number of customers arriving per unit time i.e., inter arrival rate follows a poisson distribution and the inter arrival time follows an exponential distribution. The arrival rate is denoted by λ. The size of input population is considered either to be finite or infinite. When the number of arrivals are dependent on the number of customers already in the system, the input population is finite. For example of a finite input populations is a factory having only five machine which often require service and two of them are in working condition. That at any time, there are only two machines which can possibly require service.

If a new customer's arrival is independent of the number of customers already in the system, the input population is called infinite. For examples of infinite population are the customers arriving at a bank, students arriving to get admission in the college.

(b) Queuing process : The queuing process refers to the number of queues and their respective lengths. Certain service system of the queue depends upon the operational situation such as physical space, attitude of the customers. In certain cases, a service system is unable to accomodate more than the required number of customers at a time. No further customers all allowed to enter until space becomes available to accommodate new customers. Such type of situations are referred to as finite source queue. For example, in a cinema halls, restaurants, banks etc. If a service system is able to accomodate any number of customers at a time, then. it is referred to as infinite source queue. For example, in a sales department where the customer orders are received.

(c) Queue discipline : The queue discipline is the rule, in which the manner of the customer's behaviour while waiting and the manner in which they are choosen for service. Some disciplines are

(i) **First come, first served (FCFS) :** If the customers are served in the order of their arrival, then it is known as the first come, first served. For example, at cinema ticket window, railway reservation centre etc.

(ii) **Last come, first served •(LCFS) :** Last arrival in the system is referred to the

servicing of the customer in an order just reserve of their arrival, so that the one who joins the queue in the last is serviced first. For example, in a big godown, where the last item loaded is removed first.

(iii) **Selection in random order (SIRO) :** In this rule, the customers are selected for service at random irrespective of their arrivals in the service system.

(iv) **Priority service :** Under this rule, the service is of two types (a) Pre-emptive and (b) non-pre-emptive.

In pre-emptive, the high priority customers are given service over the low priority customers and in non-pre emptive, the low priority customers are serviced before the high priority customers.

(d) Service mechanism : The service mechanism is specified that customer receive once they are selected from a queue. It is also called a server. The amount of time where a customer takes to be serviced by the server is called the service time and is denoted by μ.

Some queue models are single server, some are multiple servers. The service time is a random variable with the statistical distribution for all arrivals.

(e) Customer's behaviour : The customers generally behave in four ways..

(i) **Balking :** A customer may leave the queue because the queue is too long and it has no time to wait.

(ii) **Reneging :** The customers leave the queue if they have waited too long for service.

(iii) **Priorities :** Some customers are served before others regardless of their order of arrival.

(iv) **Jockeying:** Customers may jockey from one waiting line to another.

Transient and Steady States

When a service system is started it progresses through a number of changes. However, if attains stability after some time. Before the start of the service operations it is very much influenced by the initial conditions and the elapsed time. this period of transition is stated as transient state. However, after sufficient time has passed, the system becomes independent of the initial conditions and of the elapsed time and enters a steady-state condition, i.e., A system is said to be in "transient state" when its operating behaviour are dependent on time.

A "steady state condition" is said to prevail when the behaviour of the system becomes independent of time.

Let $P_n(t)$ denote the probability that there are *n* customers in the system at time *t*. The rate of change in the value of $P_n(t)$ with respect to time t is represented by the derivative of $P_n(t)$ *i.e.,* $P_n'(t)$. In the steady state, we have

$$\lim_{t\to\infty} P_n(t) = P_n \text{ (independent of t)}$$

$$\Rightarrow \qquad \lim_{t\to\infty} \frac{d}{dt}\{P_n(t)\} = \frac{d}{dt}\{P_n\}$$

$$\Rightarrow \qquad \lim_{t\to\infty} P_n'(t) = 0$$

In some situations, when the arrival rate of customers in the system is greater than the service rate, a steady state cannot be reached regardless of the length of the elapsed time.

A List of Symbols

The following symbols used in the queuing models are as follows next time n = Number of customers in the system (waiting and in system).

$P_n(t)$ = Probability of n customers in the system at time t.

P_n = Probability of n customers in the system.

λ = Average (expected) customers arrival rate in the queuing system.

μ = Average (expected) service rate at the place of service.

N = Maximum number of customers in the system.

s = Number of service facilities or servers.

$$\rho = \frac{\lambda}{\mu} = \frac{\text{Average service time}\left(\frac{1}{\mu}\right)}{\text{Average inter - arrial time}\left(\frac{1}{\lambda}\right)}$$

= traffic intensity or server utilization factor.

L_s = Average (expected) number of customers in the system (waiting and in service).

L_q = Average (expected) number of customers in the queue (queue length).

L_b = Average (expected) length of non-empty queue.

Ws = Average (expected) waiting time in the system (waiting and in service).

Wq = Average (expected) waiting time in the queue.

Pw = Probability that an arriving customers has to wait.

Traffic Intensity (Utilization Factor)

Traffic Intensity $(\rho) = \dfrac{\text{average arrival rate}}{\text{average service rate}}$

i.e,
$$\rho = \frac{1/\mu}{1/\lambda} \quad \frac{\text{average service time}}{\text{average inter-arrival time}}$$

The unit of traffic intensity is Erlang.

If $\rho < 1 \Rightarrow \dfrac{\lambda}{\mu} < 1 \Rightarrow \lambda < \mu$ *i.e.*, the arrival rate is less than the service rate. It is necessary condition for a system to have settled down to steady state.

If $p > 1$ *i.e.*, the arrival rate is greater than the service rate and consequently, the number of units in the queue tends to increase indefinitely as the time passes on, provided the rate of service is not attached by the length of queue.

Probability Distributions in Queuing Systems

It is assumed that the customers joining a queuing system arrive in random manner and follow a poisson distribution or equivalently the inter arrival times follow exponential distribution.

The number of arrivals and departures during an interval of time in a queuing system is controlled by the following axioms.

(i) The probability of an event occurring during the time interval $(t, t + \Delta t)$ depends on the length of time interval Δt.

(ii) The probability of more than one event occurring during the time interval $(t; t + \Delta t)$ is negligible. It is represented by $O(\Delta t)$.

(iii) Atmost one event can occur during a small time interval Δt. The probability of an arrival during the time interval $(t, t + \Delta t)$ is given by $P(\Delta t) = \lambda\Delta t + O(\Delta t)$ where λ is a constant and is independent of time t.

Distribution of Arrival (Pure Birth Process)

In this process the customers arrive.at the queuing system and never leave it. Such process is known as pure birth process.

Theorem 1 : If the arrivals are completely random, then the probability distribution of a number of arrivals in a fixed time interval follows a poisson distribution.

Proof : Let us define the following terms which are used in queuing system:

Δt = The probability of more than one customer's arrival is negligible, *i.e.*, during small interval of time Δt only one customer can arrive.

$\lambda \Delta t$ = Probability that a customer will arrive in the system during Δt time.

1- Δt = Probability that no customers will arrive in the system during Δt time.

Case I When $n > 0$. For $n > 0$, there may be two mutually exclusive ways of having **n** customers at time $t + \Delta t$.

(i) The system contains n customers at time t and there is no arrival during time interval Δt.

(ii) The system contains (n-1) customers at time t and there is one arrival and no departure during time interval Δt.

Hence, $P_n(t+\Delta t)$ = {(Probability of n customers at time t)×(Probability of no arrival during time Δt)} + {(Probability of (n-1) customer at time t)×(Probability of one arrival during time Δt)}

$$= P_n(t)\{1-\lambda\Delta t\} + P_{n-1}(t)\{\lambda\Delta t\} \quad \text{....(i)}$$

Case II When $n = 0$. If there is no customer in the system at time $t+\Delta t$, then there will be no arrival during time interval Δt.

Hence, $P_0(t+\Delta t)$ = {(Probability of *no* customers at time t)×(Probability of no arrival during time Δt)}

$$= P_0(t)\{1-\lambda\Delta t\} \quad \text{....(ii)}$$

Eqs. (i) and (ii) may be written as

$$P_n(t+\Delta t) - P_n(t) = P_n(t)\times(-\lambda\Delta t) + P_{n-1}(t)\lambda\Delta t, n > 0$$

and
$$P_0(t+\Delta t) - P_0(t) = P_0(t)(-\lambda\Delta t), n = 0$$

dividing both sides by Δt and taking limit as $\Delta t \to 0$

$$\lim_{\Delta t \to 0} \frac{P_n(t+\Delta t) - P_n(t)}{\Delta t} = -\lambda P_n(t) + \lambda P_{n-1}(t) \quad \text{....(iii)}$$

and $$\lim_{\Delta t \to 0} \frac{P_0(t+\Delta t) - P_0(t)}{\Delta t} = -\lambda P_0(t) \quad \text{...(iv)}$$

by the definition of first derivative,

$$\lim_{\Delta t \to 0} \frac{P_n(t+\Delta t) - P_n(t)}{\Delta t} = \frac{d}{dt} P_n(t) = P_n^1(t)$$

The equations (iii) and (iv) can be written as

$$P_n^1(t) = -\lambda P_n(t) + \lambda P_{n-1}(t), n > 0 \quad \text{...(v)}$$

$$P_0^1(t) = -\lambda P_0(t), n = 0 \quad \text{...(vi)}$$

Which is known as differential-difference equations.

Eqs. (vi) can be written as

$$\frac{P_0^1(t)}{P_0(t)} = -\lambda$$

Integrating both sides w.r.t. t, we get

$$\log P_0(t) = -\lambda t + A \quad \text{...(vii)}$$

The value of constant A can be determined by using the following initial conditions.

$$P_n(0) = \begin{cases} 1 \; ; n = 0 \\ 0 \; ; n > 0 \end{cases}$$

Substituting $t = 0$ in Eq. (vii), we get $P_0(0) = 1$

$\therefore \quad A = 0$

Thus, Eq. (vii) gives, $\log P_0(t) = -\lambda t$

$$\Rightarrow \quad P_0(t) = e^{-\lambda t} \quad \text{...(viii)}$$

Now, putting $n = 1$ in Eq. (v) and using Eq. (viii), we have

$$P_1'(t) = -\lambda P_1(t) + \lambda e^{-\lambda t}$$

$$\Rightarrow \quad P_1'(t) + \lambda P_1(t) = \lambda e^{-\lambda t} \quad \text{...(ix)}$$

which is a linear differential equation of first order. It can be solved by multiplying both sides of it by integrating factor.

I.F. $= e^{\int \lambda dt} = e^{\lambda t}$

Then Eq. (ix) becomes

$$e^{\lambda t}\left[P_1^1(t) + \lambda P_1(t)\right] = \lambda$$

or $$\frac{d}{dt}\left[e^{\lambda t} P_1(t)\right] = \lambda$$

Integrating both sides w.r.t. t, we get

$$e^{\lambda t} P_1(t) = \lambda t + B \qquad \text{...(x)}$$

Where B is a constant of integration.

Initially, put $t = 0$, we get

$P_1(0) = B = 0 \quad [\because P_1(0) = 0]$

Substituting $B = 0$ in Eq. (x) we have

$$e^{\lambda t} P_1(t) = \lambda t$$

or $$P_1(t) = \lambda t e^{-\lambda t} \qquad \text{...(xi)}$$

Similarly, Putting $n = 2$ in eqn. (v) and using eqn. (xi), we get

$$P_2^1(t) + \lambda P_2(t) = \lambda(\lambda t e^{-\lambda t})$$

or $$\frac{d}{dt}\left[e^{\lambda t} P_2(t)\right] = \lambda(\lambda t)$$

Integrating w.r.t. t, we get

$$e^{\lambda t} P_2(t) = \frac{\lambda^2 .\ t^2}{2!} + C$$

Where C is a constant of integration.

Putting, $t = 0$, then $C = 0$, $P_2(0) = 0$. Hence,

$$e^{\lambda t} P_2(t) = \frac{(\lambda t)^2}{2!}$$

or $$P_2(t) = \frac{(\lambda t)^2 e^{-\lambda t}}{2!} \qquad \text{...(xii)}$$

In general, we have

$$P_n(t) = \frac{(\lambda t)^n e^{-\lambda t}}{n!}, \quad \text{for } n = 0,1,2 \ldots\ldots\ldots\ldots\ldots\ldots$$

Which is a position distribution formula. This completes the proof of the theorem.

Distribution of Inter-arrival times (Exponential Process)

If the number of arrival be n in time t, follows the poisson distribution.

$$P_n(t) = \frac{(\lambda t)^n e^{-\lambda t}}{n!}, \; n = 0,1,2,3 \ldots\ldots\ldots\ldots\ldots\ldots$$

Then the inter-arrival time T follows the exponential distribution

$$f(T) = \lambda e^{-\lambda t}$$

and vice-versa.

Proof Let t_0 be the instant of an arrival initially. The random variable T represents the inter-arrival time. There is no arrival in the interval $(t_0, t_0 + T)$ and $(t_0+T, t_0+T+\Delta t)$.

Therefore, putting $t = T$ and $n = 0$ in poisson distribution.

$$P_n(t) = \frac{(\lambda t)^n}{n!} e^{-\lambda t} \qquad \ldots(i)$$

We have,

$$P_0(T) = \frac{(\lambda t)^0}{0!} e^{-\lambda T}$$

$$= e^{-\lambda T} \qquad \ldots(ii)$$

Again, putting $t = T + \Delta T$ and $n = 0$ in Eq. (i), we get

$$P_0(T + \Delta T) = \frac{[\lambda(T + \Delta T)]^0}{0!} e^{-\lambda(T+\Delta T)}$$

$$= e^{-\lambda(T+\Delta T)}$$

$$= e^{-\lambda T} . e^{-\lambda \Delta T}$$

$$= e^{-\lambda T} [1 - \lambda \Delta T + O(\Delta T)]$$

$$= P_0(T)[1 - \lambda \Delta T + O(\Delta T)]$$

$$P_0(T + \Delta T) = P_0(T) + P_0(T) [-\lambda \Delta T + O(\Delta T)] \qquad \text{[using (ii)]}$$

$$P_0(T + \Delta T) - P_0(T) = P_0(T) . [-\lambda \Delta T + O(\Delta T)]$$

Dividing both sides by Δ T, we get

$$\frac{P_0(T+\Delta T)-P_0(T)}{\Delta T}=-\lambda P_0(T)+\frac{O(\Delta T)}{\Delta T}P_0(T)$$

Taking limit on both sides as Δ T→ 0

$$\lim_{\Delta T\to 0}\frac{P_0(T+\Delta T)-P_0(T)}{\Delta T}=\lim_{\Delta T\to 0}\left[-\lambda P_0(T)+\frac{O(\Delta T)}{\Delta T}P_0(T)\right]$$

or, $$\frac{d}{dT}P_0(T)=-\lambda P_0(T) \qquad \left[\because \lim_{\Delta T\to 0}\frac{O(\Delta T)}{\Delta T}=0\right]$$

∴ Probability density function

$$f(T)=\frac{d}{dT}P_0(T)=-\lambda P_0(T)$$

or $$f(T)=-\lambda e^{-\lambda T}$$ [using Eq. (ii)]

or $$f(T)=\lambda e^{-\lambda T}$$

(neglecting negative sign because probability density function is always positive.)

which is the exponential law of probability.

Where $\frac{1}{\lambda}$ is the mean and variance is $\frac{1}{\lambda^2}$.

i.e. $$E(T)=\frac{1}{\lambda},\ var\,(T)=\frac{1}{\lambda^2}$$

The converse of this theorem can be proved in a similar way.

Distribution of Departures (Pure Death Process)

In this process when no additional customer joins the system while service is continued. Let at time t = 0, the departure occur at a rate μ per unit time. The distribution of departure can be obtained on the basis of the following three axioms.

(i) Probability of one departure during time Δt is μ Δt.

(ii) Probability of more than one departure between time t and t + Δt is negligible.

(iii) The number of departure in non-overlapping intervals are statistically independent, *i.e.* the process has independent arrivals.

Using the same arguments as in pure birth process, we obtain the differential difference equation in three ways.

Case I When $0 < n < N$, since there are N customers in the system. Then

$$P_n(t+\Delta t) = P_n(t)[1-\mu\Delta t] + P_{n+1}(t).\mu\Delta t \quad \text{...(i)}$$

Case II When $n = N$, $P_{n+1}(t) = 0$

$$\therefore \quad P_N(t+\Delta t) = P_N(t)[1-\mu\Delta t] \quad \text{...(ii)}$$

Case III When $n = 0$,

$$\therefore \quad P_0(t+\Delta t) = P_0(t) + P_1(t)\mu\Delta t \quad \text{...(iii)}$$

Rearranging the terms and dividing by the Δt with taking limit as $\Delta t \rightarrow 0$. The Eqs. (ii) (i) and (iii) becomes

$$P_N^1(t) = -\mu P_N(t), n = N \quad \text{...(iv)}$$

$$P_n^1(t) = -\mu P_n(t) + \mu P_{n+1}(t), 0 < n < N \quad \text{...(v)}$$

$$P_0^1(t) = -\mu P_1(t), n = 0 \quad \text{...(vi)}$$

The solution of Eqs. (iv), (v) and (vi) can be obtained by using iterative method.

From eq. (iv)

$$\frac{P_N'(t)}{P_N(t)} = -\mu$$

Integrating both sides w.r.t.t, we get

$$\log P_N(t) = -\mu t + A \quad \text{...(vii)}$$

Using boundary condition $P_N(0) = 1$ as $t = 0$, we get $A = 0$

Therefore eq. (vii) becomes

$$\log P_N(t) = -\mu t$$

$$\text{or} \quad P_N(t) = e^{-\mu t} \quad \text{..(viii)}$$

In eq. (v), put $n = N-1$,

$$P_{N-1}'(t) = -\mu P_{N-1}(t) + \mu P_N(t)$$

$$= -\mu P_{N-1}(t) + \mu e^{-\mu t} \quad \text{[from Eq. (viii)]}$$

$$\text{or} \quad P_{N-1}'(t) + \mu P_{N-1}(t) = \mu e^{-\mu t}$$

which is a linear differential equation.

So, integration factor I.F.= $e^{\int \mu dt} = e^{\mu t}$

Therefore, the solution of this equation is given by

$$P_{N-1}(t)e^{\mu t} = \int \mu e^{-\mu t}.e^{\mu t}\,dt + B$$
$$= \mu t + B$$

or $$P_{N-1}(t) = \mu t e^{-\mu t} + Be^{-\mu t} \quad \text{...(ix)}$$

Initially $t = 0$, $P_{N-1}(t) = 0$, then from eq. (ix) to get $B = 0$.

Therefore, $P_{N-1}(t) = \mu t e^{-\mu t}$

Again, putting $n = N-2$, in Eq.(v) and proceeding exactly as Eq. (ix)

$$P_{N-2}(t) = \frac{(\mu t)^2\, e^{-\mu t}}{2!}$$

Putting $n = N-3$, ……….. N-i, we get,

$$P_{N-i}(t) = \frac{(\mu t)^i\, e^{-\mu t}}{i!}, i = 0,1,2..........N-1$$

In general, at $n = N-n$, we have

$$P_n(t) = \frac{(\mu t)^{N-n}\, e^{-\mu t}}{(N-n)!}, n = 0,1,2..........N \quad \text{...(x)}$$

In order to find $P_0(t)$, we use the following procedure.

Since, $$1 = \sum_{n=0}^{N} P_n(t) = P_0(t) + \sum_{n=1}^{N} P_n(t)$$

$$\therefore \quad P_0(t) = 1 - \sum_{n=1}^{N} P_n(t) = 1 - \sum_{n=1}^{N} \frac{(\mu t)^{N-n}}{(N-n)!} e^{-\mu t}, \text{[using Eq.(x)]...(xi)}$$

Finally, combining the results (x) and (xi), we get

$$P_n(t) = \begin{cases} \dfrac{(\mu t)^{N-n}}{(N-n)!} e^{-\mu t}; & n = 1,2.........., N, t \geq 0 \\ 1 - \displaystyle\sum_{n=1}^{N} \dfrac{(\mu t)^{N-n}}{(N-n)!} e^{-\mu t}; & n = 0, t \geq 0 \end{cases}$$

Thus, the number of departure in time t follows the "Truncated Poisson Distribution".

Distribution of Service Times

Let $S(t)$ be the cumulative density function and $s(t)$ be the probability density function of T. We observed that the probability of no services during time 0 to t is equivalent to the probability of having no departure during the same period.

Thus, probability [service time $T \geq t$] = Probability [no departure during t] = $P_N(t)$ where N be the number of customers in the system and no arrival is allowed after N customers. Therefore,

$$P_N(t) = e^{-\mu t}$$

$$S(t) = \text{probability } [T \leq t] = 1 - \text{probability } [T \geq t]$$

or,

$$S(t) = 1 - e^{-\mu t}$$

Differentiating both sides w.r.t. t, we get

$$\frac{d}{dt}S(t) = s(t) = \begin{cases} \mu e^{-\mu t} & ; t \geq 0 \\ 0 & ; t < 0 \end{cases}$$

Hence, it is concluded that the service time distribution is 'Exponential distribution' with the mean $\frac{1}{\mu}$ and variance $\frac{1}{\mu^2}$.

Concepts of Queuing Models

The standard notations of queuing models were introduced by D.G. Kendall in 1953 in the from $(a/\ b/\ c)$. Later A.M. Lee in 1966 added the symbols d and e to the Kendall notation. Generally queuing model may be completely specified in the following symbolic form:

$$\{(a/b/c); (d/e)\}$$

where

a = arrival distribution
b = departure distribution
c = number of service channels
d = maximum number of customers allowed in the system
e = queue discipline

Solution of Queuing Models

Model I: $(M/M/1) : (\infty/\infty/FCFS)$: (Birth and Death model)

The model $(M/M/1) : (\infty/\infty/FCFS)$ represents a single server which has unlimited queue capacity and infinite calling customers both arrivals and service are poisson processed.

To determine the model, it is necessary to find an expression for the probabilities of n customers in the system at time t. The probability of n customers in the system at time $(t+\Delta t)$ can be expressed as the sum of the probability of the three cases *i.e.*

1. The system is in state n and no arrival and no departure.
2. The system is in state $n+1$ and no arrival and one departure.
3. The system is in state $n-1$ and one arrival and no departure.

We will begin with determining probability P_n of n units in the system at time t as follows.

Step I : To obtain the system of differential equation.

Let $P_n(t)$ be the probability of n units at time t in the system, the probability that the system will contain n units at time $(t+\Delta t)$ can be expressed as the sum of three probabilities as mentioned above.

for $n \geq 1$,

$$P_n(t+\Delta t) = P_n(t) \times \text{ (Probability of no arrival in time } \Delta t \text{ and no departure in } \Delta t)$$

$$+ P_{n+1}(t) \times \text{ (Probability of no arrival in } \Delta t \text{ and one departure in } \Delta t)$$

$$+ P_{n-1}(t) \times \text{ (Probability of one arrival in } \Delta t \text{ and no departure in } \Delta t)$$

$$= P_n(t).\{(\text{Probability of no arrival in } \Delta t) \times (\text{Prob. of no departure in } \Delta t)$$

$$+ P_{n+1}(t).\{(\text{Probability of no arrival in } \Delta t) \times (\text{Prob. of one departure in } \Delta t)$$

$$+ P_{n-1}(t).\{(\text{Probability of one arrival in } \Delta t) \times (\text{Prob. of no departure in } \Delta t)$$

$$= P_n(t)\{1-\lambda\Delta t\}\{1-\mu\Delta t\} + P_{n+1}(t)\{1-\lambda\Delta t\}.\mu\Delta t + P_{n-1}(t).\lambda\Delta t.\{1-\mu\Delta t\}$$

$$= P_n(t)\{1-\lambda\Delta t - \mu\Delta t + \lambda\mu(\Delta t)^2\} + P_{n+1}(t)\{\mu\Delta t - \lambda\mu(\Delta t)^2\}$$

$$+ P_{n-1}(t)\{\lambda\Delta t - \lambda\mu(\Delta t)^2\}$$

$$= P_n(t)\{1-(\lambda+\mu)\Delta t\} + P_{n+1}(t).\mu\Delta t + P_{n-1}(t).\lambda\Delta t + \text{term involving } (\Delta t)^2$$

Since, Δt is very small, hence $(\Delta t)^2$ can be neglected.

$$P_n(t+\Delta t) = P_n(t) - P_n(t)(\lambda+\mu)\Delta t + P_{n+1}(t).\mu\Delta t + P_{n-1}(t).\lambda\Delta t$$

$$P_n(t+\Delta t) - P_n(t) = \lambda P_{n-1}(t)\Delta t + \mu P_{n+1}(t)\Delta t - (\lambda+\mu)P_n(t)\Delta t$$

Dividing by Δt, we get

$$\frac{P_n(t+\Delta t) - P_n(t)}{\Delta t} = \lambda P_{n-1}(t) + \mu P_{n+1}(t) - (\lambda+\mu)P_n(t)$$

Taking limit on both sides as $\Delta t \to 0$, we have

$$\lim_{\Delta t \to 0} \frac{P_n(t+\Delta t) - P_n(t)}{\Delta t} = \frac{d}{dt}\{P_n(t)\} = \lambda P_{n-1}(t) + \mu P_{n+1}(t) - (\lambda+\mu)P_n(t)$$

$$P_n'(t) = \lambda P_{n-1}(t) + \mu P_{n+1}(t) - (\lambda+\mu)P_n(t) \quad \text{...(i)}$$

Which is the required differential equation.

In a similar way, if there will be no unit in the system at time $(t + \Delta t)$ *i.e.,* at n=0, the sum of two probabilities are as follows:-

$P_0(t + \Delta t)$ = (Probability of no unit in system at time t and no arrival in time Δt) +(Probability of one unit in system at time t and one unit serviced in time Δt and no arrival in Δt)

= {(Probability of no unit in the system at time t) × (Probability of no arrival in time Δt)}
+ {(Probability of one unit in the system at time t)}
× (Probability of one unit serviced in time Δt)
× (Probability of no arrival in time Δt)

$$= P_0(t).(1-\lambda\,\Delta t) + P_1(t).\,\mu\,\Delta t\,(1-\lambda\Delta t)$$

$$= P_0(t) - P_0(t)\lambda\,\Delta t + P_1(t)\,\mu\,\Delta t - P_1(t)\,\mu\,\lambda\,(\Delta t)^2$$

(the term involving $(\Delta t)^2$ can be neglected)

$$P_0(t + \Delta t) - P_0(t) = \mu\,P_1(t)\,\Delta t - \lambda\,P_0(t)\,\Delta t$$

Dividing by Δt and taking limit on both sides as $\Delta t \to 0$, we get

$$\lim_{\Delta t \to 0} \frac{P_0(t + \Delta t) - P_0(t)}{\Delta t} = \mu\,P_1(t) - \lambda\,P_0(t)$$

$$P_0^1(t) = \mu\,P_1(t) - \lambda\,P_0(t) \quad \text{...(ii)}$$

Step II : To obtain the system of steady- state equations

In the steady state, $P_n(t)$ is independent of time t, and the rate of change $P_n(t)$ can be considered to be zero *i.e.*

$$\lim_{n \to \infty} P_n(t) = P_n$$

and $$\lim_{t \to \infty} \frac{d}{dt}\{P_n(t)\} = 0, \text{ for } n \geq 0,$$

Thus, Eqs. (i) and (ii) becomes

$$\lambda\,P_{n-1}(t) + \mu P_{n+1}(t) - (\lambda+\mu)P_n(t) = 0, n \geq 1, \quad \text{...(iii)}$$

and $$\mu P_1(t) - \lambda P_0(t) = 0, n = 0, \quad \text{...(iv)}$$

Thus, the Eqs. (iii) and (iv) constitute the system of steady- state differential equations for the models.

Step III : To solve the system of differential equations.

from Eq. (iv)

$$\mu P_1(t) - \lambda P_0(t) = 0$$

or $$\mu P_1 - \lambda P_0$$

or $$P_1 = \frac{\lambda}{\mu} P_0; n = 0$$

putting $n = 1$ in Eq. (iii), we get

$$\lambda P_0 + \mu P_2 - (\lambda + \mu) P_1 = 0$$

or $$P_2 = \frac{(\lambda + \mu)}{\mu} P_1 - \frac{\lambda}{\mu} P_0$$

$$= \frac{(\lambda + \mu)}{\mu} . \frac{\lambda}{\mu} P_0 - \frac{\lambda}{\mu} P_0$$

$$= \left(\frac{\lambda}{\mu}\right)^2 P_0$$

Similarly, for $n = 3$,

$$P_3 = \left(\frac{\lambda + \mu}{\mu}\right) P_2 - \left(\frac{\lambda}{\mu}\right) P_1$$

$$= \left(\frac{\lambda + \mu}{\mu}\right)\left(\frac{\lambda}{\mu}\right)^2 P_0 - \left(\frac{\lambda}{\mu}\right)\left(\frac{\lambda}{\mu}\right) P_0$$

$$= \left(\frac{\lambda}{\mu}\right)^3 P_0$$

In general, we get

$$P_n = \left(\frac{\lambda}{\mu}\right)^n P_0$$

(*i.e*, n customers in the system)

Now, using the fact that $\sum_{n=0}^{\infty} P_n = 1$

$$P_0 + P_1 + P_2 \dots\dots\dots\dots + P_n + \dots\dots\dots\dots = 1$$

or $$P_0 + \left(\frac{\lambda}{\mu}\right) P_0 + \left(\frac{\lambda}{\mu}\right)^2 P_0 + \dots\dots\dots + \left(\frac{\lambda}{\mu}\right)^n P_0 \dots\dots\dots = 1$$

or $$P_0 \left[1 + \frac{\lambda}{\mu} + \left(\frac{\lambda}{\mu}\right)^2 + \dots\dots\dots + \left(\frac{\lambda}{\mu}\right)^n + \dots\dots\dots\right] = 1$$

or $$P_0 \left(1 - \frac{\lambda}{\mu}\right)^{-1} = 1$$

or $$\frac{P_0}{1 - \frac{\lambda}{\mu}} = 1$$

or $$P_0 = 1 - \frac{\lambda}{\mu} = 1 - \rho$$

and $$P_n = \left(\frac{\lambda}{\mu}\right)^n \left(1 - \frac{\lambda}{\mu}\right) = \rho^n (1 - \rho); \rho < 1, n \geq 0 \qquad \dots(v)$$

Thus, Eq. (v) is the required probability distribution of queue length.

Step IV : To obtain probability density function of waiting time (excluding service time) distribution.

Let w be the time required by the server to serve all the units present in the system in the steady state. Let $\psi_w(t)$ be the probability density function of w, *i.e.*, $\psi_w(t) = P(w \leq t)$, $0 \leq t \leq \infty$. If n arrival customer in the system, then all the customers must have been serviced by time t. Let $a_1, a_2, \dots\dots\dots\dots\dots\dots a_n$ denote service time of n customers. Thus

$$w = \begin{cases} 0 & ; n = 0 \\ \sum_{i=1}^{n} a_i & ; n \geq 1 \end{cases}$$

The distribution function of waiting time w for a customer who have to wait is given by

$$P(w \le t) = \begin{cases} P_0 = 1 - \dfrac{\lambda}{\mu} & ; n = 0 \\ P\left\{\sum_{i=1}^{n} a_i \le t\right\} & ; n \ge 1 \end{cases}$$

Since for each customer, the service time is independent, therefore its probability density function

$$\Psi_a(t) = \mu e^{-\mu t}, t > 0$$

where μ is the service rate.

Thus, $\psi_w(t) = \left\{\sum_{i=1}^{n} [P_n \text{ . Probability of } (n\text{-}1) \text{ customers got service at time } t]\right\}$

x (Probability of one customers is waiting service at time Δt)

$$= \sum_{i=1}^{n}\left(1-\frac{\lambda}{\mu}\right)\left(\frac{\lambda}{\mu}\right)^{n}\left(\frac{(\mu t)^{n-1} e^{-\mu t}}{(n-1)!}\right).\mu \ \Delta t$$

$$= \left(1-\frac{\lambda}{\mu}\right)\left(\frac{\lambda}{\mu}\right)\mu\, e^{-\mu t}\sum_{n=1}^{\infty}\frac{(\mu t)^{n-1}\left(\dfrac{\lambda}{\mu}\right)^{n-1}}{(n-1)!}\Delta t$$

$$= \lambda\left(1-\frac{\lambda}{\mu}\right)e^{-\mu t}\sum_{n=1}^{\infty}\frac{(\lambda t)^{n-1}}{(n-1)!}\Delta t$$

$$= \lambda\left(1-\frac{\lambda}{\mu}\right)e^{-\mu t}\left[1+\frac{\lambda t}{1!}+\frac{(\lambda t)^2}{2!}+\ldots\ldots\right]\Delta t$$

$$= \lambda\left(1-\frac{\lambda}{\mu}\right)e^{-\mu t}.e^{\lambda t}\Delta t$$

$$\psi_w(t) = \lambda\left(1-\frac{\lambda}{\mu}\right)e^{-(\mu-\lambda)t}\ \Delta t \qquad \ldots\text{(vi)}$$

Step V : To find busy period distribution

The statement "Customer has to wait" is meant that the service remains busy in the busy period. Let w denote the total time (waiting + service) that a customer spends in the system. Let ψ $(w : w > 0)$ is probability density function for waiting time. Then distribution is given by

$$\psi(w : w > 0) = \frac{\psi(w)}{\text{Probability } (w > 0)}$$

$$= \frac{\psi(w)}{\int_0^\infty \psi(w)\, dw}$$

$$= \frac{\lambda\left(1-\frac{\lambda}{\mu}\right)e^{-(\mu-\lambda)w}}{\int_0^\infty \lambda\left(1-\frac{\lambda}{\mu}\right)e^{-(\mu-\lambda)w}dw}. \qquad \text{[from Eq. (vi)]}$$

$$= \frac{\lambda\left(1-\frac{\lambda}{\mu}\right)e^{-(\mu-\lambda)w}}{\frac{\lambda}{\mu}}$$

$$= (\mu - \lambda)\, e^{-(\mu - \lambda)\, w};\ w = 0$$

Now, $\int_0^\infty \psi(w : w > 0)dw = \int_0^\infty (\mu-\lambda)\, e^{-(\mu-\lambda)w}\, dw = 1$

Which is the required probability density function for the busy period.

Measures of Model I

(i) Expected (average) number of customers in the system

$$L_s = \sum_{n=0}^{n} nP_n$$

$$= \sum_{n=0}^{n} n\left(\frac{\lambda}{\mu}\right)^n\left(1-\frac{\lambda}{\mu}\right)$$

$$= \left(1-\frac{\lambda}{\mu}\right)\sum_{n=0}^{n} n\left(\frac{\lambda}{\mu}\right)^n$$

$$= \left(1-\frac{\lambda}{\mu}\right)\frac{\lambda}{\mu}\sum_{n=1}^{n} n\left(\frac{\lambda}{\mu}\right)^{n-1}$$

$$= \left(1 - \frac{\lambda}{\mu}\right)\left(\frac{\lambda}{\mu}\right)\left[1 + 2\frac{\lambda}{\mu} + 3\left(\frac{\lambda}{\mu}\right)^2 + \ldots\ldots\right]$$

$$= \left(1 - \frac{\lambda}{\mu}\right) \frac{\lambda}{\mu}\left[\frac{1}{\left(1 - \frac{\lambda}{\mu}\right)^2}\right]$$

$$= \frac{\frac{\lambda}{\mu}}{1 - \frac{\lambda}{\mu}}$$

$$L_s = \frac{\lambda}{\mu - \lambda}$$

or $$L_s = \frac{\rho}{1 - \rho}, \ where\ \rho = \frac{\lambda}{\mu}$$

(ii) Expected number of customers waiting in the queue (i.e., queue length)

Since there are (n-1) customers in the queue excluding one being served, then

$$L_q = \sum_{n=1}^{\infty}(n-1)\, P_n$$

$$= \sum_{n=1}^{\infty} nP_n - \sum_{n=1}^{\infty} P_n$$

$$= \sum_{n=1}^{\infty} nP_n - \left[\sum_{n=0}^{n} P_n - P_0\right]$$

$$= L_s - (1 - P_0)$$

$$\left(\because \sum_{n=0}^{\infty} P_n \ = 1\right)$$

$$= L_s - 1 + \left(1 - \frac{\lambda}{\mu}\right)$$

$$= \frac{\lambda}{\mu - \lambda} - 1 + 1 - \frac{\lambda}{\mu}$$

$$L_q = \frac{\lambda^2}{\mu(\mu - \lambda)}$$

or $$L_q = \frac{\rho^2}{1 - \rho}$$

(iii) Expected waiting time in the queue

Since expected time an arrival spends in the queue is

$$W_q = \int_0^\infty t \left\{ \frac{d}{dt} \psi_w(t) \right\} dt$$

$$= \int_0^\infty t.\lambda \left(1 - \frac{\lambda}{\mu}\right) e^{-(\mu - \lambda)t} dt$$

Integrating by parts, we get

$$W_q = \lambda \left(1 - \frac{\lambda}{\mu}\right) \left[t.\frac{e^{-(\mu-\lambda)t}}{-(\mu - \lambda)} - \frac{1}{(\mu - \lambda)^2} e^{-(\mu-\lambda)t} \right]_0^\infty$$

$$= \lambda \left(1 - \frac{\lambda}{\mu}\right) \frac{1}{(\mu - \lambda)^2}$$

$$W_q = \frac{\lambda}{\mu(\mu - \lambda)}$$

or $$W_q = \frac{\rho}{\mu(1 - \rho)} = \frac{L_q}{\lambda}$$

(iv) Expected waiting time in the system

W_s = expected waiting time inqueue + Expected service time

$$= W_q + \frac{1}{\mu}$$

$$= \frac{\lambda}{\mu(\mu - \lambda)} + \frac{1}{\mu}$$

$$W_s = \frac{1}{\mu - \lambda}$$

or $$W_s = \frac{L_s}{\lambda}$$

(v) Expected waiting time of a customer who has to wait

The expected length of the busy period is given by

$$(w{:}w>0) = \int_0^\infty t.\left\{\frac{d}{dt}\psi_w(t)\right\}dt$$

$$= \int_0^\infty t.(\mu - \lambda)e^{-(\mu-\lambda)t}dt$$

Integrating by parts, we get

$$(w{:}w>0) = \frac{1}{\mu - \lambda} = \frac{1}{\mu(1-\rho)}$$

(vi) Expected length of non-empty queue

$$L_b = \frac{\text{Expected length of waiting time}}{\text{Probability (n>0)}}$$

$$= \frac{L_q}{1-P_0} = \frac{\frac{\lambda}{\mu}\Big/\left(1-\frac{\lambda}{\mu}\right)}{\frac{\lambda}{\mu}}$$

$$L_b = \frac{\mu}{\mu - \lambda} = \frac{1}{1-\rho}$$

(vii) Probability that the number of customers in the system is ≥ k

$$P(n \geq k) = \left(\frac{\lambda}{\mu}\right)^k, \; P(n > k) = \left(\frac{\lambda}{\mu}\right)^{k+1}$$

Proof: $P(n \geq k) = \sum_{n=k}^{\infty} P_k$

$$= \sum_{n=k}^{\infty} \left(1-\frac{\lambda}{\mu}\right)\left(\frac{\lambda}{\mu}\right)^k$$

$$= \left(1-\frac{\lambda}{\mu}\right)\left(\frac{\lambda}{\mu}\right)^k \sum_{n=k}^{\infty}\left(\frac{\lambda}{\mu}\right)^{n-k}$$

$$= \left(1-\frac{\lambda}{\mu}\right)\left(\frac{\lambda}{\mu}\right)^k \left[1+\frac{\lambda}{\mu}+\left(\frac{\lambda}{\mu}\right)^2+\ldots\ldots\ldots\ldots\right]$$

$$P(n \geq k) = \frac{\left(1-\frac{\lambda}{\mu}\right)\left(\frac{\lambda}{\mu}\right)^k}{\left(1-\frac{\lambda}{\mu}\right)} = \left(\frac{\lambda}{\mu}\right)^k = \rho^k$$

and $\qquad P(n > k) = \left(\frac{\lambda}{\mu}\right)^{k+1}$

(viii) The variance (fluctuation) of queue length

$$\text{Var}(n) = \sum_{n=1}^{\infty} n^2 P_n - \left(\sum_{n=1}^{\infty} nP_n\right)^2$$

$$= \sum_{n=1}^{\infty} n^2 P_n - (L_s)^2$$

$$= \sum_{n=1}^{\infty} n^2 (1-\rho)\rho^n - \left(\frac{\rho}{1-\rho}\right)^2$$

$$= (1\text{-}\rho)\,[1^2.\rho + 2^2.\rho^2 + 3^2\,.\,\rho^3 + \ldots\ldots\ldots\ldots] - \frac{\rho^2}{(1-\rho)^2}$$

$$= (1\text{-}\rho)\rho\,[1 + 2^2.\rho + 3^2.\rho^2 + \ldots\ldots\ldots\ldots] - \frac{\rho^2}{(1-\rho)^2}$$

$$= \rho\,(1\text{-}\rho)\frac{1+\rho}{(1-\rho)^3} - \frac{\rho^2}{(1-\rho)^2}$$

$$\text{var}(n) = \frac{\rho}{(1-\rho)^2} = \frac{\lambda\mu}{(\mu-\lambda)^2}$$

(ix) Probability that the queue is non-empty

$$P(n > 1) = 1 - P_0 - P_1$$

$$= 1 - \left(1-\frac{\lambda}{\mu}\right) - \left(\frac{\lambda}{\mu}\right)\left(1-\frac{\lambda}{\mu}\right)$$

$$= \left(\frac{\lambda}{\mu}\right)^2$$

Illustrated Examples on Model I

Example 1 *In a railway yard, goods trains arrive at a rate of 30 trains per day. Assuming that the inter-arrival time follows an exponential distribution and the service time (the time taken to hump a train) distribution is also exponential with an average 36 minutes. Calculate the following*

(i) *The average number of trains in a queue.*

(ii) *The probability that the queue size exceeds 10.*

If the input of trains increases to an average 33 per day, what will be change in (i) and (ii) ?

Solution Let $\lambda = \frac{30}{60 \times 24} = \frac{1}{48}$ trains/minute

and $\mu = \frac{1}{36}$ trains/minute

Then, traffic intensity $\rho = \frac{\lambda}{\mu} = \frac{36}{48} = 0.75$

(i) The average number of trains in a queue

$$L_s = \frac{\rho}{1-\rho}$$

$$= \frac{0.75}{1-0.75} = 3 \text{ trains}$$

(ii) The probability that the queue size exceeds 10 is

$$P(n \geq 10) = \rho^{10} = (0.75)^{10} = 0.06$$

When the input increases to an average 33 trains per day,

Then $\lambda = \frac{33}{60 \times 24} = \frac{11}{480}$ trains/minute

and $\mu = \frac{1}{36}$ trains/minute

Then traffic intensity $\rho = \dfrac{\lambda}{\mu} = \dfrac{11\times 36}{480} = 0.83$

Hence, (i) $L_s = \dfrac{\rho}{1-\rho}$

$= \dfrac{0.83}{1-0.83} = 5$ trains

(iii) Probability $(n \geq 10) = \rho^{10}$

$= (0.83)^{10} = 0.2$ (approx.)

Example 2 *A company distributes its products by trucks loaded at its only loading station. Both, company's trucks and contractor's trucks, are used for this purpose. It was fount out that on an average every five minutes, one truck arrived and the average loading time was three minutes. 50% of the trucks belong to the contractor. Find out*

(i) *The probability that a truck has to wait.*

(ii) *The waiting time of truck has to wait.*

(iii) *The expected waiting time of contractor's truck per day, assuming a 24-hours shift.*

Solution Average arrival rate of truck $\lambda = \dfrac{60}{5} = 12$ trucks/hour

and average service rate of truck $\mu = \dfrac{60}{3} = 20$ trucks/hour.

(i) The probability that a truck has to wait is

$$\rho = \frac{\lambda}{\mu} = \frac{12}{20} = 0.6$$

(ii) The waiting time of truck per day is

$$W_s = \frac{1}{\mu-\lambda} = \frac{1}{20-12} = \frac{1}{8} \text{ hour} = 7.5 \text{ minutes}$$

(iii) The expected waiting time of contractor's truck per day is

= (no. of truck per day) × (contractor's percentage) × (expected time)

$$= (12\times 24)\times\frac{50}{100}\times\frac{\lambda}{\mu(\mu-\lambda)}$$

$$= 288 \times \frac{1}{2} \times \frac{12}{20 \times 8}$$

$$= 10.8 \text{ hours}$$

Example 3 *A computer repairman finds that the time spent on his jobs has an exponential distribution with mean 30 minutes. If he repairs sets in the order in which they come in, and if the arrival of sets is poisson with an average rate of 10 per 8 hours day, what is repairman's expected idle time each day? How many jobs are ahead of the average set just brought in?*

Solution An arrival rate is $\lambda = \frac{10}{8 \times 60} = \frac{1}{48} \frac{\text{jobs}}{\text{minute}}$

and service rate $\mu = \frac{1}{30} \frac{\text{jobs}}{\text{minute}}$

Therefore, expected number of jobs are

$$L_s = \frac{\frac{\lambda}{\mu}}{1 - \frac{\lambda}{\mu}}$$

$$= \frac{\lambda}{\mu - \lambda} = \frac{\frac{1}{48}}{\frac{1}{30} - \frac{1}{48}} = \frac{12}{3} \text{ jobs}$$

The number of hours for which the repairman remains busy in 8 hours day is

$$= 8 \times \frac{\lambda}{\mu}$$

$$= 8 \times \frac{\frac{1}{48}}{\frac{1}{30}} = 8 \times \frac{30}{48} = 5 \text{hours}$$

Example 4 *Customers arrive at a sales counter manned by a single person according to a poisson process with a mean rate of 20 per hour. The time required to serve a customer has an exponential distribution with a mean of 100 seconds. Find the average waiting time of a customer.*

Solution The average rate $\lambda = 20$ per hour.

and the service rate $\mu = \frac{60 \times 60}{100} = 36$ per hour.

The average waiting time of a customer in the queue is

$$W_s = \frac{\lambda}{\mu(\mu-\lambda)}$$

$$= \frac{20}{36(36-20)} = \frac{20}{36\times 16} \text{ hours}$$

$$= \frac{20\times 3600}{36\times 16} = 125 \text{ seconds}$$

The average waiting time of a customer in the system is

$$W_s = \frac{1}{\mu-\lambda} = \frac{1}{36-20} = \frac{1}{16} \text{ hour}$$

$$= \frac{3600}{16} = 225 \text{ seconds.}$$

Example 5 *At what average rate must a salesman at a shop work in order to ensure a probability of 0.90 that the customer will not have to wait longer than 12 minutes? It is assumed that there is only one counter to which customer arrive in a poisson distribution at an average rate of 15 per hour. The length of service by the salesmen has an exponential distribution.*

Solution Here, arrival rate λ = 15 per hour = $\frac{15}{60} = \frac{1}{4}$ customer/minute

Service rate μ = ?

Probability (waiting time ≥ 12) = 1-0.90 = 0.10

Hence,
$$\int_{12}^{\infty} \lambda\left(1-\frac{\lambda}{\mu}\right) e^{-(\mu-\lambda)t}\, dt = 0.10$$

or
$$\lambda\left(1-\frac{\lambda}{\mu}\right)\left[\frac{e^{-(\mu-\lambda)t}\, dt}{-(\mu-\lambda)}\right]_{12}^{\infty} = 0.10$$

or
$$\frac{\lambda}{\mu} e^{-(\mu-\lambda)\times 12} = 0.10$$

or
$$\frac{1}{4\mu} e^{\left(12\times\frac{1}{4}-12\mu\right)} = 0.10$$

or
$$e^{(3-12\mu)} = 0.40\mu$$

$$\frac{1}{\mu} = 2.48 \frac{\text{minute}}{\text{service}}$$

$$\mu = \frac{1}{2.48} \frac{\text{minute}}{\text{service}}$$

Practice Problems

1. On an average 96 patients per 24 hours day require the service of an emergency clinic. Also on average, a patient requires 10 minutes of active attention. Assume that the facility can handle only one emergency at a time. Suppose that it costs the clinic Rs.100 per patient treated to obtain an average servicing time of 10 minutes and that each minute of decrease in this average time would cost Rs.10 per patient treated. How much would have to be budgeted by the clinic to decrease the average size of the queue from $1\frac{1}{3}$ patients to $\frac{1}{2}$ patient.

   [**Ans.** The budget should be increased from Rs.100 to Rs.125 per patient]

2. Telephone users arrive at a booth following a poisson distribution with an average time of 5 minutes between one arrival and the next. The time taken for a telephone call is on an average 3 minutes and it follows an exponential distribution. What is the probability that the booth is busy ? How many more booths should be established to reduce the waiting time to less than or equal to half of the present waiting time ?

   [**Ans.** $Wq = \frac{3}{40}$ hour, $Ws = \frac{1}{8}$ hour, waiting time is required to be reduced to half (μ') = 28 per hour. The number of booths =1.40] ,

3. A bank counter window has a mean service time of 2 minutes and customers arrive at the rate of 20 per hour. Assume that these represent rates with a poisson distribution

   (i) What percent of time will the counter be idle ?

   (ii) How long will it take the mean customer to wait in line and beserved ?

   (iii)What fraction of customers will have to wait in the line ?

   [**Ans.** (i) 33.3% of time (ii) $\frac{1}{10}$ hour (iii) $\frac{2}{3}$ hour or 66.67%]

4. A firm has single machinist in a repair shop. He works eight hours a day and on an average four machines break each day. It takes on an average one hour to repair a machine. Using poisson exponential distribution, determine

   (a) The average number of machine in the repair shop.

   (b) The average number of machine in the shop on which the machinist has not started to work.

   (c) The expected down time per machine.

(d) The average time a machine waits for service.

(e) The expected proportion of time facility will be idle.

[**Ans.** (a) 1machine (b) 0.50 machine (c) 2 hours after failing (d) 1hour (e) 0.50 the repairman will spend 50% of each working day = 4 hour.]

5. An airlines has one reservation clerk on duty in its local branch at any given time. The clerkhandle information regarding passenger reservation and flight timings. Assume that the number of customers arriving during any given period is poisson distribution with an arrival rate of 8 per hour and that the clerk can serve a customer in 6 minutes on an average, with an exponential distribution service time.

(a) What is the probability that the system is busy ?

(b) What is the expected time a customer spends in the system?

(c) What is the average length of the queue and what is the number of customers in the system?

[**Ans.** (a) 0.8 (b) 30 minutes (c) 3.2 customers and 4 customers]

Model II. (M/M/1) : (N/∞ / FCFS). Exponential Service, Finite (or Limited) Queue

The capacity of the system of this model is finite (limited). Suppose that no more than N customers can be arrive at any time in the system. Thus any customer arriving when the system is already contains N customers does not enter the system and is lost.

Step I : To obtain steady-state differential equations.

The differential equations derived in model I will also be same for this model as long as n < N. The equations are

$$P_0(t+\Delta t) = P_0(t)\,[1-\lambda\,\Delta t\,] + P_1(t)\,\mu\,\Delta t + O(\Delta t), \text{ for } n = 0 \quad \ldots\text{(i)}$$

$$P_n(t+\Delta t) = P_n(t)\,[1-(\lambda+\mu)\,\Delta t] + P_{n-1}(t)\,\lambda\,\Delta t + P_{n+1}(t)\,\mu\,\Delta t + O(\Delta t), \quad \ldots\text{(ii)}$$

for $n = 1,2 \ldots N\text{-}1$

and putting $n = N,\ P_{N+1}(t) = 0,\ \lambda = 0$, we have

$$P_N(t+\Delta t) = P_N(t)\,[1-\mu\Delta t] + P_{N-1}(t)\lambda\,\Delta t + O(\Delta t) \quad \ldots\text{(iii)}$$

Now, dividing Eqs. (i), (ii) and (iii) by at Δt and taking limit as Δt→ 0 these equations are

$$P_0'(t) = -\lambda P_0(t) + \mu P_1(t) \quad \ldots\text{(iv)}$$

$$P_n^1(t) = -(\lambda+\mu)\,P_n(t) + \lambda P_{n-1}(t) + \mu P_{n+1}(t) \quad \ldots\text{(v)}$$

and
$$P_N^1(t) = -\mu P_N(t) + \lambda P_{N-1}(t) \quad \ldots\text{(vi)}$$

For steady-state condition, $t \to \infty$, $P_n(t) \to P_n$ and $P_n^1(t) \to 0$ Eqs. (iv), (v) and (vi) are

$$0 = -\lambda P_0 + \mu P_1 \quad \text{...(vii)}$$

$$0 = -(\lambda + \mu)\, P_n + \lambda P_{n-1} + \mu P_{n+1} \quad \text{...(viii)}$$

$$0 = -\mu P_N + \lambda P_{N-1} \quad \text{...(ix)}$$

Step II : To solve the system of differential equation.

Here, $P_0 = P_0$ (initial condition)

$$P_1 = \frac{\lambda}{\mu} P_0 \quad \text{(From Eq. (vii))}$$

$$P_2 = \left(\frac{\lambda}{\mu}\right)^2 P_0 \quad \text{(putting n = 1 in Eq. (viii))}$$

Similarly, $P_3 = \left(\frac{\lambda}{\mu}\right)^3 P_0$

In general, $P_n = \left(\frac{\lambda}{\mu}\right)^n P_0,\ n < N$ ...(x)

$$P_N = \left(\frac{\lambda}{\mu}\right)^N P_0,\ n = N$$

Now, to obtain P_0, we know that

$$\sum_{n=0}^{N} P_n = 1$$

$$P_0\left[1 + \frac{\lambda}{\mu} + \left(\frac{\lambda}{\mu}\right)^2 + \dots\dots\dots + \left(\frac{\lambda}{\mu}\right)^N\right] = 1$$

or $P_0\left[\frac{1-\rho^{N+1}}{1-\rho}\right] = 1$, where $\rho = \frac{\lambda}{\mu}$

or $P_0 = \frac{1-\rho}{1-\rho^{N+1}}$ ...(xi)

Substituting the value of P_0 in Eq. (x), we have

$$P_n = \left[\frac{1-\rho}{1-\rho^{N+1}}\right]\rho^n, \text{ for n = 0,1,2 .... N} \quad \text{...(xii)}$$

Thus, Eqs. (xi) and (xii) gives the required solution of this model which do not require $\lambda < \mu$ *i.e.*, ρ may be greater than 1.

Measures for Model II

1. Expected number of customers in the system

$$L_s = \sum_{n=0}^{N} nP_n$$

$$= \sum_{n=0}^{N} n\left(\frac{1-\rho}{1-\rho^{N+1}}\right)\rho^n$$

$$= \frac{1-\rho}{1-\rho^{N+1}} \sum_{n=0}^{N} n\rho^n$$

$$= P_0 \sum_{n=0}^{N} n\rho^n$$

2. Expected queue length

$$L_q = L_s - \frac{\lambda}{\mu}$$

3. Expected waiting time of a customer in the system

$$W_s = \frac{L_s}{\lambda}$$

4. Expected waiting time of a customer in the queue

$$W_q = W_s - \frac{1}{\mu} = \frac{L_q}{\lambda}$$

5. Fraction of time system is full (Fraction of potential customers lost)

$$P_N = P_0 \rho^N$$

Effective arrival rate, $\lambda_e = \lambda(1-P_N)$

Effective traffic intensity, $\rho_e = \frac{\lambda_e}{\mu}$

Illustrated Examples on Model II

Example 6 *If for a period of 2 hours in a day (8-10 A.M.)trains arrive at the yard every 20 minutes but the service time continues to remain 36 minutes, then calculate for this period.*

(i) the probability that the yard is empty.

(ii)average queue length.

On the assumption that the line capacity of the yard is limited to 4 trains only.

Solution Here, arrival rate $\lambda = \frac{1}{20}$ train/minute

And service rate $\mu = \frac{1}{36}$ train/minute

$\therefore$ Traffic intensity $\rho = \frac{\lambda}{\mu} = \frac{36}{20} = 1.8$

and $N = 4$ trains.

(i) Probability $P_0 = \left[\frac{\rho - 1}{\rho^{N+1} - 1}\right] = \frac{1.8 - 1}{(1.8)^5 - 1} = 0.04$

(ii)Average queue length

$$= P_0 \sum_{n=0}^{N} n\rho^n$$

$$= 0.04 \sum_{n=0}^{4} n\rho^n$$

$$= 0.04\,[\rho + 2\rho^2 + 3\rho^3 + 4\rho^4]$$

$$= 0.04\,[1.8 \times 2\,(1.8)^2 + 3(1.8)^3 + 4(1.8)^4]$$

$$= 0.04 \times 72 = 29, \text{ say 3 trains}$$

Example 7 *Consider a single server queuing system with poisson input, exponential service times. Let the mean arrival rate is 3 calling units per hour, the expected service time is 0.25 hour and the maximum calling units in the system is two. Derive the steady-state probability distribution of the number of calling units in the system and then calculate the expected number in the system.*

Solution Here, arrival rate $\lambda = 3$ units per hour

Service rate $\mu = 4$ units per hour

and $N = 2$

Then traffic intensity $\rho = \dfrac{\lambda}{\mu} = \dfrac{3}{4} = 0.75$

The steady-state probability distribution of n calling units in the system is

$$P_n = \left(\frac{1-\rho}{1-\rho^{N+1}}\right)\rho^n$$

$$= \left[\frac{1-0.75}{1-(0.75)^3}\right](0.75)^n$$

$$= (0.43)\,(0.75)^n$$

and $$P_0 = \frac{1-\rho}{1-\rho^{N+1}} = \frac{1-0.75}{1-(0.75)^3} = 0.43$$

The expected number of calling units in the system is

$$L_s = \sum_{n=0}^{N} nP^n$$

$$= \sum_{n=0}^{2} n(0.43)(0.75)^n$$

$$= (0.43)\sum_{n=0}^{2} n\,(0.75)^n$$

$$= (0.43)\,[0.75 + 2(0.75)^2]$$

$$= 0.81$$

Example 8 *Assume that goods trains are coming in a yard, the service time for each train is assumed to be exponential with an average of 36 minutes. If the yard can admit of 9 trains only (these being 10 lines, one of which is car marked for the shunting engine to reverse itself from the crest of the hump to the rare of the train). Calculate the following on the assumption that 30 trains, on average are received in the yard.*

(i) *The probability that the yard is empty.*

(ii) *Average queue length.*

Solution Here, arrival rate $\lambda = \dfrac{30}{60\times 24} = \dfrac{1}{48}\,\dfrac{\text{trains}}{\text{minute}}$

Service rate $\mu = \frac{1}{36}\frac{\text{trains}}{\text{minute}}$

$\therefore$ Traffic intensity $\rho = \frac{\lambda}{\mu} = \frac{36}{48} = 0.75$

and $N = 9$

(i) The probability is $P_0 = \frac{1-\rho}{1-\rho^{N+1}}$

$$= \frac{1-0.75}{1-(0.75)} = \frac{0.25}{0.90} = 0.28$$

(ii) Average queue length is

$$L_s = \left(\frac{1-\rho}{1-\rho^{N+1}}\right)\sum_{n=0}^{N} n\rho^n$$

$$= \frac{1-0.75}{1-(0.75)^{10}}\sum_{n=0}^{9} n(0.75)^n$$

$$= 0.28 \times [0.75 + 2(0.75)^2 + \ldots\ldots\ldots + 9(0.75)^9]$$

$$= 0.28 \times 9.58$$

$$= 2.79, \text{ say, } 3 \text{ trains.}$$

Model III (m/m/s): (∞/∞/FCFS). Exponential Service – Unlimited Queue

This model is of the type multi-server queuing models. In the queuing system the customers arrive in a poisson distribution with mean arrival rate λ. There are S number of service stations arranged in parallel and the customer can go to any of the server for his service where the service time at each server is identical and follows the same exponential distribution. The mean service rate is μ per busy server. The overall service rate of server is obtained in two situations, when n customers are in the system.

(i) If $n < S$, then there will be no queue, $(S\text{-}n)$ number of server may remain idle and then

$\mu_n = n\mu \,;\, n < S$

(ii) If $n \geq S$, then all servers will be busy and the maximum number of customers in the queue will be (n -S). Then $\mu_n = S\mu \,;\, n \geq S$

Thus, to derive the results for this model, we get

$$\lambda_n = \lambda \,;\, n \geq 0$$

$$\mu_n = \begin{cases} n\mu & ; n < S \\ S\mu & ; n \geq S \end{cases}$$

The method of determining probability P_n of n customers in the system at time t are obtained below:

Step I: Obtain the system of differential-difference equations.

By the similar arguments as in model I, we get

$$P_n(t+\Delta t) = P_n(t)[1-\lambda\Delta t][1-n\mu\Delta t] + P_{n+1}(t)[1-\lambda\Delta t][(n+1)]\mu\Delta t] + P_{n-1}(t)[\lambda\Delta t][1-(n-1)\mu\Delta t]$$

$$= -(\lambda + n\mu)P_n(t)\Delta t + (n+1)\mu P_{n+1}(t)\Delta t + \lambda P_{n-1}(t)\Delta t + P_n(t) + (\Delta t)^2,\ n < s \quad \text{...(i)}$$

$$P_n(t+\Delta t) = P_n(t)[1-\lambda\Delta t][1-S\mu\Delta t] + P_{n+1}(t)[1-\lambda\Delta t][S\mu\Delta t] + P_{n-1}(t)[\lambda\Delta t][1-S\mu\Delta t]$$

$$= -(\lambda + S\mu)P_n(t)\Delta t + S\mu P_{n+1}(t)\Delta t + \lambda P_{n-1}(t)\Delta t + P_n(t) + (\Delta t)^2,\ n \geq s \quad \text{...(ii)}$$

and $$P_0(t+\Delta t) = P_0(t)[1-\lambda\Delta t] + P_1(t)\mu\Delta t;\ n=0 \quad \text{...(iii)}$$

Dividing Eqs. (i), (ii) and (iii) by Δt and taking limit as $\Delta t \to 0$, we get

$$P_n^1(t) = -(\lambda + n\mu)P_n(t) + (n+1)\mu P_{n+1}(t) + \lambda P_{n-1}(t),\ n < S \quad \text{...(iv)}$$

$$P_n^1(t) = -(\lambda + S\mu)P_n(t) + S\mu P_{n+1}(t) + \lambda P_{n-1}(t),\ n \geq S \quad \text{...(v)}$$

and $$P_0'(t) = -\lambda P_0(t) + \mu P_1(t),\ n = 0 \quad \text{...(vi)}$$

Step II : Obtain the system of steady-state equations.

In the steady-state condition, when $t \to \infty$, $P_n(t) \to P_n$ and $P_n^1(t) \to 0$, for all n, then Eqs. (iv) (v) and (vi) becomes

$$0 = -(\lambda + n\mu)P_n + (n+1)\mu P_{n+1} + \lambda P_{n-1},\ n<S \quad \text{...(vii)}$$

$$0 = -(\lambda + S\mu)P_n + S\mu P_{n+1} + \lambda P_{n-1},\ n \geq S \quad \text{...(viii)}$$

and $$0 = -\lambda P_0 + \mu P_1,\ n=0 \quad \text{...(ix)}$$

Step III : To solve the system of differential equations.

Here, $$P_1 = \frac{\lambda}{\mu}P_0 \quad \text{[from eq. (viii)]}$$

Putting $n = 1$ in Eq. (vii), we get

$$P_2 = \frac{\lambda}{2\mu}P_1 = \frac{1}{2!}\left(\frac{\lambda}{\mu}\right)^2 P_0$$

Putting $n = 2$, $$P_3 = \frac{1}{3!}\left(\frac{\lambda}{\mu}\right)^3 P_0$$

In general $$P_n = \frac{\lambda}{n\mu} P_{n-1} = \frac{1}{n!}\left(\frac{\lambda}{\mu}\right)^n P_0, n \le S$$

....

....

$$P_S = \frac{\lambda}{S\mu} P_{S-1} = \frac{1}{S!}\left(\frac{\lambda}{\mu}\right)^S P_0$$

$$P_{S+1} = \frac{\lambda}{S\mu} P_S = \frac{1}{S}\cdot\frac{1}{S!}\left(\frac{\lambda}{\mu}\right)^{S+1} P_0$$

$$P_{S+2} = \frac{\lambda}{S\mu} P_{S+1} = \frac{1}{S^2}\cdot\frac{1}{S!}\left(\frac{\lambda}{\mu}\right)^{S+2} P_0$$

In general $$P_n = P_{S+(n-S)} = \frac{1}{S^{n-S}}\cdot\frac{1}{S!}\left(\frac{\lambda}{\mu}\right)^n P_0, \text{for } n \ge S$$

Now, to find P_n by using the fact that

$$\sum_{n=0}^{\infty} P_n = 1$$

or $$\sum_{n=0}^{S-1} P_n + \sum_{n=s}^{\infty} P_n = 1 \qquad (n \le S - 1)\ (n \le S)$$

or $$\sum_{n=0}^{S-1} \frac{1}{n!}\left(\frac{\lambda}{\mu}\right)^n P_0 + \sum_{n=S}^{\infty} \frac{1}{S^{n-1}}\cdot\frac{1}{S!}\left(\frac{\lambda}{\mu}\right)^n P_0 = 1$$

or $$P_0\left(\sum_{n=0}^{S-1} \frac{S^n}{n!}\left(\frac{\lambda}{S\mu}\right)^n + \frac{1}{S!}\sum_{n=S}^{\infty} \frac{S^n}{S^{n-1}}\cdot\left(\frac{\lambda}{S\mu}\right)^n\right) = 1$$

Let traffic intensity $\rho = \lambda/S\mu$, Then

$$P_0\left(\sum_{n=0}^{S-1} \frac{(S\rho)^n}{n!} + \frac{S^S}{S!}\sum_{n=S}^{\infty} \rho^n\right) = 1$$

or $$P_0 = \left(\sum_{n=0}^{S-1} \frac{(S\rho)^n}{n!} + \frac{S^S}{S!} \frac{\rho^S}{1-\rho} \right)^{-1} \quad \text{...(x)}$$

$$\left(\because \sum_{n=s}^{\infty} \rho^n = \rho^S + \rho^{S+1} + \rho^{S+2} \text{..........} \infty = \frac{\rho^S}{1-\rho} \right)$$

Thus, steady-state distribution of n arrivals is

$$P_n = \begin{cases} \dfrac{1}{n!}\left(\dfrac{\lambda}{\mu}\right)^n P_0 & ; \ for\ n \le S-1 \\ \dfrac{1}{S!}\dfrac{1}{S^{n-s}}\left(\dfrac{\lambda}{\mu}\right)^n P_0 & ; \ for\ n \ge S \end{cases} \quad \text{...(xi)}$$

Measures for Model III

1. Expected number of customers waiting in the queue

$$L_q = \sum_{n=s}^{\infty} (n-s) P_n$$

$$= \sum_{n=s}^{\infty} (n-s) \frac{\rho^n}{s^{n-S}.s!} P_0$$

$$= \frac{\rho^s.P_0}{s!} \sum_{n=s}^{\infty} (n-s) \rho^{n-s}$$

$$= \frac{\rho^s.P_0}{s!} \sum_{a=0}^{\infty} a\rho^a, n-s=a$$

$$= \frac{\rho^s}{s!} P_0.\rho \sum_{a=0}^{\infty} a\rho^{a-1}$$

$$= \frac{\rho^s}{s!}.\rho\, P_0 \frac{d}{d\rho}\left[\sum_{a=1}^{\infty} \rho^a \right]$$

$$= \frac{\rho^s}{s!}.\rho\, P_0 \frac{1}{(1-\rho)^2}$$

$$= \frac{1}{s!}\left(\frac{\lambda}{\mu}\right)^s \frac{\lambda . s\mu}{(s\mu-\lambda)^2} P_0, \ \rho = \frac{\lambda}{\mu}$$

$$L_q = \frac{1}{(s-1)!}\left(\frac{\lambda}{\mu}\right)^s \frac{\lambda\mu}{(s\mu-\lambda)^2} P_0$$

2. **Expected number of customers in the system**

$$L_s = \sum_{n=0}^{\infty} n\,P_n = \sum_{n=0}^{s-1} n\,P_n + \sum_{n=s}^{\infty} n\,P_n$$

$$= \sum_{n=0}^{s-1} n\,P_n + \sum_{n=s}^{\infty}(n-s)\,P_n + s\sum_{n=s}^{\infty} P_n$$

$$= \sum_{n=0}^{s-1} n\,P_n + L_q + s\left[1-\sum_{n=0}^{s-1} P_n\right]$$

$$= L_q + s + \sum_{n=0}^{s-1}(n-s)\,P_n$$

$$= L_q + s + \sum_{n=0}^{s-1}(n-s)\frac{(s\rho)^n}{n!}P_0$$

$$L_s = L_q + \frac{\lambda}{\mu}$$

3. **Expected waiting time of a customer in the queue**

$$W_q = \frac{L_q}{\lambda}\frac{1}{(s-1)!}\left(\frac{\lambda}{\mu}\right)^s \frac{\mu}{(s\mu-\lambda)^2}.P_0$$

4. **Expected waiting time of a customer spends in the system**

$$W_s = W_q + \frac{1}{\mu}$$

$$= \frac{L_q}{\lambda} + \frac{1}{\mu}$$

5. **Probability that an arriving customer has to wait (i.e., busy period)**

$$\text{Prob. (busy period)} = \text{Probability } (n \geq s) = \sum_{n=s}^{\infty} \frac{1}{s!\,s^{n-s}}\left(\frac{\lambda}{\mu}\right)^n P_0$$

$$= \frac{1}{s!}\left(\frac{\lambda}{\mu}\right)^s P_0 \sum_{a=0}^{\infty}\left(\frac{\lambda}{\mu}\right)^a$$

$$\text{or Probability } (n \geq s) = \frac{1}{s!}\left(\frac{\lambda}{\mu}\right)^s P_0 \left(\frac{1}{1-\dfrac{\lambda}{\mu}}\right)$$

or $$= \frac{P_s}{1-\rho}, \text{ where } P_s = \frac{\left(\dfrac{\lambda}{\mu}\right)^s P_0}{s!}$$

Illustrated Examples on Model III

Example 9 *A bank has two tellers working on saving accounts. The first teller handles withdrawals only. The second teller handles deposits only. It has been found that the service time distribution for the deposits and withdrawals both are exponential with mean service time 3 minutes per customer. Depositors are found to arrive in a poisson distribution throughout the day with mean arrival rate 16 per hour. Withdrawals also arrive in a poisson distribution with mean arrival rate 14 per hour. What would be the effect on the average waiting time for depositors and withdrawals if each teller could handle both withdrawals and deposits. What would be the effect if this could only be accomplished by increasing the service time to 3.5 minutes?*

Solution Here, in the first case, $\lambda_1 = \frac{16}{60} = \frac{4}{15}$ per minutes

$\mu = \frac{1}{3}$ per minutes.

(i) Average waiting time in the queue is

$$W_q = \frac{\lambda_1}{\mu(\mu - \lambda_1)} = \frac{4}{15 \times \frac{1}{3}\left(\frac{1}{3} - \frac{4}{15}\right)} = 12 \text{ minutes.}$$

(ii) In the second case, $\lambda_2 = \frac{14}{60} = \frac{7}{30}$ perminutes, $\mu = \frac{1}{3}$ per minutes.

Average waiting time in the queue is

$$W_q = \frac{\lambda_2}{\mu(\mu - \lambda_2)} = \frac{7}{30 \times \frac{1}{3}\left(\frac{1}{3} - \frac{7}{30}\right)} = 7 \text{ minutes.}$$

Now, combining these two case, we get

$$\lambda = \lambda_1 + \lambda_2 = \frac{16}{60} + \frac{14}{60} = \frac{1}{2}, \mu = \frac{1}{3}$$

Since, it becomes the case of two server i.e. s=2,

Then $$\rho = \frac{\lambda}{s\mu} = \frac{\frac{1}{2}}{2\times\frac{1}{3}} = \frac{3}{4}$$

Therefore, $$P_0 = \left(\sum_{n=0}^{S-1} \frac{(s\rho)^n}{n!} + \frac{(s\rho)^S}{s!(1-\rho)}\right)^{-1}$$

$$= \left(\sum_{n=0}^{1} \frac{\left(2\times\frac{3}{4}\right)^n}{n!} + \frac{\left(2\times\frac{3}{4}\right)^2}{2!(1-\frac{3}{4})}\right)^{-1}$$

$$= \left(\sum_{n=0}^{1} \frac{1}{n!}\left(\frac{3}{2}\right)^n + \frac{1}{2}\times\frac{9}{4}\times 4\right)^{-1}$$

$$= \left(1 + \frac{3}{2} + \frac{9}{4}\right)^{-1} = (7)^{-1} = \frac{1}{7}$$

Average waiting time of arrival in the queue is

$$W_q = \frac{L_q}{\lambda} = \left(\frac{1}{(s-1)!}\left(\frac{\lambda}{\mu}\right)^s \frac{\mu}{(s\mu-\lambda)^2}\right) P_0$$

$$= \left(\frac{3}{2}\right)^2 \times \frac{\frac{1}{3}}{\left(\frac{2}{3}-\frac{1}{2}\right)^2} \times \frac{1}{7}$$

$$= \frac{27}{7} = 3.86 \text{ minutes.}$$

Again, combining waiting time with increased service time, when

$$\lambda = 30 \text{ per hour}, \ \mu = \frac{60}{3.5} = \frac{120}{7} \text{ per hour}$$

$$\frac{\lambda}{\mu}=\frac{21}{12}$$

We get, $$P_0=\left(\sum_{n=0}^{1}\frac{1}{n!}\left(\frac{21}{12}\right)^n+\frac{1}{2!}\left(\frac{21}{12}\right)^2\frac{2\times\frac{120}{7}}{2\times\frac{120}{7}-30}\right)^{-1}$$

$$=\left(1+\frac{21}{12}+\frac{1}{2}\times\frac{21\times21}{12\times12}\times8\right)^{-1}$$

$$=\left(1+\frac{7}{4}+\frac{49}{4}\right)^{-1}=(15)^{-1}=\frac{1}{15}$$

$\therefore$ $$W_q=\frac{L_q}{\lambda}=\left(\frac{1}{(s-1)!}\left(\frac{\lambda}{\mu}\right)^s\frac{\mu}{(s\mu-\lambda)^2}\right)P_0$$

$$=\left(\frac{7}{4}\right)^2\times\frac{\frac{120}{7}}{\left(\frac{30}{7}\right)^2}\times\frac{1}{15}$$

$$\frac{343}{30\times60}\text{ hour } 11.43 \text{ minutes.}$$

Example 10 *Four counters are being run to check the passport and papers of the tourist. The tourists choose a counter at random. If the arrival is poisson distribution at the rate λ and the service time is exponential with parameter $\frac{\lambda}{2}$. What is the steady average at each counter?*

Solution Given that, s = 4, service rate $\mu=\frac{\lambda}{2}$, arrival rate = λ

$\therefore$ $$\rho=\frac{\lambda}{s\mu}=\frac{\lambda}{4\times\frac{\lambda}{2}}=\frac{1}{2}$$

Therefore, $$P_0=\left(\sum_{n=0}^{S-1}\frac{(s\rho)^n}{n!}+\frac{(s\rho)^S}{s!(1-\rho)}\right)^{-1}$$

$$= \left(\sum_{n=0}^{3} \frac{2^n}{n!} + \frac{2^4}{4!} \frac{1}{\left(1 - \frac{1}{2}\right)} \right)^{-1}$$

$$= \left[1 + 2 + 2 + \frac{8}{6} + \frac{4}{3}\right]^{-1}$$

$$= \left(5 + \frac{8}{3}\right)^{-1} = \left(\frac{23}{3}\right)^{-1}$$

$$= \frac{3}{23}$$

Now, expected queue length is

$$L_q = \frac{\left(\frac{\lambda}{\mu}\right)^s}{s!} . \frac{\rho}{(1-\rho)^2} . P_0$$

$$= \frac{2^4}{4!} \times \frac{\frac{1}{2}}{\left(\frac{1}{2}\right)^2} \times \frac{3}{23} = \frac{4}{23}$$

Example 11 *A supermarket has two girls at the sales counters. If the service time for each customer is exponential with mean 4 minutes and if people arrive in a poisson distribution at the counter at the rate of 10 per hour.*

(i) *Calculate the probability that an arrival will have to wait for service.*

(ii) *Find the expected percentage of idle time for each girl.*

(iii) *If a customer has to wait, find the expected length of his waiting time.*

Solution (i) Probability of having to wait for service is

$$P(w > 0) = \frac{\left(\frac{\lambda}{\mu}\right)^s}{s!(1-\rho)} P_0$$

Here, $\lambda = \frac{1}{6}, \mu = \frac{1}{4}, s = 2 \text{ and } \rho = \frac{\lambda}{\mu s} = \frac{1}{3}$

Now, $$P_0 = \left(\sum_{n=0}^{S-1} \frac{(s\rho)^n}{n!} + \frac{(s\rho)^S}{s!(1-\rho)} \right)^{-1}$$

$$= \left(\sum_{n=0}^{1} \frac{\left(\frac{2}{3}\right)^n}{n!} + \frac{\left(\frac{2}{3}\right)^2}{2!\left(1-\frac{1}{3}\right)} \right)^{-1}$$

$$= \left(1 + \frac{2}{3} + \frac{\frac{4}{9}}{2\times\frac{2}{3}} \right)^{-1} = (2)^{-1} = \frac{1}{2}$$

Therefore, probability P (w > 0) $= \dfrac{\left(\frac{4}{6}\right)^2}{2!\left(1-\frac{1}{3}\right)} \times \dfrac{1}{2} = \dfrac{1}{6}$

(ii) The expected percentage of idle time is

$$= \left(1 - \frac{1}{3}\right) = \frac{2}{3} = 67\% \ (approx.)$$

(iii) The expected length of his waiting time is

$$= \frac{1}{1-\rho} \cdot \frac{1}{s\mu}$$

$$= \frac{1}{1-\frac{1}{3}} \times \frac{1}{2\times\frac{1}{4}} = 3 \text{ minutes.}$$

Practice Problems

Model II

1. A typist has 5 person for whom he performs typing work. Arrival rate is poisson and service times are exponential. Average arrival rate is 4 per hour with an average service time of 10 minutes. Cost of waiting is Rs. 8 per hour, while the cost of servicing is Rs. 2.50 each. Calculate

   (a) The average waiting time of an arrival.

(b) The average length of the waiting line.

(c) The average time which an arrival spends in the system.

(d) The minimum cost service rate.

[**Ans.** (a) 12.4 minutes (b) 0.79 for one typist (c) 22.4 minutes]

2. At a railway station, only one train is handled at a time. The railway yard is only for two trains to wait, while the other is given signal to leave the station. Trains arrive at the station at an average rate of 6 per hour and the railway station can handle them on an average of 12 per hour. Assuming poisson arrivals and exponential service distribution. Find the steady-state probability for the various number of trains in the system. Also find the average waiting time of a new train coming in the yard.

[**Ans.** $\lambda = 6, \mu = 12, \rho = \dfrac{\lambda}{\mu} = 0.5, N = 3, P_0 = 0.53,$

$L_s = \sum_{n=1}^{3} nP_n = 0.74, L_q = 0.24, W_q = 0.04$]

3. A barber shop has space to accommodate only 10 customers. He can serve only one person at a time. If a customer comes to his shop and finds it full he goes to the next shop. Customers randomly arrive at an average rate $\lambda = 10$ per hour and the barber's service time is negative exponential with an average of $\dfrac{1}{\mu} = 5$ minutes per customer.

(a) Write recurrence relations for the steady state queuing system for above.

(b) Determine P_0 and P_n probability of having 0 and n-customers in the shop.

[**Ans.** $N = 10; \lambda = \dfrac{10}{60}, \mu = \dfrac{1}{5}$. To find $P_0 = \dfrac{1-\rho}{1-\rho^{N+1}}, P_n = P_0\rho^n$]

4. A petrol station has a single pump and space for not more than 3 cars (2 waiting, 1 being served). A car arriving when the space is filled to capacity goes elsewhere for petrol. Cars arrive according to poisson faisson at a mean rate of one every 8 minutes. Their service time has an exponential distribution with a mean 4 minutes. The owner has the opportunity of renting an adjacent piece of land, which would provide space for an additional car to wait. The rent would be Rs.2000 per month. The expected net profit from each customer is Rs.2 and the station is open 10hours everyday. Would if be profitable to rent the additional space ?

[**Ans.** $\lambda = \dfrac{1}{48}, \mu = \dfrac{1}{36}, P_0 = 0.28, L_s = 3$ cars]

Model III

5. A telephone exchange has two long distance operators. The telephone company finds that, during the peak load, long distance calls arrive in a poisson distribution at an average rate of 15 per hour. The length of service on these calls is exponential distribution with mean length 5 minutes.

   (i) What is the probability that a subscriber will have to wait for his long distance call during the peak hours of the day ?

   (ii) If the subscribers will wait and serviced in turn, what is the expected waiting time ?

   [**Ans.** $s = 2, \lambda = \frac{5}{60}, \mu = \frac{1}{5}$ $P_0 = \frac{3}{13}$ (i) 0.48 (ii) 3.2 minutes]

6. A company has two tools cribs, each having a single clerk, in its manufacturing area. One tool crib handles only the tools for the heavy machinery, while the second one handles all other tools. It is observed that for each tool crib the arrivals follow a poisson distribution with a mean of 20 per hour and the service time distribution is negative exponential with a mean of 2 minutes.

   The tool manager feels that if tool cribs are combined in such a way that either clerk can handle any kind of tools as demand arises would be more efficient and the waiting problem could be reduced to some extent. It is believed that the mean arrival rate at the two tool cribs will be 40 per hour, while the service time will remain unchanged.

   Compare in status of queue and the proposal with respect to the total expected number of machines at the tool crib(s), the expected waiting time including service time for each mechanic and probability that he has to wait for more than five minutes.

   [**Ans.** (i) $\lambda = 20, \mu = 2, \rho = \frac{2}{3}$, $L_s = 2$, $W_s = 6$ minutes (each), probability $(t > 5) = 0.435$

   (ii) $\lambda = 40, \mu = 30, s = 2, P_0 = \frac{1}{5}$ L_s =2.4 persons, $W_s = 3.6$ minutes. $P(t > 5) = 0.25$]

7. A general insurance company has three claims adjusters in its branch office. People with claims against the company are found to arrive in poisson distribution, at an average rate of 20 per 8 hour day. The amount of time that an adjuster spends with a claimant is found to have a negative exponential distribution, with mean service time 40 minutes. Claimants are processed in the order of their appearance.

   (i) How many hours a week can an adjuster expect to spend with claimants ?

   (ii) How much time, on an average, does a claimant spend in the branch office ?

   [**Ans.** (i) 22.2 hours, (ii) 45 minutes]

Chapter - 16

Fuzzy Sets

Introduction

According to the traditional view science should strive for certainly in all its manifeslation. (Precision, specificity, sharpness, consistency etc.) hence uncertainty (imprecision nonspecificity, vagueness, inconsistency etc.,) is regarded as unscientific.

According to Morden view uncertainty is considered essential to science, it is not only on unavoidable plague, but it has, in fact, a great utility.

Fuzzy set theory exhibits immense potential for effective solving of the uncertainly is the problem fuzzyness means "vagueness". Fuzzy set theory is an extent mathematical tool to handle; the uncertainty arrising due to vagueness. Understanding human speech and recognizing hand written characters are some common instances where fuzziners manifests.

L.A.Zadeh viewed about fuzzy set theory in his seminal paper (zadel 1965). Science than, a lot of theoretical development law taken place in this field.

Fuzzy versus crisp

We consider the query, "Is sky blue?". The answer to this is a definite yes/true or no/false as warranted by the situation. If "yes" is accorded a value of 1 and "no" is accorded a value of O. Such a logic which demands a binary (0/1) type of handling is termed crisp in the domain of fuzzy set theory.

On the other way we consider the statement "Is shyam honest?"

The answer to this query need not be defiruite "yes" or "no".

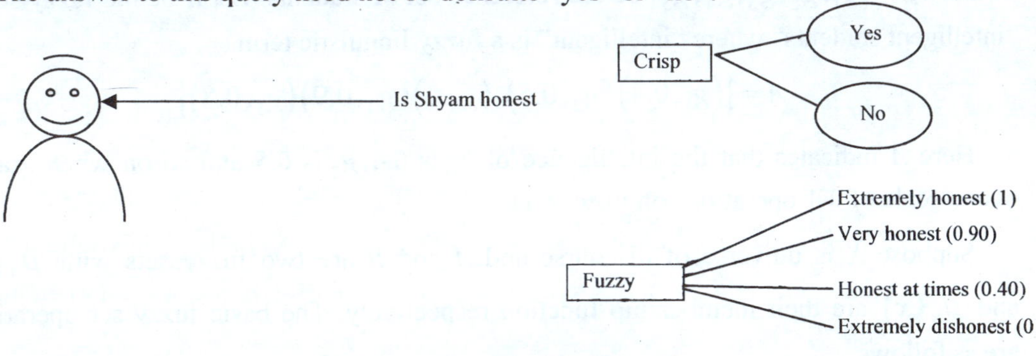

In terms of membership function considering the degree to which one knowns shyam, a variety of answers spanning a range, such as "extremely honest", extremely dishonest, "honest at times" very honest" could be generated. We gave them degrees 1,0,.4,.9 respectively. The situation is therefore so fluid that it can accept values between 0 and 1 in contrast to the earlier one which was either 0 and 1. Such a situation is termed fuzzy.

Fuzzy sets : L.A. Zadeh in 1965 introduced the concept of fuzzy sets on the basis of membership function defined as:

$$\mu : x \to \{0,1\}$$

Here, $\mu(x)=1$, means full membership

$\mu(x)=0$, means, non membership

$0<\mu(x)<1$, means intermediate membership

Fuzzy sets support a flexible sense of membership of elements to a set while in crisp set theory an element either belongs to or does not belong to a set.

Definition : Let X be a non zero set. A fuzzy set A of this set X is defined by the following set of pairs $A=\{[x,\mu_A(x)]\}:x\in X$

Where $\mu_A : X \to [0,1]$ is a function called as the membership function of A and $\mu_A(x)$ is the grade of membership or degree of belongings or degree of membership of $x\in X$ in A.

Thus a fuzzy set is a set of pairs consisting of a particular elements of the universe and its degree of membership. A can be written as

$$A=\{[x_1,\mu_A(x_1)],[x_2,\mu_A(x_2)],...[x_n,\mu_A(x_n)]\}$$

Let $X=\{g_1,g_2,g_3,g_4,g_5\}$ be the reference set of students. Let A be the fuzzy set of "intelligent students" where "intelligent" is a fuzzy linguistic term.

$$A=\{(g_1,0.4)(g_2,0.5),(g_3,1)(g_4,0.9)(g_5,0.8)\}$$

Here A indicates that the intelligence of g_1 is 0.4, g_2 is 0.5 and so on when graded over a scale of 0-1 operations on fuzzy sets.

Suppose X is universe of discourse and A and B are two fuzzy sets with $\mu_A(x)$ and $\mu_B(x)$ are their membership function respectively. The basic fuzzy set operations are as follows

(i) Union (ii) Intersection (iii) Complement (iv) Product of two fuzzy sets (v) Product by a crisp number (vi) Difference of two fuzzy sets and (vii) Disjunctive sum of two fuzzy sets.

Union : The union of two fuzzy sets A and B is a new fuzzy set denoted by $A \cup B$, which is defined on set X with a membership function

$\mu_{A\cup B}(x)$ define as

$$\mu_{A\cup B}(x) = \max\ \{\mu_A(x), \mu_B(x)\}$$

Example We write

A= {0.3 Shiv,0.9 Dev,0.6 Richa,0.4 Nidhi} for the set of Rich people.

B= {0.4 Richa,0.6 Nidhi,0.8 Dev,0.6 Shiv} for the set of industrial peoples.

Here the degree in $A \cup B$ of each element is the maximum of degree of membership of this element in A and in B.

$$A \cup B = \{0.6 \text{ Shiv}, 0.9 \text{ Dev}, 0.6 \text{ Richa}, 0.6 \text{ Nidhi}\}$$

Intersection of two fuzzy sets

The intersection of two fuzzy sets A and B is a new fuzzy set denoted by $A \cap B$ with membership function defined as $\mu_{A\cap B}(x) = \min\ \{\mu_A(x), \mu_\beta(x)\}$

Example In the above example.

$$A \cap B = \{0.3 \text{ Shiv}, 0.8 \text{ Dev}, 0.4 \text{ Richa}, 0.4 \text{ Nidhi}\}$$

Because,

$$\mu_{A\cap B}(\text{Shiv}) = \min\{\mu_A(\text{Shiv}), \mu_\beta(\text{Shiv})\}$$

$$= \min\ \{0.3, 0.6\} = 0.3$$

$$\mu_{A\cap B}(\text{Dev}) = \min\{\mu_A(\text{Dev}), \mu_\beta(\text{Dev})\}$$

$$= \min\ \{0.9, 0.8\} = 0.8$$

$$\mu_{A\cap B}(\text{Richa}) = \min\{\mu_A(\text{Richa}), \mu_\beta(\text{Richa})\}$$

$$= \min\ \{0.6, 0.4\} = 0.4$$

$$\mu_{A\cap B}(\text{Nidhi}) = \min\{\mu_A(\text{Nidhi}), \mu_\beta(\text{Nidhi})\}$$

$$= \min\ \{0.4, 0.6\} = 0.4$$

Complement of a fuzzy set: The complement of a fuzzy set A is a new fuzzy set A^c with a membership function $\mu_{A^C}(x) = 1 - \mu_A(x)$.

Example For the fuzzy set A defined as honest people as

A = {0.6 Naman, 0.5 Karan, 0.7Yash, 0.8 Rama, 0.7 Siddhi}

Then A^c = {0.4 Naman, 0.5 Karan, 0.3Yash, 0.2 Rama, 0.3 Siddhi}

because

$$\mu_{A^c}(\text{Naman}) = 1 - \mu_A(\text{Naman})$$
$$= 1\text{-}0.6 = 0.4$$
$$\mu_{A^c}(\text{Karan}) = 1 - \mu_A(\text{Karan})$$
$$= 1 - 0.5 = 0.5$$
$$\mu_{A^c}(\text{Yesh}) = 1 - \mu_A(\text{Yesh})$$
$$= 1 - 0.7 = 0.3$$
$$\mu_{A^c}(\text{Rama}) = 1 - \mu_A(\text{Rama})$$
$$= 1 - 0.8 = 0.2$$
$$\mu_{A^c}(\text{Siddhi}) = 1 - \mu_A(\text{Siddhi})$$
$$= 1 - 0.7 = 0.3$$

Product of two fuzzy sets

The product of two fuzzy sets A and B is a new fuzzy set $A.B$ whose membership function is defined as

$$\mu_{A.B}(x) = \mu_A(x)\mu_B(x)$$

Example If A and B are two fuzzy sets which are define as

$$A = \{(x_1, 0.2), (x_2, 0.8), (x_3, 0.4), (x_4, 0.3)\}$$
$$B = \{(x_1, 0.4), (x_2, 0), (x_3, 0.1), (x_4, 0.2)\}$$

Then, $A.B = \{(x_1, 0.08), (x_2, 0), (x_3, 0.04), (x_4, 0.06)\}$

Since $\mu_{AB}(x_1) = \mu_A(x_1).\mu_B(x_1)$

$$= 0.2 \times 0.4 = 0.08$$
$$\mu_{AB}(x_2) = \mu_A(x_2).\mu_B(x_2)$$
$$= 0.8 \times 0 = 0$$
$$\mu_{AB}(x_3) = \mu_A(x_3).\mu_B(x_3)$$
$$= 0.4 \times 0.1 = 0.04$$
$$\mu_{AB}(x_4) = \mu_A(x_4).\mu_B(x_4) = 0.3 \times 0.2 = 0.06$$

Difference of two fuzzy sets

The difference of two fuzzy sets A and B is a new fuzzy set A-B defined as

$$A-B=\left(A\cap B^{c}\right)$$

Example Let A and B be two fuzzy sets defined as

$$A=\{(x_1,0.2),(x_2,0.5),(x_3,0.6),(x_4,0.3)\}$$
$$B=\{(x_1,0.1),(x_2,0.4),(x_3,0.5),(x_4,0.2)\}$$

then $B^c=\{(x_1,0.9),(x_2,0.6),(x_3,0.5),(x_4,0.8)\}$

$$A-B=A\cap B^c$$
$$=\{(x_1,0.2),(x_2,0.5),(x_3,0.5),(x_4,0.8)\}$$

because

$$\mu_{A\cap B^c}(x_1)=\min\{\mu_A(x_1),\mu_{B^c}(x_1)\}$$
$$=\min\{0.2,0.9\}=0.2$$
$$\mu_{A\cap B^c}(x_2)=\min\{\mu_A(x_2),\mu_{B^c}(x_2)\}$$
$$=\min\{0.5,0.6\}=0.5$$
$$\mu_{A\cap B^c}(x_3)=\min\{\mu_A(x_3),\mu_{B^c}(x_3)\}$$
$$=\min\{0.6,0.5\}=0.5$$
$$\mu_{A\cap B}(x_4)=\min\{\mu_A(x_4),\mu_{B^c}(x_4)\}$$
$$=\min\{0.3,0.8\}=0.3$$

Disjunctive sum

The disjunctive sum of two fuzzy sets A and B is a new fuzzy set $A\oplus B$ and defined by

$$A\oplus B=\left(A^c\cap B\right)\cup\left(A\cap B^c\right)$$

Example We take two fuzzy sets A and B, such that

$$A=\{(x_1,0.4),(x_2,0.8),(x_3,0.6),(x_4,0.7)\}$$
$$B=\{(x_1,0.2),(x_2,0.6),(x_3,0.9),(x_4,0.5)\}$$
$$A^c=\{(x_1,0.6),(x_2,0.2),(x_3,0.4),(x_4,0.3)\}$$
$$B^c=\{(x_1,0.8),(x_2,0.4),(x_3,0.1),(x_4,0.5)\}$$
$$A^c\cap B=\{(x_1,0.2),(x_2,0.2),(x_3,0.4),(x_4,0.3)\}$$

$$A \cap B^c = \{(x_1, 0.4), (x_2, 0.4), (x_3, 0.1), (x_4, 0.5)\}$$

So, $$A \oplus B = (A^c \cap B) \cup (A \cap B^c)$$

$$= \{(x_1, 0.4), (x_2, 0.4), (x_3, 0.4), (x_4, 0.5)\}$$

Power of a fuzzy set : The k power of a fuzzy set A is a new fuzzy set A^k whose membership function is given by

$$\mu_{A^k}(x) = (\mu_A(x))^k$$

Example $A = \{(x_1, 0.4), (x_2, 0.2), (x_3, 0.7), (x_4, 0.3)\}$

For $k = 2$

$$\mu_{A^2}(x) = \{\mu_A(x)\}^2$$

Hence, $(A)^2 = \{(x_1, 0.16), (x_2, 0.04), (x_3, 0.49), (x_4, 0.09)\}$

Since, $$\mu_{A^2}(x_1) = [\mu_A(x_1)]^2 = (0.4)^2 = 0.16$$

$$\mu_{A^2}(x_2) = [\mu_A(x_2)]^2 = (0.2)^2 = 0.04$$

$$\mu_{A^2}(x_3) = [\mu_A(x_3)]^2 = (0.7)^2 = 0.49$$

$$\mu_{A^2}(x_4) = [\mu_A(x_4)]^2 = (0.3)^2 = 0.09$$

Product of a fuzzy set with a crisp number

Multiplying a fuzzy set A by a crisp number n, results a new fuzzy set product $n\,.A$ with the membership function.

$$\mu_{n.A}(x) = n.\mu_A(x)$$

Example $A = \{(x_1, 0.4), (x_2, 0.6), (x_3, 0.8), (x_4, 0.5)\}$

$$n = 0.5$$

$$n.A = \{(x_1, 0.20), (x_2, .30), (x_3, 0.40), (x_4, 0.25)\}$$

Equality : Two fuzzy sets are said to be equal

$(A = B)$ if $\mu_A(x) = \mu_B(x)$

Example $A=\{(x_1,0.2),(x_2,0.8),(x_3,0.4)\}$

$$B=\{(x_1,0.3),(x_2,0.8),(x_3,0.4)\}$$

$$C=\{(x_1,0.2),(x_2,0.8),(x_3,0.4)\}$$

here, $A \neq B$ because $\mu_A(x_1) \neq \mu_B(x_1)$

But $\mu_A(x_1)=\mu_C(x_1)$

$$\mu_A(x_2)=\mu_C(x_2)$$

$$\mu_A(x_3)=\mu_C(x_3)$$

So, $A=C$

Properties of fuzzy sets

As we know that any fuzzy set A is a subset of the reference sets X (such as universal set in crisp). Membership of any element belonging to null set Q is 0 and membership of an element belonging to reference set X is 1.

Fuzzy set satisfies the following properties same as the crisp.

(i) Commutativity $A \cup B = B \cup A$

$$A \cap B = B \cap A$$

(ii) Associativity $A \cup (B \cup C) = (A \cup B) \cup C$

$$A \cap (B \cap C) = (A \cap B) \cap C$$

(iii) Distributivity

$$A \cup (B \cap C) = (A \cup B) \cap (A \cup C)$$

$$A \cap (B \cup C) = (A \cap B) \cup (A \cap C)$$

(iv) Idempotent law

$$A \cup A = A$$

$$A \cap A = A$$

(v) Identity

$$A \cup Q = A$$

$$A \cup X = X$$

$$A \cap X = A$$

$$A \cap Q = Q$$

(vi) Transitivity : If $A \subseteq B \subseteq C$, then $A \subseteq C$

(vii) Involution $(A^c)^c = A$

(viii) De Morgan's laws

$$(A \cap B)^c = (A^c \cup B^c)$$

$$(A \cup B)^c = (A^c \cap B^c)$$

Note: The laws of excluded middle not necessarily true so

$$A \cup A^c \neq X$$

$$A \cap A^c \neq Q$$

Fuzzy Relations

A crisp relation represents the presence or absence of association, or interconnected ness between the elements of two or more sets. Let *A* and *B* be two sets then the crisp relation *R* is subset of $A \times B .i.e., R = \{(x, y) : x \in A, y \in B, xRy\}$.

This concept can generalized to allow for various degrees or strengths of relation or interaction between elements. Degree of association can be represented by membership grades in a fuzzy relation in the same way as degrees of set membership are represented in the fuzzy set. The crisp relation can be considered to the restricted case of the fuzzy relation.

Def : Fuzzy relation is a fuzzy set defined on the Cartesian product of crisp sets $X_1, X_2 ......X_n$, where the *n* tuples $(x_1, x_2, ...x_n)$ may have varying degrees of membership within the relation. The membership values indicate the strength of the relation between the tuples.

Example Let *R* be the fuzzy relation between two sets *X* and *Y*, where *X* is the set of diseases and *Y* is the set of symptoms.

X = {Maleria, Typhoid, Viral fever}

Y = {sichness, high temperature, shivering & feaver}

The fuzzy relation *R* may be defined as

	Sihness & fever	High temperature	Shivering
Malaria	0.1	0.9	0.8
Typhoid	0.2	0.9	0.7
Viral fever	0.9	0.4	0.6

Fuzzy Cartesian product

Let *A* be a fuzzy set defined on universe *X* and *B* be a fuzzy set defined on the universe *Y*, the Cartesian product between the fuzzy set *A* and *B* is denoted by $A \times B$. It gives a fuzzy relation *R* defined by $R = A \times B \subseteq X \times Y$

Where *R* has a membership function given by

$$\mu_R(x, y) = \mu_{A\times B}(x, y)$$
$$= \min\ [\mu_A(x), \mu_B(y)]$$

Let $X = \{x_1, x_2, x_3\}$ and $Y = \{y_1, y_2\}$ are the universes of discourse. Let $A = \{(x_1, 0.2), (x_2, 0.5), (x_3, 0.4)\}$ and $B = \{(y_1, 0.5)(y_2, 0.7)\}$ be two fuzzy sets defined on *X*. Then the fuzzy relation *R* resulting out of the fuzzy cartesion product $A \times B$ is given by

$$R = A \times B$$

$$\begin{array}{c} \\ x_1 \\ x_2 \\ x_3 \end{array} \begin{array}{c} \begin{array}{cc} Y_1 & Y_2 \end{array} \\ \begin{bmatrix} 0.2 & 0.2 \\ 0.5 & 0.5 \\ 0.4 & 0.4 \end{bmatrix} \end{array}$$

because
$$R(x_1, y_1) = \min\{\mu_A(x_1), \mu_B(y_1)\}$$
$$= \min\ \{0.2, 0.5\} = 0.2$$
$$R(x_1, y_2) = \min\{\mu_A(x_1), \mu_B(y_2)\}$$
$$= \min\ \{0.2, 0.7\} = 0.2$$
$$R(x_2, y_1) = \min\{\mu_A(x_2), \mu_B(y_1)\}$$
$$= \min\ \{0.5, 0.5\} = 0.5$$
$$R(x_2, y_2) = \min\{\mu_A(x_2), \mu_B(y_2)\}$$
$$= \min\ \{0.5, 0.7\} = 0.5$$
$$R(x_3, y_1) = \min\{\mu_A(x_3), \mu_B(y_1)\} = \min\ \{0.4, 0.5\} = 0.4$$
$$R(x_3, y_2) = \min\{\mu_A(x_3), \mu_B(y_2)\}$$
$$= \min\ \{0.4, 0.7\} = 0.4$$

Composition of relations: Suppose *R* is a fuzzy relation defined on $X_1 \times X_2$ and *S* is a fuzzy relation defined on $X_2 \times X_3$, then *ROS* is a fuzzy relation on $X_1 \times X_3$. The fuzzy max-min composion is defined as

$$\mu_{ROS} = \max\{\mu_R(x_1, x_2), \mu_S(x_2, x_3)\}$$

Example $X_1 = \{x_1, x_2, x_3\},\ X_2 = \{k_1, k_2\}, X_3 = \{z_1, z_2, z_3\} R$ be a fuzzy relation

$$\text{defined by } \begin{array}{c} \\ x_1 \\ x_2 \\ x_3 \end{array} \begin{array}{c} \begin{array}{cc} k_1 & k_2 \end{array} \\ \begin{bmatrix} 0.5 & 0.1 \\ 0.2 & 0.9 \\ 0.8 & 0.6 \end{bmatrix} \end{array}$$

$$\text{Let } S \text{ be a fuzzy relation } \begin{array}{c} \\ k_1 \\ k_2 \end{array} \begin{array}{c} \begin{array}{ccc} z_1 & z_2 & z_3 \end{array} \\ \begin{bmatrix} 0.6 & 0.4 & 0.7 \\ 0.5 & 0.8 & 0.9 \end{bmatrix} \end{array}$$

Then ROS, by max-min composition yields,

$$ROS = \begin{array}{c} \\ x_1 \\ x_2 \\ x_3 \end{array} \begin{bmatrix} z_1 & z_2 & z_3 \\ 0.5 & 0.4 & 0.5 \\ 0.5 & 0.8 & 0.9 \\ 0.6 & 0.6 & 0.7 \end{bmatrix}$$

$$\mu_{ROS}(x_1, z_1) = \max \{\min(0.5, 0.6), \min(0.1, 0.5)\}$$

$$= \max \{0.5, 0.1\} = 0.5$$

$$\mu_{ROS}(x_1, z_2) = \max \{\min(0.5, 0.4), \min(0.1, 0.8)\}$$

$$= \max \{0.4, 0.1\} = 0.4$$

$$\text{Similarly, } \mu_{ROS}(x_1, z_3) = \max \{0.5, 0.1) = 0.5$$

$$\mu_{ROS}(x_2, z_1) = \max \{0.2, 0.5) = 0.5$$

$$\mu_{ROS}(x_2, z_2) = \max \{0.2, 0.8) = 0.8$$

$$\mu_{ROS}(x_2, z_3) = \max \{0.2, 0.9) = 0.9$$

$$\mu_{ROS}(x_3, z_1) = \max \{0.6, 0.5) = 0.6$$

$$\mu_{ROS}(x_3, z_2) = \max \{0.4, 0.6) = 0.6$$

$$\mu_{ROS}(x_3, z_3) = \max \{0.7, 0.6) = 0.7$$

Binary Relations

Any relations between two sets X and Y is known as a binary relation, It is denoted by $R(X,Y)$. For membership function the sagittal diagrams are another representation for binary relation $R(X,Y)$. Each of the sets X,Y is represented by a set of nodes in the diagram nodes corresponding to one set are clearly distinguished from nodes representing the other set. Elements of $X \times Y$ with non zero membership grades in $R(X,Y)$ are represented in the diagram by lines connecting the respective nodes. These lines are labelled with the value of the membership grade $\mu_R(X,Y)$.

If R(X,Y) is a fuzzy relation, its domain in the fuzzy set dom $R(X,Y)$ whose membership function is defined by $\mu \text{ dom}_R(X) = \max_{y \in Y} \mu_R(x,y)$ for each $x \in X$. Thus, each element of set X belongs in the domain of R to the degree equal to the strength of its strongest relation to any member of set Y.

The range of fuzzy relation $R(X,Y)$ is a fuzzy set ran $R(X,Y)$ whose membership function is defined by $\mu_{ran\,R}(y) = \max_{x \in X} \mu_R(x,y)$ for each $y \in Y$. Therefore, the strength of the strongest relation that each element of Y has to an element of X is equal to the degree of that elements membership in the range of R . In addition, the height of a fuzzy relation R is a number h (R) defined by

$$h(R) = \max_{y \in Y} \max_{x \in X} \mu_R(x,y)$$

Remark : If $h(R)$ =1, then the relation is called normal. If $h(R) \neq 1$ then it is called subnormal.

Examples of two convenient representation of a fuzzy relation (i) sagittal diagram (ii) membership matrix.

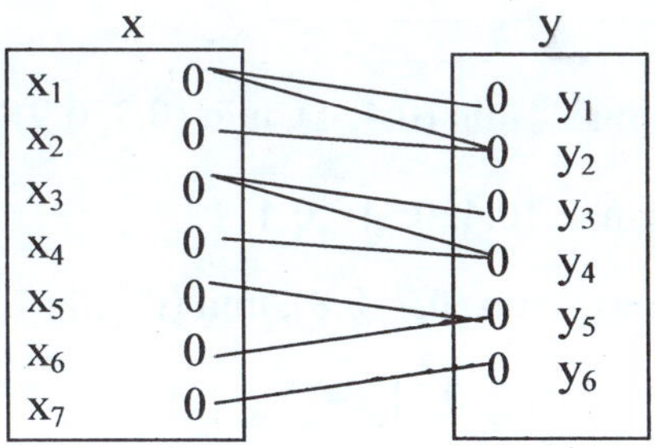

(i)

	y_1	y_2	y_3	y_4	y_5	y_6
x_1	0.9	1	0	0	0	0
x_2	0	0.4	0	0	0	0
x_3	0	0	1	0.2	0	0
x_4	0	0	0	0.4	0	0
x_5	0	0	0	0	0.5	0
x_6	0	0	0	0	0.2	0
x_7	0	0	0	0	0	0.3

(ii)

Example (composition of binary relation).

Let two binary relation $R_1(x, y)$ and $R_2(y, z)$ be specified by their sagittal diagrams.

Let $R(x,z) = R_1(x,y) 0 R_2(y,z)$. Then $R(x,z)$ is denoted by the membership function specified.

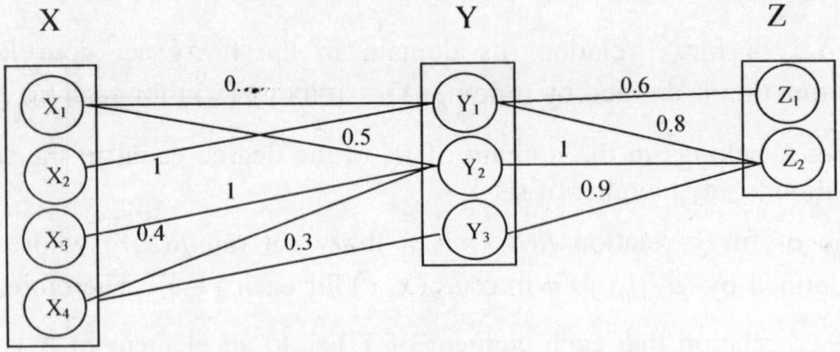

Composition : $R : R_1 \, 0 \, R_2$

x	z	$\mu_R(x,z)$
x_1	z_1	0.6
x_1	z_2	0.7
x_2	z_1	0.6
x_2	z_2	0.8
x_3	z_2	1
x_4	z_2	0.4

Because $\mu_R(x_4, z_2) =$ max {min (0.4, 1), min (0.3, 0.9)}

= max {0.4, 0.3}= 0.4

$\mu_R(x_1, z_2) =$ max {min (0.7, 0.8), min (0.5, 1) }

= max {0.7, 0.5} = 0.7 etc.

Operations on fuzzy relations

Let R_1 and R_2 be fuzzy relations on $X \times Y$ then union of two relations, intersection of two relations and complement of relation are also fuzzy sets. Their membership functions can be defined as

Union

$$\mu_{R_1 \cup R_2} = \max\{\mu_{R_1}(x, y), \mu_{R_2}(x, y)\}$$

Intersection

$$\mu_{R_1 \cap R_2} = \min\{\mu_{R_1}(x, y), \mu_{R_2}(x, y)\}$$

Complement

$$\mu_{R_1^C}(x, y) = 1 - \mu_{R_1}(x, y)$$

$$\mu_{R_2^C}(x, y) = 1 - \mu_{R_2}(x, y)$$

Fuzzy logic

Logic is the study of methods and principles of reasoning in all its possible forms. Classical logic deals with propositions that are required to be either true or false. Each propositions has its opposite, which is usually called a negation of the proposition. A proposition and its negation are required to assume opposite truth values.

In crisp logic, the truth values acquired by propositions or predicates are 2-valued. True and false which may be treated numerically equivalent to (0,1), but in fuzzy logic, truth values are multivalued such as absolutely true, partly true, absolutely false, partially false and so on and which are numerically equivalent to (0-1).

Fuzzy proposition

Fuzzy propositions are associated with fuzzy sets. In other words a fuzzy proposition is a statement which acquires a fuzzy truth value.

If given α in to be a fuzzy proposition, $T(\alpha)$ represents the truth valu (0,1) attached α. The fuzzy membership value associated with the fuzzy set A for α is treated as the fuzzy truth value $T(\alpha)$.

$i..e, T(\alpha) = \mu_A(x)$, where $0 \leq \mu_A(x) \leq 1$.

Example. α : Kanad is honest

$T(\alpha)$: 0.7, if α is partly true.

$T(\alpha) = 1$ if α is absolutely true.

$T(\alpha) = 0$ if α is absolutely wrong.

Types of fuzzy propositions

We take simple fuzzy propositions, which are classified into the following four types:

1. Unconditional and unqualified propositions.
2. Unconditional and qualified propositions.
3. Conditional and unqualified propositions.
4. Conditional and qualified propositions.

Fuzzy connectives

In fuzzy logic following connective are used

(i) Negation : $-$

(ii) Disjunction : $\vee$

(iii) Conjunction : ^

(iv) Implication : $\Rightarrow$

(i) Negation : It is denoted by $-$, and which is defined by $\overline{P} = 1 - T(P)$

(ii) Disjunction : It is denoted by $\vee$ *i.e.*, $P \vee Q$ and is defined as

$$P \vee Q = \max \{T(P), T(Q)\}$$

(iii) Conjunction : It is denoted by ^ and defined by

$$P \wedge Q = \min \{T(P), T(Q)\}$$

(iv) Implication : It is denoted by $\Rightarrow$ and defined by

$$\Rightarrow Q = \max \{1 - T(P), T(Q)\}$$

Example P : Shivansh is efficient, $T(P) = 0.75$

Q : Devansh is efficient, $T(Q) = 0.60$.

(i) $\overline{P}$: Shivansh is not efficient

$$T(\overline{P}) = 1 - T(P) = 1 - 0.75 = 0.25$$

(ii) $P \vee Q$: Either shivansh or Devansh is efficient

$$T(P \vee Q) = \max \{T(P), T(Q)\}$$

$$= \max \{0.75, 0.60\} = 0.75$$

(iii) $P \wedge Q$: Shivamsh and Devansh both efficient

$$T(P \wedge Q) = \min \{T(P), T(Q)\}$$

$$= \min \{0.75, 0.60\} = 0.60$$

Remark I P and Q, which are related by the '$\Rightarrow$' are known as antecedent and consequent respectively. Similarly 'too' represents the IF-THEN statement as

IF x_1 is A THEN y_1 is B, which is equivalent to

$$R = (A \times B) \cup (\bar{A} \times y)$$

The membership function of R is given by

$$\mu_R(x_1, y_1) = \max\{\min[\mu_A(x_1), \mu_B(y_1), 1 - \mu_A(x_1)]\}$$

Remark II For compound implication we take as IF x_1 is A then y_1 is B, otherwise y_1 is C" the relation R is equivalent to

$$R = (A \times B) \cup (A \times C)$$

The membership function R is given by

$$\mu_R(x, y) = \max\{\min[\mu_A(x_1), \mu_B(y_1)], \min[1 - \mu_A(x_1), \mu_C(y_1)]\}$$

Example Let X = {Mona, Krali, Siddhi, Chikoo}

$Y = \{1,2,3,4\}$

and A ={(Mona, 0),(Krali,0.8),(Siddhi, 0.6),(Chikoo,1)}

$B = \{(1,0.2),(2,1),(3,0.8),(4,0)\}$

$C = \{(1,0),(2,0.4),(3,1),(4,0.8)\}$

Find the implication relations

(i) If x is A THEN y is B

(ii) If x is A THEN y is B otherwise y is C

Solution To determine (i) compute

$$R = (A \times B) \cup (\bar{A} \times Y) \text{ where}$$

$$\mu_R(x, y) = \max\{\min[\mu_A(x), \mu_B(y)], 1 - \mu_A(x)\}$$

$$A \times B = \begin{array}{c} \\ a \\ b \\ c \\ d \end{array} \begin{array}{c} \begin{array}{cccc} 1 & 2 & 3 & 4 \end{array} \\ \begin{bmatrix} 0 & 0 & 0 & 0 \\ 0.2 & 0.8 & 0.8 & 0 \\ 0.2 & 0.6 & 0.6 & 0 \\ 0.2 & 1 & 0.8 & 0 \end{bmatrix} \end{array}$$

$$\bar{A}\times Y = \begin{array}{c} \\ a \\ b \\ c \\ d \end{array}\begin{array}{c} \begin{array}{cccc} 1 & 2 & 3 & 4 \end{array} \\ \begin{bmatrix} 1 & 1 & 1 & 1 \\ 0.2 & 0.2 & 0.2 & 0.2 \\ 0.4 & 0.4 & 0.4 & 0.4 \\ 0 & 0 & 0 & 0 \end{bmatrix} \end{array}$$

Here, Y the universe of discourse could be viewed as {(1,1),(2,1),(3,1),(4,1)} a fuzzy set all of whose elements x have $\mu(x)=1$.

Therefore

$$\bar{R} = \begin{array}{c} \\ a \\ b \\ c \\ d \end{array}\begin{array}{c} \begin{array}{cccc} 1 & 2 & 3 & 4 \end{array} \\ \begin{bmatrix} 1 & 1 & 1 & 1 \\ 0.2 & 0.8 & 0.8 & 0.2 \\ 0.4 & 0.6 & 0.6 & 0.4 \\ 0.2 & 0.1 & 0.8 & 0 \end{bmatrix} \end{array}$$

Which represent If x is A THEN y is B.

To determine (ii) compute

$$R = (A\times B)\cup(\bar{A}\times C)$$

$$\mu_R(x,y) = \max\{\min(\mu_A(x),\mu_B(y)), \min[1-\mu_A(x),\mu_C(y)]\}$$

$$A\times B = \begin{array}{c} \\ a \\ b \\ c \\ d \end{array}\begin{array}{c} \begin{array}{cccc} 1 & 2 & 3 & 4 \end{array} \\ \begin{bmatrix} 0 & 0 & 0 & 0 \\ 0.2 & 0.8 & 0.8 & 0 \\ 0.2 & 0.6 & 0.6 & 0 \\ 0.2 & 1 & 0.8 & 0 \end{bmatrix} \end{array}$$

$$\bar{A}\times C = \begin{array}{c} \\ a \\ b \\ c \\ d \end{array}\begin{array}{c} \begin{array}{cccc} 1 & 2 & 3 & 4 \end{array} \\ \begin{bmatrix} 0 & 0.4 & 1 & 0.8 \\ 0 & 0.2 & 0.2 & 0.2 \\ 0 & 0.4 & 0.4 & 0.4 \\ 0 & 0 & 0 & 0 \end{bmatrix} \end{array}$$

Therefore, $R = \max\{(A \times B), (\overline{A} \times C)\}$ gives

$$R = \begin{array}{c} \\ a \\ b \\ c \\ d \end{array}\begin{array}{c} \begin{array}{cccc} 1 & 2 & 3 & 4 \end{array} \\ \begin{bmatrix} 0 & 0.4 & 1 & 0.8 \\ 0.2 & 0.8 & 0.8 & 0.2 \\ 0.2 & 0.6 & 0.6 & 0.4 \\ 0.2 & 1 & 0.8 & 0 \end{bmatrix} \end{array}$$

The above R represents IF x is $\overline{A}$ THEN y is $\overline{B}$ ELSE y is C.

Fuzzy Quantifiers

Fuzzy logic propositions are quantified by fuzzy quantifiers. There are two classes of fuzzy quantifiers such as

(i) Absolute quantifiers

(ii) Relative quantifiers

Absolute quantifiers are defined over R and relative quantifiers are defined over [0-1].

Example

Absolute quantifier	Relative quantifier
round about 300	almost
much greater than 5	about
some where around 25	most

Fuzzy Inference

Fuzzy inference also referred to as approximate reasoning refers to computational procedures used for evaluating linguistic desriptions. The two important inferring procedures are follows.

(i)Generalized modus ponens (GMP)

(ii)Generalized modus tollens (GMT)

GMP is formally stated as

If x is A THEN y is B

$$\frac{\qquad x \text{ is } A'}{y \text{ is } B' \qquad}$$

Here, A, B, A' and B' are fuzzy terms.

Every fuzzy linguistic statement above the line in analytically known and what is below is analytically unknown.

To compute the membership function of B', the max-min composition of fuzzy set A' with $R(x,y)$ which is the known implication relation (IF-THEN relation) is used.

$$B' = A' O R(x, y), \text{ in terms of membership function}$$

$$\mu_{B'}(y) = \max\{\min(\mu_A(x), \mu_R(x, y))\}$$

where $\mu_A(x)$ is the membership function of A'. $\mu_R(x, y)$ is the membership function of the implication relation and $\mu_{B'}(y)$ is the membership function of B'.

GMT has the form

IF x is A THEN y is B

y is B'

---

x is A'

The membership of A' is computed on similar lines as

$$A' = B' O\ R(x, y)$$

In terms of membership function

$$\mu_{A'}(x) = \max\{\min(\mu_{B'}(y), \mu_R(x, y))\}$$

Example *Apply the fuzzy modus ponens rule to deduce Rotation is quite slow given*

(i) If the speed is high then the rotation is slow.

(ii) The speed is very high.

Solution. Let H (High), VH(very high), S (slow) and QS (Quite slow)

For $X = \{30,40,50,60,70,80,90,100\}$ the set of speed and

$Y = \{10,20,30,40,50,60\}$ the set of rotations per minute,

$H = (70,1),(80,1)(90,0.3)\}$

$VH = \{(90,0.9)(100,1)\}$

$QS = \{(30,0.8)(40,1)(50,0.6)\}$

To find $R(x,y)$ representing the implication relation (i) we compute $R(x, y) = \max\ (H \times S, \overline{H} \times Y)$

Where,

$$H \times S = \begin{array}{c} \\ 30 \\ 40 \\ 50 \\ 60 \\ 70 \\ 80 \\ 90 \\ 100 \end{array} \begin{array}{c} \begin{array}{cccccc} 10 & 20 & 30 & 40 & 50 & 60 \end{array} \\ \begin{bmatrix} 0 & 0 & 0 & 0 & 0 & 0 \\ 0 & 0 & 0 & 0 & 0 & 0 \\ 0 & 0 & 0 & 0 & 0 & 0 \\ 0 & 0 & 0 & 0 & 0 & 0 \\ 0 & 0 & 0.8 & 1 & 0.6 & 0 \\ 0 & 0 & 0.8 & 1 & 0.6 & 0 \\ 0 & 0 & 0.3 & 0.3 & 0.3 & 0 \\ 0 & 0 & 0 & 0 & 0 & 0 \end{bmatrix} \end{array}$$

$$\overline{H} \times Y = \begin{array}{c} \\ 30 \\ 40 \\ 50 \\ 60 \\ 70 \\ 80 \\ 90 \\ 100 \end{array} \begin{array}{c} \begin{array}{cccccc} 10 & 20 & 30 & 40 & 50 & 60 \end{array} \\ \begin{bmatrix} 1 & 1 & 1 & 1 & 1 & 1 \\ 1 & 1 & 1 & 1 & 1 & 1 \\ 1 & 1 & 1 & 1 & 1 & 1 \\ 1 & 1 & 1 & 1 & 1 & 1 \\ 0 & 0 & 0 & 0 & 0 & 0 \\ 0 & 0 & 0 & 0 & 0 & 0 \\ 0.7 & 0.7 & 0.7 & 0.7 & 0.7 & 0.7 \\ 1 & 1 & 1 & 1 & 1 & 1 \end{bmatrix} \end{array}$$

$$R(x, y) = \begin{array}{c} \\ 30 \\ 40 \\ 50 \\ 60 \\ 70 \\ 80 \\ 90 \\ 100 \end{array} \begin{array}{c} \begin{array}{cccccc} 10 & 20 & 30 & 40 & 50 & 60 \end{array} \\ \begin{bmatrix} 1 & 1 & 1 & 1 & 1 & 1 \\ 1 & 1 & 1 & 1 & 1 & 1 \\ 1 & 1 & 1 & 1 & 1 & 1 \\ 1 & 1 & 1 & 1 & 1 & 1 \\ 0 & 0 & 0.8 & 1 & 0.6 & 0 \\ 0 & 0 & 0.8 & 1 & 0.6 & 0 \\ 0.7 & 0.7 & 0.7 & 0.7 & 0.7 & 0.7 \\ 1 & 1 & 1 & 1 & 1 & 1 \end{bmatrix} \end{array}$$

$$QS = VH \ O \ R(x, y)$$

$$= [0\ 0\ 0\ 0\ 0\ 0\ 0.9\ 1] \times \begin{bmatrix} 1 & 1 & 1 & 1 & 1 & 1 \\ 1 & 1 & 1 & 1 & 1 & 1 \\ 1 & 1 & 1 & 1 & 1 & 1 \\ 1 & 1 & 1 & 1 & 1 & 1 \\ 0 & 0 & 0.8 & 1 & 0.6 & 0 \\ 0 & 0 & 0.8 & 1 & 0.6 & 0 \\ 0.7 & 0.7 & 0.7 & 0.7 & 0.7 & 0.7 \\ 1 & 1 & 1 & 1 & 1 & 1 \end{bmatrix}$$

$$= [1\ 1\ 1\ 1\ 1\ 1]$$

Fuzzy Relation Equations

The nation of fuzzy relation equations is associated with the concept of composition of binary relations. The composition of two fuzzy binary relation *L*(*x, y*) and *M*(y, z) can be defined, in general, in terms of an operation on the membership matrices of *L* and *M* that resembles matrix multiplication. This operation involves exactly the same combinations of matrix entries as in the regular matrix multiplication. The multiplications and additions that are applied to these combinations in the matrix multiplications are replaced with other operations, the appropriate operations of fuzzy set inter section and union respectively.

We consider three binary relations $[L(X,Y), M(Y,Z)]$ and $N(X,Z)]$ which are defined on the sets

$$X = \{xi | i \in I\}, Y = \{yj | j \in J\}, Z = \{zk \mid k \in K\}$$

Let the membership matrices of L, M and N be denoted by

$L = [P_{i,j}], M = [q_{j,k}], N = [r_{i;k}]$ respectively, where

$$P_{i,j} = \mu_L(x_i, y_J), q_{J,k} = \mu_M(y_j, z_k), r_{i,k} = \mu_N(x_i, z_k)$$

$\forall i \in I, j \in \text{J}$ and $\text{k} \in \text{K}$. It is clear that all entries in the matrices, *L*, *M* and *N* are real numbers in the internal (0, 1).

Assume now that the three relations constrain each other in such a way that

$$L\, o\, M = N \qquad \ldots\text{(i)}$$

Where 0 denotes the max-min composition.

$$i.e., \qquad \max_{j \in J} \min\,(p_{i,j}, q_{j,k}) = r_{i,k} \qquad \ldots\text{(ii)}$$

$\forall i \in I$ and $k \in K$. That is, the matrix eq. (i) encompasses $n \times s$ simultaneous equations of the forms of eq. (ii).

When two of the components in each of the equations are given and one is unknown, these equations are referred to as fuzzy relations equations.

Basic method for fuzzy relation equations :

1. Decompose eq. $LoM = N$ into equations of the form $lom=n$ ...(i)
   One for each row in L and N (l is associated with index j, m with indices j and k and n with index k).
2. For each equation (i), if
   $$\max_{j \in J} m_{j,k} < \max n_k$$
   for some k, then the equation has no solution. $N(m,n) = \phi$ and the procedure terminates, otherwise proceed to step 3.
3. Determine $\hat{l}$ as follows
   $$\hat{l} = [\hat{l}_j : j \in J] \text{ in which}$$
   $$\hat{l}_{\mathrm{j}} = \min_j \sigma(m_{j,k}, n_k)$$
   where, $$\sigma(m_{j,k}, n_k) = \begin{cases} n_k & if \quad m_{j,k} > n_k \\ 1 & \quad otherwise \end{cases}$$
4. If $\hat{l}$ is not a solution of eq. (i), then the equation has no solution $N(m,n) = \phi$ and the procedure terminates, otherwise, proceed to Step 5.
5. For each $\hat{l}_{\mathrm{j}} = 0$ and $n_k = 0$, exclude these components, as well as the corresponding rows j and columns k from matrix M in (i). This results in the reduced equation
   $$L'oM' = n' \qquad \text{...(ii)}$$
   Where we assume $j \in J', k \in K'$
6. Determine all minimal solution of the reduced equation (ii) by the following procedure this results in $N(m', n')$.
   (i) Permute elements of n' and the corresponding columns of m' appropriately to arrange them in decreasing order.

(ii) Determine sets

$$J_k(\hat{l}') = \{j \in J' / \min(\hat{l}'_j, m'_{j,k}) = n'_k\}$$

For all $k \in K$ and form

$$j(\hat{l}') = \underset{k \in K'}{X} J_k(\hat{l}').$$

(iii) For each $B \in J(\hat{l}')$ and each $j \in J$ determine the set

$$O(B, j) = \{k \in K \mid B_k = j\}$$

(iv) For each $B \in J(\hat{l}')$ generate the tuple

$$g(B) = \{g_j(B) / j \in J'\}$$

By taking $g_j(B) = \begin{cases} \max r'_k & \text{if } O(B_1, j) \neq \phi \\ 0 & \text{otherwise} \end{cases}$

(v) From all tuples g(B)generated in step 4, select only the minimal ones: this result in $\overline{N}(m', n')$

7. Determine the solution set of the reduced equation (6)

$$S(m', n') = U_{\hat{l},} < \hat{l}', \hat{l}' >,$$

Where the union is taken over all $\hat{l}' \in \hat{N}(m', r')$

8. Extend all solutions is $N(m', r')$ by zeros that were excluded in Step 5. This results in the solution set $N(m, r)$ of equation (i).

9. Repeat steps 2-8 for all equations of type (i) that are embedded in eq. $LoM = N$. This results in all matrices L that satisfied $LoM = N$.

Example Consider the matrix equation

$$\begin{bmatrix} l_{11} & l_{12} & l_{13} \\ l_{21} & l_{22} & l_{23} \end{bmatrix} o \begin{bmatrix} 0.9 & 0.5 \\ 0.7 & 0.8 \\ 1 & 0.4 \end{bmatrix} = \begin{bmatrix} 0.6 & 0.3 \\ 0.2 & 1 \end{bmatrix}$$

Whose general form is

$$[l_{i,j}] \; o \; [m_{j,k}] = [n_{i,k}]$$

determine all particular configurations of matrix first for which the equation is satisfied.

The given matrix represents the following four equations of the form of eq.

$$\max_{j \in J} \min (l_{i,j}, m_{j,k}) = n_{i,k}$$

$$\max[\min (l_{11},0.9), \min (l_{12},0.7), \min (l_{13},1)] = 0.6 \quad \ldots(i)$$

$$\max[\min (l_{11},0.5), \min (l_{12},0.8), \min (l_{13},0.4)] = 0.3 \quad \ldots(ii)$$

$$\max[\min (l_{21},0.9), \min (l_{22},0.7), \min (l_{23},1)] = 0.2 \quad \ldots(iii)$$

$$\max[\min (l_{21},0.5), \min (l_{22},0.8), \min (l_{23},0.4)] = 1 \quad \ldots(iv)$$

observe that equations (i) and (ii) contain only unknowns l_{11}, l_{12} and l_{13}, whereas equations (iii) and (iv) contain only unknowns l_{21}, l_{22} and l_{23}. This means that each of these pairs of equation can be solved independently of the other. Hence, as previously argued, the given matrix equations can be decomposed into two simple matrix equations.

$$\begin{bmatrix} l_{11} & l_{12} & l_{13} \end{bmatrix} o \begin{bmatrix} 0.9 & 0.5 \\ 0.7 & 0.8 \\ 1 & 0.4 \end{bmatrix} = \begin{bmatrix} 0.6 & 0.3 \end{bmatrix}$$

and $$\begin{bmatrix} l_{21} & l_{22} & l_{23} \end{bmatrix} o \begin{bmatrix} 0.9 & 0.5 \\ 0.7 & 0.8 \\ 1 & 0.4 \end{bmatrix} = \begin{bmatrix} 0.2 & 1 \end{bmatrix}$$

It is clear that, however, the second equation, which is associated with i = 2, satisfies the inequality.

$\max_{j \in J} m_{j,k} < n_{i,k}$ and therefore, has no solution.

Specifically max (0.5,0.8,0.4) < 1

Thus the given matrix equations has no solution.

Example Given and $m = \begin{bmatrix} 0.1 & 0.4 & 0.5 & 0.1 \\ 0.9 & 0.7 & 0.2 & 0 \\ 0.8 & 1 & 0.5 & 0 \\ 0.1 & 0.3 & 0.6 & 0 \end{bmatrix}$ $n = [0.8,\ 0.7,\ 0.5,\ 0]$

determine all solution of $l \ o \ m = n$.

First, we determine the maximum solution $\hat{l}$.

$$\hat{l}_1 = \min(1,0.1,1,0) = 0.1$$

$$\hat{l}_2 = \min(0.8,1,1,1) = 0.8$$

$$\hat{l}_3 = \min(1,0.7,1,1) = 0.7$$

$$\hat{l}_4 = \min(1,1,0.5,1) = 0.5$$

Thus, $\hat{l} = (0,0.8,0.7,0.5)$. We can easily check that $\hat{l} \in N$ *(m,r)*: hence $N(m,r) \neq \phi$

Since, $\hat{l}_1 = 0$, we may reduce the matrix equation by excluding l_1 and the first row of matrix *M*, since $n_4 = 0$ we may make a further reduction by excluding n_4 and the fourth column *M*. The reduced equation has the form

$$[l_1 \quad l_2 \quad l_3] o \begin{bmatrix} 0.9 & 0.7 & 0.2 \\ 0.8 & 1 & 0.5 \\ 0.1 & 0.3 & 0.6 \end{bmatrix} = [0.8 \quad 0.7 \quad 0.5]$$

O(B,j)	*j*	*g(B)*
B = 1 1 2	{1,2},{3},ϕ	(0.8,0,5,0)
1 1 3	{1,2},ϕ,{3}	(0.8,0,0.5)
1 2 2	{1},{2,3},ϕ	(0.8,0.7,0)
1 2 3	{1}{2},{3}	(0.8,0.7,0.5)

Next, we apply the four steps of the procedure for determining the set $\hat{N}(a,n)$ of all minimal solutions of this reduced matrix equation:

(i) Employing the maximum solution

$\hat{l} = (0.8,0.7,0.5)$ of the reduced equation, we obtain

$J_1(\hat{l}) = \{1\}, J_2(\hat{l}) = \{1,2\}, J_3(\hat{l}) = \{2,3\}$

hence, $J(\hat{l}) = \{1\} \times \{1\}, \{2\} \times \{2,3\}$

(ii) The sets *O(B,j)* that we must determine for all $B \in J(\hat{l})$ and all $j \in J$ are listed in table.

(iii) For each $B \in J(\hat{l})$ we generated the triples *g(B),* which are also listed in table.

(iv) Two of the triples *g(B)* in table are minimal : (0.8,0.5,0) and (0.8,0,0.5). By adding O as the first component to each of these triples, we obtained the minimal solutions of the original matrix equation. Hence

$\hat{N}$ (m,n)={1$\hat{l}$=(0,0.8,0.5,0), 2$\hat{l}$ = (0,0.8,0.5)

The set *N(m,n)* of all solutions of the given matrix equation is now fully captured by the maximum solution.

$\hat{l}$ = (0,0.8,0.7,0.5) and the two minimal solution.

1$\hat{l}$ = (0,0.8,0.5,0)

2$\hat{l}$ = (0,0.8,0,0.5)

So $N(m,n) = \{l \in l| \, {}^{1}\hat{l} \leq l \leq \hat{l}\}$

$\cup \{l \in l| \, {}^{2}\hat{l} \leq l \leq \hat{l}\}$

Defuzzification

The reverse process of fuzzification is called defuzzification.

In many conditions for a system whose output is fuzzy, it is easier to take a crisp decision if the output is represented as a single scaler quantity. This conversion of a fuzzy set to single crisp value is called defuzzification.

Method of Defuzzification

(i) Controid method

(ii) Center of sums (cos) method

(iii) Mean of maxima (Mom)

Centroid method : This method is also known as the center of gravity or center of area method, it obtains the center of area ⓧ occupied by the fuzzy set. It is given by

$$ⓧ = \frac{\int \mu(x) x dx}{\int \mu(x) dx}$$ for a continuous membership function

and

$$\textcircled{x} = \frac{\sum_{i=1}^{n} x_i \mu(x_i)}{\sum_{i=1}^{n} \mu(x_i)} \quad \text{for a discrete membership function}$$

Where n is the number of elements in the sample, x_i' s are the elements, $\mu(x_i)$ is its membership functin.

Center of sums (Cos) method

In center of sums method the overlapping area is counted twice. Cos builds the resultant membership function by taking the algebraic sum of outputs from each of the contributing fuzzy sets $A_1, A_2, \ldots\ldots\ldots\ldots$. The defuzzified value $\textcircled{x}$ is given by

$$\textcircled{x} = \frac{\sum_{i=1}^{n} x_i \sum_{k=1}^{n} \mu_{Ak}(x_i)}{\sum_{i=1}^{n}\sum_{k=1}^{n} \mu_{Ak}(x_i)}$$

Where n is the number of fuzzy sets and N is the number of fuzzy variables.

Mean of maxima (Mom) defuzzification

In the method to find the output of defuzzification we take the crisp value with the highest degree of membership. In cases with more then one element having the maximum value, the mean value of the maxima is taken. The equation of the difuzzified value is given by

$$\textcircled{x} = \frac{\sum_{xi \in m} x_i}{|M|}$$

Where M = {x_i : $\mu(x_i)$ is equal to the height of fuzzy set}

|M| is the cardinality of the sets M.

In the continuous case, M can be defined as M = {x∈[-c,c]/μ(x) is equal to the height of the fuzzy set}

In such a case, the mean of maxima is the arithmetic average of mean values of all intervals contained in M including zero length intervals.

Note

(i) The height of a fuzzy set *A, i.e h(A)* is the largest membership grade obtained by any element in that set.

(ii) COS is the most commonly used defuzzificaiton method.

(iii) The COS method can be implemented easily and leads to rather fast inference cycles.

(iv) In the centroid method, the overlapping area is counted once whereas in center of sums, the overlapping area is counted twice.

Chapter - 17

Decision Theory

Introduction

In any organization, we have to take proper decision i.e.management planning, controlling and motivation processes. Decisions whether to accept or reject a new jobs, to buy or not to buy shares of a particular company etc. would be made without a detailed analysis in each situation. Such decisions are the personal decisions.

Infect, the decision should be made whenever the organization or an individual faces a problem of decision making or dissatisfied with the existing decisions or when alternative selection is specified. Then it is necessary to develop rational decision making to improve decision making abilities. The Principle of decision making can be useful in deciding the best course of action from the given alternatives.

In the process of decision making generally the following steps are involved:

(a) The proper understanding of the problem.

(b) Different alternatives or acts available for the problem.

(c) The selection of proper alternative from available alternatives.

(d) The decision maker has control over these alternatives.

(e) The Estimates of payoff of different alternatives.

"A decision is defined as the selection by the decision maker of an act. Considered to be the best according to some predesignated standard form among the several available options."

Basic concept of Decision Theory

There are certain basic concepts associated with the process of decision making listed as follows:

1. **Strategies or Acts:** There is a finite number of alternatives available with the decision maker at each point in time when a decision is made. The number of possible courses of action may be large or small, but these are under the control of decision maker. e.g. for a farmer, before rain starts he has to decided about the crop to be taken.

2. **State of Nature :** These are not under the control of decision maker. The state of nature are usually not determined by the action of an individual or an organization. The state of nature mutually exclusive and collectively exhaustive with respect to any decision problem. The state of nature can have a numerical description such as demand of some (finite) units of a given item or a non-numerical like employee's strike.

3. **Pay off :** A numerical value resulting from each possible combination of a Course of action and state of nature is known as Pay off.

   It is also known as profits or conditional values because of state of nature is unknown. The pay off is measured within a specified period. Which is called decision horizon.

   The arrangement or different Pay offs against the corresponding states of nature in a tabular form is known as Payoff matrix.

   Let m states of nature are denoted by O_1, O_2, - - - - O_m with respect to n alternatives (Courses of action) S_1, S_2, - - - - S_n. For a specified combination of state of nature and course of action, the pay off matrix is represented as a_{ij}.

State of nature	Courses of action			
	S_1	S_2	-------------	S_n
O_1	a_{11}	a_{12}	-------------	a_{1n}
O_2	a_{21}	a_{22}	-------------	a_{2n}
⋮	⋮	⋮	⋮ ⋮	⋮
O_m	a_{m1}	a_{m2}	-------------	a_{mn}

Pay off matrix

Types of Decision making Environment

The decision making are classified such that the type of information which is given about the occurrence of the various states of nature as well as depending upon the decision environment. Therefore they are four types of decision environments:

1. Decision making under certainty.(e.g. to purchase NSS, Indira Vikas Patra)
2. Decision making under risk.(e.g. prob of getting head in the toss of a coin is 0.5)
3. Decision making under uncertainty(e.g. prob that Mr. x will be P.M. of the country 15 years from now is not known)
4. Decision making under conflict. (e.g. the theory of games)

Decision making under uncertainty

If probabilities associated with the occurrence of states of nature are not given, then decision maker has no way to determined the expected payoff for his alternative. There are several different criteria of decision making in this situation.

1. **Criterion of Pessimism (Maximin or Minimax):** The decision maker ensures to identify the worst possible out come in each course of action. i.e. maximum loss or minimum outcome that would occur under each decision alternative and then selecting the best out of the worst outcome i.e. maximum payoff. In order to choose the optimal course of action.

2. **Criterion of Optimism (Maximax or Minimin):** In this case, for each Conditional profit or cost value (payoff), a regret value equal to the difference between the maximum payoff under a state of nature and payoff resulting from each course of action under that state is calculated.

   regret Payoff = maximum Payoff from a course of action – Payoff.

3. **Criterion of Equal Probabilities (Laplace or Baye's principle):** According to this principle, equal probabilities are assigned to different states of nature. As states of nature are mutually exclusive and collectively exhaustive so the probability of each of these must be 1/(number of states of nature).

The working methods are as follows:

   (i) Assign equal probability value to each state of nature by 1/(number of States of nature)

   (ii) Compute the expected payoff for each course of action by adding all the payoff and dividing by the number of possible states of nature.

   (iii) Select the best expected payoff value.

4. **Criterion of Realism (Hurwicz criterion):** A decision maker should be neither completely optimistic nor pessimistic. Hurwicz introduced the idea of coefficient of optimism. Let the coefficient of optimism be $\propto \left(0 \leq \propto \leq 1\right)$. Where I represents a

complete optimistic attitude about the future and 0 a complete pessimistic attitude about the future. According to Hurwicz,

H = $\propto$ (maximum payoff in column) + (1- $\propto$) (minimum payoff in column).

Example 1 A food product company is contemplating the introduction of a revolutionary new product with new packaging to replace the existing product at much price (S_1) or a moderate change in the composition of the existing product with a new packaging at a small increase in price (S_2) or a small change in the composition of the existing except the word 'new' with a negligible increase in price (S_3). The three possible states of nature of events are: (i) high increase in sates (N_1), (ii) no change in sales (N_2), and (iii) decreases in sales (N_3).

The marketing department of the company worked out the payoffs in terms of yearly net profits for each course of action for these events. This is represented in the following table:

State of nature	Courses of action		
	S_1	S_2	S_3
N_1	7,00,000	5,00,000	3,00,000
N_2	3,00,000	4,50,000	3,00,000
N_3	1,50,000	0	3,00,000

Which strategy should the company choose on the basis of: (a) Maximin criterion, (b) Maximax criterion, (c) Minimax regret criterion, (d) Laplace criterion.

Solution (a) The payoff matrix is rewritten as follows.

State of nature	Courses of action		
	S_1	S_2	S_3
N_1	7,00,000	5,00,000	3,00,000
N_2	3,00,000	4,50,000	3,00,000
N_3	1,50,000	0	3,00,000
Minimum payoff (Column)	1,50,000	0	3,00,000

Since maximum of column minimum is Rs. 3, 00,000.

Hence, the company should adopt course of action S_3, applying the maximin criterion.

(b)

State of nature	Courses of action		
	S_1	S_2	S_3
N_1	7,00,000	5,00,000	3,00,000
N_2	3,00,000	4,50,000	3,00,000
N_3	1,50,000	0	3,00,000
Maximum payoff (Column)	7,00,000	5,00,000	3,00,000

Since maximum of column maximum is Rs. 7, 00,000.

Hence, the company should adopt course of action S_1, applying the maximax criterion.

(c) Select maximum course of action in each row and subtracting all course of action in each row.

State of nature	Courses of action		
	S_1	S_2	S_3
N_1	7,00,000-7,00,000 =0	7,00,000-5,00,000 =2,00,000	7,00,000-3,00,000 =4,00,000
N_2	4,50,000-3,00,000 =1,50,000	4,50,000-4,50,000 =0	4,50,000-3,00,000 =1,50,000
N_3	3,00,000-1,50,000 =1,50,000	3,00,000-0 =3,00,000	3,00,000-3,00,000 =0
Maximum regret	1,50,000	3,00,000	4,00,000

The minimum value among the maximum regret is Rs. 1,50,000 and this corresponds to course of action S_1, applying minimax regret criterion.

(d) Here it is assured that each course of action has a probability of occurrence equal to 1/3. Therefore expected returns can be obtained by the formula:

$$\text{Expected return} = \frac{1}{3}(\text{Sum of each course of action})$$

Course of acti on	Expected return
S_1	$\frac{1}{3}(7,00,000+ 3,00,000 + 1,50,000)= 3,83,333.33$
S_2	$\frac{1}{3}(5,00,000+ 4,50,000 +0)= 3,16,666.66$
S_3	$\frac{1}{3}(3,00,000+ 3,00,000 + 3,00,000)= 3,00,000$

Since the largest expected return is Rs. 3, 83,333.33

Hence, the executive must select the strategy S_1.

Example 2 A manufacturer makes a product, of which the principal ingredient is a chemical A. At the moment, the manufacturer spends Rs. 1,000 per year on supply of A, but there is a possibility that the price may soon increase to four times its present figure because of worldwide shortage of the chemical. There is another chemical B, which the manufacturer could use in conjunction with a third chemical C, in order to give the same effect as chemical A. chemicals B and C would together cost the manufacturer Rs. 3,000 per year, but their prices are unlikely to rise. What action should the manufacturer take? Apply the maximin and minimax criteria per decision making and give two sets of solutions. If the coefficient of optimism is 0.4, find the course of action that minimizes the cost.

Solution The data of the problem is summarized in the following table with negative figures represent profit.

States of nature	Course of action	
	S_1 (use B and C)	S_2 (use A)
N_1 (Price of A increase)	-3000	-4000
N_2 (Price of A dose not increase)	-3000	-1000

(i) Maximin criterion:

States of nature	Course of action	
	S_1	S_2
N_1	-3000	-4000
N_2	-3000	-1000
Minimum (column)	-3000	-4000

Since maximum of minimum column is -3000

Hence, the manufacturer should adopt course of action S_1

(ii) Minimax criterion:

States of nature	Course of action	
	S_1	S_2
N_1	-3000 - (-3000)= 0	-3000 - (-4000) =1000
N_2	-1000 - (-3000)= 2000	-1000 - (-1000) = 0
Maximum (column)	2000	1000

Since minimum of maximum column is 1000

Hence, the manufacturer should adopt course of action S_2.

(iii) Hurwicz criterion: - Let the coefficient of optimism is $\alpha = 0.4$

Then coefficient of Pessimism will be $1-\alpha = 1 - 0.4 = 0.6$

Then according to Hurwicz, select course of action that maximum for Profit and minimum for cost value i.e.

$H = \alpha$ (Best payoff) + $(1-\alpha)$ (worst payoff)

$= \alpha$ (maximum in column) + $(1-\alpha)$ (minimum in column)

Course of action	Best Payoff	Worst Payoff	H
S_1	-3000	-3000	-3000
S_2	-1000	-4000	-2800

For S_1, $H = 0.4(-3000) + 0.6(-3000) = -3000$

For S_2, $H = 0.4(-1000) + 0.6(-4000) = -2800$

Since course of action S_2 has the least cost (maximum profit) is Rs. 2800. Then manufacturer should adopt S_2.

Practice Problems

1. The following matrix gives the payoff of different alternatives S_1, S_2 and S_3 against conditions N_1, N_2, N_3 and N_4.

Alternative	States of Nature			
	N_1	N_2	N_3	N_4
S_1	4000	-100	6000	18000
S_2	20000	5000	400	0
S_3	20000	15000	-2000	1000

Indicate the decision taken under the following approaches:

(a) Pessimistic (b) Optimistic

(c) Equal probability (d) Regret

(e) Hurwicz criterion, the degree of optimism being 0.7.

[Ans : (a) S_2 (b) S_2 or S_3 (c) S_3 (d) S_1 (e) S_2]

2. In the toy manufacturing company, suppose that the product acceptance probabilities are not known, so that the available information is as shown below:

Product acceptance	Anticipated first year Profit (Rs. 1,000)		
	Product line		
	Full	Partial	Minimal
Good	80	70	50
Fair	50	45	40
Poor	-25	-10	0

Determine the optimal decision under each of the following decision criteria and show how you arrived in it:

(a) Maximax (b) Maximin (c) Equal likelihood (d) Minimax regret.

[**Ans :** (a) Full (b) Minimal (c) Full or Partial (d) Partial]

3. A decision matrix with cost data is given below:

Alternative	States of Nature			
	A_1	A_2	A_3	A_4.
S_1	6	8	3	5
S_2	2	5	4	7
S_3	1	3	8	5
S_4	4	6	6	3

Find the best alternative using (i) Minimax criterion (ii) Minimin criterion (iii) Minimax regret criterion.

[**Ans :** (i) S_4 (ii) S_3 (iii) S_2 or S_4]

4. Mr. X has Rs. 10,000 to invest in one of three options: P, Q or R. The return on his investment depends on whether the economy experiences inflation, recession, or no change at all. His possible returns under each economic condition are given below:

Alternative	States of Nature		
	Inflation	Recession	No change
P	2,000	1,200	1,500
Q	3,000	800	1,000
R	2,500	1,000	1,800

What should he decide using the (i) Pessimistic criterion (ii) Optimistic criterion (iii) Equally likely criterion and (iv) regret criterion.

[**Ans:** (i) A, Rs.120 (ii) B, Rs.300 (iii) C, Rs.176.6 (iv) C, Rs.50]

Decision Making Under Risk

When the probabilities of occurrence are known to decision maker to select from among several possible options. This decision maker is said take decision under risk. Each strategy will have as many payoffs as the states of nature in the situation. This procedure is suitable only if the decisions are of repetitive type. The most widely used criterion for evaluating various courses of action under risk is the Expected monetary value (EMV) or Expected Utility.

1. **Expected Monetary Value (EMV)**: The expected monetary value (EMV) for the specified course of action is the weighted sum of possible payoffs i.e. by summing the payoff for several course of action multiplied by the probability associated with each outcome.

The expected monetary value is stated as

$$EMV(S_i) = \sum_{j=1}^{n} P_{ij} \; P_j$$

Where S_i = course of action, i

P_j = Probability of occurrence of state of nature.

P_{ij} = Payoff associated with course of action i and state of nature j.

n = number of possible states of nature.

Steps for Calculating EMV: The various steps for calculating EMV are as follows:

Step 1. Construct a payoff matrix and define the state of nature and the course of action.

Step 2. List the payoff associated with each possible combination along with the Corresponding probability of each state of nature.

Step 3. Calculate the EMV for each course of action by multiplying the Conditional profits by the associated probabilities and then add these weighted values for each course of action.

Step 4. Select the course of action corresponding to the optimal EMV.

2. **The Expected value of perfect Information (EVPI)**: If the decision maker will have perfect or reliable information about the occurrence of the state of nature, then he will be able to choose the optimal coursed of action with certainity.

The expected profit with perfect information (EPPI) is the maximum attainable expected monetary value (EMV) based on perfect information about the state of nature that will occur. It is defined as the sum of the product of best state of nature corresponding to each optimal course of action and its probability.

Therefore, the expected value of perfect information may be defined as the maximum amount one would be willing to pay to obtain perfect information about the state of nature that would occur. Hence, the perfect information would increase profit from EMV up to the value of EPPI. i.e.

EVPI = EPPI – EMV*

= (Expected Profit with perfect information) – (Maximum expected monetary value).

3. **The Expected opportunity Loss (EOL)**: An alternative way of maximizing expected monetary value is to minimize the expected opportunity loss or expected value of regret. The expected opportunity loss is determined as the difference between the highest profit for a state of nature and the actual profit obtained for the particular course of action taken. The expected opportunity loss for i th course of action S_i is Stated as follows:

$$\text{EOL (State of nature } N_i) = \sum_{i=1}^{n} l_{ij} P_i$$

Where l_{ij} = opportunity loss due to state of state of nature N_i and course of action S_j.

P_i = Probability of occurrence of state of nature N_i.

n = number of possible states of nature.

Steps for calculating EOL: The various steps for calculation EOL are as follows:

Step 1. Construct a conditional profit table for each course of action and state of nature combining along with the associated probabilities.

Step 2. Calculate the conditional opportunity loss for each state of nature by subtracting each payoff from the maximum payoff for that outcome.

Step 3. Calculate EOL for each course of action by multiplying the probability of each state of nature with the COL value and then adding the values.

Step 4. Select a course of action for which the EOL value is minimum.

Example 3 Based on Past experience it is known that the probability of weather being cold is 0.4 then (a) Set up the opportunity loss table and compute opportunity loss of each alternative course of action. The payoff matrix is given below:

Alternative	States of nature	
	Cold weather	Warm weather
Sell cold drinks	110	50
Sell ice cremes	80	90

Show that the decisions using EMV and EOL are the same.

Solution Given that the probability of cold weather is 0.4 Hence probability of warm weather is 1- 0.4 = 0.6

EMV Table

Alternative	EMV (Rs.)
Sell cold drinks	110 x 0.4 + 50 x 0.6 = 74
Sell ice cremes	80 x 0.4 + 90 x 0.6 = 86

Since EMV is more for selling ice cremes i.e. Rs. 86

Opportunity loss table

Alternative	States of nature		EOL
	Cold weather	Warm weather	
Sell cold drinks	110 - 110=0	90 – 50 = 40	0x 0.4+40x0.6 =24
Sell ice cremes	110 – 80 =30	90 – 90 = 0	30x0.4+0x0.6 =12

Since EOL is minimum for selling ice cremes i.e. Rs. 12

Thus, the decision obtained by EMV and EOL are the same.

Example 4 Find EVPI for the following payoff table:

Event	Acts			Probability
	A_1	A_2	A_3	
S_1	40	-10	-100	0.20
S_2	400	440	400	0.70
S_3	650	720	760	0.10

Solution (i) for expected payoff (EP) multiply eact act by probability.

EP for each act

Event	Probability	Acts		
		A_1	A_2	A_3
S_1	0.20	40 x 0.20 = 8.0	-10 x 0.20 = -2.0	-100 x 0.20 = -20.0
S_2	0.70	400 x 0.70 = 280.0	440 x 0.70 = 308.0	400 x 0.70 = 280.0
S_3	0.10	650 x 0.10 = 65.0	720 x 0.10 = 72.0	760 x 0.10 = 76.0
Total		353.0	378.0	336.0

The maximum payoff under uncertainity corresponds to act A_2 i.e.

EP = 378

Table for EPPI

(ii)

Events	Probability	Acts			Maximum payoff for each events	EPPI
		A_1	A_2	A_3		
S_1	0.20	40	-10	-100	40	40 x 0.2 = 8
S_2	0.70	400	440	400	440	440 x 0.7= 308
S_3	0.10	650	720	760	760	760 x 0.1 = 76
					Total	392

$\therefore$ EVPI = EPPI- EP

= 392 - 378 = 14

Example 5 A company manufactures goods for a market in which the technology of the products is changing rapidly. The research and development department has produced a new product which appears to have potential for commercial exploitation. A further Rs. 60,000 is required for development testing. The company has 100 customers and each customer might purchase, at the most, one unit of the product. Market research suggests a selling price of Rs. 6,000 for each unit with total variable costs of manufacture and selling estimated at Rs. 2,000 for each unit.

As a result of previous experience of this type of market, it has been possible to derive probability distribution relating to the proportion of customers who will buy the product, as follows:

Proportion of customers :	0.04	0.08	0.12	0.16	0.20
Probability :	0.1	0.1	0.2	0.4	0.2

Determine the expected opportunity losses, given no further information than that stated above and state, whether or not, the company should develop the product.

Solution Let the proportion of customers be p, who purchase the new product.

Then the conditional profit will be

= (6000 - 2000) (p x 100) - 60000

= Rs 1000 (400p - 60) -----(i)

Now, the company has two alternatives i.e. to develop the product or not to develop the product. The conditional profit table can now be constructed.

Proportion of customers	Probability	Alternatives	
		Do not develop	Develop
0.04	0.1	0	-44000
0.08	0.1	0	-28000
0.12	0.2	0	-12000
0.16	0.4	0	4000
0.20	0.2	0	20000

(Note: Putting values of p in eqn (i) and find the develop alternative.)

From the conditional profit table, the opportunity loss table can be derived and expected opportunity losses can be found.

Proportion of custome rs	Probability	Alternatives	
		Do not develop	Develop
0.04	0.1	0	44000
0.08	0.1	0	28000
0.12	0.2	0	12000
0.16	0.4	4000	0
0.20	0.2	20000	0

∴ EOL (Do not develop) = 4000 x 0.4 + 20000 x 0.2 = Rs. 5600

and EOL (Develop) = 44000 x 0.1 + 28000 x 0.1+12000 x 0.2 = Rs 9600

Since do not develop is the lower EOL of Rs. 5600.

Hence the best decision is not to develop the product.

Example 6 A company manufactures parts of passenger cars and sells them in lots of 10000 parts each. The company has a policy of inspecting of inspecting each lot before it is actually shipped to the retailer. Five inspection categories established for quality control represent the percentage of defective items contained in each lot. These are given in the following table. The daily inspection chart for past 100 inspections. Shows the following rating or break down inspection:

The management is considering two possible course of action:

(a) Shut – down the entire plant operations and thoroughly inspect each machine.

Rating	Proportion of defective items	Frequency
Excellent(A)	0.02	25
Good(B)	0.05	30
Acceptable(C)	0.10	20
Fair(D)	0.15	20
Poor(E)	0.20	5
	Total	100

(b) Continue production as it now exists but offer the customer a refund for defective items that are discovered and subsequently returned. The first alternative will cost Rs. 600 while the second alternative will cost the company Rs. 1 for each defective item that is returned.

What is the optimum decision for the company? Find the EVPI.

Solution Calculation of inspection and refund cost table are as follows:

Rating	Defective rate	Probability	Expected cost		Opportunity loss	
			Inspect	Refund	Inspect	Refund
A	0.02	0.25	600	200	400	0
B	0.05	0.30	600	500	100	0
C	0.10	0.20	600	1000	0	400
D	0.15	0.20	600	1500	0	900
E	0.20	0.05	600	2000	0	1400
Total =			600	800	130	330

The cost of refund is computed as follows:

For A : 10000 x0.02 x 1.00 = Rs. 200.

Similarly the cost of refund for other rating is calculated.

Expected cost of inspective is:

600 x 0.25+ 600 x 0.30 + 600 x 0.20 + 600 x 0.20 + 600 x 0.05 = Rs. 600

Expected cost of refund is:

200 x 0.25 + 500 x 0.30 + 1000 x 0.20 + 1500 x 0.20 + 2000 x 0.05 = Rs. 800

Since the cost of refund is more than the cost of inspection.

Hence the plant should be shut down for inspection

and EVPI = EOL of inspection = Rs. 130

Example 7 A certain piece of equipment is to be purchased for a construction project at a remote location. This equipment contains as expensive part which is subject to random failure. Spares of this past can be purchased at the same time the equipment is purchased. Their unit cost is Rs. 1500 and they have no scrap value. If the part fails on the job and no spare is available, the part will have to be manufactured on a special order basis. If this is required, the total cost, including down time of the equipment is estimated as Rs. 9000 for each such occurrence. Based on previous experience with similar parts, the following probability estimates of the number of failures expected over the duration of the product are provided as given below:

Failure	:	0	1	2	3 and above
Probability	:	0.80	0.15	0.05	0.00

(a) Determine optimal EMV* and optimal number of spares to purchase initially.

(b) Based on opportunity losses, determine the optimal course of action and optimal value of EOL.

(c) Determine expected profit with perfect information and EVPI.

Solution (a) Let A_1(no failure), A_2(1 failure) and A_3(2 failure) be the possible states of nature. Similarly let S_1 (no spare purchased), S_2 (1 spare purchased) and S_3 (2 spare purchased) be possible course of action.

Then conditional costs table are as follows.

State of nature (spare required)	Course of action (no. of spare purchased)	Purchase cost	Emergency cost	Total cost
0	0	0	0	0
0	1	1500	0	1500
0	2	3000	0	3000
1	0	0	9000	9000
1	1	1500	0	1500
1	2	3000	0	3000
2	0	0	18000	18000
2	1	1500	9000	10500
2	2	3000	0	3000

By using conditional costs and the probability of states of nature, the expected monetary value can be computed for each of three states of nature as shown in table below:

State of nature	Probability	Conditional cost			Weighted cost due to course of action		
		S_1	S_2	S_3	S_1	S_2	S_3
A_1	0.80	0	1500	3000	.80 x 0 = 0	.80 x1500 = 1200	.80 x 3000 =2400
A_2	0.15	9000	1500	3000	.15 x 9000 =1350	.15x1500 = 225	.15x3000 = 450
A_3	0.05	18000	10500	3000	.05x18000 = 900	.05x10500 = 525	.05x3000 = 150
				EMV	2250	1950	3000

Since the weighted cost is Rs. 1950 is lowest due to course of action S_2.

Hence S_2 should be chosen. If EMV is expressed in terms of profit.

Then $EMV^* = EMV$ of S_2 = - Rs. 1950

Hence, the optimal number of spares to be purchased initially should be one.

(b) To calculate EOL, we first calculate COL. The computation for COL are shown in table given below:

State of nature	Conditional cost			Conditional opportunity loss		
	S_1	S_2	S_3	S_1	S_2	S_3
A_1	0	1500	3000	0	1500	3000
A_2	9000	1500	3000	9000-1500 = 7500	1500-1500 = 0	3000-1500 = 1500
A_3	18000	10500	3000	18000-3000 = 15000	10500-3000 = 7500	3000-3000 = 0

(Since we are dealing with the condition cost rather than conditional Profit, lower value for each state of nature shall be considered for computing opportunity losses)

The calculations for expected opportunity loss are shown in table below:

State of nature	Prob-ability	COL			Weighted opportunity loss		
		S_1	S_2	S_3	S_1	S_2	S_3
A_1	0.80	0	1500	3000	.80 x 0 = 0	.80 x 1500 = 1200	.80 x 3000 = 2400
A_2	0.15	7500	0	1500	.15 x 7500 = 1125	.15 x 0 = 0	.15 x 1500 = 225
A_3	0.05	15000	7500	0	.05 x 15000 = 750	.05 x 7500 = 375	.05 x 0 = 0
				Total =	1875	1575	2625

Since $EOL^* = EOL(S_2)$ = Rs. 1575 (the lowest cost), Hence it is adopt course of action S_2 and to purchase one spare.

(c) The EPPI can be computed by choosing optimal course of action for each state of nature, multiplying its conditional cost by the probability and then summing these products. The EPPI calculations are as follows:

State of nature	Probability	Optimal cause of action	Cost of optimal course of action	
			Conditional cost	Weighted opportunity loss
A_1	0.80	S_1	0	0.80 x 0 = 0
A_2	0.15	S_2	1500	0.15 x 1500= 225
A_3	0.05	S_3	3000	0.05 x 3000= 150
			Total =	375

Since we are dealing with the expected cost, EPPI = - Rs. 375

Now, the expected value of perfect information is

$$EVPI = EPPI - EMV^*$$

$$= -375 - (-1950)$$

$$= \text{Rs. } 1575$$

Here, it can be observed that

$$EVPI = EOL^* = \text{Rs. } 1575$$

Practice Problem

1. An investor is given the following investment alternatives and percentage rates of return.

	State of nature (Market Conditions)→		
	Low	Medium	High
Regular shares	7%	10%	15%
Risky shares	-10%	12%	25%
Property	-12%	18%	30%

Over the past 300 days, 150 days have been medium market conditions and 60 days have had high market increases.

On the basis of these data, state the optimum investment strategy for the investment.

[**Ans :** Expected return 23%, highest for Property, then investor should invest in this alternatives.]

2. An electric manufacturing company has seen its business expanded to the point where it needs to increase production beyond its existing capacity. It has narrowed the alternatives to two approaches to increase the maximum production capacity. (a) Expansion, at a cost of Rs. 8 million or (b) modernization at a cost of Rs. 5 million. Both approaches would require the same amount of time for implementation. Management believes that over the required payback period, demand will either be high or moderate. Since high demand is considered to be some what less likely than moderate demand, the probability of high demand has been setup at 0.35, if the demand is high, expansion would gross an estimated additional Rs. 12 million but modernization only an additional Rs. 6 million, due to lower maximum production capacity. On the other hand, if the demand is moderate, the comparable figures would be Rs. 7 million for expansion and Rs. 5 million for modernization.

   (a) Calculate conditional profit in relation to various action and outcome combinations and state of nature.

   (b) If company wishes to maximize its expected monetary value, then should it modernize or expand.

   (c) Calculate the EVPI.

   (d) Construct the conditional opportunity loss table and also calculate EOL.

[**Ans :** (a) S_1 is .4, -1 and S_2 is 1,0

(b) EMV (S_1) = Rs. 0.75 Million, EMV (S_2) = Rs. 0.35,

EMV*= EMV (S_1) = Rs. 0.75 million and must expand S_1.

(c) EVPI = Rs. 0.65 million

(d) EOL (S_1) = Rs. 0.65, EOL (S_2) = Rs. 1.05 million]

3. A TV dealer find that the cost of holding a TV in stock for a week is RS. 50. Customers who can not obtain new TV immediately tend to go to other dealers and he estimates that for every customer who can not get immediate delivery he loses an average of Rs. 200. For one particular model of TV, the probabilities of a demand of 0,1,2,3,4 and 5 TVs in a week are 0.05,0.10,0.20,0.30.0.20 and 0.15 respectively.

(a) How many TV per week should the dealer order? Assume that there is no time log between ordering and delivery.

(b) Compute EVPI.

(c) The dealer is thinking of spending on a small market survey to obtain additional information regarding the demand levels. How much should be willing to spend on such a survey.

[Hint: Cost function = $\begin{cases} 50\text{ S} + 200\text{ (D-S), D} \geq \text{S} \\ 50\text{ D} + 50\text{ (S-D), D<S} \end{cases}$

where D is demand and S is number of TV stored.

Ans: (i) Expected cost = Rs. 230, dealer must order 4 TV sets per week.

(ii) EVPI = EC – ECPI = 230- 147.5= Rs. 82.5

(iii) Dealer should not willing to spend more than Rs. 82.5 for the market survey]

4. A group of volunteers of a service organization rises money each year by selling gift articles outside the stadium after a hockey match between team A and B. They can buy any of the three different types of gift articles from a dealer. Their sales are mostly dependent on which team wins the match.

A conditional payoff table is given below:

Events	Type of gifts articles(Rs.)		
	I	II	III
Team A wins	400	500	800
Team B wins	1000	900	600

(i) Construct the opportunity loss table

(ii) Which type of gift article should be volunteers buy if the probability of team A winning is 0.8?

[Ans: (i)

Events	Type of gifts articles(Rs.)			Probability
	I	II	III	
Team A wins	800-400= 400	300	0	0.8
Team B wins	1000-1000 = 0	100	400	0.2

(ii) EOL = 80, the lowest value.

So the volunteer should buy type III gift article.]

Chapter - 18

Calculus of Variation

Introduction

The calculus of variations is a powerful technique which was developed from the theory of maxima and minima of function of a single or more variable. The calculation of variation is concerned with the problem of extre-mising "functionals" *i.e* with maximizing or minimizing functionals. The solution of the basic problems of the calculus of variations are of the following types:

(a) Find the shortest path between two given points on a surface.

(b) Find the curve between two given points in a plane.

(c) Find the curve along with a particle will slide in the shortest time.

Functionals

A variable quantity $I[y(x)]$ is a functional dependent on a function $y(x)$ if to each function $f(x)$ belongs to certain class of functions. Consider the problem of finding a curve through two paints (x_o, y_o) and (x_1, y_1) whose length l is minimum, on the curve $y = y(x)$. Such that

$$I[y(x)] = \int_{x_0}^{x_1} \sqrt{1+\left(\frac{dy}{dx}\right)^2}\, dx.$$

Variational problem- The mathematical problem which investigates extrema of a functional is known as variational problem. The laws which govern functionals to attain extrema are the variational principles.

Variation of a function - The variation or increment δy of the function y(x) of the functional $I\left[y(x)\right]$ is the difference between two functions $y(x)$ and $y_1(x)$ belonging to the eertain class of functions: $\delta y = y(x) - y_1(x)$.

Linear functionals – The functional $I\left[y(x)\right]$ defined in the class of functions M is called linear if it satisfies.

(i) $I\left[cy(x)\right] = cI\left[y(x)\right]$, where c is an arbitrary constant

(ii) $I\left[y_1(x)+y_2(x)\right] = I\left[y_1(x)\right] + I\left[y_2(x)\right]$, where $y_1(x) \in M$ & $y_2(x) \in M$.

Example 1 Show that the functional

$$I_1\left[y(x)\right] = \int_a^b \left[y^1(x) + y(x)\right]dx$$

is linear in $c\left[a,b\right]$,but the functional

$$I_2\left[y(x)\right] = \int_a^b \left[p(x)\left(y^1(x)\right)^2 + q(x)\left(y(x)\right)^2\right]dx$$

is non linear.

Solution Let c be a constant then.

(i) $I_1[cy(x)] = \int_a^b [c\{y^1(x) + y(x)\}]dx$

$$= \int_a^b [cy^1(x) + cy(x)]dx$$

$$= c\int_a^b [y^1(x) + y(x)]dx$$

$$= cI_1[y(x)]$$

(ii) $I_1[y_1(x) + y_2(x)] = \int_a^b [\{y_1(x) + y_2(x)\}^1 + \{y_1(x) + y_2(x)\}]dx$

$$= \int_a^b [y_1^1(x) + y_2^1(x) + y_1(x) + y_2(x)]dx$$

$$= \int_a^b [y_1^1(x) + y_1(x)]dx + \int_a^b [y_2^1(x) + y_2(x)]dx$$

$$= I_1[y_1(x)] + I_2[y_2(x)]$$

Hence, the functional $I_1[y(x)]$ is Linear.

Now, $I_1[cy(x)] = \int_a^b [p(x)(cy^1(x))^2 + q(x)(cy(x))^2]dx$

$$= c^2 \int_a^b [p(x)(y^1(x))^2 + q(x)(y(x))^2]dx$$

$$= c^2 I_2[y(x)]$$

Thus, $I_2[y(x)]$ is non linear.

Increment of a Functional – Given a functional $I[y(x)]$ on the class M and let the increment be $\delta y = y(x) - y_1(x)$ of the function. Then the expression

$$\Delta I = I[y(x) + \delta y] - I[y(x)]$$

is called the increment of $I[y(x)]$ corresponding to the increment δy.

Variation of the Functional – The increment of functional $I[y(x)]$.

$$\Delta I = I[y(x) + \delta y] - I[y(x)]$$

Can be represented as

$$\Delta I = L[y(x), \delta y] + \beta[y(x), \delta y] \ \max|\delta y|$$

Where $L[y(x), \delta(y)]$ is a linear functional w.r.t. δy and $\beta[y(x),\delta y] \to 0$ as $|\delta y| \to 0$. Then $L[y(x),\delta y]$ is called the variation of the functional and is denoted by δI.

Example 2 If a functional $I[y(x)] = \int_0^1 \sqrt{1 + (y^1(x))^2}\, dx$

is defined on a class $C^1[0,1]$ then prove that.

$$I[1] = 1,\ I[x] = \sqrt{2},\ I[x^2] = \frac{\sqrt{5}}{2} + \frac{1}{4}\sinh^{-1} 4.$$

Solution Let the functional be

$$I[y(x)] = \int_0^1 \sqrt{1 + [y^1(x)]^2}\, dx$$

Now, $I[1] = \int_0^1 1.dx \qquad \left[\because y^1(x) = \frac{d}{dx} y(x) = \frac{d}{dx}.1 = 0\right]$

$$= [x]_0^1$$

$$= 1$$

and $$I[x] = \int_0^1 \sqrt{1+\left(\frac{d}{dx}x\right)^2}\,dx$$

$$= \int_0^1 \sqrt{1+1}\,dx$$

$$= \sqrt{2}\,[x]_0^1 = \sqrt{2}$$

Also, $$I[x^2] = \int_0^1 \sqrt{1+\left(\frac{d}{dx}x^2\right)^2}\,dx$$

$$= \int_0^1 \sqrt{1+(2x)^2}\,dx$$

$$= \int_0^1 \sqrt{1+4x^2}\,dx$$

$$= 2\int_0^1 \sqrt{\frac{1}{4}+x^2}\,dx$$

$$= 2\left[\frac{1}{2}\left\{x\sqrt{\frac{1}{4}+x^2}\right\}+\frac{1}{2}.\frac{1}{4}\sinh^{-1}4x\right]_0^1$$

$$= \frac{\sqrt{5}}{2}+\frac{1}{4}\sinh^{-1}4$$

Euler's Equation – Let us examine a function $y = y(x)$ for which the extremum of the function.

$$I[y(x)] = \int_a^b F(x, y(x), y^1(x))dx \quad \ldots(1)$$

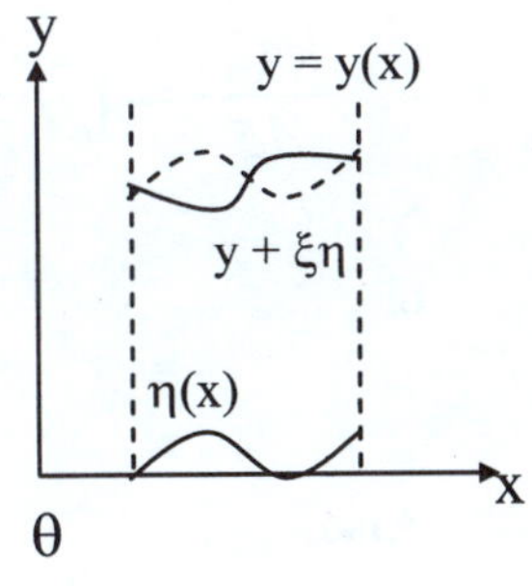

Subject to the boundary conditions.

$$y(a) = y_1 \quad \ldots(2)$$

and $$y(b) = y_2 \quad \ldots(3)$$

Where y_1 and y_2 are prescribed at the fixed boundary points *a* and *b*.

Let us choose a function η(x) such that η(a) = η(b) = 0. Then for an arbitrary constant ξ, however small the sum $y + \xi n$ will satisfy the boundary condition (2) and (3).

Now, replacing y by $y+\xi\eta$, and $\dfrac{dy}{dx}$ by $\dfrac{dy}{dx}+\xi\dfrac{d\eta}{dx}$ then equation (1) becomes,

$$I[y(x)] = \int_a^b F\left(x, y+\xi n, \frac{dy}{dx}+\xi\frac{d\eta}{dx}\right)dx$$

If we denote this integral by $\phi(\xi)$.

Which being a function of ξ, is a maxima or minima at $\xi = 0$.

When $$\frac{d\phi}{d\xi} = 0 \text{ at } \xi = 0 \qquad \text{...(4)}$$

Where $$\phi(\xi) = \int_a^b F\left(x, y+\xi\eta, \frac{dy}{dx}+\xi\frac{d\eta}{dx}\right)dx \qquad \text{...(5)}$$

Expanding by $F(x, y+\xi\eta, y^1+\xi\eta^1)$ Taylor's series, we have

$F(x, y+\xi\eta, y^1+\xi\eta^1) = F(x, y, y^1)+\xi\eta\dfrac{\partial F}{\partial y}+\xi\eta^1\dfrac{\partial F}{\partial y^1}+$ terms containing higher power of ξ.

$$\therefore \phi(\xi) = \int_a^b [F(x, y, y^1)+\xi\eta\frac{\partial F}{\partial y}+\xi\eta^1\frac{\partial F}{\partial y^1} + \text{terms containing higher power of } \xi]\, dx$$

Now, differentiating $\phi(\xi)$ w.r.t. ξ, we get

$$\frac{d\phi(\xi)}{d\xi} = \int_a^b \eta\frac{\partial F}{\partial y}+\eta^1\frac{\partial F}{\partial y^1} + \text{terms containing higher power of } \xi]\, dx \qquad \text{...(6)}$$

or $$\int_a^b \left(\eta\frac{\partial F}{\partial y}+\eta^1\frac{\partial F}{\partial y^1}\right)dx = 0 \text{ at } \xi = 0 \text{ [from eqn. (4)]} \qquad \text{...(7)}$$

Also, $$\int_a^b \eta^1\frac{\partial F}{\partial y^1}dx = \left[\eta\frac{\partial F}{\partial y^1}\right]_a^b - \int_a^b \frac{d}{dx}\left(\frac{\partial F}{\partial y^1}\right)\eta dx \text{ (by parts)}$$

Hence, eqn (7) becomes

$$\int_a^b \left[\eta\frac{\partial F}{\partial y}-\eta\frac{d}{dx}\left(\frac{\partial F}{\partial y^1}\right)\right]dx + \left[\eta\frac{\partial F}{\partial y^1}\right]_a^b = 0$$

But η(a) = η (b) = 0 thus, we have

$$\int_a^b \left[\eta \frac{\partial F}{\partial y} - \frac{d}{dx}\left(\frac{\partial F}{\partial y^1}\right)\right] dx + 0 = 0$$

Since η being an arbitrary quantity, there fore

$$\frac{\partial F}{\partial y} - \frac{d}{dx}\left(\frac{\partial F}{\partial y^1}\right) = 0$$

This equation is known an Euler's equation.

Other forms of the Euler's equation – Let F be a function of x, y, y'.

Then
$$\frac{dF}{dx} = \frac{\partial F}{\partial x} + \frac{\partial F}{\partial y}\frac{dy}{dx} + \frac{\partial F}{\partial y'}\frac{dy'}{dx}$$

$$= \frac{\partial F}{\partial x} + y'\frac{\partial F}{\partial y} + y''\frac{\partial F}{\partial y^1} \qquad \text{...(1)}$$

and
$$\frac{d}{dx}\left(y'\frac{\partial F}{\partial y^1}\right) = y''\frac{\partial F}{\partial y^1} + y'\frac{d}{dx}\left(\frac{\partial F}{\partial y'}\right)$$

or
$$y''\frac{\partial F}{\partial y'} = \frac{d}{dx}\left(y'\frac{\partial F}{\partial y'}\right) - y'\frac{d}{dx}\left(\frac{\partial F}{\partial y'}\right) \qquad \text{...(2)}$$

From eqn.(1) and (2), we have

$$\frac{dF}{dx} = \frac{\partial F}{\partial x} + y'\frac{\partial F}{\partial y} + \frac{d}{dx}\left(y'\frac{\partial F}{\partial y'}\right) - y'\frac{d}{dx}\left(\frac{\partial F}{\partial y'}\right)$$

or
$$\frac{d}{dx}\left[F - y'\frac{\partial F}{\partial y'}\right] = \frac{\partial F}{\partial x} + y'\left[\frac{\partial F}{\partial y} - \frac{d}{dx}\left(\frac{\partial F}{\partial y'}\right)\right]$$

or
$$\frac{1}{y'}\left[\frac{d}{dx}\left(F - y'\frac{\partial F}{\partial y'}\right) - \frac{\partial F}{\partial x}\right] = \frac{\partial F}{\partial y} - \frac{d}{dx}\left(\frac{\partial F}{\partial y'}\right)$$

or
$$\frac{1}{y'}\left[\frac{d}{dx}\left(F - y'\frac{\partial F}{\partial y'}\right) - \frac{\partial F}{\partial x}\right] = 0 \qquad \left[\because \frac{\partial F}{\partial y} - \frac{d}{dx}\left(\frac{\partial F}{\partial y'}\right) = 0\right]$$

or
$$\frac{d}{dx}\left(F - y'\frac{\partial F}{\partial y'}\right) - \frac{\partial F}{\partial x} = 0 \qquad \text{...(3)}$$

Solution of Euler's Equation – Every solution of the Euler's equation is called an extremal of the variational problems.

Case I If F be a function of y and y' and independent of x.

$$\Rightarrow \quad \frac{\partial F}{\partial x} = 0$$

Hence, eqn (3) becomes $\frac{d}{dx}\left(F - y'\frac{\partial F}{\partial y'}\right) = 0$

Integrating, we get $F - y'\frac{\partial F}{\partial y'} = C$

Case II If F be a function of x and y' and independent of y.

$$\Rightarrow \quad \frac{\partial F}{\partial y} = 0$$

Hence, eqn (3) becomes $\frac{d}{dx}\left(\frac{\partial F}{\partial y'}\right) = 0$

Integrating, we get $\frac{\partial F}{\partial y'} = C$

Case III If F be a function of x and y and independent of y'.

$$\Rightarrow \quad \frac{\partial F}{\partial y'} = 0$$

Hence, eqn (3) becomes $\frac{\partial F}{\partial y} = 0$

The solution of this equation does not involve any arbitrary constants. Which is not possible to find y and no extremum can be obtained.

Example 3 Find the extremals of the functional $\int_{x_0}^{x_1}\left(\frac{y'^2}{x^3}\right)dx$

Solution Let the function

$$F(x, y, y') = \frac{y'^2}{x^3}$$

Which is independent of y *i.e.* $\frac{\partial F}{\partial y} = 0$

$\therefore$ Euler's equation reduces to $\frac{d}{dx}\left(\frac{\partial F}{\partial y'}\right)=0$

or $$\frac{d}{dx}\left(\frac{2y'}{x^3}\right)=0$$

or $$\frac{x^3.2y''-2y'.3x^2}{x^6}=0$$

or $$\frac{2x^2(xy''-3y')}{x^6}=0$$

or $$xy''-3y'=0$$

Integrating, we get $\int\frac{y''}{y'}=\int\frac{3}{x}+C$

$\log y' = 3 \log x + \log C$

$= \log Cx^3$

or $y' = Cx^3$.

Again integrating, we get

$$y=\frac{Cx^4}{4}+C_1$$

Which is the required extremal

Example 4 Test for extremum the functional.

$$I[y(x)]=\int_0^{\frac{\pi}{2}}(y'^2-y^2)dx$$

$$y(0) = 0,\ y\left(\frac{\pi}{2}\right)=1$$

Solution Let the function $f\ (x, y, y') = y'^2 - y^2$

Hence, The Euler's equation is

$$\frac{\partial F}{\partial y}-\frac{d}{dx}\left(\frac{\partial F}{\partial y'}\right)=0$$

or $$(-2y)-\frac{d}{dx}(2y')=0$$

or $$y+\frac{d}{dx}y'=0$$

or $$y+\frac{d^2y}{dx^2}=0$$

Which is a higher order differential equation.

The auxiliary equation is $1+m^2=0$

$\Rightarrow \quad m=\pm i$

The general solution is $y = C_1 \cos x + C_2 \sin x$ ...(1)

Now, applying initial condition, when $x = 0, y = 0$, then $C_1 = 0$

When, $x=\frac{\pi}{2}, y=1$ then $C_2 = 1$

Substituting the values of C_1 and C_2 in eqn. (1), we have,

$$y = \sin x$$

This is the required extremal.

Example 5 Solve the variational Problem

$$\int_1^2 [x^2y'^2+2y(x+y)]\,dx \text{ given } y(1)=y(2)=0$$

Solution Let the function $f(x, y, y') = x^2y'^2 + 2xy + 2y^2$

Hence, Euler's equation is $\frac{\partial F}{\partial y}-\frac{d}{dx}\left(\frac{\partial F}{\partial y'}\right)=0$

or $$(+2x+4y)-\frac{d}{dx}(2x^2y')=0$$

or $$2x+4y-(2x^2y''+4xy')=0$$

or $$x^2y''+2xy'-2y=x$$

Which is a linear homogeneous differential equation.

Putting $x=e^z \Rightarrow z=\log x$

Then $(D^2 + D - 2)y = e^z$

Auxiliary equation is $m^2 + m - 2 = 0$

$$m = 1, -2$$

Complementary function is $C_1 e^z + C_2 e^{-2z}$

Particular integral is $= \dfrac{1}{D^2+D-2}e^z$

$$= \left[\frac{1}{3(D-1)}+\frac{1}{(D+2)}\right]e^z$$

$$= \frac{1}{3}e^z(z+1)$$

Its solution is $y = C_1 e^z + C_2 e^{-2z} + \frac{1}{3}e^z(z+1)$

or $$y = C_1 x + \frac{C_2}{x_2} + \frac{1}{3}x(\log x+1) \quad \text{...(i)}$$

Since $y(1) = 0$ then $C_1 + C_2 + \frac{1}{3} = 0$

When $y(2) = 0$ then $2C_1 + \frac{1}{4}C_2 + \frac{2}{3}(\log 2+1) = 0$

By solving these two equations, we get $C_2 = \frac{8}{21}\log 2$, $C_1 = -\frac{1}{21}(7+8\log 2)$

Putting these values of C_1 and C_2 in eqn (i) , we have

$$y = -\frac{1}{21}(7+8\log 2)x + \frac{8\log 2}{21}.\frac{1}{x^2}+\frac{1}{3}x(\log x+1)$$

or $$y = \frac{1}{21}\left[-7x+8\log 2\left(\frac{1}{x^2}-x\right)+7x(\log x+1)\right]$$

or $$y = \frac{1}{21}\left[8\log 2\left(x^{-2}-x\right)+7x\log x\right]$$

Which is the required extremal.

Functional Dependent on Higher order Derivatives-

Consider the extremum of the functional

$$I[y(x)] = \int_a^b F(x, y(x), y'(x), y''(x))dx \quad \text{...(1)}$$

Subject to the boundary conditions.

$$y(a) = y'(a) = 0$$
$$y(b) = y'(b) = 0 \quad \text{...(2)}$$

Let $y(x,\xi) = y(x) + \xi\delta\, y(x)$

and $\phi(\xi) = I\,[y(x,\xi)]$

Sothat $\phi(0) = I[y(x)]$

But we know that

$$\left(\frac{\partial \phi}{\partial \xi}\right)_{\xi=0} = \int_a^b \left[\frac{\partial F}{\partial y}\delta y + \frac{\partial F}{\partial y'}\delta y' + \frac{\partial F}{\partial y''}\delta y''\right] dx \quad \ldots(3)$$

Now $$\int_a^b \frac{\partial F}{\partial y'}\delta y' dx = \left[\frac{\partial F}{\partial y'}\delta y\right]_a^b - \int_a^b \frac{d}{dx}\left(\frac{\partial F}{\partial y'}\right)\delta y dx$$

Using boundary conditions δy (a) = δy (b) = 0, then

$$\int_a^b \frac{\partial F}{\partial y'}\delta y' dx = -\int_a^b \frac{d}{dx}\left(\frac{\partial F}{\partial y'}\right)\delta y dx \quad \ldots(4)$$

and $$\int_a^b \frac{\partial F}{\partial y''}\delta y'' dx = \left[\frac{\partial F}{\partial y''}\delta y'\right]_a^b - \int_a^b \frac{d}{dx}\left(\frac{\partial F}{\partial y''}\right)\delta y' dx$$

Using boundary conditions δy '(a) = δy ' (b) = 0, then

$$\int_a^b \frac{\partial F}{\partial y''}\delta y'' dx = -\int_a^b \frac{d}{dx}\left(\frac{\partial F}{\partial y''}\right)\delta y' dx = \int_a^b \frac{d^2}{dx^2}\left(\frac{\partial F}{\partial y''}\right)\delta y dx \quad \ldots(5)$$

Substituting the values of eqn (4) and (5) in eqn (3), we get

$$\left(\frac{\partial \phi}{\partial \xi}\right)_{\xi=0} = \int_a^b \left[\frac{\partial F}{\partial y} - \frac{d}{dx}\left(\frac{\partial F}{\partial y'}\right) + \frac{d^2}{dx^2}\left(\frac{\partial F}{\partial y''}\right)\right]\delta y dx$$

For maxima or minima,

$$\left(\frac{\partial \phi}{\partial \xi}\right)_{\xi=0} = 0$$

∴ $$\int_a^b \left[\frac{\partial F}{\partial y} - \frac{d}{dx}\left(\frac{\partial F}{\partial y'}\right) + \frac{d^2}{dx^2}\left(\frac{\partial F}{\partial y''}\right)\right]\delta y dx = 0$$

or $$\frac{\partial F}{\partial y} - \frac{d}{dx}\left(\frac{\partial F}{\partial y'}\right) + \frac{d^2}{dx^2}\left(\frac{\partial F}{\partial y''}\right) = 0 \quad \ldots(6)$$

This differential equation is called the **Euler – Poisson Equation** and its solutions are called extremals.

In general, for the functional

$$I[y(x)] = \int_a^b F(x, y, y', y'', \ldots\ldots\ldots\ldots y^{(n)})dx$$

The Euler- Poisson equation is

$$\frac{\partial F}{\partial y} - \frac{d}{dx}\left(\frac{\partial F}{\partial y'}\right) + \frac{d^2}{dx^2}\left(\frac{\partial F}{\partial y''}\right) \ldots\ldots\ldots\ldots\ldots\ldots$$

$$+ (-1)^n \frac{d^n}{dx^n}\left(\frac{\partial F}{\partial y^{(n)}}\right) = 0$$

Example 6 Show that the curve which extremizes the functional

$$I[y(x)] = \int_0^{\frac{\pi}{4}} (y''^2 - y^2 + x^2)dx \text{ under the conditions.}$$

$$y(0) = 0, y'(0) = 1, y\left(\frac{\pi}{4}\right) = y'\left(\frac{\pi}{4}\right) = \frac{1}{\sqrt{2}} \text{ is } y = \sin x.$$

Solutions The Euler Poisson equation is

$$\frac{\partial F}{\partial y} - \frac{d}{dx}\left(\frac{\partial F}{\partial y'}\right) + \frac{d^2}{dx^2}\left(\frac{\partial F}{\partial y''}\right) = 0 \qquad \ldots(1)$$

The given function be $F = y''^2 - y^2 + x^2$

$\therefore$ equation (1) becomes

$$-2y + \frac{d^2}{dx^2}(2y'') = 0$$

or $$y^{iv} - y = 0$$

or $$(D^4 - 1)y = 0$$

Its solution is $y(x) = C_1 e^x + C_2 e^{-x} + C_3 \cos x + C_4 \sin x$ ...(2)

and $$y'(x) = C_1 e^x - C_2 e^{-x} - C_3 \sin x + C_4 \cos x \qquad \ldots(3)$$

Appling the boundary conditions.

$$y(0) = 0 = C_1 + C_2 + C_3 \qquad \ldots(4)$$

$$y'(0) = 1 = C_1 - C_2 + C_4 \qquad \ldots(5)$$

$$y\left(\frac{\pi}{4}\right)=\frac{1}{\sqrt{2}}=C_1e^{\frac{\pi}{4}}+C_2e^{-\frac{\pi}{4}}+\frac{1}{\sqrt{2}}C_3+\frac{1}{\sqrt{2}}C_4 \quad ...(6)$$

$$y'\left(\frac{\pi}{4}\right)=\frac{1}{\sqrt{2}}=C_1e^{\frac{\pi}{4}}-C_2e^{-\frac{\pi}{4}}-\frac{1}{\sqrt{2}}C_3+\frac{1}{\sqrt{2}}C_4 \quad ...(7)$$

By solving equations (4), (5), (6) and (7), we have

$$C_1 = C_2 = C_3 = 0 \text{ and } C_4 = 1$$

Hence, equation (2) becomes

$$y(x) = \sin x$$

Example 7 Determine the extremal of the functional

$$I[y(x)]=\int_{-l}^{l}\left[\frac{1}{2}\mu y''^2+\ell y\right]dx$$

Subject to $y(-l) = 0, y'(-l) = 0, y(l) = 0, y'(l) = 0$

Solutions The Euler- Poisson equation is

$$\frac{\partial F}{\partial y}-\frac{d}{dx}\left(\frac{\partial F}{\partial y'}\right)+\frac{d^2}{dx^2}\left(\frac{\partial F}{\partial y''}\right)=0 \quad ...(1)$$

The given funtion is $F=\frac{1}{2}\mu y''^2+\ell y$

$\therefore$ equation (1) becomes,

$$\ell+\frac{d^2}{dx^2}(\mu y'')=0$$

or $$\ell+\mu y^{iv}=0$$

or $$\mu\frac{d^4y}{dx^4}+\ell=0$$

or $$\frac{d^4y}{dx^4}=\frac{-\ell}{\mu}$$

Its solution is $y(x)=-\frac{\ell}{\mu}\frac{x^4}{24}+C_1\frac{x^3}{6}+C_2\frac{x^2}{2}+C_3x+C_4$ ...(2)

and $y'(x)=-\frac{\ell}{\mu}\frac{x^3}{6}+C_1\frac{x^2}{2}+C_2x+C_3$ ...(3)

Applying the boundary conditions

$$y(-l)=0=\frac{-\ell l^4}{24\mu}-\frac{C_1l^3}{6}+\frac{C_2l^2}{2}-C_3l+C_4 \qquad \ldots(4)$$

$$y'(-l)=0=\frac{+\ell l^3}{\mu.6}+\frac{C_1l^2}{2}-C_2l+C_3 \qquad \ldots(5)$$

$$y(+l)=0=\frac{-\ell l^4}{\mu.24}+\frac{C_1l^3}{6}+C_2\frac{l^2}{2}+C_3l+C_4 \qquad \ldots(6)$$

$$y'(+l)=0=\frac{-\ell}{\mu}\frac{l^3}{6}+C_1\frac{l^2}{2}+C_2l+C_3 \qquad \ldots(7)$$

By solving equations (4), (5), (6) and (7), we get

$$C_1=0, C_2=\frac{\ell l^2}{6\mu}, C_3=0, C_4=\frac{-\ell l^4}{24\mu}$$

Hence, equation (2) becomes

$$y(x)=\frac{-\ell x^4}{24\mu}+\frac{\ell l^2x^4}{12\mu}-\frac{\ell l^4}{24\mu}$$

or $$y(x)=\frac{-\ell}{24\mu}\left[x^4-2l^2x^2+l^4\right]$$

Functional dependent on more than one Independent variable –

Consider the functional $I[z(x,y)]=\iint\limits_D F(x,y,z,p,q)dxdy$ ...(1)

Where $$p=\frac{\partial z}{\partial x}, q=\frac{\partial z}{\partial y}$$

Let the extremizing surface be $z = z(x,y)$. So that one parameter

Surface is $z(x, y, \xi) = z(x, y) + \xi\, \delta z$

$p(x, y, \xi) = p(x, y) + \xi\, \delta p$

$q(x, y, \xi) = q(x, y) + \xi\, \delta q$

and $$\phi(\xi)=\iint\limits_{\infty} F(x,y,z(x,y,\xi),p(x,y,\xi),q(x,y,\xi))dxdy$$

Then $$\left(\frac{\partial\phi}{\partial\xi}\right)_{\xi=0}=\iint\limits_{\infty}\left(\frac{\partial F}{\partial z}.\delta z+\frac{\partial F}{\partial p}.\delta p+\frac{\partial F}{\partial q}.\delta q\right)dxdy \qquad \ldots(2)$$

Since $$\frac{\partial}{\partial x}\left(\frac{\partial F}{\partial p}.\delta z\right)=\frac{\partial}{\partial x}\left(\frac{\partial F}{\partial p}\right)\delta z+\frac{\partial F}{\partial p}\delta p \qquad ...(3)$$

and $$\frac{\partial}{\partial y}\left(\frac{\partial F}{\partial q}.\delta z\right)=\frac{\partial}{\partial y}\left(\frac{\partial F}{\partial q}\right)\delta z+\frac{\partial F}{\partial q}\delta q \qquad ...(4)$$

Now, $$\iint_{\infty}\left(\frac{\partial F}{\partial p}.\delta p+\frac{\partial F}{\partial q}.\delta q\right)dxdy$$

$$=\iint_{\infty}\left[\frac{\partial}{\partial x}\left(\frac{\partial F}{\partial p}.\delta z\right)+\frac{\partial}{\partial y}\left(\frac{\partial F}{\partial q}.\delta z\right)\right]dxdy$$

$$-\iint_{\infty}\left[\frac{\partial}{\partial x}\left(\frac{\partial F}{\partial p}\right)+\frac{\partial}{\partial y}\left(\frac{\partial F}{\partial q}\right)\right]\delta z dxdy \qquad ...(5)$$

Also, by Green's Theorem

$$\iint_{\infty}\left[\frac{\partial}{\partial x}\left(\frac{\partial F}{\partial p}\delta z\right)+\frac{\partial}{\partial y}\left(\frac{\partial F}{\partial q}\delta z\right)\right]dxdy$$

$$=\int_{G}\left[\frac{\partial F}{\partial p}dy-\frac{\partial F}{\partial q}dx\right]\delta z=0$$

Since on G, the function z is fixed and hence $\delta z = 0$
Substituting the above value in equ (5), we have

$$\iint_{\infty}\left[\frac{\partial F}{\partial p}\delta p+\frac{\partial F}{\partial q}\delta q\right]dxdy$$

$$=-\iint_{\infty}\left[\frac{\partial}{\partial x}\left(\frac{\partial F}{\partial p}\right)+\frac{\partial}{\partial y}\left(\frac{\partial F}{\partial q}\right)\right]\delta z dxdy$$

Substituting this value in eqn. (2), we have

$$\left(\frac{\partial \phi}{\partial \xi}\right)_{\xi=0}=\iint_{\infty}\left[\frac{\partial F}{\partial z}-\frac{\partial}{\partial x}\left(\frac{\partial F}{\partial p}\right)-\frac{\partial}{\partial y}\left(\frac{\partial F}{\partial q}\right)\right]\delta z dxdy$$

For maxima or minima.

$$\left(\frac{\partial \phi}{\partial \xi}\right)_{\xi=0}=0$$

Hence, $$\iint_{\infty}\left[\frac{\partial F}{\partial z}-\frac{\partial}{\partial x}\left(\frac{\partial F}{\partial p}\right)-\frac{\partial}{\partial y}\left(\frac{\partial F}{\partial q}\right)\right]\delta z dxdy=0$$

Then, $$\frac{\partial F}{\partial z}-\frac{\partial}{\partial x}\left(\frac{\partial F}{\partial p}\right)-\frac{\partial}{\partial y}\left(\frac{\partial F}{\partial q}\right)=0 \qquad \ldots(6)$$

This differential equation is called the **Euler – Ostrogradsky equation** and its solutions are called extremals.

Example 8 Find the Euler – ostrogradsky equation for

$$I[u(x,y)]=\iint_{\infty}\left[\left(\frac{\partial u}{\partial x}\right)^2+\left(\frac{\partial u}{\partial y}\right)^2\right]dxdy$$

Where the values of u are prescribed on the boundary Γ of the domain D.

Solution Let the function be

$$F(x,y,u,u_x,u_y) = u_x^2+u_y^2$$

where $U_x=\dfrac{\partial u}{\partial x}$ and $U_y=\dfrac{\partial u}{\partial y}$

Then $$\frac{\partial F}{\partial u}=0,\frac{\partial}{\partial x}\left(\frac{\partial F}{\partial u_x}\right)=\frac{\partial}{\partial x}(2u_x)=2\frac{\partial^2 u}{\partial x^2}$$

and $$\frac{\partial}{\partial x}\left(\frac{\partial F}{\partial u_y}\right)=\frac{\partial}{\partial y}(2u_y)=2\frac{\partial^2 u}{\partial y^2}$$

Now, The Euler ostrogradsky equation is

$$\frac{\partial F}{\partial u}-\frac{\partial}{\partial x}\left(\frac{\partial F}{\partial u_x}\right)-\frac{\partial}{\partial y}\left(\frac{\partial F}{\partial u_y}\right)=0$$

or $$0-2\frac{\partial^2 u}{\partial x^2}-2\frac{\partial^2 u}{\partial y^2}=0$$

or $$\frac{\partial^2 u}{\partial x^2}+\frac{\partial^2 u}{\partial y^2}=0$$

Which is a laplace's equation in the domain D, where function is harmonic because it satisfies the laplace's equation.

Example 9 Find the surface of a minimum area, stretches over a given closed space curve C, enclosing the domain D in the xy-plane.

Solution Here the problem is to determine the extremals of the functional

$$S[z(x,y)]=\iint_D\sqrt{1+z_x^2+z_y^2}\,dxdy$$

where the function be

$$f(x,y,z,z_x,z_y)=\sqrt{1+z_x^2+z_y^2}$$

Now, $$\frac{\partial F}{\partial z}=0,\frac{\partial}{\partial x}\left(\frac{\partial F}{\partial z_x}\right)=\frac{\partial}{\partial x}\left(\left(1+z_x^2+z_y^2\right)^{-\frac{1}{2}}.z_x\right)$$

$$=\left(1+z_x^2+z_y^2\right)^{-\frac{1}{2}}.z_{xx}-z_x\left(1+z_x^2+z_y^2\right)^{-\frac{3}{2}}\left(z_xz_{xx}+z_y.z_{xy}\right)$$

and $$\frac{\partial}{\partial y}\left(\frac{\partial F}{\partial z_y}\right)=\frac{\partial}{\partial y}\left(\left(1+z_x^2+z_y^2\right)^{-\frac{1}{2}}.z_y\right)$$

$$=\left(1+z_x^2+z_y^2\right)^{-\frac{1}{2}}z_{yy}-z_y\left(1+z_x^2+z_y^2\right)^{-\frac{3}{2}}\left(z_xz_{xy}+z_y.z_{yy}\right)$$

Hence, the Euler - Ostrogradsky equation is

$$\frac{\partial F}{\partial z}-\frac{\partial}{\partial x}\left(\frac{\partial F}{\partial z_x}\right)-\frac{\partial}{\partial y}\left(\frac{\partial F}{\partial z_y}\right)=0 \quad \text{...(i)}$$

Substituting the above values in eqn (i), we have

$$\left(1+z_x^2+z_y^2\right)^{-\frac{3}{2}}\left[\left(1+z_x^2+z_y^2\right)z_{xx}-z_x\left(z_xz_{xx}+z_y.z_{xy}\right)+\left(1+z_x^2+z_y^2\right)z_{yy}\right.$$
$$\left.-z_y\left(z_xz_{xy}+z_y.z_{yy}\right)\right]=0$$

or $$\left(1+z_x^2+z_y^2\right)z_{xx}-z_x^2z_{xx}-2z_xz_yz_{xy}+(1+z_x^2+z_y^2)z_{yy}-z_y^2z_{yy}=0$$

or $$(1+z_y^2)z_{xx}-2z_xz_yz_{xy}+(1+z_x^2)z_{yy}=0$$

Which is the required equation of minimal surfaces.

Variation Problems in Parametric form-

Consider the parametric form of a line be

$$x=\phi(t),\quad y=\psi(t)$$

Let us consider the functional

$$I[x(t),y(t)]=\int_a^b F(t,x(t),y(t),\dot{x}(t),\dot{y}(t))\,dt \quad \text{...(1)}$$

Where $$\dot{x}(t)=\frac{dx}{dt}\quad and\quad \dot{y}(t)=\frac{dy}{dt}$$

and $\qquad x(t), y(t) \in C'[a,b]$

If $\qquad \phi(\alpha, \beta) = I[x(t) + \alpha\delta x, y(t) + \beta\delta y]$

Then for an extremum, we have

$$\left[\frac{\partial \phi}{\partial \alpha}\right]_{\alpha=\beta=0} = 0 \;\; and \;\; \left[\frac{\partial \phi}{\partial \beta}\right]_{\alpha=\beta=0} = 0$$

Now, Let $\qquad \bar{x}(t) = x_1(t), x_2(t) \ldots\ldots\ldots\ldots x_n(t)$

Then $\qquad \dot{\bar{x}}(t) = \dot{x}_1(t), \dot{x}_2(t) \ldots\ldots\ldots\ldots \dot{x}_n(t)$

Let L $\left(t\,\bar{x}\,\dot{\bar{x}}\right)$ be any function of variable $t\,\bar{x}\,\dot{\bar{x}}$ and the functional be

$$J[\bar{x}(t)] = \int_a^b L\left(t\bar{x}(t)\,\dot{\bar{x}}(t)\right)dt \qquad \ldots(2)$$

The problem is to minimize the functional J [$\bar{x}(t)$] by extremals

Which satisfy the boundary conditions

$$\bar{x}\left(\bar{a}\right) = \bar{\alpha} \text{ and } \bar{x}\left(\bar{b}\right) = \bar{\beta} \qquad \ldots(3)$$

Where $\bar{a} = (a_1, a_2, \ldots\ldots\ldots a_n)$ and $\bar{\alpha} = (\alpha_1, \alpha_2, \ldots\ldots\ldots \alpha_n)$ etc. Then L must

Satisfy the n second order Euler equations

$$\frac{\partial L}{\partial x_i} - \frac{d}{dt}\left(\frac{\partial L}{\partial \dot{x}_i}\right) = 0, \text{ for i} = 1,2,3 \ldots\ldots\ldots\ldots n \qquad \ldots(4)$$

This equations is called **Euler-Langrange Equation** and their

Solutions are called extremals.

Note: Let the potential energy V and T be the kinetic energy. Then Lagrangian function be

L = T-V

Example 10 Find the extremal of the functional

$$V(x) = \int_0^{\frac{\pi}{2}} \left(x_1'^2 + x_2'^2 + 2x_1x_2\right)dt$$

Subject to the boundary conditions

$$x_1(0)=0,\ x_1\left(\frac{\pi}{2}\right)=1, x_2(0)=0,\ and\ x_2\left(\frac{\pi}{2}\right)=-1$$

Solutions The given function be

$$F(x_1,x_2,x_1^1,x_2^1)=x_1^{1^2}+x_2^{1^2}+2x_1x_2$$

For extrema

$$\frac{\partial F}{\partial x_1}-\frac{d}{dt}\left(\frac{\partial F}{\partial x_1^1}\right)=0$$

and
$$\frac{\partial F}{\partial x_2}-\frac{d}{dt}\left(\frac{\partial F}{\partial x_2^1}\right)=0$$

or
$$x_2-\frac{d}{dt}\left(x_1^1\right)=0 \quad and \quad x_1-\frac{d}{dt}\left(x_2^1\right)=0$$

or
$$x_2-\frac{d^2x_1}{dt^2}=0 \quad and \quad x_1-\frac{d^2x_2}{dt^2}=0$$

Combining the above equations, we have

$$\frac{d^4x_1}{dt^4}=\frac{d^2x_2}{dt^2}=x_1$$

Its solution is $x_1(t)=C_1e^t+C_2e^{-t}+C_3\sin t+C_4\cos t$ ....(i)

Then $x_2(t)=\dfrac{d^2x_1}{dt^2}=C_1e^t+C_2e^{-t}-C_3\sin t-C_4\cos t$ ...(ii)

Applying the boundary conditions.

$$x_1(0)=0=C_1+C_2+C_4 \qquad \text{...(iii)}$$

$$x_1\left(\frac{\pi}{2}\right)=1=C_1e^{\frac{\pi}{2}}+C_2e^{-\frac{\pi}{2}}+C_3 \qquad \text{...(iv)}$$

$$x_2(0)=0=C_1+C_2-C_4 \qquad \text{...(v)}$$

$$x_2\left(\frac{\pi}{2}\right)=-1=C_1e^{\frac{\pi}{2}}+C_2e^{-\frac{\pi}{2}}-C_3 \qquad \text{...(vi)}$$

By solving equations (iii),(iv),(v), and (vi), we get

$$C_1=C_2=C_4=0 \quad \text{and} \quad C_3=1$$

Substituting these values in (i) and (ii), we have

$$x_1(t)=\sin t \quad \text{and} \quad x_2(t)=-\sin t$$

Which are the required extremals.

Example 11 A mass, suspended at the end of a light spring having spring constant k, is set into vertical motion. Use Lagrange's equation, to find the equation of motion of the mass.

Solution Let the displacement of m from the equilibrium position 0 be x at any time t. Then the kinetic energy of P is

$$T=\frac{1}{2}m\left(\frac{dx}{dt}\right)^2=\frac{1}{2}m\dot{x}^2$$

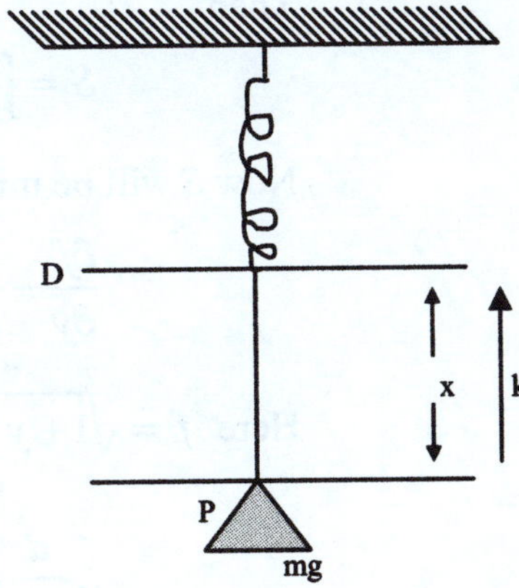

Also the work done during its fall from 0 to P is

$$W=\int_0^x (mg-kx)\,dx$$

$$=mgx-\frac{kx^2}{2}$$

If V is the potential energy of the mass at P, then

$$V=-W=\frac{1}{2}kx^2-mgx$$

$\therefore$ Lagrange function L = T – V

$$=\frac{1}{2}m\dot{x}^2+mgx-\frac{1}{2}kx^2$$

Thus, the Lagrange's equation

$$\frac{\partial L}{\partial x}-\frac{d}{dt}\left(\frac{\partial L}{\partial \dot{x}}\right)=0$$

or $$(mg-kx)-\frac{d}{dt}\left(m\dot{x}\right)=0$$

or $$mg-kx-m\frac{d^2x}{dt^2}=0$$

or $$m\frac{d^2x}{dt^2}=mg-kx$$

Which is the required equation of motion.

Example 12 Prove that the shortest distance between two points in a plane is a straight line.

Solution Let $A(x_1, y_1)$ and $B(x_2, y_2)$ be the given two points and S the arc length of a curve connecting them.

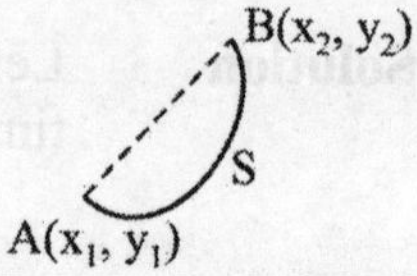

Then

$$S = \int_{x_1}^{x_2} ds = \int_{x_1}^{x_2} \sqrt{1 + y'^2}\, dx$$

Now S will be minimum if it satisfies Euler's equation.

$$\frac{\partial f}{\partial y} - \frac{d}{dx}\left(\frac{\partial f}{\partial y'}\right) = 0$$

Here $f = \sqrt{1 + y'^2}$, which is independent of y i.e., $\frac{\partial f}{\partial y} = 0$

$$\therefore \frac{d}{dx}\left\{\frac{\partial}{\partial y'}\sqrt{1 + y'^2}\right\} = 0$$

or $$\frac{d}{dx}\left\{\frac{y'}{\sqrt{1 + y'^2}}\right\} = 0$$

on integration, we have $\frac{y'}{\sqrt{1 + y'^2}} = \text{constant}$

$\Rightarrow$ $y' = \text{constant}.$

$y = mx + c.$

Which is a straight line. The constants m and c area determined from the fact that the straight line passes through A and B.

Example 13 Find the curve passing through the points (x_1, y_1) and (x_2, y_2) which when rotated about the x-axis gives a minimum surface area.

Solution The surface area $= \int_{x_1}^{x_2} 2\pi y\, ds$

$$= 2\pi \int_{x_1}^{x_2} y\sqrt{1 + y'^2}\, dx$$

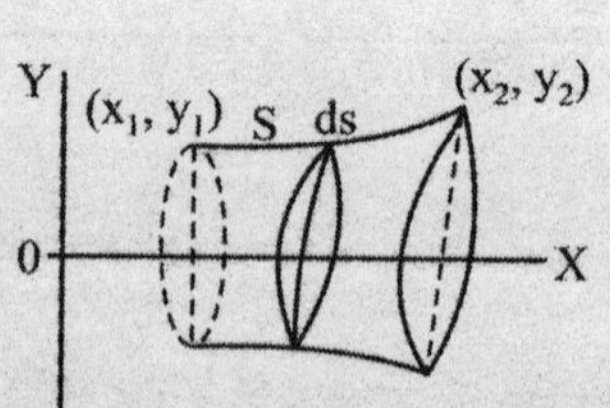

This has to be minimum.

Since $\sqrt{1 + y'^2}$ is independent of x, therefore,

Euler's equation reduces to

$$f - y'\frac{\partial f}{\partial y'} = C$$

$$\therefore\ \sqrt{1+y'^2} - y'\frac{\partial}{\partial y'}\left\{y\sqrt{1+y'^2}\right\} = C$$

$$\Rightarrow \quad y\sqrt{1+y'^2} - y'\left\{\frac{y}{2}(1+y'^2)^{\frac{-1}{2}}.2y'\right\} = C$$

$$\Rightarrow \quad \frac{y}{\sqrt{1+y'^2}} = C \text{ or } y' = \frac{dy}{dx} = \frac{\sqrt{y^2-c^2}}{c}$$

Separating the variables and integrating, we have

$$\int \frac{dy}{\sqrt{y^2-c^2}} = \int \frac{dx}{c} + c_1$$

$$\Rightarrow \quad \cos h^{-1}\left(\frac{y}{c}\right) = \frac{x+a}{c}$$

$$\Rightarrow \quad y = C\cos h\left(\frac{x+a}{c}\right)$$

Which is a catenary.

Example 14 On which curve the function $\int_0^{\frac{\pi}{2}} (y'^2 - y^2 + 2xy)dy$ with $y(0) = 0$ and $y\left(\frac{\pi}{2}\right) = 0$ be extremized?

Solution Let $f(x) = y'^2 - y^2 + 2xy$

So that $\frac{\partial f}{\partial y} = 0 - 2y + 2x$

and $\frac{d}{dx}\left(\frac{\partial f}{\partial y}\right) = \frac{d}{dx}(2y') = 2y''$

$\therefore$ Euler's equation $\frac{\partial f}{\partial y} - \frac{d}{dx}\left(\frac{\partial f}{\partial y'}\right) = 0$ becomes

$$-2y + 2x - 2y'' = 0$$

$$\Rightarrow \quad y'' + y = x$$

$\Rightarrow \qquad (D^2+1)y = x$

It's A.E. gives $D = \pm i$

$\therefore \qquad C.F. = C_1 \cos x + c_2 \sin x$

$$\text{E} P.I. = \frac{1}{D^2+1}x = [1+D^2]^{-1}x = [1-D^2]x = x.$$

Thus $\qquad y = C_1 \cos x + C_2 \sin x + x \qquad$ .....(i)

Using boundary conditions :

When $\qquad x = 0, C_1 = 0$

When $\qquad x = \dfrac{\pi}{2}, C_2 = -\dfrac{\pi}{2}$

Hence equation (i) reduces to

$$y = x - \frac{\pi}{2}\sin x.$$

Which is the only curve on which the given functional can be extremized.

Example 15 Find the extremals of $\int_0^1 yy'^3 dx, y(0) = 0, y(1) = 1.$

Solution Here $f = yy'^3$, which is free from x.

$\therefore$ Euler's equation becomes

$$f - y'\frac{\partial f}{\partial y'} = C \text{ (Constant)}$$

$\Rightarrow \qquad yy'^3 - y^1.3yy'^2 = C$

$\Rightarrow \qquad yy'^3 = K \text{ (constant)}$

$\Rightarrow \qquad y^{1/3}y' = K^{1/3}$

or $\qquad y^{1/3}dy = K^{1/3}dx$

On integration, we have

$$y^{4/3} = \frac{4}{3}K^{1/3}x + c_1$$

Using the boundary conditions, we get

$$y = x^{3/4}.$$

Example 16 Find the extremals of $I(y)=\int_1^2 \frac{\sqrt{1+y'^2}}{x}dx, y(1)=0, y(2)=1.$

Solution Here $f\frac{\sqrt{1+y'^2}}{x}$, which is free from y.

Euler's equation be $\frac{\partial f}{\partial y'}=C$ (constant)

$$\Rightarrow \quad \frac{y'}{x\sqrt{1+x'^2}}=C$$

$$\Rightarrow \quad y'^2(1-c^2x^2)=C^2x^2$$

$$\Rightarrow \quad y'=\frac{cx}{\sqrt{1-c^2x^2}}$$

On integration, we get $y=-\frac{1}{c}\sqrt{1-c^2x^2}+c_1$

$$(y-c_1)^2+x^2=\frac{1}{c^2}$$

So that the final solution is

$$(y-2)^2+x^2=5.$$

Example 17 Prove that the sphere is the solid figure of revolution which, for a given surface area, has maximum volume.

Solution Consider the arc OPA of the curve which rotates about the x-axis as shown in the figure.

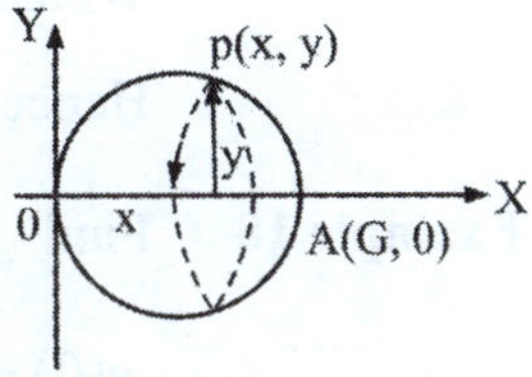

Then surface area $S=\int_0^a 2\pi y ds$

$$=\int_0^a 2\pi y\sqrt{1+y'^2}dx$$

and volume of the solid $V=\int_0^a \pi y^2 dx$

Here we have to maximize V subject to fixed S.

Taking $f=\pi y^2$ and $g=2\pi y\sqrt{1+y'^2}$

Consider the function $H=f+\lambda g$

$$= \pi y^2 + 2\pi\lambda y\sqrt{1+y'^2}$$

Now H has to satisfy Euler's equation, which is independent of x.

$$\therefore \quad \pi y^2 + 2\pi\lambda y\sqrt{1+y'^2} - y'.2\pi\lambda y\frac{y'}{\sqrt{1+y'^2}} = C$$

or $$\pi y^2 + \frac{2\pi\lambda y}{\sqrt{1+y'^2}} = C \qquad .....(i)$$

Since the curve passes through O and A for which y = 0,

Equation (i) gives C = 0.

$$\therefore \quad y + \frac{2\lambda}{\sqrt{1+y'^2}} = 0$$

$$\Rightarrow \quad y' = \frac{\sqrt{4\lambda^2 - y^2}}{y}$$

On integration, we get

$$\int dx = \int \frac{y\,dy}{4\lambda^2 - y^2} + C_1$$

or $$x = C_1 - \sqrt{4\lambda^2 - y^2} \qquad .....(ii)$$

When $x = 0, y = 0 \;\therefore C_1 = 2\lambda$

$\therefore$ Equation (ii) becomes

$$(x - 2\lambda)^2 + y^2 = 4\lambda^2$$

Which is a circle with centre $(2\lambda, 0)$ and radius 2λ.

Hence the figure formed by the revolutions of given are is a sphere.

Example 18 Find the extremals of $\int_0^{\pi} y'^2 dx,$ by boundary conditions $y(0) = 0, y(\pi) = 0$ with the constraint $\int_0^{\pi} y^2 dx = 1.$

Solution Taking $f = y'^2$ and $g = y^2$

Consider the function $H = f + \lambda g$

$$= y'^2 + \lambda y^2$$

Which gives the Euler's equation $y'' + \lambda y = 0$

Its solution is $y = C_1 \cos\sqrt{\lambda}x + C_2 \sin\sqrt{\lambda}x$

Using boundary conditions.

$$C_1 = 0,\ \lambda = n^2, n \text{ is non-zero integer.}$$

$$\therefore \qquad y = C_2 \sin nx$$

Now $\quad \int_0^\pi y^2 dx = 1$

$$\Rightarrow \qquad \int_0^\pi c_2^2 \sin^2 nx\, dx = 1$$

On solving, we have $c_2 = \sqrt{\dfrac{2}{\pi}}$

$\therefore$ The extremal is $y = \sqrt{\dfrac{2}{\pi}} \sin x.$

Example 19 Find the extremal of the function.

$$\int_0^{\pi/2} (y'^2 + z'^2 + 2yz)dx, y(0) = 0, y\left(\frac{\pi}{2}\right) = 1,\ z(0) = 0, z\left(\frac{\pi}{2}\right) = -1.$$

Solution The Euler's equations are

$$\frac{\partial f}{\partial y} - \frac{d}{dx}\left(\frac{\partial f}{\partial y'}\right) = 0 \text{ and } \frac{\partial f}{\partial z} - \frac{d}{dx}\left(\frac{\partial f}{dz'}\right) = 0$$

i.e., $y'' - z = 0$ and $z'' - y = 0$

Solving these two equations, we get

$g'' - y = 0$

A.E. is $\qquad (D^4 - 1)y = 0$

$D = \pm 1, \pm i$

$\therefore$ C.F. $y = c_1 e^x + c_2 e^{-x} + c_3 \cos x + c_4 \sin x$

And $z = y'' = c_1 e^x + c_2 e^x - c_3 \cos x - c_4 \sin x$

Using boundary conditions, we have

$c_1 = c_2 = c_3 = 0,\ c_4 = 1$

$\therefore$ The extremal of the functional is

$y = \sin x$ and $z = -\sin x.$

Example 20 Find the extremal of the functional

$$\int_0^{\pi}(2yz-2y^2+y'^2)dx, y(0)=0, y(\pi)=1, z(0)=0, z(\pi 0=-1.$$

Solution The Euler's equations are

$$\frac{\partial f}{\partial y}-\frac{d}{dx}\left(\frac{\partial f}{\partial y'}\right)=0 \text{ and } \frac{\partial f}{\partial z}-\frac{d}{dx}\left(\frac{\partial f}{\partial z^1}\right)=0$$

i.e., $y''+2y-z=0$ and $z''+y=0$

Solving these two equation two equations, we get

$$y^{iv}+2y^{ii}+y=0$$

Its A.E. is $(D^2+1)^2 y=0$

$$D=\pm i, \pm i$$

C.F. is $y=c_1\cos x+c_2\sin x+c_3 x\cos x+c_4 x\sin x$

and

$$z=c_1\cos x+c_2\sin x-2c_3\sin x+c_3 x\cos x+2c_4\cos x+c_4 x\sin x$$

using boundary conditions, we have

$$c_1=c_4=0, c_3=\frac{1}{\pi}$$

Thus, the extremal of the functional be

$$y=c_2\sin x+\frac{1}{\pi}x\cos x$$

And $z=c_2\sin x-\frac{1}{\pi}(2\sin x-x\cos x)$

Example 21 Find the shortest distance between the parabola $y=x^2$ and the straight line $x=y=5$.

Solution Consider $I=\int_{x_1}^{x_2}\sqrt{1+y'^2}\,ds$

Where $f=\sqrt{1+y'^2}$, which is independent of x and y.

Then Euler's equation reduces to $y''=0$, which gives a solution of the form $y=ax+b$.

The transversality condition at the line

$x-y=5$ given $a=-1$

$\therefore y=b-x$

The transversality condition at the parabola $y=x^2$ be

$x_1=\frac{1}{2}, y_1=\frac{1}{4}.$

As the line $y=b-x$ passes through the point $\left(\frac{1}{2},\frac{1}{4}\right)$

$\therefore b=\frac{3}{4}.$

Therefore the shortest path is the line, that is

$y=\frac{3}{4}-x$ through the point $\left(\frac{1}{2},\frac{1}{4}\right).$

$\Rightarrow x+y=\frac{3}{4}$

At the point (x_2, y_2), we get the equations.

$x_2+y_2=\frac{3}{5}$ and $x_2-y_2=5$

$\Rightarrow x_2=\frac{23}{8}, y_2=\frac{-17}{8}$

Shortest distance between $\left(\frac{1}{2},\frac{1}{4}\right)$ and $\left(\frac{23}{8},\frac{-17}{8}\right)$ be $=\frac{19}{8}\sqrt{2}.$

Rayleigh – Ritz Method- To solve the boundary value problem by Rayleigh – Ritz method, we take the differential equation as the Euler's equation of some variational Problem. Then we reduce this problem to a minimizing problem assuming an approximate solution in the form.

$$\bar{y}(x)=y_0(x)+\sum C_i\psi_i(x) \qquad \text{...(i)}$$

Where $\psi_i(x)$ the trial functions satisfy the boundary conditions and $\psi_i(x)=0$ on the boundary C of its region R.

Let the integral to be extermised be $I=\int_a^b f(y,y',x)dx$ ...(ii)

such that y(a) = R and y(b) = Q.

Substituting (i) in eqn (ii) by replacing y by $\bar{y}$ in I, giving $\bar{I}$ as a function of the unknown. Which requires

$$\frac{\partial \bar{I}}{\partial C_i} = 0, \ i = 1,2,\ldots\ldots..$$

By solving equations, we get the values of C_i which when substituting in eqn (i), gives the required solution.

Example 22 Solve the boundary value problem $y'' - y + x = 0 \quad (0 \le x \le 1), y(0) = 0,$ $y(1) = 0$ by Rayleigh-Ritz method.

Solution The given differential equation be

$$y'' - y + x = 0 \qquad \ldots(i)$$

Its solution is equivalent to extermising the integral

$$I = \int_0^1 F(x, y, y')dx$$

Where $\quad F(x, y, y') = 2xy - y^2 - y'^2 \qquad \ldots(ii)$

Since the Euler's equation $\frac{\partial F}{\partial y} - \frac{d}{dx}\left(\frac{\partial F}{\partial y'}\right) = 0$ gives (i)

Consider the trial function be

$$\bar{y}(x) = C_0 + C_1 x + C_2 x^2 \qquad \ldots(iii)$$

To satisfy the boundary conditions y(0) = y(1) = 0

We get $\quad C_0 = 0, C_2 = - C_1$

Hence, eqn (iii) becomes

$$\bar{y}(x) = C_1 x(1 - x) \qquad \ldots(iv)$$

Substituting $\bar{y}$ and $\bar{y}'$ in I, we have

$$I = \int_0^1 \left[2x\bar{y} - \bar{y}^2 - \bar{y}'\right] dx$$

$$= \int_0^1 \left[2C_1(x^2 - x^3) - C_1^2(x - x^2)^2 - C_1^2(1 - x)^2\right] dx$$

$$= \frac{1}{6}C_1 - \frac{11}{30}C_1^2$$

Its stationary values are given by

$$\frac{dI}{dC_1}=0$$

$$\therefore \qquad \frac{1}{6}-\frac{11}{30}\times 2C_1=0 \quad \Rightarrow C_1=\frac{5}{22}$$

Hence, the approximate solution is

$$\bar{y}(x)=\frac{5}{22}x(1-x)$$

Example 23 Find the approximate solution of $y''+xy=-x,\ y(0)=y(1)=0$

Solution The given equation is the Euler's equation of the variational problem of the functional

$$I=\int_0^1\left(y'^2-xy^2-2xy\right)dx$$

Choosing the solution $y(x)=C(x-x^2)$ and $y'(x)=C(1-2x)$

$$\therefore \qquad I=\int_0^1\left[C^2(1-2x)^2-C^2\left(x-x^2\right)^2x-2Cx\left(x-x^2\right)\right]dx$$

Its stationary values are $\dfrac{dI}{dC}=0$

$$\Rightarrow \qquad \int_0^1\left\{C\left[(1-2x)^2-x\left(x-x^2\right)^2\right]-x\left(x-x^2\right)\right\}dx=0$$

$$\therefore \qquad 19C=5 \text{ or } C=\frac{5}{19}$$

The approximate solution is

$$y(x)=\frac{5}{19}\left(x-x^2\right)$$

and better approximation

$$y(x)=\left(x-x^2\right)\left(C_1+C_2x\right)$$

where $C_1=0.177,\ C_2=0.173$

Then the approximation becomes

$$y(x)=(0.177+0.173x)(x-x^2)$$

Example 24 Find the approximate solution of $y''+y=-x,\ y(0)=y(1)=0$

Solution The given equation is the Euler's equation of the variational problem of the functional

$$I=\int_0^1\left(y'^2-y^2-2xy\right)dx$$

Choosing solution $y(x)=C(x-x^2)$ and $y'(x)=C(1-2x)$

$$\therefore \quad I=\int_0^1\left[C^2(1-2x)^2-C^2\left(x-x^2\right)^2-2Cx\left(x-x^2\right)\right]dx$$

$$=\frac{3}{10}C^2-\frac{C}{6}$$

Its stationary values are $\dfrac{dI}{dC}=0\Rightarrow C=\dfrac{5}{18}$

The approximate solution is

$$y(x)=\frac{5}{18}\left(x-x^2\right)$$

Example 25 Solve $y''+y=x,\ y'(0)=1,\ y(1)=1$

Solution The given equation is the Euler equation of the variational problem of the functional

$$I=\int_0^1\left(y'^2-y^2+2xy\right)dx+2y(0)$$

Choosing solution $y(x)=a+(1-a)x$ and $y'(x)=(1-a)$

$$\therefore \quad I=\int_0^1\left[(1-a)^2-a^2-(1-a)^2x^2-2a(1-a)x+2ax\right.$$
$$\left.+2(1-a)x^2\right]dx+2a$$

Its stationary conditions are $\dfrac{dI}{da}=0$

$$\Rightarrow \quad \frac{4a}{3}=0\Rightarrow a=0$$

$\therefore$ The required solution is $y(x)=x$

Example 26 Solve $y''+y=x,\ y(0)=0,\ y(1)-y'(1)=0$

Solution Here $$I=\int_0^1\left(y'^2-y^2+2xy\right)dx-y^2(1)$$

Choosing solution $y(x)=Cx$ and $y'(x)=C$

$$\therefore \quad I=\int_0^1\left(C^2-C^2x^2+2Cx^2\right)dx-C^2$$

$$=\frac{2C-C^2}{3}$$

The stationary values are $\dfrac{dI}{dC}=0$

$$\Rightarrow C=1$$

$\therefore$ The required solution is $y=x$

Example 27 Find the approximate solution of $y''=1,\ y(0)=y(1)=0$

Solution The given equation is the Euler's equation of the variational problem of the functional

$$I=\int_0^1\left(y'^2+2y\right)dx$$

Choosing solution $y(x)=C(x-x^2)$ and $y'(x)=C(1-2x)$

$$\therefore \quad I=\int_0^1\left[C^2(1-2x)^2+2C\left(x-x^2\right)\right]dx$$

The stationary values are $\dfrac{dI}{dC}=0$

$$\Rightarrow C=-\frac{1}{2}$$

The required solution is $y(x)=\dfrac{1}{2}\left(x^2-x\right)$

Example 28 Find the approximately the smallest eigen value λ of

$$y''+\lambda y=0,\ y(0)=y(1)=0$$

Solution Here $$I=\int_0^1\left(y'^2-\lambda y^2\right)dx$$

Choosing solution $y(x)=C(x-x^2)$ and $y'(x)=C(1-2x)$

$$\therefore \quad I=\int_0^1\left[C^2(1-2x)^2-\lambda C^2(x-x^2)\right]dx$$

The stationary values are $\dfrac{dI}{dC}=0$

$$\Rightarrow \quad \lambda=\frac{\int_0^1(1-2x)^2\,dx}{\int_0^1(x-x^2)^2\,dx}=0$$

The exact eigen values are $\lambda_n=n^2\pi^2$ and least eigen value $\lambda_1=\pi^2=9.870$

Example 29 Find the smallest equation value of $y''+\lambda y=0,\ y(-1)=0,\ y(1)=0$

Solution Here $I=\int_{-1}^1\left(y'^2-\lambda y^2\right)dx$

Choosing solution $y(x)=C(1-x^2)$ and $y'(x)=-2Cx$

$$\therefore \quad I=\int_{-1}^1\left[4C^2x^2-\lambda C^2(1-x^2)^2\right]dx$$

The stationary values are $\dfrac{dI}{dC}=0$

$$\Rightarrow \quad \lambda=\frac{\int_{-1}^1 4x^2\,dx}{\int_{-1}^1(1-x^2)^2\,dx}=2.5$$

The exact value $=\dfrac{\pi^2}{4}=2.467$

Example 30 Find the least eigen value of $y''+\lambda y=0,\ y'(0)=0,\ y(1)=0$

Solution Here $I=\int_0^1\left(y'^2-\lambda y^2\right)dx$

Choosing solution in one term approximately

$$y(x)=C(x-1) \quad \text{and} \quad y'(x)=C$$

$$\therefore \quad I=\int_0^1\left[C^2-\lambda C^2(x-1)^2\right]dx$$

The stationary values are $\dfrac{dI}{dC}=0$

$$\Rightarrow \qquad \lambda=\frac{1}{\int_0^1 (x-1)^2\ dx}=3$$

Now, choosing solution in two term approximately

$$y(x)=(x-1)(C_1+C_2x)$$

and $$y'(x)=(C_1-C_2)+2C_2x$$

$$\therefore I=\int_0^1\Big[(C_1-C_2)^2+4C_2^2x^2+4C_2(C_1-C_2)x^2-\lambda\big\{C_1^2+(C_1-C_2)^2x^2$$
$$+C_2^2x^4-2C_1(C_1-C_2)x+2C_2(C_1-C_2)x^3-2C_1C_2x^2\big\}\Big]dx$$

$$=(C_1-C_2)^2+\frac{4C_2^2}{3}+2C_2(C_1-C_2)-\lambda\left\{C_1^2+\frac{(C_1-C_2)^2}{3}+\frac{C_2^2}{5}\right.$$
$$\left.-C_1(C_1-C_2)+\frac{C_2(C_1-C_2)}{2}-\frac{2C_1C_2}{3}\right\}$$

The stationary values are $\dfrac{\partial I}{\partial C_1}=0,\ \dfrac{\partial I}{\partial C_2}=0$ gives

$$\frac{2}{3}(3-\lambda)\ C_1-\frac{\lambda}{6}C_2=0$$

and $$-\frac{\lambda}{6}C_1+\frac{20-\lambda}{30}C_2=0$$

for non trivial solution

$$\begin{vmatrix}\frac{2}{3}(3-\lambda) & -\frac{\lambda}{6}\\ -\frac{\lambda}{6} & \frac{20-\lambda}{30}\end{vmatrix}=0$$

$$\Rightarrow \qquad \lambda^2+92\lambda-240=0$$

or $\lambda = \dfrac{-92+\sqrt{(92)^2-4\times 240}}{2} > 0$

or $\lambda = -46+\sqrt{2256}$

or $\lambda = 2.539$

Galerkin's method – Consider the boundary value problem

$$y'' = f(y, y', x) \text{ with } y(a) = P \text{ and } y(b) = Q \quad \text{...(i)}$$

Where the differential equation as $R = \overline{y}'' - f(\overline{y}, \overline{y}', x)$ ...(ii)

Here, R is the residual of the equation.

Consider the trial solution as $\overline{y}(x) = C_1\psi_1(x) + C_2\psi_2(x) + ...$

Then trial solution is differentiated twice and substituting in eqn (ii)

Now, to find the values of C_1, C_2, ... and weight the residual by trial functions $\psi_1(x), \psi_2(x), ...$ and then set the integrals to zero.

Thus, we have

$$\int_c R\psi_1(x) = 0 \text{ and } \int_c R\psi_2(x) = 0$$

These lead to simultaneous equations in the unknowns.

Having found C_1, C_2, ... the approximate solution is obtained.

Example 31 Solve the boundary value problem $y'' - y + x = 0 \; (0 \le x \le 1)$, y(0) = 0, y(1) = 0 by Galerkin's method. Also compare your approximate solution with the exact solution.

Solutions The given residual be

$$R = \overline{y}'' - \overline{y} + x \quad \text{...(i)}$$

To find the trial solution which satisfies the boundary conditions, we derive the Euler's Lagrange equation which passes through the points. $(0,0)$, $(\frac{1}{2},C)$ and $(1,0)$

The resulting equation be $\bar{y}(x) = 4Cx(1-x)$

So that $\psi(x) = x(1 - x)$

Substituting $\bar{y}(x)$ and $\bar{y}''(x)$ in eqn (i), we have

$$R = 4Cx^2 + (1-4C)x - 8C$$

Thus, $\int R\psi(x)dx = 0$

$\Rightarrow \int_0^1 \left[4Cx^2 + (1-4C)x - 8C\right]x(1-x)dx = 0$, where $C = \frac{5}{88}$

Hence, the approximate solution is $\bar{y}(x) = \frac{5}{22}x(1-x)$

Exact Solution: The given differential equation can be written as

$$(D^2 - 1)y = -x$$

its solution is $y = C_1e^x + C_2e^{-x} + x$

using boundary conditions, $y(0) = 0$ and $y(1) = 0$, we get

$$C_2 = -C_1 = \frac{1}{e - e^{-1}}$$

Hence, the exact solution is $y = x - \frac{e^x - e^{-x}}{e - e^{-1}}$

The approximate and the exact solutions for some values of x one given below for comparison:

x	Approximate Solution	Exact solution
0.25	0.043	0.035
0.50	0.057	0.057
0.75	0.043	0.05

Discretization: Discretization concerns the process of transferring continuous models and equations into discrete counterparts. This process is usually carried out as a first step toward making them suitable for numerical evaluation. In other words "The process of replacing a problem defined on a continuum, an interval [0, 1] by an approximating problem on a finite set of points, nh, n = 0, 1, 2, ...N, where $h = \frac{1}{N}$", problems aries in many branches of numerical analysis, i.e., ordinary and partial differential equations, where the finite difference method and the finite element method are common forms of discretization.

For ordinary differential equation

$$y' = f(x, y),\ 0 < x < 1,\ y(0) = y_0$$

A simple discretization is given by Euler's method

$$\frac{1}{h}(y_{n+1} - y_n) = f(x_n, y_n)$$

Where $x_n = nh$, $n = 0, 1, 2, \ldots N$ and $h = \frac{1}{N}$, y_n denotes the approximation to the true solution $y(x)$ at the point x_n.

Discretization is also related to discrete mathematics. It is also refer to modification of variable of category granularity.

Finite Element method: The finite element method (FEM) is a numerical technique for finding approximate solution of partial differential equations as well as of integral equations. Solution of these equations gives us the approximate behaviour of the continuum. These facts suggest that we need to keep pace with the developments by under-standing the basic theory, modeling techniques and computational aspects of the finite element method. Finite element method is widely used in all fields of Science and engineering. Such as, fluid flows, aero elasticity, thermodynamics, structural engineering, machine foundation, lubrication, electrical engineering, textile engineering etc. FEM allows entire designs to be constructed, refined and optimized before the design is manufactured.

Variational Formulation: Let the function $f(x) = u''(x)$ in the interval (0, 1) and the boundary conditions are u(0)=0, u(1)=0 where f is given, u is an unknown function of x and u'' is the second derivative of u w.r.t. x.

To solve the problem, the first step is to convert it into their variational equivalents. Let for any function v that satisfies the displacement boundary conditions *i.e.* v = 0 at $x = 0$ and $x = 1$. We have

$$\int_0^1 f(x)v(x)\,dx = \int_0^1 u''(x)v(x)\,dx \qquad \text{.. (i)}$$

Integrating by parts on the right hand side of equation (i), we have

$$\int_0^1 f(x)v(x)\,dx = \left[u'(x)v(x)\right]_0^1 - \int_0^1 u'(x)v'(x)\,dx$$

Using boundary condition that $v(0)=v(1)=0$, we have

$$\int_0^1 f(x)v(x)\,dx = -\int_0^1 u'(x)v'(x)\,dx$$

$$= -\phi(u,v)$$

Practice Problems

1. If a functional $I[y(x)] = \int_0^1 y(x)dx$ is defined on class C [0,1] then prove that $I[1] = 1$ and $I[\sin \pi x] = \dfrac{2}{\pi}$

2. Test for an extremum the functional

$$I[y(x)] = \int_0^1 \left(xy + y^2 - 2y^2 y'\right) dx$$

$y(0) = 1,\ y(1) = 2.$

[**Ans**: when $x = 0$, $y(0) = 0$, when $x = 1$, $y(1) = -1/2$
Hence $I[y(x)]$ can not attain an extremum].

3. Find the extremals of the functional

$$I[y(x)] = \int x(dx^2 + dy^2)^{\frac{1}{2}}$$

[**Ans.** $y = a \cosh^{-1}\dfrac{x}{a} + b$]

4. Test the extremum the functional

$$I[y(x)] = \int_0^4 (xy' - y'^2)dx,\ y(0) = 0,\ y(4) = 3.$$

[**Ans.** $\dfrac{1}{4}(x^2 - 1)$]

5. Find the extremum of the functional

$$I[y(x)] = \int_{x_0}^{x_1} \frac{\left(1 + y'^2\right)^{\frac{1}{2}}}{x} dx$$

[**Ans.** $x^2 + (y - C_2)^2 = C_1^2$]

6. Find the extremizing function for the functional.

$$I[z(x,y)] = \iint_D \left[\left(\frac{\partial^2 z}{\partial x^2} \right)^2 + \left(\frac{\partial^2 z}{\partial y^2} \right)^2 + \left(\frac{\partial^2 z}{\partial x \partial y} \right)^2 - 2zf(x,y) \right] dxdy$$

where f(x,y) is a known function.

[**Ans.** $\frac{\partial^4 z}{\partial x^4} + 2\frac{\partial^4 z}{\partial x^2 \partial y^2} + \frac{\partial^4 z}{\partial y^4} = f(x,y)$]

7. Prove that the extremals of the functional.

$I[y(x)] = \int_a^b \sqrt{1+y'^2}\, dx$ are straight line.

8. Prove that the Euler-Ostrogradsky equation of the functional.

$I[z(x,y)] = \iint_D (p^2 + q^2 + 2zf(x,y))dxdy$ is the Poisson

equation $\nabla^2 z = f(x,y)$

9. Find the extremals of the functional.

$$I[y(x)] = \int_{x_0}^{x_1} \left(2xy + y'''^2\right) dx$$

[**Ans.** $y = \frac{x^7}{7!} + C_1x^5 + C_2x^4 + C_3x^3 + C_4x^2 + C_5x + C_6$]

10. Find the extremals of the functional.

$$I[y(x)] = \int_{x_0}^{x_1} \frac{1+y^2}{y'^2}\, dx$$

[**Ans.** y = sinh (C_1x + C_2)]

11. Solve the boundary value problem $y''+y+x=0$ $(0 \le x \le 1)$, y(0) = y(1) = 0 by (i) Rayleigh – Ritz method (ii) Galerkin's method compare your solution with the exact solution.

[**Ans.** (i) & (ii) $\bar{y} = \frac{5}{18}x(1-x)$]

12. Using Galerkin's method solve the boundary value problem $y'' = 3x + 4y$, *y(0) =0, y(1)= 1.*

[**Ans.** $\bar{y} = \frac{x}{4}(5x-1)$]

13. Using Ritz's method, find the approximate solutions of the problem $y'' - y + 4xe^x = 0$, *y'(0) – y(0) =1, y'(1) + y(1)= – e*

[**Ans.** $\bar{y}$ = 0.58 + 0.27x]

14. Given the boundary value problem $y''+\pi^2 y = x, y(0) = 1, y(1) = -0.9$. use Galerkin's method to estimate y(0.5), taking the trial solution:
$y = 1 - 1.9x + C_1 x(1 - x) + C_2 x^2(1 - x)$

[Ans. 0.05]

15. Two masses m_1 and m_2 are connected by an inextensible string which passes over a fixed pulley. Using Euler's Lagrange equations, show that the acceleration of either mass is

$$(m_1 - m_2)g/(m_1 + m_2)$$

Chapter - 19

Theory of Reliability and Fault Tolerance

Introduction

In daily life we use the term reliable. A product is said to be reliable if its quality gives dependable, maintenance free results *i.e.* the possibility of failure can not ruled out. In recent year, a fundamental problem is estimating reliability is whether a system will function in a prescribed manner in a given environment for a given period of time. This, of course, depends on many factors such as the design of system, the parts and components used and the environment. The performance of a system under specified conditions for a specified period of time can be considered as a probabilistic event.

Definition of Reliability: Reliability has been defined in various ways: " Reliability is defined as the probability of a device performing its purpose adequately for the period of times intended under the operating conditions." Or "Reliability is a characteristic of an item, expressed by the probability that the item will perform its required function under given conditions for a given time interval." It is denoted by **R(t)** . Sometimes, we use the word 'unit' for device as the unit is composed of several elements and parts which contributes partly to the unreliable factors.

Failure : When a system or unit does not perform satisfactorily, it is said to be failed. The failure is a function of time. In other words "A failure occurs when an item stops performing its required function." If T is the time till the failure of a unit occurs, then the probability that it will not fail before time t is the reliability of the unit $R(t)$, where $R(t) = P(T>t)$.

Failure Rate : The failure rate plays an important role in reliability analysis. Let us consider $S(t)$ be the number of surviving components after time t and δt is the span of time, then the failure rate.

$$Z(t) = \frac{\text{No. of failures during a particular span of time}}{\text{No. of surviving components at the beginning of interval of time}}$$

$$= \lim_{\delta t \to 0} \frac{S(t) - S(t+\delta t)}{S(t)\delta t}$$

$$= -\lim_{\delta t \to 0} \frac{S(t+\delta t) - S(t)}{\delta t} \cdot \frac{1}{S(t)}$$

$$= -\frac{d}{dt} S(t) \cdot \frac{1}{S(t)}$$

$$= -\frac{1}{S(t)} \cdot \frac{d}{dt} S(t)$$

The failure rate is also known as **Hazard rate**.

Reliability Functions: Let us consider a set of N identical components of a sample. Let $S(t)$ be the number of surviving components and $F(t)$ be the number of components that have failed after a time t, where $S(t) + F(t) = N$.

Let $R(t)$ be the reliability of probability of survival and $Q(t)$ be the probability of failure, so that $R(t) = \dfrac{S(t)}{S(t)+F(t)} = \dfrac{S(t)}{N}$ and $Q(t) = \dfrac{F(t)}{N}$ ...(i)

where $R(t) + Q(t) = 1$ or $Q(t) = 1 - R(t)$.

Therefore $R(t) = 1 - Q(t)$

or
$$R(t) = 1 - \frac{F(t)}{N} \quad \text{...(ii)}$$

Differentiating both sides *w.r.t. t*, we get

$$\frac{dR(t)}{dt} = -\frac{1}{N}\frac{d}{dt}F(t) \quad \text{...(iii)}$$

or
$$\frac{d}{dt}F(t) = -N\frac{dR(t)}{dt} \quad \text{...(iv)}$$

Also,
$$\frac{d}{dt}S(t) = \frac{d}{dt}(N - F(t)) = \frac{-d}{dt}F(t) \quad \text{...(v)}$$

Which shows that the rate of survival of components is the negative of the rate at which the component fail.

From equation (V), We get

$$\frac{1}{S(t)}\frac{d}{dt}F(t) = -\frac{N}{S(t)}\frac{dR(t)}{dt}$$

or $$\frac{1}{S(t)}\frac{dF(t)}{dt} = -\frac{1}{R(t)}\frac{dR(t)}{dt} \qquad [\because \text{ using equation (i)}]$$

Let the failure rate is denoted by λ *i.e.* $\lambda = \frac{1}{S(t)}\frac{dF(t)}{dt}$, then

$$\lambda = -\frac{1}{R(t)}\frac{dR(t)}{dt} \qquad \text{...(vi)}$$

Integrating, we get

$$\int_0^t \lambda\, dt = -\int_1^{R(t)} \frac{dR(t)}{R(t)}$$

or $$\lambda[t]_0^t = -[\log R(t)]_1^{R(t)}$$

or $$\lambda t = -[\log R(t) - \log 1]$$

or $$\lambda t = -\log R(t)$$

or $$\log R(t) = -\lambda t$$

or $$R(t) = e^{-\lambda t} \qquad \text{...(vii)}$$

This relationship is generally known as the exponential failure law.

Properties of Reliability:

(i) Reliability is associated with probability its numerical value lies between 0 and 1 *i.e.* $0 \le R(t) \le 1$.

(ii) When $t = 0$, then $R(0) = 1$ and when $t = \infty$ then $R(\infty) = 0$.

(iii) The reliability function is an exponential distribution.

(iv) Reliability is a decreasing function.

Mean Failure Rate: If $z_1, z_2, ...z_k$ are failure rates, then mean failure rate is

$$m = \frac{z_1 + z_2 + \cdots\cdots + z_k}{k} = \frac{1}{k}\sum_{i=1}^{k} z_i$$

Mean Time To Failure (MTTF):

(i) Let the number of devices but into test are N items with failure times. $t_1, t_2, ...t_k$, then mean time to failure is defined as

$$\text{MTTF} = \frac{1}{N}(t_1 + t_2 + \cdots\cdots + t_k)$$

$$= \frac{1}{N}\sum_{i=1}^{k} t_i$$

(ii) Let the components is described by its reliability function, then mean time to failure is given by

$$\text{MTTF} = -\int_0^\infty t\,dR(t)$$

$$= -[tR(t)]_0^\infty + \int_0^\infty R(t)\,dt$$

$$= \int_0^\infty R(t)\,dt \quad [\because \text{R}(\infty) = 0]$$

Mean Time Between Failure (MTBF): Since the probability that a system will perform successfully depends upon the conditions under which it is operating and the time of operation. The average time where a system will run between failures. This time is known as mean time between failure.

$$\text{MTBT} = \int_0^\infty R(t)dt$$

$$= \int_0^\infty e^{-\lambda t} dt$$

$$= \left[\frac{e^{-\lambda t}}{-\lambda}\right]_0^\infty$$

$$= \frac{-1}{\lambda}[0-1] = \frac{1}{\lambda}$$

Note: MTTF is the mean time to first failure, whereas MTBF is the mean time between two successive component failures. The components need not be identical, while MTTF and MTBF are identical.

Relation Between Reliability and Mean Time Between Failures:

Let the reliability equation be

$$R(t) = e^{-\lambda t} \qquad \text{...(i)}$$

and the mean time between failure be

$$\text{MTBF} = \frac{1}{\lambda} \qquad \text{...(ii)}$$

$\therefore$ From equation (i) & (ii), we have

$$R(t) = e^{-t/\text{MTBF}} \qquad \text{...(iii)}$$

When the product λt is small, then reliability

$$R(t) = 1 - \lambda t \qquad \text{...(iv)}$$

∴ From equation (ii) & (iv), we have

$$R(t) = 1 - \frac{t}{\text{MTBF}}$$

Therefore $$\text{MTBF} = \frac{t}{1 - R(t)} \qquad \text{...(v)}$$

Maintainability: The probability that a failed system will be repaired failure to working order within a specified time is called maintainability of the system. In other words, maintainability of the system is the probability of isolating and repairing a fault within a given specified time.

It is denoted by M(t).

Hence the relationship between maintainability are repair rate (μ) with mean time to repair (MTTR) be

$$\mu = \frac{1}{\text{MTTR}}$$

$$\therefore M(t) = \int_0^t \mu e^{-\mu t} dt$$

$$= 1 - e^{-\mu t}$$

$$M(t) = 1 - e^{-\frac{t}{\text{MTTR}}}$$

Where t is the permissible time constraint for the maintenance action.

Availability: The availability of a system is the probability that the system will be up *i.e.* functioning according to expectations at any time during its scheduled working period.

$$\text{Availability} = \frac{\text{System up-time}}{\text{System up-time + System down time}}$$

$$= \frac{\text{System up-time}}{\text{System up-time + (number of failure} \times \text{MTTR)}}$$

$$= \frac{\text{System up-time}}{\text{System up-time + System up-time} \times \lambda \times \text{MTTR)}}$$

$$= \frac{\text{System up-time}}{\text{System up-time}(1 + \lambda \times \text{MTTR})}$$

$$= \frac{1}{1 + \lambda \times \text{MTTR}}$$

Since $\lambda = \dfrac{1}{\text{MTBF}}$, then

$$\text{Availability} = \frac{\text{MTBF}}{\text{MTBF+MTTR}}$$

or $$\text{Availability} = \frac{\mu}{\lambda+\mu} + \frac{\lambda}{(\lambda+\mu)^2 \cdot T}\left[1-e^{-(\lambda+\mu)T}\right] \text{ in } (0,T)$$

System Reliability: The concept of system reliability is a measure of how well the system performs or meets its design objectives.

Two limiting cases of system design are frequently met in practice:

(i) Series System

(ii) Parallel System

(i) **Series System:** A series system is one in which all components are connected that the entire system will fail if any of its components fails.

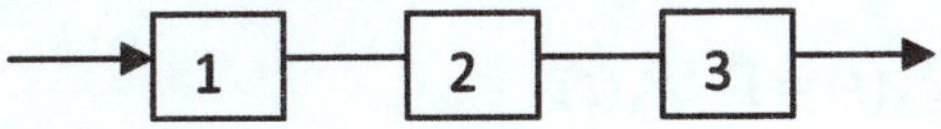

Series System

Consider $R_S(t)$ be the system reliability and all the n components are independent, the system will servive for operating time t, then

$$R_S(t) = R_1(t) \cdot R_2(t) \cdot \cdots\cdots\cdot R_n(t)$$

or $$R_S(t) = \prod_{i=1}^{n} R_i(t) \qquad \text{...(i)}$$

Which is known as the product rule in reliability.

But $$R_i(t) = e^{-\lambda_i t} \qquad \text{...(ii)}$$

Then from equation (i) & (ii), we have

$$R_S(t) = \prod_{i=1}^{n} e^{-\lambda_i t}$$

$$R_S(t) = \exp\left(-\sum_{i=1}^{n} \lambda_i t\right)$$

and $$\text{MTBF} = \frac{1}{n\lambda}$$

(ii) Parallel System: A parallel system is a system that is not considered to have failed unless all components have failed.

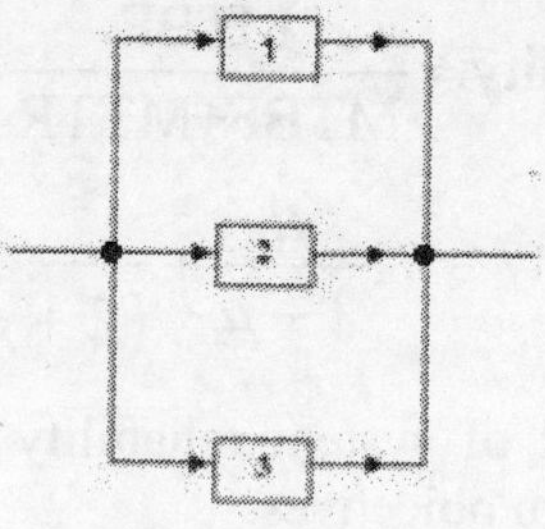

Parallel System

Consider $R_1(t)$, $R_2(t)$ and $R_3(t)$ be the reliability of the subsystem respectively.

Then $$F_1(t) = 1 - R_1(t)$$

$$F_2(t) = 1 - R_2(t)$$

and $$F_3(t) = 1 - R_3(t)$$

If $F_P(t)$ be the failure distribution of the system, then

$$F_P(t) = [1 - R_1(t)][1 - R_2(t)][1 - R_3(t)]$$

If $R_P(t)$ be the system reliability, then

$$R_P(t) = 1 - F_P(t)$$

$$= 1 - [1 - R_1(t)][1 - R_2(t)][1 - R_3(t)]$$

or $$R_P(t) = 1 - \prod_{i=1}^{n}[1 - R_i(t)]$$

But, if all components have equal reliability, say $r(t)$, then

$$R_P(t) = 1 - [1 - r(t)]^n$$

$$R_P(t) = 1 - \left[1 - e^{-\lambda t}\right]^n$$

Illustrative Examples

Example 1 Any circuit has an MTBF of 0.8 years. Assuming random failures,

(i) Calculate the probability that the circuit will survive 1 year without failure.

(ii) What is the probability that there will be more than 2 failures in the first year?

(iii) What is the expected number of failure per year?

Solution: Here, MTBF = 0.8 year

$$\therefore \qquad \lambda = \frac{1}{\text{MTBF}} = \frac{1}{0.8} = 1.25 \text{ per year}$$

Since the failures are random events, the number of failures in an interval of length t follows a Poisson process, given by

$$P[N(t)=r] = \frac{e^{-\lambda t}\cdot(\lambda t)^r}{r!}$$

(i) $$P[N(t)=0] = \frac{e^{-\lambda}\cdot\lambda^0}{0!} = e^{-1.25} = 0.2865$$

(ii) $$P[N(t)>2] = 1 - e^{-\lambda}\left[\frac{\lambda^0}{0!}+\frac{\lambda^1}{1!}+\frac{\lambda^2}{2!}\right]$$

$$= 1 - 0.2865\left[1 + 1.25 + \frac{(1.25)^2}{2}\right]$$

$$= 0.1315$$

(iii) Expected number of failure per year, $\lambda = 1.25$

Example 2. The density function of the time to failure is $f(t) = \dfrac{4}{(t+2)^3}$, $t > 0$ is in years.

Find (i) The reliability function $R(t)$. (ii) The failure rate λ and (iii) MTTF.

Solution: Given the density function be $f(t) = \dfrac{4}{(t+2)^3}$

(i) Reliability $$R(t) = \int_t^{\infty} f(t)\, dt$$

$$= \int_t^{\infty} \frac{4}{(t+2)^3}\, dt$$

$$= 4\left[\frac{-1}{2(t+2)^2}\right]_t^\infty$$

$$= -2\left[0 - \frac{1}{(t+2)^2}\right] = \frac{2}{(t+2)^2}$$

(ii) Failure rate $\lambda = \dfrac{f(t)}{R(t)}$

$$\lambda = \frac{4/(t+2)^3}{2/(t+2)^2}$$

$$\lambda = \frac{2}{t+2}$$

(iii) $\text{MTTF} = \int_0^\infty R(t)\,dt$

$$= \int_0^\infty \frac{2}{(t+2)^2}\,dt$$

$$= 2\left[\frac{-1}{t+2}\right]_0^\infty = -2\left[0 - \frac{1}{2}\right] = 1$$

Example 3 A product has an MTBF of 1500 hours. What is the reliability for a 500 hours mission?

Solution: Here $t = 500$ hours and $\text{MTBF} = \dfrac{1}{\lambda} = 1500$ hours

$\therefore$ $\lambda = \dfrac{1}{1500}$ per hour

$\therefore$ Reliability $R(t) = e^{-\lambda t}$

$$= e^{\frac{-1}{1500}\times 500}$$

$$= e^{-\frac{1}{3}} = 0.717$$

Example 4 A component has MTBF = 200 hours and MTTR = 10 hours with both failure and repair exponential distribution. Find the availability of the component in the interval (0,10) hours, at the end of 10 hours and after a long time.

Solution : Here, $\text{MTBF} = \frac{1}{\lambda} = 200$

$$\therefore \quad \lambda = \frac{1}{200} = 0.005$$

and $\text{MTTR} = \frac{1}{\mu} = 10$

$$\therefore \quad \mu = \frac{1}{10} = 0.1$$

$$\therefore \quad \text{Availability} = \frac{\text{MTBF}}{\text{MTBF} + \text{MTTR}}$$

or $$\text{Availability} = \frac{\mu}{\lambda+\mu} + \frac{\lambda}{(\lambda+\mu)^2 \cdot T}\left[1 - e^{-(\lambda+\mu)T}\right] \text{ in } (0,T)$$

In the interval (0, 10),

$$\text{Availability} = \frac{0.1}{0.005+0.1} + \frac{0.005}{(0.005+0.1)^2 \cdot 10}\left[1 - e^{-(0.005+0.1)\times 10}\right]$$

$$= \frac{0.1}{0.105} + \frac{0.005}{(0.105)^2 \times 10}\left(1 - e^{-1.05}\right)$$

$$= 0.952 + \frac{0.005}{0.11}(1 - 0.035)$$

$$= 0.9812$$

At the end of $t = 10$

$$\text{Availability} = \frac{\mu}{\lambda+\mu} + \frac{\lambda}{\lambda+\mu} e^{-(\lambda+\mu)t}$$

$$= \frac{0.1}{(0.005+0.1)} + \frac{0.005}{(0.005+0.1)} e^{-(0.005+0.1)\times 10}$$

$$= 0.952 + 0.047 \times 0.035$$

$$= 0.9684$$

After a time mean $t = \infty$,

$$\text{Availability} = \frac{\text{MTBF}}{\text{MTBF+MTTR}}$$

$$= \frac{200}{200+10} = 0.9524$$

Example 5 Time to repair a power generator is best described by its density function

$$f(t) = \frac{t^2}{333}, \quad 1 \le t \le 10 \text{ hours}.$$

(i) Calculate MTTF

(ii) Find the repair rate

(iii) Find the probability that a repair will be completed in 6 hours.

Solution: Given the density function be $f(t) = \dfrac{t^2}{333}$

(i) $\text{MTTR} = \int_0^{10} t\, f(t)\, dt$

$$= \int_0^{10} t \cdot \frac{t^2}{333} dt$$

$$= \frac{1}{333} \int_0^{10} t^3 dt$$

$$= \frac{1}{333} \left[\frac{t^4}{4} \right]_0^{10} = \frac{1}{1332} \times 10^4 = 7.5 \text{ hours}$$

(ii) Repair rate $\mu(t) = \dfrac{f(t)}{1 - F(t)}$

$$= \frac{\dfrac{t^2}{333}}{\displaystyle\int_t^{10} \frac{t^2}{333} dt}$$

$$= \frac{\frac{t^2}{333}}{\frac{1}{333}\left[\frac{t^3}{3}\right]_t^{10}} = \frac{\frac{t^2}{333}}{\frac{10^3 - t^3}{999}}$$

$$= \frac{3t^2}{1000 - t^3} \text{ per hour}$$

(iii) $P(T < 6) = \int_1^6 f(t)\, dt$

$$= \int_1^6 \frac{t^2}{333} dt = \left[\frac{t^3}{3 \times 333}\right]_1^6 = \frac{6^3 - 1^3}{999} = 0.2152$$

Example 6 To make a prediction regarding the weather, a computer requires statistical data from three sources. Assuming the three sources posses the respective hazard function $\lambda_1(t) = 4t^6, \lambda_2(t) = 0.8t^2, \lambda_3(t) = 0.6t$, where t is measured in months. Find the reliability function. Assume source failures are independent.

Solution: Here $\lambda_1(t) = 4t^6, \lambda_2(t) = 0.8t^2, \lambda_3(t) = 0.6t$

Since the failures are independent, then reliability is

$$R(t) = e^{-\int_0^t \lambda_1(t)dt} \times e^{-\int_0^t \lambda_2(t)dt} \times e^{-\int_0^t \lambda_3(t)dt}$$

$$= e^{-\int_0^t 4t^6 dt} \times e^{-\int_0^t 0.8t^2 dt} \times e^{-\int_0^t 0.6t\, dt}$$

$$= e^{-\frac{4}{7}t^7} \times e^{-\frac{0.8}{3}t^3} \times e^{-\frac{0.6}{2}t^2}$$

$$= e^{-\left[\frac{4}{7}t^7 + \frac{0.8}{3}t^3 + 0.3t^2\right]}$$

Example 7 A hybrid electronic circuit consists of 20 valves, 30 resistors and 10 capacitors all connected in series. The components in each category are identical and were found to have the following failure rate:

		Failure rate (No. of failures per 1000 hours/unit)
Valves	:	0.10
Resistors	:	0.05
Capacitors	:	0.05

What is the MTBF of the system? What is the reliability for 100 hours?

Solution: In series system, the failure rate is given by

$$\lambda(t) = N_1\lambda_1 + N_2\lambda_2 + N_3\lambda_3$$

given that $N_1 = 20, N_2 = 30, N_3 = 10$

and $\lambda_1 = 0.10, \lambda_2 = 0.05, \lambda_3 = 0.05$

$$\therefore \quad \lambda(t) = 20\times0.10 + 30\times0.05 + 10\times0.05$$

$$= 2 + 1.5 + 0.5$$

Failure rate $\lambda(t) = 4.0$

Since the failure rate corresponds to an operating time of 1000 hours.

Therefore, $\text{MTBF} = \dfrac{1000}{4.0} = 250$

Reliability $R(t) = e^{-\lambda t}$

$$= e^{-t/\text{MTBF}}$$

$$= e^{-100/250} = e^{-0.4} = 0.671$$

Example 8 Calculate the reliability of the following series system, where each component has the identical reliability.

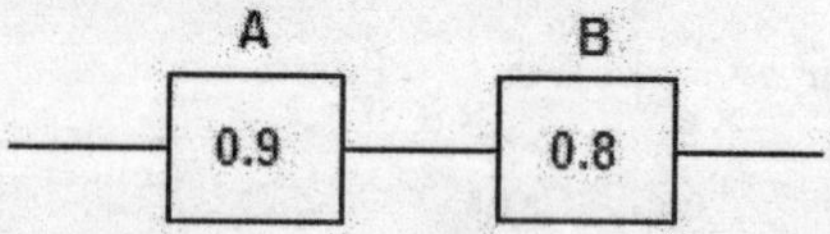

Solution: The reliability of components are $R_A = 0.9$ and $R_B = 0.8$

Then, the reliability of the series system is

$$R_{AB} = R_A \times R_B$$

$$= 0.9\times0.8 = 0.72$$

Example 9 Calculate the reliability of the following parallel system, where each component has the identical reliability.

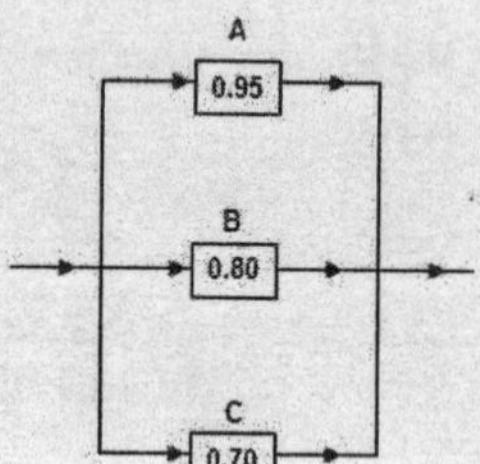

Solution: The reliability of the components are $R_A = 0.95$, $R_B = 0.80$, $R_C = 0.70$ The parallel system reliability is

$$\begin{aligned} R_P &= 1-(1-R_A)(1-R_B)(1-R_C) \\ &= 1-(1-0.95)(1-0.80)(1-0.70) \\ &= 1-(0.05)(0.20)(0.30) \\ &= 1-0.003 \\ &= 0.997 \end{aligned}$$

Example 10 Obtain the system reliability for the following system.

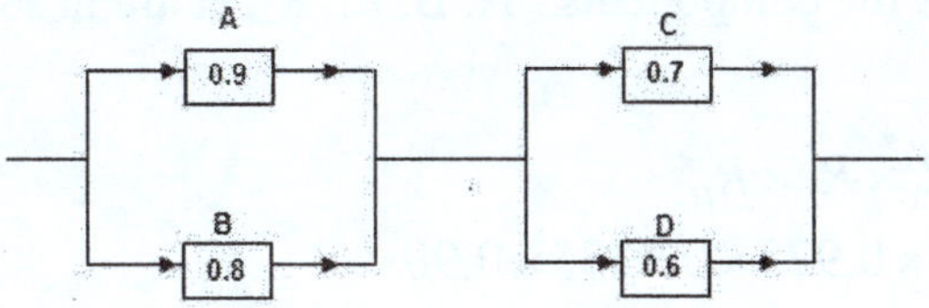

Solution : Here $R_A = 0.9$, $R_B = 0.8$, $R_C = 0.7$ and $R_D = 0.6$

$$\begin{aligned} \therefore \quad R_{AB} &= 1-(1-R_A)(1-R_B) \\ &= 1-(1-0.9)(1-0.8) \\ &= 1-0.1\times 0.2 \quad = \ 0.98 \end{aligned}$$

and

$$\begin{aligned} R_{CD} &= 1-(1-R_C)(1-R_D) \\ &= 1-(1-0.7)(1-0.6) \\ &= 1-0.3\times 0.4 \quad = \ 0.88 \end{aligned}$$

$$\begin{aligned} \therefore \quad R_S &= R_{AB}\times R_{CD} \\ &= 0.98\times 0.88 \\ &= 0.8624 \end{aligned}$$

Example 11 Compute the reliability of the system for the connection given in the following figure.

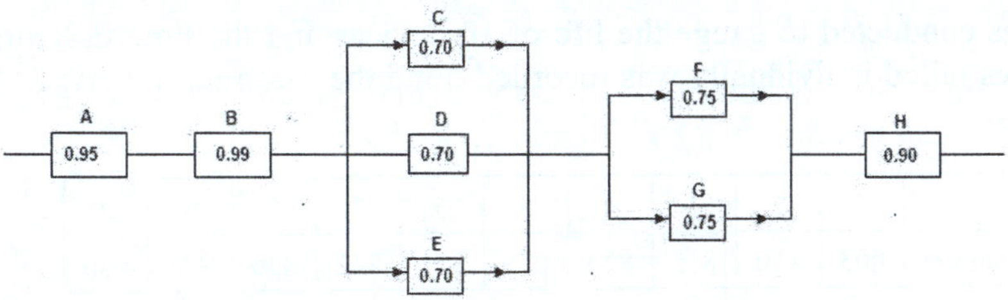

Solution: The components C, D, E are parallel, then reliability

$$R_1 = 1-(1-R_C)(1-R_D)(1-R_E)$$
$$= 1-(1-0.70)^3 = 0.973$$

Again, the components F, G are parallel, then reliability

$$R_2 = 1-(1-R_F)(1-R_G)$$
$$= 1-(1-0.75)^2 = 0.9375$$

Now, in the system, the components A, B, R_1, R_2, H are in series,

The reliability is

$$R_S = R_A \times R_B \times R_1 \times R_2 \times R_H$$
$$= 0.95 \times 0.99 \times 0.973 \times 0.9375 \times 0.90$$
$$= 0.772$$

Practice Problem

1. A density function of the time to failure in years of the gizmos manufactured by a certain company is given by $f(t) = \dfrac{200}{(t+10)^3}$, $t \geq 0$.

   (a) Derive the reliability function and determine the reliability for the first year of operation.

   (b) Find the MTTF.

   [**Ans.** (a) $R(t) = \dfrac{100}{(t+10)^2}$, $R(1) = 0.8264$ (b) 10 years]

2. A product with constant hazard rate has been found to have a reliability of 90% from operating time of 100 hours. Find the failure rate of the product.

   [**Ans.** $\lambda = 0.001$ failure/hour]

3. A test was conducted to gauge the life of 10 devices and the time in hours before the devices failed individually was recorded. Find the mean failure rate and MTTF of devices.

Device :	1	2	3	4	5	6	7	8	9	10
Time to failure	805	810	815	820	825	832	842	856	875	900

[**Ans.** 0.0011, MTTF = 838 hours]

4. A computer has an MTTF = 34 hours and MTTR = 2.5 hours. What is the steady state availability? If the MTTR is reduced to 1.5 hours, what MTTF can be tolerated without decreasing the steady state availability of the computer?

[**Ans.** 0.931, 20.239 hours]

5. A system consists of 6 components connected as in the following figure:

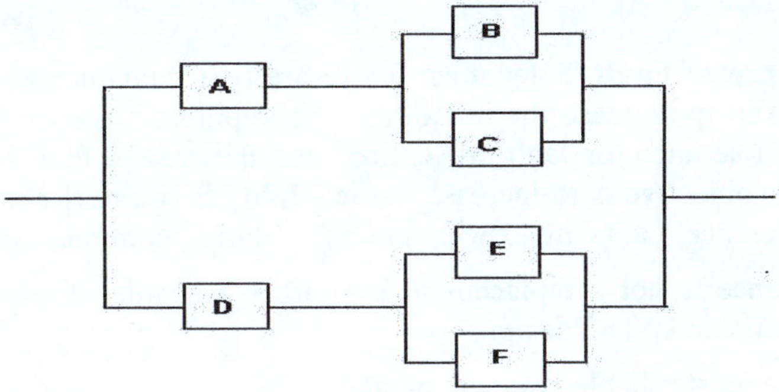

Find the overall reliability of the system given that the reliabilities of A, B, C, D, E and F are respectively 0.95, 0.80, 0.90, 0.99, 0.90 and 0.85.

[**Ans.** 0.9983]

6. Compute the reliability of the system for the connection given in the following figure, if the reliabilities of A, B, C, D are 0.90, 0.95, 0.96 and 0.98 respectively:

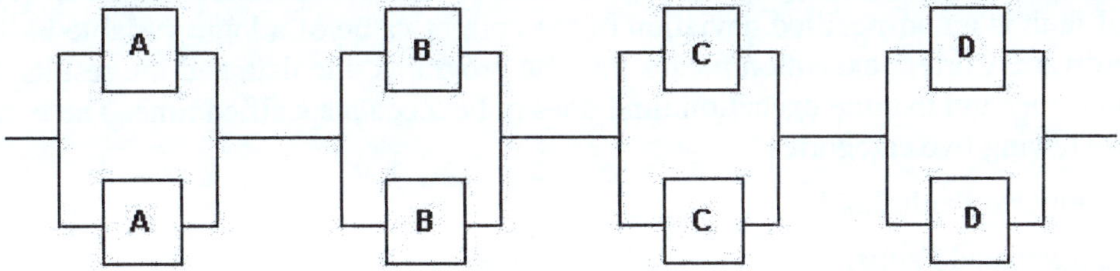

[**Ans.** 0.9856]

7. The reliability of a component with constant rate of failure $\alpha = 0.002$ is specified to be 0.95. Find the number of hours of operation under given conditions.

[**Ans.** 25.65 hours]

8. An electronic circuit consists of 5 silicon transistors, 6 silicon diodes, 7 composition resistors and 8 ceramic capacitors connected in series configuration. The hourly failure rate of each component is given below:

Silicon transistors : $\lambda_1 = 2\times10^{-5}$

Silicon diodes : $\lambda_2 = 3\times10^{-5}$

Composition resistors : $\lambda_3 = 4\times10^{-5}$

Ceramic capacitors : $\lambda_4 = 5\times10^{-5}$

Calculate the reliability of the circuit for 10 hours when the component follows exponential distribution.

[**Ans.** $R(10) = e^{-0.0096}$]

The Importance of Fault Tolerance: There are two fundamental different approaches that can be taken to increase the reliability of computing systems. The first approach is called fault in tolerance (or fault prevention) and the second fault tolerance. In the fault prevention, the objective is to increase the reliability by a priori elimination of all faults. In the fault tolerance, faults are expected to occur during computation.

Fault tolerance is not a replacement but rather a supplement to the most important principles of reliable system design.

(i) Use the most reliable component, and

(ii) Keep the system as simple as possible.

While fault tolerance can be used to increase significantly the reliability of an already reliable system.

Basic concept of Fault Tolerance: Fault tolerance is defined as, "It is a system which has the built-in capability to preserve the continued correct execution of its programs and input/output functions in the presence of a certain set of operational faults." An operational fault is an unspecified deviation of the correct value of a logic variable in the system hardware. Correct execution means that the programs, the data and the results do not contain errors and that the execution time does not exceed a specified time. The types of faults are having two categories:

(i) Anticipated Faults, and

(ii) Unanticipated Faults.

Anticipated faults are those whose occurrence in an operational system can be foreseen. Unanticipated faults on the other hand cannot be foreseen but affect the operation of a system.

Hence, tolerance of both anticipated and unanticipated faults must be considered in high-reliability applications.

Chapter - 20

Goal Programming

Introduction: Goal programming is an approach used for solving a multi objective optimization problem. The relationship of G.P. within the fields of management science and operation research and multiple criteria decision making will be discussed. A problem is modelled into a GP model in a manner similar to that of the LP model.

Concept of Goal Programming: The concept of GP was first introduced by Charnes, Cooper and Ferguson in 1955. The basic idea of GP has been traced by some other mathematicians. Like Charnes and Cooper in 1961, they suggested a method for solving an infeasible LP problem arising from various conflicting resource constraints. Ijiri (1965) developed the concept of pre-emptive priority factors, assigning different priority levels to goals and different weights for the goals at the same priority level. Lee (1972) and Ignizio (1976) have discussed the subject of GP which is an extension of LP. Then the basic idea of GP has been traced by Romero in 1992 to a study by Charnes and Cooper on executive compensation.

Basically, the method of GP consists of formulating an objective function in which optimization comes as close as possible to the specified goals.

Goal Programming Model Formulation: The formulation of GP model is similar to that of LP model. Therefore all specified goals are ranked in order of their priority.

Let us consider m goals are expressed by an m-component column b_i, a_{ij} be the coefficient for the jth decision variable in the ith constraint, x_i be a decision variable and w_i, the weights of each goal. Then the general GP model can be stated as follows:

minimize $$Z = \sum_{j=1}^{m} a_{ij} x_j = b_i$$

or $$Z = \sum_{i=1}^{m} w_i \left(\overset{-}{d}_i + \overset{+}{d}_i \right)$$

Subject to $$\sum_{j=1}^{m} a_{ij} x_j + \overset{-}{d}_i - \overset{+}{d}_i = b_i, \quad i = 1, 2, ----m$$

and $$x_j,\ \bar{d}_i,\ \overset{+}{d}_i \geq 0, \quad \text{for all } i, j$$

Where $\bar{d}_i$ = negative deviation from the goal (under achievement)

$\overset{+}{d}_i$ = positive deviation from the goal (over achievement)

Since both under and over achievement of a goal cannot be achieved simultaneously, one or both may be zero in the solution *i.e.* $\bar{d}_i \times \overset{+}{d}_i = 0$. Now we shall discuss the goal model extensions and their solutions more systematically as follows:

Single-Goal Models

Example 1. A firm produces two products X and Y. The unit profit from product X is Rs. 100 and that of product Y is Rs. 50. The goal of the firm is to earn a total profit of exactly Rs. 700 in the next week.

Solution: Let x_1 and x_2 be the number of units of the product X and Y respectively.

The single goal of profit is stated as

Maximize $$Z = 100x_1 + 50x_2$$

Since the goal of the firm is to earn a target profit of Rs. 700 per week, the profit goal can be restated to allow for under or over-achievement as:

$$100x_1 + 50x_2 + \bar{d}_1 - \overset{+}{d}_1 = 700$$

The goal programming formulation of the problem is

Minimize $$Z = \bar{d}_1 + \overset{+}{d}_1$$

Subject to $$100x_1 + 50x_2 + \bar{d}_1 - \overset{+}{d}_1 = 700$$

and $$x_1,\ x_2,\ \bar{d}_1,\ \overset{+}{d}_1 \geq 0$$

If the profit goal is not achieved completely, the slack in the profit goal will be expressed by negative deviation variable, $\bar{d}_1$ from the goal. But if the solution shows a profit in excess of Rs. 700, the surplus in the profit will be expressed by positive deviational variable, $\overset{+}{d}_1$ from the goal. If the profit goal of Rs. 700 is exactly achieved, both $\bar{d}_1$ and $\overset{+}{d}_1$ will be zero.

The required solution will be any linear combination of x_1 and x_2 between two points. $x_1 = 7,\ x_2 = 0$ and $x_1 = 0,\ x_2 = 14$

Multiple Goal Models

Example 2. In the previous example 1, in addition to the profit goal. Management consider to achieve sales

volume for products X and Y close to 5 and 4 respectively. Formulation this problem as a GP model.

Solution: The constraints of this problem can be stated as:

$$100x_1 + 50x_2 = 700$$

Subject to $x_1 \le 5$

$$x_2 \le 4$$

The profit goal and the sales goals are expressed as

$$100x_1 + 50x_2 + \bar{d}_1 - \overset{+}{d}_1 = 700$$

$$x_1 + \bar{d}_2 = 5$$

$$x_2 + \bar{d}_3 = 4$$

The GP model as follows.

$$\text{Minimize } 100x_1 + 50x_2 + \bar{d}_1 - \overset{+}{d}_1 = 700$$

Subject to $x_1 + \bar{d}_2 = 5$

$$x_2 + \bar{d}_3 = 4$$

and $x_1,\ x_2,\ \bar{d}_1,\ \overset{+}{d}_1,\ \bar{d}_2,\ \bar{d}_3 \ge 0$

where $\bar{d}_2$ and $\bar{d}_3$ represent under achievement of sales volume for products X and Y respectively. Since sales target are maximum possible sales volume. Hence $\overset{+}{d}_2$ and $\overset{+}{d}_3$ are not included in the sales constrained.

The solution of this problem is that if $x_1 = 5,\ x_2 = 4$ then all targets will be completely achieved.

Thus $\quad \overset{-}{d}_1 = \overset{-}{d}_2 = \overset{-}{d}_3 = \overset{+}{d}_1 = 0$

The General Goal Programming Model: The general GP model may be stated as :

Minimize
$$Z = \sum_{i=1}^{m}\sum_{k=1}^{k} p_k \left(\overset{+}{w}_{i,k}\, \overset{+}{d}_i + \overset{-}{w}_{i,k}\, \overset{-}{d}_i \right)$$

Subject to
$$\sum_{j=1}^{n} a_{ij}x_j + \overset{-}{d}_i - \overset{+}{d}_i = b_i, \quad \text{for } i = 1, 2, ---m$$

$$\sum_{j=1}^{n} a_{ij}x_j \,(\leq = \geq)\; b_i, \quad \text{for } i = m+1, m+2, ---m+p$$

and
$$x_j, \overset{-}{d}_i, \overset{+}{d}_i \geq 0, \quad \text{for } j = 1, 2, ---n, \;\; i = 1, 2, ---m$$

where P_k = The priority coefficient for the kth priority.

$\overset{+}{w}_{i,k}$ = The relative weight of the $\overset{+}{d}_i$ variable in the kth priority level.

$\overset{-}{w}_{i,k}$ = The relative weight of the $\overset{-}{d}_i$ variable in the kth priority level.

Z = The sum of the derivations from all desired goals.

a_{ij} = The constants attached to each decision variable.

b_i = The right hand side values of each constraint.

Graphical Solution of GP Problems: A graphical solution for GP problems is similar to graphical solution of LP. The following steps are:

Step 1. Graph all system constraints and identify the feasible region. If no structural constraints exist, then the feasible region is the first quadrant *i.e.* both x_1 and x_2 are ≥ 0. If no feasible region exists, there is no solution to the problem.

Step 2. Graph the lines corresponding to the goal constraints and set the deviational variables.

Step 3. Identify the top-priority solution in the feasible region, determine the points that satisfy the highest priority goal.

Step4. Consider the remaining goal having the next highest priority and determine the best solutions.

Such that this best solution does not degrade the solutions already achieved for goals of highest priority.

Step 5. Repeat step4 until all priority have been investigated.

Example 3: A company produces two products P and Q. According to past experience, production of either product P or Q requires an average of one hour in the plant. The plant has a normal production capacity of 400 hours a month. The marketing department of the firm reports that because of limited market, the maximum number of product P and Q that can be sold in a month are 240 and 300 respectively. The net profit from the sale of product P and Q are Rs. 800 and Rs. 400 respectively. The manager of the company has set the following goals arranged in the order of importance.

P₁: He wants of avoid any under-utilization of normal production capacity.

P₂: He wants to sale maximum possible units of products P and Q. Since the net profit from the sale of product P is twice the amount from product Q. Therefore, the manager has twice as much desire to achieve sales for product P as for product Q.

P₃: He wants to minimize the overtime operation of the plant as much as possible.

Formulate and solve the given problem by graphical method of GP.

Solution: Let x_1 and x_2 be the units of product P and Q. Since overtime operation is allowed, the plant capacity constraint can be expressed as

$$x_1 + x_2 + \overset{-}{d_1} - \overset{+}{d_1} = 400$$

Where $\overset{-}{d_1}$ is the under-utilization and $\overset{+}{d_1}$ is the overtime operation.

Since the sales goals are the maximum possible sales, positive deviations will not appear in the sales. Thus, the sales constraints are

$$x_1 + \overset{-}{d_2} = 240$$

$$x_2 + \overset{-}{d_3} = 300$$

Where $\overset{-}{d_2}$ and $\overset{-}{d_3}$ are the under-achievements of the sales goal. Thus, the GP model becomes:

Minimize: $$z = P_1 \overset{-}{d_1} + P_2\left(2\overset{-}{d_2} + \overset{-}{d_3}\right) + P_3 \overset{+}{d_1}$$

Subject to : $$x_1 + x_2 + \overset{-}{d_1} - \overset{+}{d_1} = 400$$

$$x_1 + \overset{-}{d_2} = 240$$

$$x_2 + \overset{-}{d_3} = 300$$

and $\quad x_1, x_2, \overset{-}{d_1}, \overset{-}{d_2}, \overset{-}{d_3}, \overset{+}{d_1} \geq 0.$

Step 1. First step to plot all the goal constraints on a graph. Since the under-utilization and over-utilization of the plant capacity are permissible, both the deviational variables $\overset{-}{d_1}$ and $\overset{+}{d_1}$ are indicated by the arrows. Similarly, $\overset{-}{d_2}$ and $\overset{-}{d_3}$ also indicated by arrows. Once all the goal constraints have been plotted on the graph, the feasible region is represented by the shaded region O A B C.

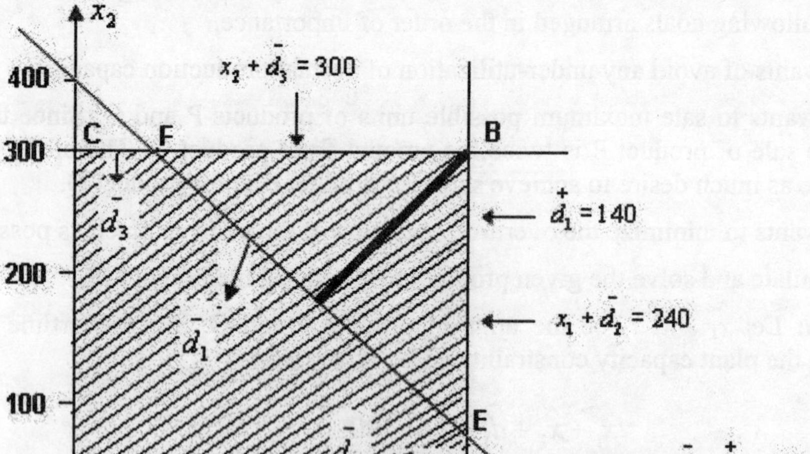

Step 2. The first goal is completely achieved by the line EF in the reduced feasible region EBF. Then we achieve the sales goal of product P. Since at the second priority level, the sales goal constraint of product P has more weight than the product Q, the manager would like to achieve first the sales goal of product P which can be completely achieved, and shown by the line EB in the region EBF. Also the second priority given to product Q is also achieved and shown by the line FB in the region EBF. Thus, the first two goals are completely achieved at point B in the region EBF. The third priority is given to the overtime operation which cannot be achieved at the expense of first two goals.

Now, the solution of the GP problem becomes.

$$x_1 = 240, x_2 = 300, \overset{-}{d_1} = 0, \overset{-}{d_2} = 0, \overset{-}{d_3} = 0, \overset{+}{d_1} = 140$$

Example 4. A company produces two products X and Y. Each product must be processed through two parts. Part one has 30 hours of production capacity per day, and part two has 60 hours. Each units of product X requires 2 hours in part one and 6 hours in part two.

Each unit of product Y requires 3 hours in part one and 4 hours in part two. Manager has rank ordered the following goals it would like to achieve in determining the daily product mix:

P_1 : Minimize the under achievement of joint total production of 10 units.

P_2 : Minimize the under achievement of producing 7 units of product Y.

P_3 : Minimize the under achievement of producing 8 units of product X.

Formulate this problem as a GP model and then solve it by using the graphical method.

Solution: The GP model for the problem is

Minimize $Z = P_1 \bar{d_1} + P_2 \bar{d_3} + P_3 \bar{d_2}$

Subject to $2x_1 + 3x_2 = 30$

$6x_1 + 4x_2 = 60$

$x_1 + x_2 + \bar{d_1} - \overset{+}{d_1} = 10$

$x_1 + \bar{d_2} - \overset{+}{d_2} = 8$

$x_2 + \bar{d_3} - \overset{+}{d_3} = 7$

and $x_1, x_2, \bar{d_1}, \bar{d_2}, \bar{d_3}, \overset{+}{d_1}, \overset{+}{d_2}, \overset{+}{d_3}, \geq 0$

Now, we can proceed with graphical method:

Step 1. The first two constraints are system constraint and the last three are goal constraints in GP model. The feasible region is the non-negative quadrant.

Step 2. Put the deviational variables to zero in each goal constraint, we can plot the straight line.

Step 3. The highest priority is to minimize $\bar{d_1}$, which represents under achievement of joint total production of 10 units. Thus all possible points that have positive values of $\bar{d_1}$ have been eliminated. To achieve second priority goal, all points having positive values of $\bar{d_3}$ have been eliminated. The points eliminated are those that lie below the second goal constraint line. These represent combination of X and Y that fall short of the production goal of 7 units for product Y.

Step 4. The third priority goal is to minimize d_2 , which represent under achievement of the production goal of 8 units of product X.

In the line is ABC each point involves under achievement of third goal. The optimal solution occurs at corner point C where $\bar{d}_3$ is made as small as possible. At point D or E,

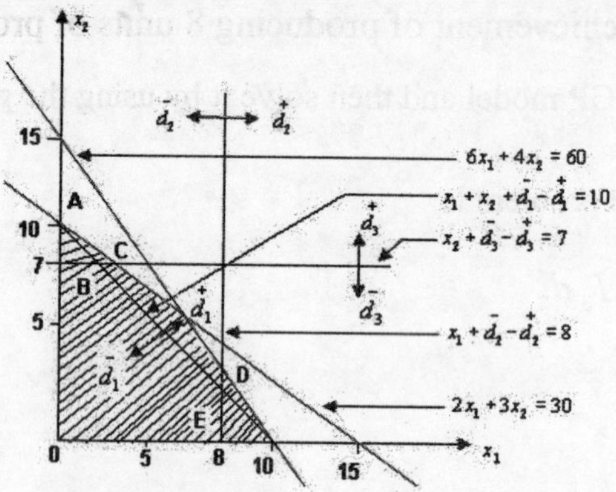

$\bar{d}_2$ would become zero, and positive value for $\bar{d}_3$. Sacrificing a higher priority goal, this is not allowed.

Since goal equation 1 and 5 intersect at point C, we can solve these equations to find the final solution for decision variables $x_1 = 4.5$ and $x_2 = 7$ in the feasible region.

The values for deviational variables are

$$\bar{d}_1 = 0,\ \bar{d}_2 = 0,\ \bar{d}_3 = 0,\ \overset{+}{d}_1 = 1.5,\ \overset{+}{d}_2 = 0,\ \overset{+}{d}_3 = 0,$$

Simplex Method of GP : The simplex method for GP is similar to the LP problem. The following steps of simplex method are :

Step 1. The values of Z_j and $Z_j - C_j$ are computed for each priority level. Take $K = 1$, where K is a pointer which represent ($Z_j - C_j$) row associated with P_K priority.

Step 2. Check the value of $Z_j - C_j$ in P_1 row first. If zero exists then P_K priority goal has been met.

Then go to step 6. If a zero values do not exist proceed to step 3.

Step 3. To determine the variable to enter into the new solution, start examine $Z_j - C_j$ row of highest priority P_K and select the largest positive value. If no positive value exists in the P_K row that meets the above conditions. Then go to step 6, otherwise go to step 4.

Step 4. Determine the out coming variable by using the standard procedure from the simplex method.

Step 5. Develop the new table. The new $Z_j - C_j$ rows are computed in the same manner used in developing the initial table. Go to step2.

Step 6. Compute the next lower priority level. Any positive value in the $Z_j - C_j$ row which has negative $Z_j - C_j$ value under any lower priority rows are ignored. It is because deviations from the highest priority goal would be increased with the entry of this variable in the solution.

Example 5: Use modified Simplex method to solve the following GP problem.

Minimize $\quad Z = P_1 d_1^- + 2P_2 d_2^- + P_2 d_3^- + P_3 d_1^+$

Subject to $\quad x_1 + x_2 + d_1^- - d_1^+ = 400$

$$x_1 + d_2^- = 240$$

$$x_2 + d_3^- = 300$$

and $\quad x_1, x_2, d_1^-, d_2^-, d_3^-, d_1^+ \geq 0$

Solution: The initial Simplex table for this problem is given in the table.

Basic Variable	C_B	x_1	x_2	d_1^-	d_2^-	d_3^-	d_1^+	x_B	Min ratio
d_1^-	P_1	1	1	1	0	0	-1	400	$\frac{400}{1} = 400$
d_2^-	$2P_2$	(1)	0	0	1	0	0	240	$\frac{240}{1} = 240 \leftarrow$
d_3^-	P_2	0	1	0	0	1	0	300	-
C_j		0	0	P_1	$2P_2$	P_2	P_3		
Z_j		$P_1 + 2P_2$	$P_1 + P_2$	P_1	$2P_2$	P_2	$-P_1$		
	P_3	0	0	0	0	0	-1	0	
$\Delta_j = Z_j - C_j$	P_2	2	1	0	0	0	0	780	
	P_1	$1\uparrow$	1	0	0	0	-1	400	

$$Z = 400 \times P_1 + 240 \times 2P_2 + 300 \times P_2 = 400P_1 + 780P_2$$

Hence $P_1 = 400, P_2 = 780, P_3 = 0$

We have already said that C_j values represent the priority factors assigned to deviational variables and Z_j values represent the sum of the product of entries in C_B column with column coefficient matrix. Thus

$$Z_1 - C_1 = (P_1 \times 1 + 2P_2 \times 1 + P_2 \times 0) - 0 = P_1 + 2P_2 \quad i.e. \; P_1 = 1, P_2 = 2, P_3 = 0$$

$$Z_2 - C_2 = (P_1 \times 1 + 2P_2 \times 0 + P_2 \times 1) - 0 = P_1 + P_2 \quad i.e. \; P_1 = 1, P_2 = 1, P_3 = 0$$

and so on.

Now, to select the key column, we must select the largest positive element in the highest priority level P_1. The largest positive element 1 in P_1 row occurs at two places. To remove tie, check the next lower priority and select the key column that has a greater value at the lower priority level. So, we select the variable x_1. Next in choosing the key row, we find the minimum ratio and select that row. We select row 2 as the key row. The intersection between the key row and key column is called the key element and is squared.

Which is the standard simplex method, the solution is revised to obtain the second improved solution in the next table.

We notice that $\bar{d}_2$ has been replaced by x_1. Proceeding in the same manner.

Basic Variable	C_B	x_1	x_2	$\bar{d}_1$	$\bar{d}_2$	$\bar{d}_3$	$\overset{+}{d}_1$	x_B	**Min ratio**
$\bar{d}_1$	P_1	0	(1)	1	-1	0	-1	160	$\frac{160}{1} = 160 \leftarrow$
x_1	0	1	0	0	1	0	0	240	-
$\bar{d}_3$	P_2	0	1	0	0	1	0	300	$\frac{300}{1} = 300$
C_j		0	0	P_1	$2P_2$	P_2	P_3		
Z_j		0	$P_1 + P_2$	P_1	$-P_1$	P_2	$-P_1$		
	P_3	0	0	0	0	0	-1	0	
$\Delta_j = Z_j - C_j$	P_2	0	1	0	-2	0	0	300	
	P_1	0	$1\uparrow$	0	-1	0	-1	160	

$$Z = P_1 \times 160 + 0 \times 240 + P_2 \times 300 = 160P_1 + 300P_2$$

i.e. $\quad P_1 = 160, P_2 = 300, P_3 = 0$

Here, we notice that $\bar{d}_1$ has been replaced by x_2. We improve the solution are shown in the next table.

Basic Variable	C_B	x_1	x_2	$\bar{d}_1$	$\bar{d}_2$	$\bar{d}_3$	$\overset{+}{d}_1$	x_B	Min ratio
x_2	0	0	1	1	-1	0	-1	160	$\frac{160}{-1} = -160$
x_1	0	1	0	0	1	0	0	240	-
$\bar{d}_3$	P_2	0	0	-1	1	1	(1)	140	$\frac{140}{1} = 140 \leftarrow$
C_j		0	0	P_1	$2P_2$	P_2	P_3		
Z_j		0	0	$-P_2$	P_2	P_2	P_2		
$\Delta_j = Z_j - C_j$	P_3	0	0	0	0	0	-1	0	
	P_2	0	0	-1	-1	0	1	140	
	P_1	0	0	-1	0	0	$0\uparrow$	0	

$$Z = 0 \times 160 + 0 \times 240 + P_2 \times 140 = 140P_2$$

i.e. $\quad P_1 = 0, P_2 = 140, P_3 = 0$

Here, $\bar{d}_3$ has been replaced by $\overset{+}{d}_1$. The revised solution is shown in the next table.

Basic Variable	C_B	x_1	x_2	$\bar{d}_1$	$\bar{d}_2$	$\bar{d}_3$	$\overset{+}{d}_1$	x_B
x_2	0	0	1	0	0	1	0	300
x_1	0	1	0	0	1	0	0	240
$\overset{+}{d}_1$	P_3	0	0	-1	1	1	1	140
C_j		0	0	P_1	$2P_2$	P_2	P_3	
Z_j		0	0	$-P_3$	P_3	P_3	P_3	
$\Delta_j = Z_j - C_j$	P_3	0	0	-1	1	1	0	140
	P_2	0	0	0	-2	-1	0	0
	P_1	0	0	-1	0	0	0	0

$$Z = 0\times300 + 0\times240 + P_3\times140 = 140P_3$$

We observe that all the elements are either zero or negative in the higher priority goal P_2 *i.e.* P_2 is also fully achieved. In the attempt to satisfy the third goal P_3 , we see that there are two positive elements 1 and 1. However, these positive elements will not be considered, because there is already a negative element at a higher priority level. Hence the solution can not be improved further.

The optimal solution is

$$x_1 = 240, x_2 = 300, \overset{-}{d_1} = 0, \overset{-}{d_2} = 0, \overset{-}{d_3} = 0, \overset{+}{d_1} = 140$$

Practice Problems

1. A company produces two products. Each product requires time in two production departments : product X requires 20 hours in department 1 and 10 hours in department 2. Product Y requires 10 hours in department 1 and 10 hours in department 2. Production time is limited in department 1 to 60 hours and in department 2 to 40 hours. Contribution to profits for the two products is Rs. 40 and Rs. 80 respectively. Management's goal is to maximize profits.
2. A company is considering the allocation of Rs. 1, 50,000 advertising budget to two magazines P and Q. Rated exposures per hundred rupees of advertising expenditures are 1000 and 750 respectively for the two magazines, and it has been forecast that on the average Rs. 10 in sales results from each advertisement exposure. Management has decided that no more than 75% of the advertising budget can be expended in magazine P. The company has indicated that it would like to achieve exactly 1.5 million exposures form its advertising program. Management's goal is to allocate its money to advertising in such a way that sales are maximized.
3. A company had a single goal of achieving 1.5 million exposures but because of budget constraints was unable to achieve the goal. If we modify the problem so that the budget constraint is set as a goal, we can determine the budget requirement necessary to meet the exposure goal. Specifically the goals are:
   P_1: To achieve an exposure goal of exactly 1.5 million exposures.
   P_2: To minimize the budget requirements.
4. An office equipment manufacturer produces two types of products: Chair and Lamp. Production of either a chair or lamp requires one hour of production capacity in the plant. The plant has a maximum production capacity of 50 hours per week. Because of the limited sales capacity, the maximum number of chairs and lamps that can be sold are 6 and 8 per week, respectively. The gross margin from the sale of a chair is Rs. 90 and Rs. 60 for a lamp.
   The plant manager desires to determine the number of units of each product that should be produced per week in consideration of the following set of goals:

P_1: Available production capacity should be utilized as much as possible but not exceeded 50 hours per week.

P_2: Sales of two products should be as much as possible.

P_3: Overtime should not exceed 20% of available production time.

Formulate and solve this problem as a GP model so that the plant manager may achieve his goals as closely as possible.

5. A TV company manufactures two types of 35mm TVs. The production process for manufacturing the TVs is such that two departmental operations are required. To produce their standard TV requires 2 hours of production time in department 1 and 3 hours in department 2. To produce their deluxe model requires 4 hours of production time in department 1 and 3 hours in department 2. Currently, 80 hours of labours are available each week in each of the departments. This labour time is a somewhat restrictive factor since the company has a general policy of avoiding overtime, if possible. The manufacturer's profit on each standard TV is Rs. 30, while the profit on the deluxe model is Rs. 40. Management has set the following goals:

   P_1: Avoid overtime operations in each department.

   P_2: Prior sales-records indicated that, on the average, a minimum of 10 standard and 10 deluxe TVs can be sold weekly. Management would like to meet these sales goals. Since production time may limit producing this number of each TV, and since the deluxe TV has higher profit margin, the sales goals should be weighted by the profit contribution for the respective TVs, *i.e.* Rs. 30 for the standard TV, Rs. 40 for the deluxe TV. (We could also use weights of 3 and 4 since they have the same ratios as the profit contributions).

   P_3: Maximize Profits.

   Formulate as a GP model and solve it by using the graphical method.

6. A company produces Car seats. The company has two production lines. The production rate for line 1 is 50 seats per hour and for line 2 it is 60 seats per hour. The company has entered into a contract to supply 1200 seats daily to another company. Currently, the normal operation period for each line is 8 hours. The production manager of the company is trying to determine the best daily operation hours for the two lines. He sets the priorities to achieve his goals as given below:

   P_1: Produce and deliver 1200 seats daily.

   P_2: Limit the daily overtime operation hours of line 2 to 3 hours.

   P_3: Minimize underutilization of the regular daily operation hours of each line. Assign differential weights based on the relative productivity rate.

   P_4: Minimize the daily overtime operation hours of each line as much as possible. Assign differential weight based on the relative cost of overtime. It is assumed that the cost of operation is identical for the two production lines.

   Formulate this problem as a GP model and solve it by using the graphical method.

7. A computer company produces three different types of computers: Epic, Galaxie and Utopia. The production of all computers is conducted in a complex and modern assembly line. The production of an Epic requires 5 hours in the assembly line, a

Galaxie requires 8 hours and a Utopia requires12 hours. The normal operation hours of assembly line are 170 per month. The marketing and accounting departments have estimated that profits per unit for the three types of computers are Rs. 1, 00, 000 to the Epic, Rs. 1, 44, 000 for the Galaxie, and Rs. 2, 52, 000 for the Utopia. The marketing department further reports that the demand is such that the firm can expect to sell all the computers it produces in the month. The manager of the company has established the following goals, listed according to their importance:

P_1: Avoid underutilization of capacity in terms of regular hours of operation of the assembly line.

P_2: Meet the demand of the north-eastern sales district for five Epics, five Galaxie, and 8 Utopias (differential weights should be assigned according to the net profit ratios among the three types of computer.).

P_3: Limit overtime operation of the assembly line to 20 hours.

P_4: Meet the sales goal for each type of computers: Epic 10, Galaxie 12, and Utopia 10 (again assign weight according to the relative profit function for each computer).

P_5: Minimize that total overtime operation of the assembly line.

Formulate a GP model and solve the problem through two iterations by Simplex method of goal programming.

8. A company plans to schedule its annual advertising campaign. The total advertising budget is set at Rs. 10, 00, 000. The firm can purchase local radio spots at Rs. 2,000 per spot, local TV spots at Rs. 12, 000 per spot or magazine advertising at Rs. 4, 000 per insertion. The payoff from each advertising medium is a function of its audience size and audience characteristics. Let this payoff be defined as audience points. Audience points for the three advertising vehicles are:

Radio	**TV**	**Magazine**
50 points per spot	250 points per spot	200 points per insertion

The advertising manager of the company has established the following goals for the advertising campaign, listed in the order of importance:

(i) The total budget should not exceed Rs. 10, 00, 000.

(ii) The contract with the Radio and TV station requires that the firm spend at least Rs. 3,00,000 for TV and Radio ads.

(iii) The company does not wish to spend more than Rs. 2, 00, 000 for magazine ads.

(iv) Audience points for the advertising campaign should be maximized.

Formulate a GP model and solve this problem as a goal programming problem by the Simplex method.

Chapter - 21

MATLAB

Introduction

What is MATLAB?

MATLAB is the most dominant commercial computing environment and is used extensively in industry for computing, analysis, and visualization. A numerical analyst called Cleve Moler wrote the first version of Matlab in the 1970s. It has since evolved into a successful commercial software package.MATLAB stands for MATrix LABoratory and the software is built up around doing operations on vectors and matrices. MATLAB is not only useful for linear algebra, but it is also a great tool for analyzing data, optimization, signal and image processing, numerical integration, and numerically solving differential equations. MATLAB has very powerful graphics tools and can produce nice pictures in both two and three dimensions. Finally, MATLAB is also a programming language, and is one of the easiest programming languages for rapidly implementing algorithms for scientifically oriented applications.

Starting MATLAB

- Windows Systems

  On Windows systems MATLAB is started by double clicking the MATLAB icon on the desktop or by selecting MATLAB from the start menu. The starting procedure takes the user to the Command window where the Command line is indicated with '>>'. Used in the calculator mode all Matlab commands are entered to the command line from the keyboard. Matlab can be used in a number of different ways or modes; as an advanced calculator in the calculator mode, in a high level programming language mode and as a subroutine called from a C-program. Help and information on Matlab commands can be found in several ways, from the command line by using the 'help topic' command

- Unix Systems
  - o You should have a directory reserved for saving files associated with Matlab. Create such a directory (mkdir) if you do not have one. Change into this directory (cd).
  - o Start up a new xterm window (do xterm & in the existing xterm window).
  - o Launch Matlab in one of the xterm windows with the command **matlab** .

After a short pause, the logo will be shown followed by a window containing the Matlab interface. Should you wish to run Matlab in an xterm window, use the command **matlab –nojvm** and, following dislpay of the logo, the Matlab prompt >> will appear. Type quit at any time to exit from Matlab.

You can quit MATLAB by typing exit or quit in the command window.

The MATLAB environment

Note: From now on an instruction to press a certain key will be denoted by < >, e.g., pressing the enter key will be denoted as <enter>. Commands that should be typed at the MATLAB prompt, will be written in courier font.

The MATLAB environment (on most computer systems) consists of menus, buttons and a writing area similar to an ordinary word processor. There are plenty of help functions that you are encouraged to use. The writing area that you will see when you start MATLAB, is called the *command window*. In this window you give the commands to MATLAB. For example, when you want to run a program you have written for MATLAB you start the program in the command window by typing its name at the prompt and hitting enter. The command window is also useful if you just want to use MATLAB as a scientific calculator or as a graphing tool. If you write longer programs, you will find it more convenient to write the program code in a separate window, and then run it in the command window. MATLAB provides a very nice editor for writing code, however, any standard text editor can be used.

In the command window you will see a prompt that looks like >> . You type your commands immediately after this prompt. Once you have typed the command you wish MATLAB to perform, press <enter>. If you want to *interrupt a command* that MATLAB is running, type <ctrl> + <c>.

The commands you type in the command window are stored by MATLAB and can be viewed in the *Command History* window. To repeat any command you have already input, you can simply use the <up arrow> and <down arrow> at the command prompt to process through your commands until you reach the one you desire. You can also double click on the command you wish to repeat in the *Command History* window or drag-and-drop that command from that window to the command window.

Useful functions and operations in MATLAB

The following table lists some useful operations, functions and constants in MATLAB.

Operation, function or constant	MATLAB command
+ (addition)	+
- (subtraction)	-
* (multiplication)	*
/ (division)	/
^ (power)	^
|x| (absolute value of x)	abs(x)
square root of x	sqrt(x)
e^x	exp(x)
ln x (natural log)	log(x)
$\log_{10}$ x (base 10 log)	log10(x)
sin x	sin(x)
cos x	cos(x)
tan x	tan(x)
cot x	cot(x)
arcsin x	asin(x)
arccos x	acos(x)
arctan x	atan(x)
arccot x	acot(x)
n! (n factorial)	gamma(n+1)
e (2.71828...)	exp(1)
π(3.14159265...)	Pi
i (imaginary unit, sqrt(-1))	i or j

MATLAB is *case sensitive*. This means that MATLAB knows a difference between letters written as lower and upper case letters. For example, MATLAB will understand sin(2) but will not understand Sin(2).

Obtaining Help on MATLAB commands

To obtain help on any of the MATLAB commands, you simply need to type

```
help <command>
```

at the command prompt. For example, to obtain help on the gamma function, we type at the command prompt:

```
help gamma
```

Try this now. You may also get help about commands using the "Help Desk", which can be accessed by selecting the MATLAB Help option under the Help menu. This is particularly helpful if you want to search for commands for which you do not know the name of.

Note that the description MATLAB returns about the command you requested help on contains the command name in ALL CAPS. This does not mean that you use this command by typing it in ALL CAPS. In MATLAB, you almost always use all lower case letters when using a command.

Using MATLAB as a Calculator

The easiest way to use MATLAB is as a calculator:

Example: Compute 5 $\sin(2.5^{3-pi})$+1/75. In MATLAB this is done by simply typing

```
5*sin(2.5^(3-pi))+1/75
```

at the prompt. Be careful with parentheses and don't forget to type * whenever you multiply!

This example illustrates the order of precedence followed by MATLAB, which is similar to most calculators:

Precedence	Mathematical operation
First	Parentheses; the inner most parenthesis is executed first
Second	Exponentiation or raising quantities to a power
Third	Multiplication and division
Fourth	Addition and subtraction

For expressions containing multiple operations, the highest-precedence ones are executed first followed by the next highest, etc. Operations with the same precedence (e.g. addition/multiplication or multiplication/division) are executed from left to right.

Exercises

Compute the following expressions using MATLAB:

- $3\cos(\pi)$
- 1+1+1/2+1/6+1/24-e
- $\ln(1000+2^{pi-2})$
- 3^4/2
- 3/2^4
- (3/2)^4
- $e^{i\pi}$

- The number of combinations in which 12 persons can stand in line. (Hint: Use factorials. You may want to use the help menu for this!)

Formatting Output

MATLAB allows the user to control how numerical results are formatted and displayed to the screen with the format command. Type help format and enter at the command prompt to list all the possible options for formatting output. The two main options that you will probably use the most are format short and format long.

At the MATLAB prompt type:

```
format long

(1+sqrt(5))/2-1

format short

(1+sqrt(5))/2-1
```

Observe how many decimal digits are displayed after each of the numerical calculations. Note that MATLAB did both of the above calculations using the same number of digits, the only difference is in how the results are displayed.

Another useful option for the format command is compact. This command surpresses empty lines and allows more information to be displayed on the screen at one time. At the MATLAB prompt type:

```
format compact
```

and repeat the set of four MATLAB commands listed immediately above. To return the display to printing the extra line feeds type the command format loose

Variables in MATLAB

We can easily define our own variables in MATLAB. Let's say we need to use the value of 3.5sin(2.9) repeatedly. Instead of typing 3.5*sin(2.9) over and over again, we can denote this variable as x by typing the following:

```
x=3.5*sin(2.9)
```

(Please try this in MATLAB.) Now type

```
x+1
```

and observe what happens. Note that we did not need to declare x as a variable that is supposed to hold a floating point number as we would need to do in most programming languages.

In many cases we want to know what variables we have declared. We can do this by typing whos. Alternatively, we can view the values by openning the "Workspace" window. This is done by selecting the Workspace option from the View menu. If you want to erase all variables from the MATLAB memory, type clear. To erase a specific variable, say x, type clear x. To clear two specific variables, say xand y, type clear x y, that is separate the different variables with a space. Variables can also be cleared by selecting them in the Workspace window and selecting the delete option.

Suppressing output

Often, we may not want to have the result of a calculation printed-out to the command window. To supress the output of a command, you can put a semi-colon at the end of the command. Again letting

```
x=3.5*sin(2.9)
```

but instead typing the command

```
y=2*x;
```

You should see that the result has not been displayed on the screen. Note that MATLAB still performs the command in "the background", it just doesn't display the result. If you want to see the value stored in y type

```
y
```

and observe what happens.

When the percent symbol % is typed, all text following that symbol (but before a line feed) is suppressed *and not executed.* The % symbol is useful for commenting your programs and examples of using it will be given later.

Vectors and matrices (or arrays) in MATLAB

Vectors are created in MATLAB by putting the elements within [] brackets.

Example: `x=[1 2 3 4 5 6 7 8 9 10]`

We can also create this vector by typing `x=1:10`. The vector `(1 1.1 1.2 1.3 1.4 1.5)` can be created by typing `x=[ 1 1.1 1.2 1.3 1.4 1.5 ]` or by typing `x=1:0.1:1.5`.

Matrices can be created according to the following example. The matrix $A=\begin{pmatrix}1 & 2 & 3\\4 & 5 & 6\\7 & 8 & 9\end{pmatrix}$ is created by typing

```
A=[1 2 3 ; 4 5 6; 7 8 9],
```

i.e., rows are separated with semi-colons. If you want to use a specific element in a vector or a matrix, study the following example:

Example:

```
x=[10 20 30]

A=[ 1 2 3 ; 4 5 6 ; 7 8 9]

x(2)

A(3,1)
```

Here the second element of the vector is extracted by typing the variable and the position within parantheses. The same principle holds for matrices; the first number specifies the row of the matrix, and the second number specifies the column of the matrix. **Note that in MATLAB the first index of a vector or matrix starts at 1, not 0 as is common with other programming languages.**

If the matrices or vectors (which are special cases of a matrices) are of the same dimensions then matrix addition, matrix subtraction and scalar multiplication works just like they are supposed to.

Example: Type

```
x=[1 2 3]

y =[4 5 6]

a=2

x+y

x-y

a*x
```

and observe what happens.

If you want to apply an operation such as squaring each element in a matrix you have to use a dot . before the operation you wish to apply. Type the following commands in MATLAB.

```
x=1:10

x.^2

A=[1 2 3 ; 4 5 6 ; 7 8 9 ]

A.^2

A^2
```

and observe the result. **The dot allows elementwise operations to do performed all at once..** All built-in functions such as `sin`, `cos`, `exp` and so on automatically act elementwise on a matrix. Type

```
y=[0 1/4 1/2 3/4 1]

y=pi*y
sin(y)
```

and observe the result.

How to plot with MATLAB

There are different ways of plotting in MATLAB. The following two techniques, illustrated by examples, are probably the most useful ones.

Example 1: Plot sin(x^2) on the interval [-5,5]. To do this, type the following:

```
x=-5:0.01:5;

y=sin(x.^2);

plot(x,y)
```

and observe what happens.

Example 2: Plot exp(sin(x)) on the interval [-&pi,&pi]. To do this, type the following:

```
x=linspace(-pi,pi,101);
```

```
y=exp(sin(x));

plot(x,y)
```

and observe what happens. The command linspace creates a vector of 101 equally spaced values between -&pi and &pi (inclusive).

Ocassionally, you may need to plot values that vary quite differently in magnitude. In this case, the regular plot command fails to give us an adequate graphical picture of our data. Instead, you need a command that plots values on a log scale. MATLAB has 3 such commands: loglog,semilogx, and semilogy. Use the help command to see a description of each function. As an example of where we may want to use one of these plotting routines, consider the following problem:

Example 3: **Plot $x^{5/2}$ for $x = 10^{-5}$ to 10^5. To do this, type the following:**

```
x=logspace(-5,5,101);

y=x.^(5/2);

plot(x,y)
```

and observe what happens. Now type the following command:

```
loglog(x,y)
```

The command logspace is similar to linspace, however it creates a vector of 101 points lograthmically equally distributed between 10^{-5} and 10^5.

The following commands are useful when plotting:

Graphing functions	MATLAB command
Label the horizontal axis.	xlabel('text')
Label the vertical axis.	ylabel('text')
Attach a title to the plot.	title('text')
Change the limits on the x and y axis.	axis([xmin xmax ymin ymax])
"Keep plotting in the same window."	hold on
Turn off the "keep-plotting-in-the-same-window-command".	hold off

Note that all text must be put within ' '. The last two commands (hold on and hold off) are best explained by trying them next time you plot.

Exercises

- Plot sin(x) on the interval [-π,π] using spacing 0.5, 0.1 and 0.01 between the points where you will sample the function. (This will change the resolution). Experiment with the hold on command.
- Attach labels to the axis of the previous plot and give a title to the graph.
- Plot 5 $\cos(x^2+1)$ on [-2π,2π]. Note that the squaring operation will require you to use the dot . in order for the squaring operation to act on each element individually. However, the addition operation (+) automatically acts on elements individually.

All the commands were executed in the Command Window. The problem is that the commands entered in the Command Window cannot be save and executed again for several times. Therefore, a different way of executing repeatedly commands with MATLAB is:

1. to *create* a file with a list of commands,
2. *save* the file, and
3. *run* the file.

If needed, corrections or changes can be made to the commands in the file. The files that are used for this purpose are called script files or *scripts* for short. This section covers the following topics:

1. M-File Scripts
2. M-File Functions

MATLAB Programming I: Script

The purpose of this to give a basic introduction to programming in MATLAB using "scripts". It requires no programming experience, but some familiarity with MATLAB is recommended. All background needed can be found on the Introduction to MATLAB. Upon completing the this, you should know how to

- Create "scripts" in MATLAB.
- Use the MATLAB editor to write, save, and run your programs.
- Use looping mechanisms like for and while.
- Use logical mechanisms like if
- Use the logical operators *and, or, not.*
- Use the relational operators *equal, not equal, less than, greater than.*

Creating and using functions in MATLAB is covered in the next section MATLAB Programming II: Functions.

A number of examples will be given below. No output is given since it is expected that you implement the examples yourself and then run them in the MATLAB command window to get the output.

Disclaimer: Many of the examples and exercises given below may seem to be irrelevant and even a little stupid. However, by understanding them you will have all the background to write programs to solve all the homework problems.

Background on MATLAB scripts

- A MATLAB script is just a sequence of MATLAB commands, which most likely represent a program.
- Running a script is just like inputing each line of the script on the MATLAB command window and pressing enter.
- Variables created in scripts are accessible from the command window and any other script until they are cleared by the user or the MATLAB session is terminated.
- Output from script files that are generated from executing a command without a trailing semicolon are displayed on the MATLAB command window.
- To run a script it must be saved in the current working directory or it must be saved in a directory on the MATLAB path.

How to edit and run a MATLAB script

To write scripts (or programs) in MATLAB you should use a text editor. MATLAB has an integrated text editor which is very good and is highly recommended since it provides syntax highlighting and programs can be run and debugged directly from it. To open the editor, go to the "File" menu and choose either the "New...Blank M-file", if you want to create a new program, or "Open", if you want to open a existing M-file. In the editor you can now type in your MATLAB commands similar to how you input them on the command line. However, the main difference is when you press enter after typing a command, that command is not executed. Menus and short-cut keys for editing text (e.g. cut/copy/paste) are similar many other text editors and be customized through the "File/Preferences" menu. In order to run a script do the following:

1. If the script is not saved,then save it as *<filename>.m*, where *<filename>* is anything you wish to name the file. **It is important to add ".m" at the end of the filename.** Otherwise MATLAB may not understand it.
2. Use one of the following techniques to run the saved script:
   (a) If the script is currently displayed as the top window in the editor, hit the <F5> key.
   (b) If the script is currently displayed as the top window in the editor, press the "Play" button on the toolbar with the mouse.
   (c) Go to the command window and type the name of the file containing your script at the prompt and press enter. **When typing the filename in the command window do not include ".m"**. Note that for the script to run, it must be in the current working directory or on the MATLAB path.

In case your code has errors, MATLAB will complain when you try to run the program in the command window. When this happens, MATLAB will indicate what error occurred. You should try to interpret the error message and make necessary changes to the script in the editor. The error that is reported by MATLAB is hyperlinked to the line in the file that caused the problem. Using the mouse you can thus jump right to the line in your program that has caused the error. **After you have made the changes, make sure you save your file before trying to run the program again in the command window.**

General programming advice

The best programming advice, which applies to even the most experienced programmers, is to **never try to write all the code at once**. Programming should be done in small steps with each step tested to ensure it works before proceeding to program the next step. Implementing a longer program should always start with an outline using pen and paper before one line of code is ever typed into the editor.

Basic programming structures

Variables

There is no need to declare variable types anywhere in your script (or functions). However, it may be necessary to initialize some variables before they are used, and it is good MATLAB programming practice to initialize vectors and matrices to their proper sizes at the beginning programs. Furthermore, to avoid simple bugs that can be very easy to find, you should use the clear at the beginning of your script (but not function) to erase any values that may have previously been stored in these variables.

Output

To make MATLAB give an output, such as return the value of a variable, you type the name of the variable without any semi-colon (;) following the variable. In many cases, there is no need to see the value of a variable every single time MATLAB uses it. If MATLAB re-computes the value of a variable 1000 times, we probably don't want to see the result every single time. To surpress the output of a value, just add a semi-colon after the variable. Then MATLAB will perform the command, but will not show it on the screen.

As a first example, consider the following program that takes a temperature value in Kelvin and outpus that value in Farenheit and Celsius.

Example 1:

```
clear
temp_kelvin = 233.15;
temp_celsius = temp_kelvin-273.15
temp_farenheit = 9/5*temp_celsius + 32
```

Not that values of temp_celsius and temp_farenheit are displayed in the command window since no semicolon is placed after these lines.

Logical and relational operators

Logical and relational operators are fundamental for making decisions and controlling the flow of a program. These are listed below together with their MATLAB syntax.

Logical operators

Operation:	**MATLAB syntax:**
Logical and	&
Logical or	|
Negate	~

Relational operators

Operation:	**MATLAB syntax:**
Strictly less than	<
Less than or equal to	<=
Strictly greater than	>
Greater than or equal to	>=
Equal to	==
Not equal to	~=

Each of these commands results in one of two values, a logical 1 or a logical 0. 1 means the true, whilde zero means false. Additionally, the commands apply to scalar or matrices. When matrices are used with these operators the comparison is done element-by-element.

It is important to know the difference between = and ==. The former is used when assigning a number to a variable, e.g., x=3;. The latter is used to check if two expressions are equal. The examples below illustrate this difference.

Example 2: At the command prompt type the following:

```
a = 5
b = 6
a == b
a ~= b
a < b
a > b
c = 4
(a < b) & (c < b)
(a < b) & (c > b)
```

```
(a < b) | (c > b)
c = 1:10
a <= c
a >= c
a = [0 1 4 5 8 3 4 9 10 2];
a <= c
```

Conditional statements

Often in programs we want the computer to check whether a statement is true or false and perform different operations depending on the result of this test. This can be done using a so-called if-statement. The syntax is given below.

```
if logical expression
commands
end
or
if logical expression
commands
else
commands
end
or
if logical expression
commands
elseif logical expression
commands
else
commands
end
```

Note that for each if, you need to "close" the if-statement with an end.

The use of this conditional statement is best illustrated by studying the following three examples and doing Exercise 1.

Example 3: Computing the absolute value.

```
clear
x = -1;
if x < 0
x = -1*x;
end
x
```

In the next example, we make MATLAB write something depending on which test our month "passes". To make MATLAB write text as output, use single quote ' around the text.

Example 4:

```
clear
month = 3;
if month ==1 || month==3 || month ==5 || month==7 ||
month==10 || month==12 num_days = 31;
elseif month == 2
num_days = 28;
else
num_days = 30;
end
['Month ' num2str(month) ' has ' num2str(num_days) '
days']
```

The last line of this example shows how to concatenate strings. For more advanced ways to do this, see the sprintf command.

In the next example we use the command rem (remainder), which returns the remainder after the division of integer x by y. The syntax is rem(x,y).

Example 5:

```
clear
number = 38;
remainder2 = rem(number,2);
remainder3 = rem(number,3);
if remainder2==0 && remainder3==0  result = 'Your number is
divisible by both 2 and 3';
elseif remainder2==0
      result = 'Your number is divisible by 2 but not by 3';
elseif remainder3==0
          result = 'Your number is divisible by 3 but not by
2';
else
         result = 'Your number is not divisible by 2 or 3';
end
result
```

Exercise

1. Write a program similar to Example 4, which in addition to displaying how many days are in a given month, also determines the name of the month given the month number. For example, if month = 2, the text that should be output is "February has 28 days".

Repetitive operations (loops)

The power of computers is that they can repeat seemling mundane operations over and over (without getting bored!). An operation that is performed repeatedly is called a repetitive operation or, more common, a loop. There are different kinds of loops but the most common one is the for-loop. The syntax for a for-loop is:

```
for loop_variable = start_value : end_value
commands
end
```

This loop will initate *loop_variable* as *start_value*, increment *loop_variable* by 1 each step until *end_value* is reached. Below follows a few examples of how to use for-loops.

This example fills a vector with the first 20 Fibonacci then plots the results.

Example 6:

```
clear
x = zeros(20,1);
x(1) = 1;
x(2) = 1;
for i=3:20
  x(i)= x(i-1) + x(i-2);
end
semilogy(1:20,x,'xb-');
xlabel('n'), ylabel('nth Fibinocci number');
```

In the following example we see a so-called nested for-loop. This is nothing else then a "loop within a loop". Note how we must "close" each for with an end. Make sure you understand how this example works!

Example 7:

```
clear
for i=1:5
       for j=1:5
          A(i,j)=10*i+j;
        end
end
A
```

In the following example, make sure you understand the purpose of the variable mysum. This is a common way of performing summations when programming.

Example 8:

```
clear
N = 10;
mysum = 0;
for k=0:N
        mysum = mysum+1/factorial(k);
end
e_approximation = mysum
e_exact = exp(1)
```

In the next example, notice how we have to "shift" the indexing of the vector. **MATLAB must have positive integers for indexing a matrix or vector!** One of the most common mistakes when programming in MATLAB is that your program begins indexing at zero instead of one. Also note how by typing a percent sign (%) before text in the code, MATLAB does not interpret this text as code. It just serves as a comment for any person using the code. **Commenting your code is essential when writing programs.**

Example 9:

```
clear
N = 10;
for k=0:70
x(k+1)=0.1*k;    %    Indices    of    vectors    must    be    NON-ZERO!
sum = 0;
for m=0:10
sum  =  sum+(x(k+1)^m)/factorial(m);  %  Approx  exp(x)  using  its
Taylor series
end

e(k+1)  =  sum;  %  e(k+1)  contains  the  Taylor  series  approx.  for
x=x(k+1);
end
semilogy(x,e)
title('Approximation of e^x for x between 0 and 7')
xlabel('x')
ylabel('e^x')
```

Exercise

2. Write a program that approximates PI by computing the sum $\pi \approx 4\sum_{k=0}^{In}\frac{(-1)^y}{2k+1}$.

   The more terms you keep in the summation, the more accurate your answer will be. (In fact, the series converges to PI as m goes to infinity.) See how many terms you need to approximate PI with 5 decimals. (Note: This is by no means the most efficient way to approximate PI, but the formula is quite beautiful...)

MATLAB Programming II: Functions

Anatomy of a MATLAB function

MATLAB functions are similar to C functions or Fortran subroutines.MATLAB programs are stored as plain text in files having names that end with the extension ``.m''. These files are called, not surprisingly, m-files. Each m-file contains exactly one MATLAB function. Thus, a collection of MATLAB functions can lead to a large number of relatively small files.

One nifty difference between MATLAB and traditional high level languages is that MATLAB functions can be used interactively. In addition to providing the obvious support for interactive calculation, it also is a very convenient way to debug functions that are part of a bigger project.

MATLAB functions have two parameter lists, one for input and one for output. This supports one of the cardinal rules of MATLAB programming: *don't change the input parameters of a function*. Like all cardinal rules, this one is broken at times. My free advice, however, is to stick to the rule. This will require you to make some slight adjustments in the way you program. In the end this shift will help you write better MATLAB code.

You have already encountered functions that are built in to MATLAB. The sin() is a function that takes an argument and returns an output. As you start to write your own programs in MATLAB you will need to create your own functions that take one or more input arguments, operate on them, and return the result. For example, when you write your own non-linear root finder (e.g. Newton's method), you will most certainly need a loop that calls the non-linear function and it's derivative several times. It is certainly possible to write this function explicitly in your loop. However, that is not good programming practice since you would have to modify your general solver for every new specific function you want to input.

To introduce programming with functions in MATLAB, we will use the following example: Create a general program for graphing a given function over a given range.

Preliminaries

Before we begin, we must first look at the syntax for a function in MATLAB.

```
function y=<function_name>(argumentlist)
        commands
```

The code above must be written in a separate m-file! The name of the file should coincide with the name of the function, i.e. *<function_name>.m* (remember you must save the file with a .m after the file name.We illustrate the syntax for the body of a MATLAB function by implementing the mathematical function

$f(x)= x^2 + (1 - \exp(x))^2-4$

in a MATLAB function.

Example 1.

```
% Filename: f.m

function y = f(x)

y = x.^2 + (1 - exp(x)).^2 - 4; % Note the use of the dot
in the formula
```

This code must be saved in your working directory as a separate m-file with the name f.m. Note how we have not used the clear command. This is because all variables inside functions are local to that function (unless specified otherwise by using the GLOBAL command).

Programming with Functions

Now, we illustrate how we can actually program with functions by solving the problem posed by the above example. We want a general MATLAB function that plots a given mathematical function over the domain [a,b]. This can be accomplished with the following MATLAB function.

```
% Filename: fplotter.m
function fplotter(g,a,
x = linspace(a,b,501); % x-values over the range [a,b] at
which to evaluate g
y = feval(g,x); % evaluate the function g at all the values
in x
plot(x,y); % plot the function g
xlabel('x'); % label the x-axis
ylabel('f(x)'); % label the y-axis
```

This function must be saved as a separate m-file with the name fplotter.m. It must also be saved in the same directory as f.m. Note that this is a very naive implementation of the function plotter. If we were going to make it more robust, we would want to put in code for checking the range of (a,b) and checking whether or not the function g exists. Also, it would be advantageous to base the number of points that are used to divide up the plot interval on the values of a and b.

To plot the function in f.m using **fplotter**, first go to the MATLAB command window then change to the working directory that corresponds to the directory containing the two functions. Next, type the command

```
fplotter('f',-1.5,1.5)
```

This will call the function **fplotter** with the function f and plot it over the range [-1.5,1.5]. Make sure you understand how this example works! It will help you to write a non-linear root finder.

Some simple examples of function

Here is a trivial function, **addtwo.m**

```
function addtwo(x,y)
% addtwo(x,y)  Adds two numbers, vectors, whatever, and
%  print the result = x + y
     x+y
```

The **addtwo** function has no output parameters so the square brackets and the equal sign are omitted. There is only one MATLAB statement, x+y, in the function. Since this line does not end with a semicolon the results of the calculation are printed to the command window.

The first two lines after the function definition are comment statements. Not only do these statements describe the statements in the file, their position in the file supports the on-line help facility in MATLAB. If the first line of a MATLAB function definition is immediately followed by non-blank comment statements, then those comment statements are printed to the command window when you type ``help **function_name**". Try it with the **addtwo** function. MATLAB will print up until a blank line or an executable statement, whichever comes first.

To test your understanding of input and output variables, pass the following definitions of x and y to the addtwo function. (To save space the x and y variables are defined on the same line. You can enter these variables on the same line, as shown, or use separate lines.)

```
>> x = 2;               y = 3;
>> x = [2 3];           y = [4; 5];
>> x = eye (3,3);       y = -2*eye(3,3);
```

```
>> x = 'Sue';          y = 'Bob'
>> x = 'Jane';         y = 'Bob'
```

Here is another simple function, **traparea.m**, with three input parameters and one output parameter. Since there is only one output parameter the square brackets may be omitted.

```
function area = traparea(a,b,h)
% traparea(a,b,h)  Computes the area of a trapezoid given
%                  the dimensions a, b and h, where a and b
%                  are the lengths of the parallel sides and
%                  h is the distance between these sides
% Compute the area, but suppress printing of the result
area = 0.5*(a+b)*h;
```

Note that there is a blank line between the comment statements that describe the function and the single comment statement that describes the calculation of area. The comment statement beginning ``Compute the area..." will not be printed if you type ``help traparea". Also note that there is no ``return" statement needed. (There is a MATLAB return statement, but it is not needed here.) The output variable, area, is defined in the first line of the file. The value assigned to area is returned to the calling function or the command window.

Finally, here is another simple function, **cart2plr.m,** with two input parameters and two output parameters.

```
function [r,theta] = cart2plr(x,y)
% cart2plr  Convert Cartesian coordinates to polar coordinates
%
% [r,theta] = cart2plr(x,y) computes r and theta with
%
%       r = sqrt(x^2 + y^2);
%       theta = atan2(y,x);
r = sqrt(x^2 + y^2);
theta = atan2(y,x);
```

The comment statements have empty lines, but these will be printed if you type ``help cart2plr".

Exercise

1. Make a plot of the function *exp(-x^2)*cos(x)* for *x* between -4 and 4. (Hint: Do you need to change the file fplotter.m at all?)

Note that it is possible to have functions return more than one value

Types of Functions

MATLAB offers several different types of functions to use in your programming.

Anonymous Functions

An *anonymous function* is a simple form of MATLAB function that does not require an M-file. It consists of a single MATLAB expression and any number of input and output arguments. You can define an anonymous function right at the MATLAB command line, or within an M-file function or script. This gives you a quick means of creating simple functions without having to create M-files each time.The syntax for creating an anonymous function from an expression is

f = @(arglist)expression

The statement below creates an anonymous function that finds the square of a number. When you call this function, MATLAB assigns the value you pass in to variable x, and then uses x in the equation x.^2: sqr = @(x) x.^2;

To execute the sqr function defined above, type a = sqr(5)

a =25

Primary and Sub functions

All functions that are not anonymous must be defined within an M-file. Each M-file has a required *primary function* that appears first in the file, and any number of *subfunctions* that follow the primary. Primary functions have aider scope than sub functions. That is, primary functions can be invoked from outside of their M-file (from the MATLAB command line or from functions in other M-files) while sub functions cannot. Sub functions are visible only to the primary function and other sub functions within their own M-file.

Private Functions

A *private function* is a type of primary M-file function. Its unique characteristic is that it is visible only to a limited group of other functions. This type of function can be useful if you want to limit access to a function, or when you choose not to expose the implementation of a function. Private functions reside in subdirectories with the special name private. They are visible only to functions in the parent directory. For example, assume the directory newmath is on the MATLAB search path. A subdirectory of newmath called private can contain functions that only the functions in newmath can call. Because private functions are invisible outside the parent directory, they can use the same names as functions in other directories. This is useful if you want to create your own version of a particular function while retaining the original in another directory. Because MATLAB looks for private functions before standard M-file functions, it will find a private function named test.m before a nonprivate M-file named test.m.

Nested Functions

You can define functions within the body of any MATLAB M-file function. These are said to be *nested* within the outer function. A nested function contains any or all of the components of any other M-file function. In this example, function B is nested in function A:

```
function x = A(p1, p2)
...
B(p2)
function y = B(p3)
...
end
...
End
```

Like other functions, a nested function has its own workspace where variables used by the function are stored. But it also has access to the workspaces of all functions in which it is nested. So, for example, a variable that has a value assigned to it by the primary function can be read or overwritten by a function nested at any level within the primary. Similarly, a variable that is assigned in a nested function can be read or overwritten by any of the functions containing that function.

Function Overloading

Overloaded functions act the same way as overloaded functions in most computer languages. Overloaded functions are useful when you need to create a function that responds to different types of inputs accordingly. For instance, you might want one of your functions to accept both double-precision and integer input, but to handle each type somewhat differently. You can make this difference invisible to the user by creating two separate functions having the same name, and designating one to handle double types and one to handle integers. When you call the function, MATLAB chooses which M-file to dispatch to based on the type of the input arguments.

Global Variables

If you want more than one function to share a single copy of a variable, simply declare the variable as global in all the functions. Do the same thing at the command line if you want the base workspace to access the variable. The global declaration must occur before the variable is actually used in a function. Although it is not required, using capital letters for the names of global variables helps distinguish them from other variables. For example, create an M-file called falling.m:

```
function h = falling(t)
global GRAVITY
h = 1/2*GRAVITY*t.^2;
```

Then interactively enter the statements global GRAVITY

```
GRAVITY = 32;
y = falling ((0:.1:5)');
```

The two global statements make the value assigned to GRAVITY at the command prompt available inside the function. You can then modify GRAVITY interactively and obtain new solutions without editing any files.

Passing String Arguments to Functions

You can write MATLAB functions that accept string arguments without the parentheses and quotes. That is, MATLAB interprets

```
foo a b c
```

as

```
foo('a','b','c')
```

However, when you use the unquoted form, MATLAB cannot return output arguments. For example, legend apples oranges creates a legend on a plot using the strings apples and oranges as labels. If you want the legend command to return its output arguments, then you must use the quoted form: [legh,objh] = legend('apples','oranges');In addition, you cannot use the unquoted form if any of the arguments is not a string.

Constructing String Arguments in Code

The quoted form enables you to construct string arguments within the code. The following example processes multiple data files, August1.dat, August2.dat, and so on. It uses the function int2str, which converts an integer to a character, to build the filename:

```
for d = 1:31
s = ['August' int2str(d) '.dat'];
load(s)
% Code to process the contents of the d-th file
End
```

A Cautionary Note

While the unquoted syntax is convenient, it can be used incorrectly without causing MATLAB to generate an error. For example, given a matrix A,

A =0 -6 -1 6 2 -16 -5 20 -10

The eig command returns the eigenvalues of A: eig(A)ans = -3.0710

-2.4645+17.6008i

-2.4645-17.6008i

The following statement is not allowed because A is not a string; however, MATLAB does not generate an error:

```
eig A
ans = 65
```

MATLAB actually takes the eigenvalue of the ASCII numeric equivalent of the letter A (which is the number 65).

The eval Function

The eval function works with text variables to implement a powerful text macro facility. The expression or statementeval(s) uses the MATLAB interpreter to evaluate the expression or execute the statement contained in the text string s.The example of the previous section could also be done with the following code,although this would be somewhat less efficient because it involves the fullinterpreter, not just a function call:

```
for d = 1:31
s = ['load August' int2str(d) '.dat'];
eval(s)
% Process the contents of the d-th file
end
```

Function Handles

You can create a handle to any MATLAB function and then use that handle as a means of referencing the function. A function handle is typically passed in an argument list to other functions, which can then execute, or *evaluate*, the function using the handle. Construct a function handle in MATLAB using the *at* sign, @, before the function name. The following example creates a function handle for the sin function and assigns it to the variable fhandle:

```
fhandle = @sin;
```

You can call a function by means of its handle in the same way that you would call the function using its name. The syntax isfhandle(arg1, arg2, ...);

The function plot_fhandle, shown below, receives a function handle and data, generates y-axis data using the function handle, and plots it:

```
function x = plot_fhandle(fhandle, data)
plot(data, fhandle(data))
```

When you call plot_fhandle with a handle to the sin function and the argument shown below, the resulting evaluation produces a sine wave plot: plot_fhandle(@sin, -pi:0.01:pi)

Function Functions

A class of functions called "function functions" works with nonlinear functions of a scalar variable. That is, one function works on another function. The function functions include

- Zero finding
- Optimization
- Quadrature
- Ordinary differential equations

MATLAB represents the nonlinear function by a function M-file. For example,here is a simplified version of the function humps from the matlab/demos directory:

```
function y = humps(x)
y = 1./((x-.3).^2 + .01) + 1./((x-.9).^2 + .04) - 6;
```

Evaluate this function at a set of points in the interval $0 \leq x \leq 1$ with

```
x = 0:.002:1;
y = humps(x);
```

Then plot the function with

```
plot(x,y)
```

The graph shows that the function has a local minimum near $x = 0.6$. The function fminsearch finds the *minimizer*, the value of x where the function takes on this minimum. The first argument to fminsearch is a function handle to the function being minimized and the second argument is a rough guess at the location of the minimum:

```
p = fminsearch(@humps,.5)
p =
0.6370
```

To evaluate the function at the minimizer,

```
humps(p)
ans =
11.2528
```

Numerical analysts use the terms *quadrature* and *integration* to distinguish between numerical approximation of definite integrals and numerical integration of ordinary differential equations. MATLAB quadrature routines are quad and quadl. The statement

```
Q = quadl(@humps,0,1)
```

computes the area under the curve in the graph and produces

```
Q =
29.8583
```

Finally, the graph shows that the function is never zero on this interval. So, if

you search for a zero with

```
z = fzero(@humps,.5)
```

you will find one outside the interval

```
z =
-0.1316
```

Scripts versus Functions

Scripts are m-files containing MATLAB statements. MATLAB ``functions" are another type of m-file. The biggest difference between scripts and functions is that functions have input and output parameters. Script files can only operate on the variables that are hard-coded into their m-file. As you can see, functions much more flexible. They are therefore more suitable for general purpose tasks that will be applied to different data. Scripts are useful for tasks that don't change. They are also a way to document a specific sequence of actions, say a function call with special parameter values, that may be hard to remember.

There are more subtle differences between scripts and functions. A script can be thought of as a keyboard macro: when you type the name of the script, all of the commands contained in it are executed just as if you had typed these commands into the command window. Thus, all variables created in the script are added to the workspace for the current session. Furthermore, if any of the variables in the script file have the same name as the ones in your current workspace, the values of those variables in the workspace are changed by the actions in the script. This can be used to your advantage. It can also cause unwanted side effects.

In contrast, function variables are local to the function. (The exception is that it's possible to declare and use global variables, but that requires and explicit action by the user.) The local scope of function variables gives you greater security and flexibility. The only way (besides explicitly declared global variables) to get information into and out of a function is through through the variables in the parameter lists.

Difference between scripts and functions

Scripts	Functions
Do not accept input arguments or return output arguments.	Can accept input arguments and return output arguments.
Store variables in a workspace that is shared with other scripts	Store variables in a workspace internal to the function.
Are useful for automating a series of commands	Are useful for extending the MATLAB language for your application

Different Application of MATLAB:

MATLAB have a wide range of applications, including signal and image processing, communications, control design, *test and measurement*, financial modeling and analysis, and *computational biology*. Add-on toolboxes (collections of special-purpose MATLAB functions) extend the MATLAB environment to solve particular classes of problems in these application areas.

1. Developing Algorithms and Applications
2. Analyzing and Accessing Data
3. Visualizing Data
4. Performing Numeric Computation
5. Publishing Results and Deploying Applications

1. **Developing Algorithms and Applications:** MATLAB provides a high-level language and development tools that let you quickly develop and analyze your algorithms and applications.

   The MATLAB language supports the vector and matrix operations that are fundamental to engineering and scientific problems. It enables fast development and execution.

   With the MATLAB language, you can program and develop algorithms faster than with traditional languages because you do not need to perform low-level administrative tasks, such as declaring variables, specifying data types, and allocating memory. In many cases, MATLAB eliminates the need for 'for' loops. As a result, one line of MATLAB code can often replace several lines of C or C++ code.

   At the same time, MATLAB provides all the features of a traditional programming language, including arithmetic operators, flow control, data structures, data types, *object-oriented programming* (OOP), and debugging features

2. **Designing Graphical User Interfaces:** You can use the interactive tool GUIDE (Graphical User Interface Development Environment) to lay out, design, and edit user interfaces. GUIDE lets you include list boxes, pull-down menus, push buttons, radio buttons, and sliders, as well as MATLAB plots and ActiveX controls.

Alternatively, you can create GUIs programmatically using MATLAB functions.

Analyzing and Accessing Data: MATLAB supports the entire data analysis process, from acquiring data from external devices and databases, through preprocessing, visualization, and numerical analysis, to producing presentation-quality output.

Data Analysis: MATLAB provides interactive tools and command-line functions for data analysis operations, including:

- Interpolating and decimating
- Extracting sections of data, scaling, and averaging
- Thresholding and smoothing
- Correlation, Fourier analysis, and filtering
- 1-D peak, valley, and zero finding
- Basic statistics and curve fitting
- Matrix analysis

Data Access: MATLAB is an efficient platform for accessing data from files, other applications, databases, and external devices. You can read data from popular file formats, such as Microsoft Excel; ASCII text or binary files; image, sound, and video files; and scientific files, such as HDF and HDF5. Low-level binary file I/O functions let you work with data files in any format. Additional functions let you read data from Web pages and XML. You can call other applications and languages, such as C, C++, COM objects, DLLs, Java, Fortran, and Microsoft Excel, and access FTP sites and Web services. Using *Database Toolbox™*, you can also access data from ODBC/JDBC-compliant databases.

3. **Visualizing Data:** All the graphics features that are required to visualize engineering and scientific data are available in MATLAB. These include 2-D and 3-D plotting functions, 3-D volume visualization functions, tools for interactively creating plots, and the ability to export results to all popular graphics formats. You can customize plots by adding multiple axes; changing line colors and markers; adding annotation, LaTEX equations, and legends; and drawing shapes.

2-D Plotting: You can visualize vectors of data with 2-D plotting functions that create:

- Line, area, bar, and pie charts
- Direction and velocity plots
- Histograms
- Polygons and surfaces
- Scatter/bubble plots
- Animations

3-D Plotting and Volume Visualization: MATLAB provides functions for visualizing 2-D matrices, 3-D scalar, and 3-D vector data. You can use these functions to visualize and understand large, often complex, multidimensional data. You can specify plot characteristics, such as camera viewing angle, perspective, lighting effect, light source locations, and transparency. 3-D plotting functions include:

- Surface, contour, and mesh
- Image plots
- Cone, slice, stream, and isosurface

Creating and Editing Plots Interactively

MATLAB provides interactive tools for designing and modifying graphics. From a MATLAB figure window, you can perform the following tasks:

- Drag and drop new data sets onto the figure
- Change the properties of any object on the figure
- Zoom, rotate, pan, and change camera angle and lighting
- Add annotations and data tips
- Draw shapes
- Generate a function that can be reused with different data

Importing and Exporting Graphic Files

MATLAB lets you read and write common graphical and data file formats, such as GIF, JPEG, BMP, EPS, TIFF, PNG, HDF, AVI, and PCX. As a result, you can export MATLAB plots to other applications, such as Microsoft Word and Microsoft PowerPoint, or to desktop publishing software. Before exporting, you can create and apply style templates, covering characteristics such as layout, font, and line thickness, to meet publication specifications.

4. **Performing Numeric Computation:** MATLAB contains mathematical, statistical, and engineering functions to support all common engineering and science operations. These functions, developed by experts in mathematics, are the foundation of the MATLAB language. The core math functions use the LAPACK and BLAS linear algebra subroutine libraries and the FFTW Discrete Fourier Transform library. Because these processor-dependent libraries are optimized to the different platforms that MATLAB supports, they execute faster than the equivalent C or C++ code.

   MATLAB provides the following types of functions for performing mathematical operations and analyzing data:

   - Matrix manipulation and linear algebra
   - Polynomials and interpolation

- Fourier analysis and filtering
- Data analysis and statistics
- Optimization and numerical integration
- Ordinary differential equations (ODEs)
- Partial differential equations (PDEs)
- Sparse matrix operations

MATLAB can perform arithmetic on a wide range of data types, including doubles, singles, and integers.

Add-on toolboxes provide specialized mathematical computing functions for areas including signal processing, optimization, statistics, symbolic math, partial differential equation solving, and curve fitting.

5. **Publishing Results and Deploying Applications:** MATLAB provides a number of features for documenting and sharing your work. You can integrate your MATLAB code with other languages and applications and deploy your MATLAB algorithms and applications as standalone programs or software modules.

   Publishing Results: MATLAB lets you export your results as plots or as complete reports. You can export plots to all popular graphics file formats and then import the plots into other packages, such as Microsoft Word or Microsoft PowerPoint. Using the MATLAB Editor, you can automatically publish your MATLAB code in HTML, Word, LaTEX, and other formats.

   Integrating MATLAB Code with Other Languages and Applications: MATLAB provides functions for integrating C and C++ code, Fortran code, COM objects, and Java code with your applications. You can call DLLs, Java classes, and ActiveX controls. Using the MATLAB engine library, you can also call MATLAB from C, C++, or Fortran code.

   Deploying Applications: You can create your algorithm in MATLAB and distribute it to other MATLAB users directly as MATLAB code. Using MATLAB Compiler™, you can deploy your algorithm, as a standalone application or as a software module that you include in your project, to users who do not have MATLAB.

CHAPTER 22

Hankel and Mellin Transforms

Hankel Transforms

The Hankel transform of the function f(x) in the interval (− ∞, ∞) is defined as

$$H_n\{f(x)\} = \int_{-\infty}^{\infty} f(x).x\, J_n(px)\, dx$$

where $J_n(px)$ is the Bessel function of the first kind of order n and is denoted by

$$H_n\{f(x)\} \text{ or } H\{f(x)\} \text{ or } \tilde{f}(p)$$

Inversion Hankel Transform

If $\tilde{f}(p)$ is a Hankel transform of the function f(r) i.e.,

$$\tilde{f}(p) = \int_{-\infty}^{\infty} f(r).\, r\, J_n(pr)\, dr$$

then $$f(r) = \int_0^{\infty} \tilde{f}(p).\, p\, J_n(pr)\, dp$$

is called the inversion formula for the Hankel transform and is written as

$$f(r) = H^{-1}\left\{\tilde{f}(p)\right\}.$$

Proof: If $\tilde{f}$(s) be the complex fourier transform of the function *f(x)*

i.e., $$\tilde{f}(s) = \int_{-\infty}^{\infty} f(x)\, e^{isx}\, dx$$

then $$f(x) = \frac{1}{2\pi}\int_{-\infty}^{\infty} \tilde{f}(s)\, e^{-isx}\, ds$$

For the function of two variables, we have

If $$\tilde{f}(s,t)=\int_{-\infty}^{\infty}\int_{-\infty}^{\infty} f(x,y)\, e^{-i(sx+ty)}dxdy \quad \ldots(1)$$

then $$f(x,y)=\frac{1}{4\pi^2}\int_{-\infty}^{\infty}\int_{-\infty}^{\infty} \tilde{f}(s,t)\, e^{-i(sx+ty)}dsdt \quad \ldots(2)$$

Now, putting x = r cosθ, y = r sinθ, s = p cosα, t = p sinα in equation (1) and (2), we get

If $$\tilde{f}(p,\alpha)=\int_0^{\infty} r\, dr\int_0^{2\pi} f(r,\theta)\, e^{irp\cos(\theta-\alpha)}d\theta \quad \ldots(3)$$

then $$f(r,\theta)=\frac{1}{4\pi^2}\int_0^{\infty} p\, dp\int_0^{2\pi}\tilde{f}(p,\alpha)\, e^{-irp\cos(\theta-\alpha)}d\alpha \quad \ldots(4)$$

consider $f(r,\theta)=f(r).e^{-n\theta}$ be the special function, then equation (3) becomes

$$\tilde{f}(p,\alpha)=\int_0^{\infty} f(r).rdr\int_0^{2\pi} e^{i\{-n\theta+rp\cos(\theta-\alpha)\}}d\theta$$

Again, consider $\phi=\alpha-\theta-\dfrac{\pi}{2}$, then

$$\tilde{f}(p,\alpha)=\int_0^{\infty} f(r).rdr\int_0^{2\pi} e^{i\left\{n(\phi-\alpha+\frac{\pi}{2})+rp\cos(\phi+\frac{\pi}{2})\right\}}d\phi$$

$$=\int_0^{\infty} rf(r)\, dr.e^{in(\pi/2-\alpha)}\int_0^{2\pi} e^{i(n\phi-rp\sin\phi)}d\phi$$

$$=\int_0^{\infty} r.f(r)\, e^{in(\pi/2-\alpha)}.2\pi\, J_n(pr)\, dr$$

$\because$ $$\left[J_n(pr)=\frac{1}{2\pi}\int_0^{2\pi} e^{i(n\phi-rp\sin\phi)}d\phi\right]$$

$$=2\pi\, e^{in(\pi/2-\alpha)}\int_0^{\infty} rf(r)\, J_n(pr)\, dr$$

$$=2\pi\, e^{in(\pi/2-\alpha)}\tilde{f}(p)$$

Now, putting $f(r,\theta)=f(r).e^{-in\theta}$ and $\tilde{f}(p,\alpha)=2\pi e^{in(\pi/2-\alpha)}\tilde{f}(p)$ in equation (4), we get

$$f(r)e^{-in\theta}=\frac{1}{4\pi^2}\int_0^{\infty} pdp\int_0^{2\pi} 2\pi e^{in(\pi/2-\alpha)}\tilde{f}(p)e^{-irp\cos(\theta-\alpha)}d\alpha$$

$$=\frac{1}{2\pi}\int_0^{\infty} p\tilde{f}(p)dp\int_0^{2\pi} e^{i\{n(\pi/2-\alpha)\,-\,rp\cos(\theta-\alpha)\}}d\alpha$$

Let us assume $\psi = \theta - \alpha + \frac{\pi}{2}$

$$= \frac{1}{2\pi}\int_0^{\infty} p\tilde{f}(p)dp\int_0^{2\pi} e^{i\{n(\psi-\theta) - rp\cos(\pi/2-\psi)\}}d\psi$$

$$= \frac{1}{2\pi}\int_0^{\infty} p\tilde{f}(p)dp.e^{-ni\theta}\int_0^{2\pi} e^{i(n\psi - rp\sin\psi)}d\psi$$

$$= \frac{1}{2\pi}\int_0^{\infty} p\tilde{f}(p)dp.e^{-ni\theta}.2\pi\, J_n(pr)$$

$$\therefore \qquad f(r) = \int_0^{\infty} \tilde{f}(p).p\, J_n(pr)\, dp$$

Linear Property

If $f(x)$ and $g(x)$ be the two functions then

$$H_n\{af(x) + bg(x)\} = aH_n\{f(x)\} + bH_n\{g(x)\}$$

where a, b are two constants.

Proof: We have

$$H_n\{af(x) + bg(x)\} = \int_0^{\infty} x\{af(x) + bg(x)\}\, J_n(px)\, dx$$

$$= a\int_0^{\infty} xf(x)J_n(px)\, dx + b\int_0^{\infty} xg(x)J_n(px)\, dx$$

$$= a\, H_n\{f(x)\} + b\, H_n\{g(x)\}$$

Some useful results:

(i) $J_n(x) = \sum_{r=0}^{\infty} \frac{(-1)^r}{r!\,\lfloor n+r+1}\left(\frac{x}{2}\right)^{n+2r}$ (First kind of Bessel function)

(ii) $x\, J_n^1(x) = n\, J_n(x) - x\, J_{n+1}(x)$

(iii) $x\, J_n^1(x) = -n\, J_n(x) + x\, J_{n-1}(x)$

(iv) $2\, J_n^1(x) = J_{n-1}(x) - J_{n+1}(x)$

(v) $2n\, J_n(x) = x\left[J_{n-1}(x) + J_{n+1}(x)\right]$

(vi) $\frac{d}{dx}\left[x^{-n}J_n(x)\right] = -x^{-n}J_{n+1}(x)$

(vii) $\frac{d}{dx}\left[x^n J_n(x)\right] = x^n . J_{n-1}(x)$

(viii) $\int_0^\infty e^{-ax} J_o(px)\, dx = (a^2 + b^2)^{-\frac{1}{2}}$

(ix) $\int_0^\infty e^{-ax} J_1(px)\, dx = \frac{1}{p} - \frac{a}{p(a^2 + p^2)^{1/2}}$

(x) $\int_0^\infty x\, e^{-ax} J_0(px)\, dx = a\, (a^2 + p^2)^{-3/2}$

(xi) $\int_0^\infty x\, e^{-ax} J_1(px)\, dx = p\, (a^2 + p^2)^{-3/2}$

(xii) $\int_0^\infty \frac{1}{x}\, e^{-ax} J_1(px)\, dx = \frac{(a^2 + p^2)^{1/2} - a}{p}$

Note: $x\, J_n(px)$ is said to be the Kernel of the transformation.

Illustrative Examples

Example 1

Find the Hankel transform of e^{-ax} taking $x\, J_0(px)$ as the Kernel of the transform.

Solution: we have

$$H\{e^{-ax}\} = \int_0^\infty e^{-ax} . xJ_0(px)\, dx$$
$$= a(a^2 + p^2)^{-3/2}$$

Example 2

Find the Hankel transform of $\frac{e^{-ax}}{x}$ taking $x\, J_0(px)$ as the Kernel of the transform.

Solution:

$$H\left\{\frac{e^{-ax}}{x}\right\} = \int_0^\infty \frac{e^{-ax}}{x} .x\, J_0(px)\, dx$$
$$= \int_0^\infty e^{-ax} J_0(px)\, dx$$
$$= \left(a^2 + p^2\right)^{-1/2}$$

Example 3

Find the Hankel transform of e^{-ax} taking $x\, J_1(px)$ as the Kernel of the transform.

Solution:

$$H\{e^{-ax}\} = \int_0^\infty e^{-ax} . x\, J_1(px)\, dx$$
$$= p(a^2 + p^2)^{-3/2}$$

Example 4

Find the Hankel transform of $\dfrac{\sin ax}{x}$ taking $x\,J_0(px)$ as the Kernel of the transform.

Solution:

$$H\left\{\frac{\sin ax}{x}\right\}=\int_0^\infty \frac{\sin ax}{x}.\,x\,J_0(px)\,dx$$

$$=-\text{I. P of }\int_0^\infty e^{-iax}J_0(px)\,dx$$

$$=-\text{I. P of }(i^2a^2+p^2)^{-1/2}$$

$$=-\text{I. P of }(-a^2+p^2)^{-1/2}$$

$$=\begin{cases}0 & \text{if } p>a\\ (a^2-p^2)^{-1/2} & \text{if } o<p<a\end{cases}$$

Example 5

Find the Hankel transform of

$$f(x)=\begin{cases}1, & 0<x<a \quad n=0\\ 0, & x>a \quad n=0\end{cases}$$

Solution:

$$H_n\{f(x)\}=\int_0^\infty x.f(x)\,J_n(px)\,dx$$

$$=\int_0^a x.1.J_0(px)\,dx$$

Now, we have

$$\frac{d}{dx}\left[x^nJ_n(x)\right]=x^nJ_{n-1}(x)$$

For $n = 1$,

$$\frac{d}{dx}[x\,J_1(x)]=x\,J_0(x)$$

writing px for x, we get

$$\frac{1}{p}\frac{d}{dx}[px\,J_1(px)]=px\,J_0(px)$$

integrating

$$\int_0^a \frac{d}{dx}[x\,J_1(px)]=\int_0^a px\,J_0(px)\,dx$$

or

$$\int_0^a [x\,J_0(px)dx]=\left[\frac{x}{p}J_1(px)\right]_0^a=\frac{a}{p}J_1(pa)$$

Example 6

Find the Hankel transform of

$$f(x)=\begin{cases} a^2-x^2, & o<x<a \quad n=0 \\ 0 \quad , & x>a \quad\quad n=0 \end{cases}$$

Solution:

$$H_n\{f(x)\}=\int_0^\infty x\,f(x)\,J_n(px)\,dx$$

$$=\int_0^a x(a^2-x^2)\,J_0(px)\,dx$$

$$=a^2\int_0^a x\,J_0(px)\,dx-\int_0^a x^3 J_0(px)\,dx \quad \text{...(i)}$$

Now,

$$\int_0^a x\,J_0(px)\,dx=\frac{a}{p}J_1(pa)$$

again, for $n = 1$,

$$\frac{d}{dx}\left[x\,J_1(px)\right]=px\,J_0(px)$$

$$\therefore \quad \int_0^a x^3 J_0(px)\,dx=\int_0^a x^2.\frac{1}{p}\frac{d}{dx}[x\,J_1(px)]\,dx$$

$$=\left[x^2.\frac{x}{p}J_1(px)\right]_0^a-\frac{2}{p}\int_0^a x.x.J_1(px)\,dx$$

writing px for x and $n = 2$ in formula (vii)

$$\frac{1}{p}\frac{d}{dx}\left[x^2 J_2(px)\right]=x^2 J_1(px)$$

$$\therefore \quad \int_0^a x^3 J_0(px)\,dx=\frac{a^3}{p}J_1(pa)-\frac{2}{p}\int_0^a \frac{1}{p}\frac{d}{dx}\left[x^2 J_2(px)\right]dx$$

$$=\frac{a^3}{p}J_1(pa)-\frac{2}{p^2}a^2 J_2(pa) \quad \text{...(ii)}$$

writing $n = 1$ and ap for x in formula (v), we have

$$2J_1(pa)=pa\left[J_0(pa)+J_2(pa)\right]$$

$$\therefore \quad J_2(pa)=\frac{2}{pa}J_1(pa)-J_0(pa)$$

Putting the value of $J_2(pa)$ in (ii), we have

$$\int_0^a x^3 J_0(px)dx = \frac{a^3}{p}J_1(pa) - \frac{2}{p^2}a^2\left[\frac{2}{pa}J_1(pa) - J_0(pa)\right]$$

$$= \frac{a^3}{p}J_1(pa) - \frac{4a}{p^3}J_1(pa) + \frac{2a^2}{p^2}J_0(pa)$$

Hence from (i), we have

$$H_n\{f(x)\} = a^2.\frac{a}{p}J_1(pa) - \frac{a^3}{p}J_1(pa) + \frac{4a}{p^3}J_1(pa) - \frac{2a^2}{p^2}J_0(pa)$$

$$= \frac{4a}{p^3}J_1(pa) - \frac{2a^2}{p^2}J_0(pa)$$

Example 7

Find $H^{-1}\{p^{-2}e^{-ap}\}$, taking *n =1*.

Solution: Applying inversion formula

$$H^{-1}\{\tilde{f}(p)\} = \int_0^\infty \tilde{f}(p).p.J_n(px)\,dp$$

$$\therefore \quad H^{-1}\{p^{-2}e^{-ap}\} = \int_0^\infty p^{-2}e^{-ap}.p.J_1(px)\,dx$$

$$= \int_0^\infty \frac{1}{p}e^{-ap}J_1(px)\,dp$$

$$= \frac{(a^2+x^2)^{1/2} - a}{p}$$

Hankel Transform of the Derivatives

The Hankel transform of order *n* of the function *f(x)* be

$$\tilde{f}_n(p) = \int_0^\infty x.f(x)\,J_n(px)\,dx \quad \text{...(i)}$$

$\therefore$ derivative of Hankel transform $\left(i.e., \frac{df}{dx}\right)$ is

$$\tilde{f}'_n(p) = \int_0^\infty x.\frac{df}{dx}J_n(px)\,dx \quad \text{...(ii)}$$

Integrating by parts taking $xJ_n(px)$ as the first function, we get

$$\tilde{f}'_n(p) = [x\,f(x)\,J_n(px)]_0^\infty - \int_0^\infty f(x).\frac{d}{dx}[xJ_n(px)]dx$$

$$= -\int_0^\infty f(x)\{J_n(px) + xpJ_n^1(px)\}\,dx \quad \text{...(iii)}$$

using formula (ii) i.e., $xJ_n^1(x) = -nJ_n(x) + xJ_{n-1}(x)$

replacing *x* by *px*

$$pxJ_n^1(px) = -nJ_n(px) + pxJ_{n-1}(px)$$

$\therefore$ equation (iii) becomes

$$\tilde{f}_n'(p) = -\int_0^\infty f(x)[(1-n)J_n(px) + pxJ_{n-1}(px)]dx$$

$$= (n-1)\int_0^\infty f(x).J_n(px)\,dx - \int_0^\infty px\,f(x)J_{n-1}(px)\,dx$$

$$= (n-1)\int_0^\infty f(x)\left\{\frac{px}{2n}[J_{n-1}(px) + J_{n+1}(px)]\right\}dx - p\int_0^\infty x\,f(x)J_{n-1}(px)\,dx$$

[$\because$ using formula (v)]

$$= \frac{-p(n+1)}{2n}\int_0^\infty x\,f(x)J_{n-1}(px)dx + \frac{p(n-1)}{2n}\int_0^\infty x\,f(x)J_{n+1}(px)dx$$

Hence

$$\tilde{f}_n'(p) = -p\left[\frac{n+1}{2n}\tilde{f}_{n-1}(p) - \frac{(n-1)}{2n}\tilde{f}_{n+1}(p)\right] \quad \ldots\text{(iv)}$$

Proceeding in a similar way, we have

$$\tilde{f}_n''(p) = \frac{p^4}{4}\left[\frac{n+1}{n-1}\tilde{f}_{n-2}(p) - 2\frac{(n^2-3)}{(n^2-1)}\tilde{f}_n(p) + \frac{n+1}{n-1}\tilde{f}_{n+2}(p)\right]$$

Deductinos: Writing n = 1, 2, 3 … in equation (iv), we have

1. $\tilde{f}_1'(p) = -p\tilde{f}_0(p)$
2. $\tilde{f}_2'(p) = -p\left[\frac{3}{4}\tilde{f}_1(p) - \frac{1}{4}\tilde{f}_3(p)\right]$
3. $\tilde{f}_3'(p) = -p\left[\frac{2}{3}\tilde{f}_2(p) - \frac{1}{3}\tilde{f}_4(p)\right]$

The Finite Hankel Transforms

The finite Hankel transform of the function *f(r)*, which satisfies Dirichlet's conditions in the closed interval o ≤ r ≤ a is defined by

$$\tilde{f}(p_i) = \int_0^a r.f(r)\,J_n(p_i r)\,dr$$

where p_i is a root of the transcendental equation

$$J_n(p_i a) = 0$$

and at any point of the closed interval o ≤ r ≤ a at which the function *f(r)* is continuous, we have

$$f(r)=\frac{2}{a^2}\sum_i \tilde{f}(p_i)\frac{J_n(p_i r)}{\left[J_n'(p_i a)\right]^2}$$

Another form of Hankel Transform

The finite Hankel transform of f(r) is

$$\tilde{f}(p_i)=\int_0^1 r\,f(r)\,J_n(p_i r)\,dr \qquad \text{... (i)}$$

where p_i is a root of the transcendental equation

$$pJ_n'(p)\pm hJ_n(p)=0 \qquad \text{... (ii)}$$

Then, at any point of the closed interval o ≤ r ≤ 1 at which *f(r)* is continuous

$$f(r)=2\sum_p \frac{p^2\tilde{f}(p)}{n^2+p^2-n^2}\,\frac{J_n(pr)}{J_n^{\,2}(p)} \qquad \text{... (iii)}$$

where the sum is considered over all positive roots of (ii)

Illustrative Examples

Examples 1

Show that $\int_0^a rJ_0(pr)dr=\frac{a}{p}J_1(pa)$

Solution: From formula (vii), we have

$$\frac{d}{dr}\left[r^n J_n(r)\right]=r^n J_{n-1}(r)$$

replacing $n = 1$, and r by pr, we get

$$\frac{1}{p}\frac{d}{dr}[prJ_1(pr)]=prJ_0(pr)$$

$$\therefore \qquad \frac{1}{p}\frac{d}{dr}[rJ_1(pr)]=rJ_0(pr)$$

$$\therefore \qquad \int_0^a rj_0(pr)dr=\frac{1}{p}[rJ_1(pr)]_0^a$$

$$=\frac{a}{p}J_1(pa)$$

Example 2

Find the Hankel transform of r^{n-1}, if $rJ_{n-1}(p_i r)$ is the Kernal of the transform.

Solution: We have

$$\tilde{f}(p_i) = \int_0^a r.f(r)J_{n-1}(p_i r)dr$$

$$= \int_0^a r.r^{n-1}J_{n-1}(p_i r)dr \quad (\text{Given } f(r) = r^{n-1})$$

$$= \int_0^a r^n J_{n-1}(p_i r)dr$$

Now, using formula (vii) and replacing r by $p_i r$, we have

$$\frac{1}{p_i}\frac{d}{dr}\left[r^n J_n(p_i r)\right] = r^n J^{n-1}(p_i r)$$

$$\therefore \qquad \tilde{f}(p_i) = \frac{1}{p_i}\int_0^a \frac{d}{dr}\left[r^n J_n(p_i r)\right]dr$$

$$= \frac{1}{p_i}\left[r^n J_n(p_i r)\right]_0^a = \frac{a^n}{p_i}J_n(p_i a)$$

Example 3

Find the Hankel transform of $(1-r^2)$, taking $rJ_0(pr)$ as the Kernel.

Solution: We have

$$H(1-r^2) = \int_0^a (1-r^2).rJ_0(pr)dr$$

$$= \int_0^a rJ_0(pr)dr - \int_0^a r^3 J_0(pr)dr$$

$$= \frac{a}{p}J_1(ap) - \frac{a^2}{p^2}\left[2J_0(pa) + \left(ap - \frac{4}{ap}\right)J_1(pa)\right]$$

Example 4

Show that $$\int_0^a r^3 J_0(pr)dr = \frac{a^2}{p^2}\left[2J_0(pa) + \left(ap - \frac{4}{ap}\right)J_1(pa)\right]$$

Solution: From formula (vii), we have

$$\frac{d}{dr}\left[r^n J_n(r)\right] = r^n J_{n-1}(r) \qquad \text{...(i)}$$

replacing $n = 1$ and $r = pr$, we get

$$\frac{1}{p}\frac{d}{dr}[rJ_1(pr)] = rJ_0(pr)$$

$$\therefore \quad \int_0^a r^3 J_0(pr)dr = \int_0^a r^2 . \frac{1}{p}\frac{d}{dr}[rJ_1(pr)]dr$$

$$= \left[\frac{r^2}{p}.rJ_1(pr)\right]_0^a - \int_0^a \frac{2r}{p}.rJ_1(pr)dr$$

$$= \frac{a^3}{p}J_1(pa) - \frac{2}{p}\int_0^a \frac{1}{p}\frac{d}{dr}\left[r^2 J_2(pr)\right]dr$$

(by replacing n =2 and $r = pr$ in equation (i)

$$= \frac{a^3}{p}J_1(pa) - \frac{2}{p^2}\left[r^2 J_2(pr)\right]_0^a$$

$$= \frac{a^3}{p}J_1(pa) - \frac{2a^2}{p^2}J_2(pa) \quad \text{...(ii)}$$

Now, from formula (v), we have

$$2nJ_n(r) = r\left[J_{n-1}(r) + J_{n+1}(r)\right]$$

Replacing $n = 1$ and $r = pa$, we get

$$\frac{2}{pa}J_1(pa) = J_0(pa) + J_2(pa)$$

$$\therefore \quad J_2(pa) = \frac{2}{pa}J_1(pa) - J_0(pa)$$

$\therefore$ from (ii), we have

$$\int_0^a r^3 J_0(pr)dr = \frac{a^3}{p}J_1(pa) - \frac{2a^2}{p^2}\left[\frac{2}{pa}J_1(pa) - J_0(pa)\right]$$

$$= \frac{a^2}{p^2}\left[2J_0(pa) + \left(ap - \frac{4}{pa}\right)J_1(pa)\right]$$

Example 5

Prove that the finite Hankel transform of $f(x)$, $0 \le x \le 1$ is $p^{n-m}J_m(p)$, where

$$f(x) = \frac{2^{1+n-m}}{\overline{|m-n}}.x^n(1-x^2)^{m-n-1}$$

Solution: We have

$$H_n\{f(x)\} = \int_0^1 f(x).xJ_n(px)dx$$

$$= \int_0^1 \frac{2^{1+n-m}}{\overline{|m-n}} . x^n (1-x^2)^{m-n-1} . x . \sum_{r=0}^{\infty} \frac{(-1)^r}{r! \overline{|n+r+1}} \left(\frac{px}{2} \right)^{n+2r} dx$$

$$= \frac{1}{\overline{|m-n}} . \sum_{r=0}^{\infty} \frac{(-1)^r}{r! \overline{|n+r+1}} . \frac{p^{n+2r}}{2^{-1+m+2r}} \int_0^1 x^{2(n+r)} . x . (1-x^2)^{m-n-1} dx$$

Putting $x^2 = y$ so that $xdx = \frac{dy}{2}$

$$= \frac{1}{\overline{|m-n}} \sum_{r=0}^{\infty} \frac{(-1)^r p^{n+2r}}{r! \overline{|n+r+1} \,.\, 2^{m+2r}} \int_0^1 y^{(n+r+1)-1} . (1-y)^{(m-n)-1} dy$$

$$= \frac{1}{\overline{|m-n}} \sum_{r=0}^{\infty} \frac{(-1)^r p^{n+2r}}{r! \overline{|n+r+1} \,.\, 2^{m+2r}} \cdot \frac{\overline{|n+r+1} \,.\, \overline{|m-n}}{\overline{|n+r+1+m-n}}$$

$$\because \qquad \int_0^1 t^{m-1} (1-t)^{n-1} dt = \beta(m,n) = \frac{\overline{|m} \, \overline{|n}}{\overline{|m+n}}$$

$$= p^{n-m} \sum_{r=0}^{\infty} \frac{(-1)^r}{r! \overline{|m+r+1}} \left(\frac{p}{2} \right)^{m+2r}$$

$$= p^{n-m} . J_m(p)$$

Hankel Transform of $\frac{d^2 f}{dx^2} + \frac{1}{x} \frac{df}{dx} - \frac{n^2}{x^2} f$

We have

$$H\left\{ \frac{d^2 f}{dx^2} \right\} = \int_0^{\infty} \frac{d^2 f}{dx^2} . x \, J_n(px) dx$$

$$= \left[x J_n(px) . \frac{df}{dx} \right]_0^{\infty} - \int_0^{\infty} \frac{d}{dx} [x J_n(px)] . \frac{df}{dx} dx$$

(taking $xJ_n(px)$ as first function)

$$= -\int_0^{\infty} \frac{df}{dx} \left[J_n(px) + px \, J_n^1(px) \right] dx$$

or $$\int_0^{\infty} \left(\frac{d^2 f}{dx^2} + \frac{1}{x} \frac{df}{dx} \right) . x J_n(px) dx = -p \int_0^{\infty} \frac{df}{dx} . x J_n^1(px) dx$$

$$= -p \left[x J_n^1(px) . f(x) \right]_0^{\infty} + p \int_0^{\infty} \frac{d}{dx} \left\{ x J_n^1(px) \right\} . f(x) dx$$

(taking $xJ_n^1(px)$ as first function)

$$= p \int_0^{\infty} f(x) \frac{d}{dx} \{ \; x J_n^1(px) \; \} \; dx \qquad \ldots \text{(i)}$$

Now, since $J_n(x)$ satisfies the Bessel's differential equation

$$\frac{d}{dx}\left(x\frac{dy}{dx}\right)+\left(1-\frac{n^2}{x^2}\right)xy=0$$

$$\therefore \quad \frac{d}{dx}\left(x\frac{d}{dx}J_n(x)\right)+\left(1-\frac{n^2}{x^2}\right)xJ_n(x)=0$$

Replacing x by px, we have

$$\frac{1}{p}\frac{d}{dx}\left(px.J_n^1(px)\right)=-\left(p^2-\frac{n^2}{x^2}\right).\frac{x}{p}.J_n(px)$$

or

$$\frac{d}{dx}\left[xJ_n^1(px)\right]=-\left(p^2-\frac{n^2}{x^2}\right)\frac{x}{p}J_n(px)$$

Hence from (i), we have

$$\int_0^\infty\left(\frac{d^2f}{dx^2}+\frac{1}{x}\frac{df}{dx}\right)xJ_n(px)dx=-p\int_0^\infty f(x)\left(p^2-\frac{n^2}{x^2}\right)\frac{x}{p}J_n(px)dx$$

or

$$\int_0^\infty\left(\frac{d^2f}{dx^2}+\frac{1}{x}\frac{df}{dx}-\frac{n^2}{x^2}f\right)xJ_n(px)dx=-p^2\int_0^\infty f(x).xJ_n(px)dx$$

$$=-p^2\tilde{f}n(p)$$

Deduction:

1. When $n = 0$,

$$\int_0^\infty\left(\frac{d^2f}{dx^2}+\frac{1}{x}\frac{df}{dx}\right)xJ_0(px)dx=-p^2\tilde{f}_0(p)$$

2. When $n = 1$,

$$\int_0^\infty\left(\frac{d^2f}{dx^2}+\frac{1}{x}\frac{df}{dx}-\frac{1}{x^2}f\right)xJ_1(px)dx=-p^2\tilde{f}_1(p)$$

Applications of Hankel Transform

(Initial and Boundary Value problems)

We take a differential equation in which the range of variable from 0 to ∞.

Example 1

Find the potential $v(r, z)$ of a field due to a flat circular disc of unit radius with the centre at the origin, and axis along the z-axis, satisfying the differential equation

$$\frac{\partial^2 v}{\partial r^2}+\frac{1}{r}\frac{\partial v}{\partial r}+\frac{\partial^2 v}{\partial z^2}=0\ , \quad 0\le r<\infty,\ z\ge 0$$

and the boundary conditions $v = v_0$, when z = 0, 0 ≤ r < 1 and $\frac{\partial v}{\partial z} = 0$, when z = 0, r > 1.

Solution: Taking the Hankel transform for n = 0 of both the sides of the given differential equation

$$\int_0^\infty \left(\frac{\partial^2 v}{\partial r^2} + \frac{1}{r}\frac{\partial v}{\partial r}\right) rJ_0(pr)dr = -\int_0^\infty \frac{\partial^2 v}{\partial z^2}.rJ_0(pr)dr$$

or $$-p^2\tilde{v} = -\frac{d^2\tilde{v}}{dz^2} \quad \text{where } \tilde{v} = \int_0^\infty v.r.J_0(pr)dr$$

or $$\frac{d^2\tilde{v}}{dz^2} - p^2\tilde{v} = 0$$

Its solution is $\tilde{v}(p,z) = Ae^{pz} + Be^{-pz}$

Now, v must vanish as z tends to infinity.

∴ $\tilde{v}$ must also vanish as z tends to infinity,

∴ A = 0

Hence $\tilde{v}(p,z) = Be^{-pz}$

Applying the inversion formula, we have

$$v(r,z) = \int_0^\infty Be^{-pz}.pJ_0(pr)dp, \text{ B is a function of p only.} \quad \text{...(i)}$$

Now, applying the boundary conditions, we have

when z = 0, $$v = \int_0^\infty p.B\,J_0(pr)dp = v_0, \quad 0 \le r < 1$$

and $$\left(\frac{\partial v}{\partial z}\right)_{z=0} = \int_0^\infty -p^2BJ_0(pr)dp = 0, \quad r > 1$$

$$\left.\begin{aligned} &\text{or} && \int_0^\infty p\,BJ_0(pr)dp = v_0, && 0 \le r < 1 \\ &\text{and} && \int_0^\infty p^2BJ_0(pr)dp = 0, && r > 1 \end{aligned}\right\} \quad \text{...(ii)}$$

Now, the well known integrals

$$\left.\begin{aligned} & && \int_0^\infty J_0(pr)\frac{\sin p}{p}dp = \frac{\pi}{2}, && 0 \le r < 1 \\ &\text{and} && \int_0^\infty J_0(pr).\sin p\,dp = 0, && r > 1 \end{aligned}\right\} \quad \text{...(iii)}$$

Comparing (ii) and (iii), we have

$$B = \frac{2}{\pi} v_0 \frac{\sin p}{p^2}$$

Hence from (i), we have

$$v(r,z) = \frac{2v_0}{\pi} \int_0^{\infty} e^{-pz} \cdot \frac{\sin p}{p} J_0(pr)dp$$

Example 2

The vibrations of a very large membrane are governed by the equation

$$\frac{\partial^2 v}{\partial r^2} + \frac{1}{r}\frac{\partial v}{\partial r} = \frac{1}{c^2}\frac{\partial^2 v}{\partial t^2}, \qquad r \geq 0, t \geq 0$$

with $v = f(r), \dfrac{\partial v}{\partial r} = g(r)$, when $t = 0$

Show that, for $t > 0$

$$v(r,t) = \int_0^{\infty} p\tilde{f}(p)\cos(pct)J_0(pr)dp + \frac{1}{c}\int_0^{\infty} \tilde{g}(p)\sin(pct)J_0(pr)dp$$

where $\tilde{f}(p)$ and $\tilde{g}(p)$ are the zero order Hankel transforms of $f(r)$, $g(r)$ respectively.

Solution: Taking the Hankel transform for n = 0 of both the sides of the given differential equation

$$\int_0^{\infty} \left(\frac{\partial^2 v}{\partial r^2} + \frac{1}{r}\frac{\partial v}{\partial r}\right) .rJ_0(pr)dr = \frac{1}{c^2}\int_0^{\infty} \frac{\partial^2 v}{\partial t^2} .rJ_0(pr)dr$$

or $$-p^2\tilde{v} = \frac{1}{c^2}\frac{d^2\tilde{v}}{dt^2}, \qquad \text{where } \tilde{v} = \int_0^{\infty} v.r.J_0(pr)dr$$

or $$\frac{d^2\tilde{v}}{dt^2} + c^2p^2\tilde{v} = 0 \qquad \text{...(i)}$$

Its solution is $\tilde{v} = A\cos(cpt) + B\sin(cpt)$ ...(ii)

Also when t = 0

$$\tilde{v} = \int_0^{\infty} f(r).rJ_0(pr)dr = \tilde{f}(p) \qquad \text{...(iii)}$$

and $$\int_0^{\infty} \frac{\partial v}{\partial r}.r\, J_0(pr)dr = \int_0^{\infty} g(r).r.J_0(pr)dr$$

or $$\frac{d\tilde{v}}{dt} = \tilde{g}(p) \qquad \text{... (iv)}$$

$\therefore$ from (ii) and (iii), $\tilde{f}(p) = A$ and from (ii) and (iv)

$$-Apc\ \sin(cpt) + Bcp\ \cos(cpt) = \tilde{g}(p)$$

$$\therefore \qquad B = \frac{\tilde{g}(p)}{pc}$$

Putting the value of A and B in (ii), we have

$$\tilde{v} = \tilde{f}(p)\cos(cpt) + \frac{\tilde{g}(p)}{cp}\sin(cpt)$$

Applying the inversion formula, we have

$$\tilde{v}(r,t) = \int_0^\infty p\tilde{f}(p)\cos(cpt)J_0(pr)dp + \frac{1}{c}\int_0^\infty \tilde{g}(p)\sin(cpt)J_0(pr)dp$$

Example 3

Use the Hankel transforms to solve the differential equation.

$$\frac{\partial^2 v}{\partial r^2} + \frac{1}{r}\frac{\partial v}{\partial r} = \frac{1}{k}\frac{\partial v}{\partial t},\ \mathrm{t} > 0, \qquad 0 \le \mathrm{r} < 1$$

where $\qquad \dfrac{\partial v}{\partial r} + hv = 0$, when r =1, $\qquad$ t > 0

and $v = 1$, when $t = 0,\ 0 \le r < 1$.

Solution: Taking the finite Hankel transform for n = 0 of both the sides of the given differential equation

$$H_0\left\{\frac{\partial^2 v}{\partial r^2} + \frac{1}{r}\frac{\partial v}{\partial r}\right\} = \frac{1}{k}\int_0^1 \frac{\partial v}{\partial t}.rJ_0(pr)dr$$

or $$-p^2 H_o(v) = \frac{1}{k}\frac{\partial}{\partial t}\int_0^1 v.r.J_0(pr)dr$$

or $$\frac{d\tilde{v}}{dt} = -kp^2\tilde{v}, \qquad \text{where } \tilde{v} = H_o(v) = \int_0^1 v.r.J_0(pr)dr$$

Its solution is $\qquad \tilde{v} = Ae^{-kp^2 t}$ ....(i)

Now, when t = 0, 0 ≤ r < 1, v = 1

$$\therefore \qquad \tilde{v} = \int_0^1 1.r.J_0(pr)dr$$

$$= \frac{1}{p}J_1(p)$$

Hence, from (i), we have

when t = 0, $A = \frac{1}{p} J_1(p)$

$$\therefore \quad \tilde{v} = \frac{1}{p} e^{-kp^2 t} . J_1(p)$$

Applying inversion formula,

$$\tilde{v} = 2 \sum_p \frac{p^2}{p^2 + h^2} \frac{J_1(p)}{p} e^{-kp^2 t} . \frac{J_0(pr)}{J_0^{\,2}(p)}$$

or
$$\tilde{v} = 2 \sum_p \frac{pJ_1(p)J_0(pr)}{(p^2 + h^2)J_0^{\,2}(p)} e^{-kp^2 t}$$

where the summation being taken over all the positive roots of

$$pJ_0^1(p) + hJ_0(p) = 0$$

Note:
$$H_0\left\{\frac{\partial^2 v}{\partial r^2} + \frac{1}{r}\frac{\partial v}{\partial r}\right\} = -p^2 H_0(v) + \left\{\frac{\partial v}{\partial r} + hv(r)\right\}_{r=1} J_0(p)$$

By given condition $\frac{\partial v}{\partial r} + hv = 0$ when r = 1

$$\therefore \quad H_0\left\{\frac{\partial^2 v}{\partial r^2} + \frac{1}{r}\frac{\partial v}{\partial r}\right\} = -p^2 H_0(v)$$

Example 4

Use the Hankel transform to solve the differential equation

$$\frac{\partial^2 v}{\partial r^2} + \frac{1}{r}\frac{\partial v}{\partial r} = \frac{1}{k}\frac{\partial v}{\partial t}, \qquad 0 \le r < 1, t > 0$$

with $v = v_0$, when r = 1, t > 0

and $v = 0$ when t = 0, ≤ r < 1

Solution: Taking the finite Hankel transform for n = 0 of both sides of the given differential equation

$$H_0\left\{\frac{\partial^2 v}{\partial r^2} + \frac{1}{r}\frac{\partial v}{\partial r}\right\} = \frac{1}{k}\int_0^1 \frac{\partial v}{\partial t} .rJ_0(pr)dr$$

$$-p^2 H_0\{v\} - p\{v\}_{r=1} J_0^1(p) = \frac{1}{k}\frac{\partial}{\partial t}\int_0^1 v.r.J_0(pr)dr$$

or
$$-p^2\tilde{v} - p\, v_0 J_0^1(p) = \frac{1}{k}\frac{d\tilde{v}}{dt} \qquad \text{where } \tilde{v} = H_0(v) = \int_0^1 v.r.J_0(pr)dr$$

or $$\frac{d\tilde{v}}{dt}+kp^2\tilde{v}=-kpv_0J_0^1(p)$$

which is a linear differential equation of first order, whose I. F is $e^{\int kp^2 dt}=e^{\int kp^2 t}$

$\therefore$ $$\tilde{v}\,e^{kp^2t}=-\int kpv_0J_0^1(p).e^{kp^2t}dt+c$$

$$=-kpv_0J_0^1(p).\frac{e^{kp^2t}}{kp^2}+c$$

or $$\tilde{v}=c\,e^{-kp^2t}-\frac{v_0}{p}J_0^1(p) \qquad \text{...(i)}$$

Now since v = 0 when t = 0, 0 ≤ r < 1

then $$\tilde{v}=\int_0^1 v.r.J_0(pr)dr=0$$

$\therefore$ from (i),

$$0=c-\frac{v_0}{p}J_0^1(p)$$

or $$c=\frac{v_0}{p}J_0^1(p)$$

Hence, from (i), $\tilde{v}=\frac{v_0}{p}J_0^1(p)(e^{-kp^2t}-1)$

Applying the inversion formula

$$v=2\sum_p \frac{v_0}{p}J_0^1(p)(e^{-kp^2t}-1)\frac{J_0(pr)}{\left[J_0^1(p)\right]^2}$$

$$=2\,v_0\sum_p (1-e^{-kp^2t})\frac{J_0(pr)}{pJ_1(p)} \qquad \because \left[J_0^1(p)=-J_1(p)\right]$$

where the summation being taken over all positive roots of $J_0(p)=0$.

Practice Problems

1. Find the Hankel transform of (i) e^{-x} (ii) $\frac{e^{-x}}{x}$, taking $xJ_0(px)$ as the Kernel of the transformation.

[Ans: (i) $(1+p^2)^{-3/2}$ (ii) $(1+p^2)^{-1/2}$]

2. Find the Hankel transform of $x^{-2}\, e^{-x}$, taking $xJ_1(px)$ as the Kernel.

[**Ans:** $\dfrac{(1+p^2)^{1/2}-1}{p}$]

3. Find $H^{-1}\{\tilde{f}(p)\}$, if $\tilde{f}(p)=\dfrac{e^{-ap}}{p}$, where $\tilde{f}(p)$ is the Hankel transform of $f(x)$ and $n=0$.

[**Ans:** $(a^2+x^2)^{-1/2}$]

4. Find Hankel transform of $f(x)=\begin{cases} xn & 0<x<a \\ o & x>a \end{cases}$, $n>-1$ taking $xJ_n(px)$ as the Kernel.

[**Ans:** $\dfrac{a^{n+1}}{p}J_{n+1}(pa)$]

5. Find the Hankel transform of $\dfrac{\cos ax}{x}$, taking Kernel of transform $xJ_0(px)$.

[Ans: 0 if $p<a$ and $(p^2-a^2)^{-1/2}$ if $p>a$]

6. Find the Hankel transform of order zero of $f(x)=\begin{cases} 4-x^2 & 0<x<2 \\ 0 & x>2 \end{cases}$

[**Ans:** $\dfrac{8}{p^3}J1(2p)-\dfrac{8}{p^2}J_0(2p)$]

7. Find the Hankel transform of $\dfrac{df}{dx}$ where $f=\dfrac{e^{-ax}}{x}$ and $n=1$.

[**Ans:** $-p(a^2+p^2)^{-1/2}$]

8. Evaluate $\int_0^{\infty} r\left(\dfrac{d^2f}{dr^2}+\dfrac{1}{r}\dfrac{df}{dr}\right)J_0(pr)dr$, where $f(r)=\dfrac{e^{-ar}}{r}$.

[**Ans:** $-\dfrac{p2}{(a^2+p^2)^{1/2}}$]

9. Evaluate $\int_0^{\infty} r\left(\dfrac{d^2f}{dr^2}+\dfrac{1}{r}\dfrac{df}{dr}\right)J_0(pr)dr$, where $f=e^{-ar}$

[**Ans:** $-\dfrac{ap^2}{(a^2+p^2)^{3/2}}$]

10. Find the Hankel transform of c, if $rJ_0(p_i r)$ be the Kernel.

[**Ans:** $\frac{ca}{p_i} J_1(p_i a)$]

11. Find the Hankel transform of $r^n, (n > -1)$, if $rJ_n(pr)$ is the Kernel of the transform.

[**Ans:** $\frac{a^{n+1}}{p_i} J_{n+1}(p_i a)$]

12. Find the Hankel transform of $(a^2 - r^2)$, if $rJ_0(pr)$ is the Kernel.

[**Ans:** $\frac{4a}{p^3} J_1(ap) - \frac{2a^2}{p^2} J_0(pa)$]

13. Solve the differential equation.

$$\frac{\partial^2 v}{\partial r^2} + \frac{1}{r}\frac{\partial v}{\partial r} + \frac{\partial^2 v}{\partial z^2} = 0, \; r \geq 0 \; z \geq 0, \text{ with the conditions}$$

(i) $v \to 0$ a s $z \to \infty$ and $r \to \infty$ and (ii) $v = f(r)$, on $z = 0$, $r \geq 0$.

[**Ans:** $v(r,z) = \int_0^\infty \tilde{f}(p).e^{-pz} pJ_0(pr)dr$]

14. Find the potential v(r,z) of a field due to a flat circular disc of unit radius with the centre at the origin and axis along the z-axis, satisfying the differential equation.

$$\frac{\partial^2 v}{\partial r^2} + \frac{1}{r}\frac{\partial v}{dr} + \frac{\partial^2 v}{\partial z^2} = 0, \; 0 \leq r < \infty, \; z \geq 0 \text{ and the boundary conditions } v = v_o \text{ when}$$

$z = 0$, $0 \leq r < 1$ and $\frac{\partial v}{dz} = 0$ when $z = 0$, $r > 1$.

[**Ans:** $v(r,z) = \frac{2v_0}{\pi} \int_0^\infty e^{-pz} \frac{\sin p}{p} J_0(pr)dp$]

15. Heat is supplied at a constant rate Q per unit area per unti time over circular area of radius a in the plane z = 0 to an infinite solid of conductivity K. Show that the steady temperature at a point distant r from the axis of the circular area and distant z from the plane z = 0 is given by

$$\frac{Qa}{2k} \int_0^\infty \frac{e^{-pz}}{p} J_0(pr) J_1(pa)dp$$

16. The magnetic potential v for a circular disc of radius a and strength w magnetized parallel to its axis, satisfy Laplace's equation, is equal to $2\pi\omega$ on the disc itself and vanishes at exterior points in the plane of the disc. Show that at the point (r,z), z = 0.

$$v = 2\pi a\omega \int_0^\infty e^{-pz} J_0(rp) J_1(ap)dp$$

17. Viscous fluid is contained between two infinitely long concentric circular cylinders of radii a and b. The inner cylinder is kept at rest and outer cylinder suddenly starts rotating with uniform angular velocity ω. Find the velocity v of the fluid if the equation of motion is $\frac{\partial^2 v}{\partial r^2}+\frac{1}{r}\frac{\partial v}{\partial r}-\frac{v}{r^2}=\frac{1}{\nu}\frac{\partial v}{\partial t}$, a < r < b, t > 0, ν being the kinematic viscocity.

$$\textbf{[Ans: } v=\pi b\omega\sum_p \frac{1-e^{-\nu p^2 t}}{J_1^2(pa)-J_1^2(pb)}.J_1(pa)J_1(pb)B_1(pr),$$

$$\text{where } B_1(pr)=J_1(pr)Y_1(pa)-Y_1(pr)J_1(pa)\textbf{]}$$

Mellin Transforms

The mellin transform of the function f(x) over the interval 0 < x < ∞ is defined as

$$M\{f(x)\}=f^*(p)=\int_0^\infty x^{p-1}f(x)dx$$

It is also denoted by $M[f(x);p]$

Inversion Mellin Transform

$$f^*(p)=\int_0^\infty x^{p-1}f(x)dx$$

Then its inverse transform is

$$f(x)=\frac{1}{2\pi i}\int_{c-i\infty}^{c+i\infty} x^{-p}f^*(p)dp$$

Proof: If $\tilde{f}(s)$ be the complex fourier transform of the function f(x)

i.e.,
$$\tilde{f}(s)=\int_{-\infty}^{\infty} e^{ist}f(t)dt \quad \text{...(1)}$$

then
$$f(t)=\frac{1}{2\pi}\int_{-\infty}^{\infty}\tilde{f}(s)e^{-ist}ds \quad \text{...(2)}$$

Now, putting t = log x i.e., x = e^t and p = c+is in equation (1) and (2), we get

$$\tilde{f}(ic-ip)=\int_0^\infty x^{p-c}f(\log x).\frac{1}{x}dx$$

or
$$\tilde{f}(ic-ip)=\int_0^\infty x^{p-1}.x^{-c}f(\log x)\,dx \quad \text{...(3)}$$

and
$$f(\log x)=\frac{1}{2\pi i}\int_{c-i\infty}^{c+i\infty} x^{c-p}\tilde{f}(ic-ip)dp \quad \text{...(4)}$$

putting $x^{-c} f(\log x) = f(x)$ and $\tilde{f}(ic - ip) = f^*(p)$ in equation (3) and (4), we get

$$f^*(p) = \int_0^\infty x^{p-1} f(x)dx$$

and
$$f(x) = \frac{1}{2\pi i}\int_{c-i\infty}^{c+i\infty} x^{-p} f^*(p)dp$$

Linear Property

If f(x) and g(x) be the two functions then

$$M[af(x) + bg(x); p] = a\, M[f(x); p] + b\, M[g(x); p]$$

where a, b are two constants.

Proof:
$$M[af(x) + bg(x); p] = \int_0^\infty x^{p-1}\{af(x) + bg(x)\}\, dx$$

$$= a\int_0^\infty x^{p-1} f(x)dx + b\int_0^\infty x^{p-1} g(x)dx$$

$$= a\, M[f(x); p] + b\, M[g(x); p]$$

Some Elementary Properties

1. $M[f(ax); p] = a^{-p} f^*(p)$

Proof: We have $M[f(ax); p] = \int_0^\infty x^{p-1} f(ax)dx$

putting ax = t   ∴ adx = dt

$$\therefore \quad M[f(t); p] = \int_0^\infty \left(\frac{t}{a}\right)^{p-1} f(t).\frac{1}{a}dt$$

$$= a^{-p}\int_0^\infty t^{p-1} f(t)\, dt$$

or
$$M[f(x); p] = a^{-p}\int_0^\infty x^{p-1} f(x)dx = a^{-p} f^*(p)$$

2. $M[f(x^a); p] = a^{-1} f^*\left(\frac{p}{a}\right)$, a > 0

Proof: We have

$$M[f(x^a); p] = \int_0^\infty x^{p-1} f(x^a)dx$$

putting $x^a = t$ or $x = t^{\frac{1}{a}}$, $dx = \frac{1}{a}\, t^{\frac{1}{a}-1} dt$

$$\therefore \quad M[f(t);p] = \int_0^\infty \left(t^{\frac{1}{a}}\right)^{p-1} f(t).\frac{1}{a}\, t^{\frac{1}{a}-1}\, dt$$

$$= a^{-1}\int_0^\infty t^{\frac{p}{a}-1} f(t)\, dt$$

or $M[f(x);p] = a^{-1}\int_0^\infty x^{\frac{p}{a}-1}.f(x)dx = a^{-1}f^*(p/a)$

3. $M\left[x^a f(x);p\right] = f^*(p+a)$

Proof: We have

$$M\left[x^a f(x);p\right] = \int_0^\infty x^{p-1}.x^a f(x)dx$$

$$= \int_0^\infty x^{(p+a)-1} f(x)dx$$

$$= f^*(p+a)$$

4. $M\left[\frac{1}{x} f\left(\frac{1}{x}\right);p\right] = f^*(1-p)$

Proof: We have

$$M\left[\frac{1}{x} f\left(\frac{1}{x}\right);p\right] = \int_0^\infty x^{p-1}.\frac{1}{x} f\left(\frac{1}{x}\right)dx$$

putting $\frac{1}{x} = u$ or $x = \frac{1}{u}$, $dx = -\frac{1}{u^2}du$

$$= \int_\infty^0 \left(\frac{1}{u}\right)^{p-2} f(u)\left(-\frac{1}{u^2}\right)du$$

$$= \int_\infty^0 u^{-p} f(u)du$$

$$= \int_0^\infty x^{(1-p)-1} f(x)dx$$

$$= f^*(1-p)$$

5. $M[\log x\, f(x);p] = \frac{d}{dp} f^*(p)$

Proof: We have

$$M[\log x\, f(x);p] = \int_0^\infty x^{p-1}\log x\, f(x)dx \qquad \ldots\text{(i)}$$

and $f^*(p)=\int_0^\infty x^{p-1}f(x)dx$

$$\therefore \quad \frac{d}{dp}f^*(p)=\int_0^\infty x^{p-1}\log x\, f(x)dx \qquad \text{... (ii)}$$

$\therefore$ from equations (i) and (ii), we get

$$M[\log x\, f(x);p]=\frac{d}{dp}f^*(p)$$

Mellin Transform of Derivatives

$$M[f'(x);p]=\int_0^\infty x^{p-1}f'(x)dx$$

$$=\left[x^{p-1}f(x)\right]_0^\infty-\int_0^\infty (p-1)\,x^{p-2}f(x)dx$$

$$=0-(p-1)\int_0^\infty x^{(p-1)-1}f(x)\,dx$$

$$=-\,(p-1)\,M[f(x);p-1]$$

or $\quad M[f'(x);p]=-\,(p-1)f^*(p-1)$ ...(1)

Again

$$M[f''(x);p]=-\,(p-1)\,M[f'(x);p-1]$$

$$=-\,(p-1)-(p-2)\,M[f(x);p-2]$$

or $\quad M[f''(x);p]=(-1)^2(p-1)\,(p-2)f^*(p-2)$ ...(2)

Proceeding similar in a way, we get

$$M\left[f^n(x);p\right]=(-1)^n(p-1)\,(p-2)\,...\,(p-n)f^*(p-n)$$

or $\quad M\left[f^n(x);p\right]=(-1)^n\dfrac{\Gamma p}{\Gamma p-n}f^*(p-n)$ ...(3)

By property 3, we have

$$M\left[x^a f(x);p\right]=f^*(p+a)=M[f(x);p+a]$$

$$\therefore \quad M[x\,f'(x);p]=M[f'(x);p+1]$$

$$=-\{(p+1)-1\}f^*\{(p+1)-1\} \quad \text{from (1)}$$

or $\quad M\left[\left(x\dfrac{d}{dx}\right)f(x);p\right]=-pf^*(p)$ ...(4)

Now let $x\frac{d}{dx}f(x)=g(x)$

$$\therefore \quad M[x\,g'(x);p]=M[g'(x);p+1]$$

$$=-\{(p+1)-1\}\,M[g(x);(p+1)-1]$$

$$=-p\,M\left[x\frac{d}{dx}f(x);p\right]$$

$$=-p\left(-p\,f^*(p)\right) \qquad \text{from (4)}$$

or $$M\left[x\frac{d}{dx}\left\{x\frac{d}{dx}f(x)\right\};p\right]=(-p)^2 f^*(p)$$

or $$M\left[\left(x\frac{d}{dx}\right)^2 f(x);p\right]=(-p)^2 f^*(p) \qquad \ldots(5)$$

Proceeding similar in a way, we get

$$M\left[\left(x\frac{d}{dx}\right)^n f(x);p\right]=(-p)^n f^*(p) \qquad \ldots(6)$$

Mellin Transform of Integrals

Let $g(x)=\int_0^x f(u)\,du$

Now $$M[g'(x);p]=-(p-1)g^*(p-1)$$

$$=-(p-1)\,M[g(x);p-1]$$

or $$M[f(x);p]=-(p-1)\,M\left[\int_0^x f(u)\,du;p-1\right]$$

or $$M\left[\int_0^x f(u)du;p-1\right]=-\frac{1}{p-1}M[f(x);p]$$

replacing p by p + 1, we get

$$M\left[\int_0^x f(u)du;p\right]=-\frac{1}{p}M[f(x);p+1]$$

or $$M\left[\int_0^x f(u)du;p\right]=-\frac{1}{p}f^*(p+1)$$

Illustrative Examples

Example 1

Find $M\{\sin x\}$

Solution: We have $M\{\sin x\} = \int_0^\infty x^{p-1}.\sin x\, dx$

we know that

$$\int_0^\infty e^{-xt} \sin px\, dx = \left[\frac{e^{-xt}}{t^2+p^2}(-t\sin px - p\cos px)\right]_0^\infty$$

$$= \frac{p}{t^2+p^2}$$

$$= \frac{1}{2i}\left[(t-ip)^{-1} - (t+ip)^{-1}\right]$$

Differentiating both sides w.r.t.t,(m–1) times, we have

$$(-1)^{m-1}\int_0^\infty x^{m-1}e^{-xt}\sin px\, dx = \frac{1}{2i}(-1)^{m-1}(m-1)!\left[(t-ip)^{-m}(t+ip)^{-m}\right]$$

putting $t = r\cos\phi, p = r\sin\phi$ and on simplification

$$= (-1)^{m-1}(m-1)!\; r^{-m}\sin m\phi$$

$$= (-1)^{m-1}(m-1)!\; \frac{1}{(t^2+p^2)^{m/2}}\left\{\sin\left(m\, tan^{-1}\frac{p}{t}\right)\right\}$$

or $$\int_0^\infty x^{m-1}e^{-xt}\sin px\, dx = \frac{(m-1)!}{(t^2+p^2)^{m/2}}\sin\left\{m\, tan^{-1}\left(\frac{p}{t}\right)\right\}$$

replacing m = p, t = 0 and p = 1, we get

$$\int_0^\infty x^{p-1}\sin x\, dx = \Gamma p.\sin\left(\frac{p\pi}{2}\right)$$

Example 2

Find $M\{e^{-x}\}$

Solution: We have

$$M\{e^{-x}\} = \int_0^\infty x^{p-1}.e^{-x}dx = \Gamma p$$

Example 3

Find $M\left\{(1+x)^{-\alpha}\right\}$

Solution: We have

$$M\left\{(1+x)^{-\alpha}\right\}=\int_0^\infty x^{p-1}(1+x)^{-\alpha}\,dx \qquad \ldots(1)$$

we know that

$$B(m,n)=\int_0^\infty \frac{x^{n-1}}{(1+x)^{m+n}}\,dx, \quad m, n > 0$$

or

$$\int_0^\infty x^{n-1}(1+x)^{-(m+n)}\,dx = B(m,n)=\frac{\Gamma m\,\Gamma n}{\Gamma(m+n)}$$

putting m + n = α, m = α – n, we get

$$\int_0^\infty x^{n-1}(1+x)^{-\alpha}\,dx=\frac{\Gamma(\alpha-n)\,\Gamma n}{\Gamma\alpha}$$

Replacing n by p, we have

$$\int_0^\infty x^{p-1}(1+x)^{-\alpha}\,dx=\frac{\Gamma(\alpha-p)\,\Gamma p}{\Gamma\alpha}$$

Example 4

Prove that $M\left\{(1+x^a)^{-\alpha}\right\}=\dfrac{\Gamma(p/a)\,\Gamma(\alpha-p/a)}{a\,\Gamma\alpha}$

Solution: From example 3, we have

$$M\left\{(1+x)^{-\alpha}\right\}=\frac{\Gamma(\alpha-p)\,\Gamma p}{\Gamma\alpha}=f^*(p)$$

From property 2, we have

$$M\left\{f(x^a)\right\}=a^{-1}f^*(p/a)$$

$$\therefore \quad M\left\{(1+x^a)^{-\alpha}\right\}=a^{-1}f^*(p/a)$$

$$=\frac{1}{a}\cdot\frac{\Gamma(\alpha-p/a)\,\Gamma(p/a)}{\Gamma\alpha}$$

Example 5

Prove that $M\left[x^n f^n(x);p\right]=(-1)^n\dfrac{\Gamma(p+n)}{\Gamma p}f^*(p)$

Solution: By property 3, we have

$$M\left[x^n f(x);p\right]=f^*(p+n)$$

$$=M\left[f(x);p+n\right]$$

$$\therefore \qquad M\left[x^n f^n(x);p\right]=M\left[f^n(x);p+n\right]$$

$$=(-1)^n \frac{\Gamma(p+n)}{\Gamma(p+n-n)} f^*(p+n-n)$$

replacing p by p + n

$$=(-1)^n \frac{\Gamma(p+n)}{\Gamma(p)} f^*(p)$$

Convolution (or Falting) Theorem

If $f^*(p)$ and $g^*(p)$ be the Mellin transform of the functions f(x) and g(x) respectively, then

$$M\left[f(x)g(x);p\right]=\frac{1}{2\pi i}\int_{c-i\infty}^{c+i\infty} f^*(u)\, g^*(p-u)\, du$$

Proof: Let

$$M\left[f(x)g(x);p\right]=\int_0^{\infty} x^{p-1} f(x)\, g(x)\, dx$$

$$=\int_0^{\infty} x^{p-1} g(x)\left\{\frac{1}{2\pi i}\int_{c-i\infty}^{c+i\infty} f^*(u)\, x^{-u} du\right\} dx$$

$$=\frac{1}{2\pi i}\int_{c-i\infty}^{c+i\infty} f^*(u)\left\{\int_0^{\infty} x^{(p-u)-1} g(x) dx\right\} du$$

changing the order of integration

$$=\frac{1}{2\pi i}\int_{c-i\infty}^{c+i\infty} f^*(u) g^*(p-u) du$$

Practice Problems

Find the Mellin transforms of

1. $(1+x)^{-1}$

   [Ans: $\dfrac{\pi}{\sin(\pi p)}$]

2. $\cos x$

   [Ans: $\Gamma(p)\ \cos\left(\dfrac{p\pi}{2}\right)$]

3. $(1+x^9)^{-1}$, a > 0

   [Ans: $\dfrac{\pi}{a}\ \text{cosec}\left(\dfrac{\pi p}{a}\right)$]

4. Prove that $M\left[f(x)g(x);1\right]=\dfrac{1}{2\pi i}\displaystyle\int_{c-i\infty}^{c+i\infty} f^*(u) g^*(1-u) du$

5. Prove that $M\left[x^2\dfrac{d^2 f}{dx^2}+x\dfrac{df}{dx};p\right]=p^2 f^*(p)$

Chapter - 23

Integral Equation

Introduction : Integral equation is one of most important branches of mathematical analysis, particularly on account of its importance in boundary value problems in the ordinary and partial differential equations. It is one of the most useful techniques in many branches of pure analysis, such as stochastic process, functional analysis, fields of mechanics, mathematical physics and many more. The differential equations can be replaced by an integral equation with the help of initial and boundary conditions.

Integral Equation : An equation in which an unknown function, to be determined, appears under one or more integral sings, is called integral equation.

Integro Differential Equation : If the derivatives of the function are involved, it is called integro-differential equation.

Linear Integral Equation : An integral equation upon the known functions.

An equation of the form

$$a(x)\phi(x)+F(x)+\lambda\int_{\Omega}K(x,\xi)\phi(\xi)d\xi=0$$

is called a linear integral equation, where $\phi(x)$ is the unknown function. $\alpha(x)$, $F(x)$ and $K(x,\xi)$ are known functions.

λ is a non-zero real or complex parameter and the integration extends over the domain Ω of the auxiliary variable ξ.

Note: $K(x,\xi)$ is called Kernel of the integral equation.

There are two basic types of linear integral equation.

(i) Volterra integral equation

(ii) Fredholm integral equation

I. Volterra Integral Equation : If the upper limit of integration is a variable, then it is said to be a Volterra integral equation

$$a(x)\phi(x) = F(x) + \lambda\int_a^x K(x,\xi)\phi(\xi)d\xi$$

(i) When $\alpha = 0$, then

$$F(x) = \lambda\int_a^x K(x,\xi)\phi(\xi)d\xi,\ a > -\infty$$

is called the Volterra integral equation of first kind.

(ii) When $\alpha = 1$, then

$$\phi(x) = F(x) + \lambda\int_a^x K(x,\xi)\phi(\xi)d\xi$$

is called the Volterra non-homogeneous integral equation of Second Kind.

(iii) When $\alpha = 1$, and $F(x) = 0$, then

$$\phi(x) = \lambda\int_a^x K(x,\xi)\phi(\xi)d\xi$$

is called the homogeneous Volterra integral equation of Second Kind.

II. Fredholm Integral Equation : If both the limits of integration are constant, then it is said to be fredholm integral equation.

$$a(x)\phi(x) = F(x) + \lambda\int_a^b K(x,\xi)\phi(\xi)d\xi$$

(i) When $\alpha = 0$, then

$$F(x) = \lambda\int_a^b K(x,\xi)\phi(\xi)d\xi, a \le x \le b$$

is called the Fredholm integral equation of First Kind.

(ii) When $\alpha = 1$, then

$$\phi(x) = F(x) + \lambda\int_a^b K(x,\xi)\phi(\xi)d\xi, a \le x \le b.$$

is called the Fredholm non-homogeneous integral equation of Second Kind.

(iii) When $\alpha = 1$, and $F(x) = 0$, then

$$\phi(x) = \lambda\int_a^b K(x,\xi)\phi(\xi)d\xi, a \le x \le b.$$

is called the Fredholm homogeneous integral equation of Second Kind.

Non-Linear Integral Equation : If the unknown function appears under an integral sign to a power $n(>1)$, then the equation is said to be a non-linear integral equation.

$$\phi(x)=F(x)+\lambda\int_a^b K(x,\xi)\phi^n(\xi)d\xi.$$

Singular Integral Equation : When Kernel becomes infinite at one or more points within the range of integration or one or both limits of integration becomes infinite, the integral equation is called singular integral equation e.g.

$$\phi(x)=f(x)+\lambda\int_{-\infty}^{\infty} e^{-|x-\xi|}\phi(\xi)d\xi$$

and $$f(x)=\int_a^x (x-\xi)^{-\alpha}\phi(\xi)d\xi,\ 0<\alpha<1$$

Note : If $\alpha=-\frac{1}{2}$, then second equation is said to be the Abel's integral equation.

Convolution Integral : If the Kernal of the integral equation is defined as a function of the difference, i.e.

$$K(x,\xi)=K(x-\xi).$$

Then the integral equation

$$\phi(x)=F(x)+\lambda\int_a^x K(x-\xi)\phi(\xi)d\xi$$

and the corresponding Fredholm equation

$$\phi(x)=xe^x$$

are called integral equation of the convolution type.

The function defined by the integral

$$\int_0^x K(x-\xi)\phi(\xi)d\xi=\int_0^x K(\xi)\phi(x-\xi)d\xi$$

is called the Convolution or the Faltung of the two functions K and ϕ and are known as the Convolution integrals.

In general, the convolution integrals may be defined as

$$\int_{-\infty}^{\infty} K(x-\xi)\phi(\xi)d\xi=\int_{-\infty}^{\infty} K(\xi)\phi(x-\xi)d\xi$$

Solution of Integral Equation :

Example 1 Show that the function $\phi(x)=xe^x$ is a solution of the Volterra integral equation

$$\phi(x)=\sin x+2\int_0^x \cos(x-\xi)\phi(\xi)d\xi.$$

Solution : Put $\phi(x)=xe^x$ in the given integral equation, we get

$$=\sin x+2\int_0^x \cos(x-\xi).\xi e^\xi\, d\xi$$

$$=\sin x+2\left[cosx\int_0^x \xi e^\xi \cos\xi\, d\xi+\sin x\int_0^x \xi e^\xi \sin\xi\, d\xi\right]$$

$$=\sin x+2\left[\cos x\left\{\frac{\xi e^\xi}{2}(\cos\xi+\sin\xi)\Big|_0^x-\frac{1}{2}\int_0^x e^\xi(\cos\xi+\sin\xi)\,d\xi\right\}\right.$$

$$\left.+\sin x\left\{\frac{\xi e^\xi}{2}(\sin\xi-\cos\xi)\Big|_0^x-\frac{1}{2}\int_0^x e^\xi(\sin\xi-\cos\xi)\,d\xi\right\}\right]$$

Note : $\int e^{ax}\cos bx dx=\frac{e^{ax}}{a^2+b^2}(a\cos bx+b\sin bx)$ and

$$\int e^{ax}\sin bx\, dx=\frac{e^{ax}}{a^2+b^2}(a\sin bx-b\cos bx)$$

$$=\sin x+2\left[cox\left\{\frac{xe^x}{2}(\cos x+\sin x)\right\}-\frac{1}{2}\cos x\left\{\frac{1}{2}e^\xi(\cos\xi+\sin\xi+\sin\xi-\cos\xi)\right\}_0^x\right.$$

$$\left.+\sin x\left\{\frac{xe^x}{2}(\sin x-\cos x)\right\}-\frac{1}{2}\sin x\left\{\frac{1}{2}e^\xi(\sin\xi-\cos\xi-\cos\xi-\sin\xi)\right\}_0^x\right]$$

$$=\sin x+xe^x(\cos^2 x+\sin^2 x)-\cos x(e^\xi\sin\xi)_0^x+\sin x(e^\xi\cos\xi)_0^x$$

$$=\sin x+xe^x-\cos x(e^x\sin x)+\sin x(e^x\cos x-1)$$

$$=xe^x$$

Hence $\phi(x)=xe^x$ is a solution of the given integral equation.

Example 2 Show that the function $\phi(x)=1$ is a solution of the Fredholm integral equation

$$\phi(x)+\int_0^1 x(e^{x\xi}-1)\phi(\xi)d\xi = e^x - x.$$

Solution Put $\phi(x)=1$ in the given integral equation, we get

$$\text{L.H.S.} = 1+\int_0^1 x(e^{x\xi}-1)d\xi$$

$$=1+\left\{x\left(\frac{e^{x\xi}}{x}-\xi\right)\right\}_0^1$$

$$=1+\left(e^{x\xi}-x\xi\right)_0^1$$

$$=1+e^x-x-1$$

$$=e^x-x=\text{R.H.S.}$$

Thus $\phi(x)=1$ is solution of the given integral equation.

Example 3 Show that the function $\phi(x)=1-x$ is a solution of the integral equation.

$$\int_0^x e^{x-\xi}\phi(\xi)d\xi = x.$$

Solution Put $\phi(x)=1-x$ in the given equation, we get

$$\text{L.H.S.} = \int_0^x e^{x-\xi}(1-\xi)d\xi$$

$$=e^x\int_0^x e^{-\xi}d\xi - e^x\int_0^x \xi e^{-\xi}\,d\xi$$

$$=-e^x(e^{-\xi})_0^x + e^x(\xi e^{-\xi}+e^{-\xi})_0^x$$

$$=-e^x(e^{-x}-1)+e^x(xe^{-x}+e^{-x}-1)$$

$$=-1+e^x+x+1-e^x$$

$$=x=\text{R.H.S.}$$

Hence $\phi(x)=1-x$ is a solution of the given integral equation.

Example 4 Show that $\phi(x) = \cos 2x$ is a solution of the equation

$$\phi(x) = \cos x + 3\int_0^{\pi} K(x,\xi)\,\phi(\xi)d\xi$$

Where $K(x,\xi) = \begin{cases} \sin x \cos \xi, & 0 \le x \le \xi \\ \cos x \sin \xi, & \xi \le x \le \pi \end{cases}$

Solution Given

$$\phi(x) = \cos x + 3\int_0^{\pi} K(x,\xi)\,\phi(\xi)d\xi$$

$$\phi(x) = \cos x + 3\left[\int_0^{x} K(x,\xi)\phi(\xi)d\xi + \int_x^{\pi} K(x,\xi)\phi(\xi)d\xi\right]$$

$$= \cos x + 3\left[\int_0^{x} \cos x \sin \xi\phi(\xi)d\xi + \int_x^{\pi} \sin x \cos \xi\phi(\xi)d\xi\right] \qquad .....(1)$$

$\left[\because \right.$ $t > 0$ $\quad t \le x$ $\quad t > x$; $t = 0$ $\quad t = x$ $\quad t = \pi$ $\left.\right]$

Since $\phi(x) = \cos 2x$ is a solution of the equation, so it must satisfy equation (1).

Hence

$$\phi(x) = \cos + 3\left[\int_0^{x} \cos x \sin \xi \cos 2\xi\, d\xi + \int_x^{\pi} \sin x \cos \xi\, \cos 2\xi\, d\xi\right]$$

$$= \cos x + \frac{3}{2}\left[\cos x\int_0^{x} (\sin 3\xi - \sin \xi)d\xi + \sin x\int_x^{\pi} (\cos 3\xi + \cos \xi)d\xi\right]$$

$$= \cos x + \frac{3}{2}\left[\cos x\left(-\frac{\cos 3\xi}{3} + \cos \xi\right)_0^{x} + \sin x\left(\frac{\sin 3\xi}{3} + \sin \xi\right)_x^{\pi}\right]$$

$$= \cos x + \frac{3}{2}\left[\left(-\frac{1}{3}\cos x \cos 3x + \cos^2 x + \frac{\cos x}{3} - \cos x\right) + \sin x\left(-\frac{1}{3}\sin 3x - \sin x\right)\right]$$

$$= \cos x + \frac{3}{2}\left[\frac{2}{3}\cos 2x - \frac{2}{3}\cos x\right]$$

$\Rightarrow \phi(x) = \cos 2x$ is a solution of the given integral equation.

Example 5 Show that the function $\phi(x) = e^x\left[2x - \frac{2}{3}\right]$ is a solution of the Fredholm integral equation $\phi(x) + \lambda\int_0^1 e^{x-\xi}\phi(\xi)d\xi = 2xe^x, \lambda = 2.$

Solution Put $\phi(x) = e^x\left[2x - \frac{2}{3}\right]$ in the L.H.S. of the equation, we get

$$\text{L.H.S.} = e^x\left[2x - \frac{2}{3}\right] + 2\int_0^1 e^x\left(2\xi - \frac{2}{3}\right)d\xi$$

$$= e^x\left[2x - \frac{2}{3}\right] + 2e^x\left(\xi^2 - \frac{2}{3}\xi\right)_0^1$$

$$= e^x\left[2x - \frac{2}{3}\right] + 2e^x\left(1 - \frac{2}{3}\right)$$

$$= e^x\left[2x - \frac{2}{3} + 2\times\frac{1}{3}\right]$$

$$= 2xe^x = \text{R.H.S.}$$

Hence $\phi(x) = e^x\left[2x - \frac{2}{3}\right]$ is a solution of the given integral equation.

Example 6 Show that the function $\phi(x) = 2 - x$ is a solution of the integral equation.

$$\int_0^x e^{x-\xi}\phi(\xi)d\xi = e^x + x - 1.$$

Solution Put $\phi(x) = 2 - x$ in the L.H.S. of the equation, we get

$$\text{L.H.S.} = \int_0^x e^{x-\xi}(2 - \xi)d\xi$$

$$= e^x\left[2\int_0^x e^{-\xi}d\xi - \int_0^x \xi e^{-\xi}\,d\xi\right]$$

$$= e^x\left[2(-e^{-\xi})_0^x - \left\{-\xi e^{-\xi} - e^{-\xi}\right\}_0^x\right]$$

$$= e^x\left[-2e^{-x} + 2 + xe^{-x} + e^{-x} - 1\right]$$

$$= -2 + 2e^x + x + 1 - e^x$$

$$= e^x + x - 1 = \text{R.H.S.}$$

Hence, $\phi(x) = 2 - x$ is a solution of the given integral equation.

Example 7 Show that the function $\phi(x) = (1+x^2)^{-3/2}$ is a solution of the Volterra integral equation $\phi(x) = \dfrac{1}{1+x^2} - \displaystyle\int_0^x \frac{\xi}{1+x^2}\phi(\xi)d\xi.$

Solution Put $\phi(x) = (1+x^2)^{-\frac{3}{2}}$ in the given equation, we get

$$= \frac{1}{1+x^2} - \int_0^x \frac{\xi}{1+x^2}(1+\xi^2)^{-\frac{3}{2}} d\xi$$

$$= \frac{1}{1+x^2} - \frac{1}{1+x^2}\int_0^x \frac{\xi}{(1+\xi^2)^{\frac{3}{2}}} d\xi$$

$$= \frac{1}{1+x^2} - \frac{1}{1+x^2}\left(\frac{-1}{(1+\xi^2)^{\frac{1}{2}}}\right)_0^x$$

$$= \frac{1}{1+x^2} + \frac{1}{1+x^2}\left(\frac{1}{(1+x^2)^{1/2}} - 1\right)$$

$$= \frac{1}{(1+x^2)^{\frac{3}{2}}}$$

Hence, $\phi(x) = (1+x^2)^{-\frac{3}{2}}$ is a solution of the given integral equation.

Example 8 Verify that the given function $\phi(x) = \dfrac{1}{2}$ is the solution of the integral equation

$$\int_0^x \frac{\phi(\xi)}{\sqrt{x-\xi}} d\xi = \sqrt{x}.$$

Solution Put $\phi(x) = \dfrac{1}{2}$ in the given equation, we have

$$= \int_0^x \frac{1/2}{\sqrt{x-\xi}} d\xi$$

$$= \frac{1}{2}\left(-2\sqrt{x-\xi}\right)_0^x$$

$$= -(0-\sqrt{x}) = \sqrt{x}$$

Hence, $\phi = \frac{1}{2}$ is a solution of the integral equation.

Example 9 For what value of λ the function $\phi(x) = 1 + \lambda x$ is a solution of the integral equation $x = \int_0^x e^{x-\xi} \phi(\xi) d\xi$.

Solution The integral equation is $x = \int_0^x e^{x-\xi} \phi(\xi) d\xi$

Put $\phi(x) = 1 + \lambda x$, we get

$$x = \int_0^x e^{x-\xi}(1+\lambda\xi) d\xi$$

$$\Rightarrow \quad x = e^x \int_0^x e^{-\xi}(1+\lambda\xi) d\xi$$

$$\Rightarrow \quad x = e^x\{-e^{-\xi} + \lambda(-\xi e^{-\xi} - e^{-\xi})\}_0^x$$

$$\Rightarrow \quad x = -e^x\{e^{-x} + \lambda(xe^{-x} + e^{-x}) - 1 - \lambda\}$$

$$\Rightarrow \quad x = -1 - \lambda x - \lambda + (1+\lambda)e^x$$

$$\Rightarrow \quad x(1+\lambda) + (1+\lambda) - (1+\lambda)e^x = 0$$

$$\Rightarrow \quad (1+\lambda)(x+1-e^x) = 0$$

Since $x + 1 - e^x \neq 0 \quad \therefore 1 + \lambda = 0$

$$\Rightarrow \quad \lambda = -1.$$

Relation between differential and integral equation :

Consider the differential equation of nth order as

$$\frac{d^n y}{dx^n} + a_1(x)\frac{d^{n-1} y}{dx^{n-1}} + a_2(x)\frac{d^{n-2} y}{dx^{n-2}} + \ldots + a_n(x) y = F(x) \qquad \ldots..(1)$$

With initial conditions

$$y(0) = C_0, y'(0) = C_1, y''(0) = C_2, \ldots\ldots, y^{n+1}(0) = C_{n+1} \qquad \ldots..(2)$$

Now consider $\dfrac{d^n y}{dx^n} = \phi(x)$

By integrating and using the initial condition (2), we have

$$\frac{d^{n-1}y}{dx^{n-1}} = \int_0^x \phi\,(\xi)d\xi + C_{n-1},$$

$$\frac{d^{n-2}y}{dx^{n-2}} = \int_0^x \phi(\xi)d\xi^2 + C_{n-1}x + C_{n-2}$$

....

$$y = \int_0^x \phi(\xi)d\xi^n + C_{n-1}\frac{x^{n-1}}{(n-1)!} + C_{n-2}\frac{x^{n-2}}{(n-2)!} + \ldots.. + C_0 \qquad \ldots..(3)$$

From equation (3) and (1), we obtain

$$\phi(x) + a_1(x)\int_0^x \phi(\xi)d\xi + a_2(x)\int_0^x \phi(\xi)d\xi^2 + \ldots.. + a_n(x)\int_0^x \phi(\xi)d\xi^n$$

$$= F(x) + \sum_{i=1}^{n} iC_i X_i(x) \qquad \ldots..(4)$$

Where $X_i(x) = a_i(x) + \dfrac{x}{1!}a_{i+1}(x) + \ldots.. + a_n(x)\dfrac{x^{n-i}}{(n-i)!}$ .....(5)

$$\phi(x) + \int_0^x \left[a_1(x) + a_2(x)(x-\xi) + \ldots.. + a_n(x)\frac{(x-\xi)^{n-1}}{(n-1)!} \right]\phi(\xi)d\xi = G_i(x) \qquad \ldots..(6)$$

Where $G_i(x) = F(x) + \sum_{i=1}^{n} iC_i X_i(x)$.

The equation (6) represents the non-homogeneous Volterra integral equation of second kind.

Example 10 Form an integral equation corresponding to the differential equation.

$$\frac{d^2y}{dx^2}+x\frac{dy}{dx}+y=0$$

With the initial conditions $y(0)=1, y'(0)=0$.

Solution Consider $\frac{d^2y}{dx^2}=\phi(x)$ .....(1)

By integrating w.r.t. x, we have

$$\frac{dy}{dx}=\int_0^x \phi(\xi)d\xi + A$$

Since $y'(0)=0$, then $A=0$.

$$\therefore \frac{dy}{dx}=\int_0^x \phi(\xi)d\xi \quad .....(2)$$

and $y=\int_0^x \phi(\xi)d\xi^2+B$

'or' $y=\int_0^x (x-\xi)\phi(\xi)\,d\xi+1$, since $y(0)=1$. .....(3)

From the relations (1), (2) and (3), the given differential equation reduces to

$$\phi(x)+x\int_0^x \phi(\xi)d\xi+\int_0^x (x-\xi)\phi(\xi)\,d\xi+1=0$$

$$\phi(x)=-1-\int_0^x (2x-\xi)\phi(\xi)\,d\xi$$

represents a Volterra's integral equation of second kind.

Example 11 Form an integral equation corresponding to the differential equation

$$\frac{d^3y}{dx^3}+x\frac{d^2y}{dx^2}+(x^2-x)y=xe^x+1$$

with initial conditions

$$y(0)=1=y'(0), y''(0)=0.$$

Solution Consider $\frac{d^3y}{dx^3}=\phi(x)$ .....(1)

Then $\dfrac{d^2y}{dx^2} = \int_0^x \phi(\xi)d\xi + A = \int_0^x \phi(\xi)d\xi$, since $y''(0) = 0$ .....(2)

$$\frac{dy}{dx} = \int_0^x (x-\xi)\phi(\xi)d\xi + B = \int_0^x (x-\xi)\phi(\xi)d\xi + 1, \text{ since } y'(0) = 1$$

and $\quad y = \dfrac{1}{2!}\int_0^x (x-\xi)^2\phi(\xi)d\xi + x + c$

$\therefore \quad y = \dfrac{1}{2!}\int_0^x (x-\xi)^2\phi(\xi)d\xi + x + 1$, since $y(0) = 1$ .....(3)

From the relations (1), (2) and (3), the given differential equation reduces to

$$\phi(x) + x\int_0^x \phi(\xi)d\xi + (x^2 - x)\left[\frac{1}{2!}\int_0^x (x-\xi)^2\phi(\xi)d\xi + x + 1\right]$$

$$= xe^x + 1$$

or $\quad \phi(x) = xe^x + 1 - x(x^2 - 1) - \displaystyle\int_0^x \left[x + \frac{1}{2}(x^2 - x)(x-\xi)^2\right]\phi(\xi)d\xi$

represents a Volterra's integral equation of Second Kind.

Example 12 Convert the differential equation

$y''(x) - 3y'(x) + 2y(x) = 4\sin x,$

With the initial condition $y(0) = 1, y'(0) = -2$ into Volterra's integral equation of Second Kind.

Solution Consider $\dfrac{d^2y}{dx^2} = \phi(x)$ .....(1)

$\dfrac{dy}{dx} = \displaystyle\int_0^x \phi(\xi)d\xi - 2$ .....(2)

and $\quad y = \displaystyle\int_0^x (x-\xi)\phi(\xi)d\xi - 2x + 1$ .....(3)

Substituting the relations (1), (2) and (3) in the given differential equation, we obtain

$$\phi(x) = -3\left[\int_0^x \phi(\xi)d\xi - 2\right] + 2\left[\int_0^x (x-\xi)\phi(\xi)\,d\xi - 2x + 1\right] = \sin x$$

$$\phi(x) = 4(\sin x + x - 2) + \int_0^x [3 - 2(x-\xi)]\phi(\xi)\,d\xi$$

where $F(x) = 4(\sin x + x - 2), \lambda = 1,$
and $K(x,\xi) = 3 - 2(x - \xi)$

Which represents the Volterra's integral equation of Second Kind.

Example 13 Reduce initial value problem

$$\phi''(x) + \lambda\phi(x) = F(x)$$

with $\phi(0) = 1, \phi'(0) = 0.$

Solution The differential equation is given as

$$\phi''(x) + \lambda\phi(x) = F(x)$$

$$\Rightarrow \quad \phi''(x) = F(x) - \lambda\phi(x)$$

Integrating both the sides with regard to x, we have

$$\int_0^x \phi''(x)dx = \int_0^x [F(x) - \lambda\phi(x)]dx$$

$$\Rightarrow \quad [\phi'(x)]_0^x = \int_0^x [F(x) - \lambda\phi(x)]dx$$

$$\Rightarrow \quad \phi'(x) - \phi'(0) = \int_0^x [F(x) - \lambda\phi(x)]dx$$

$$\Rightarrow \quad \phi'(x) = \int_0^x [F(x) - \lambda\phi(x)]dx \quad \text{since } \phi'(0) = 0$$

Integrating both the sides with regard to x, we have

$$\int_0^x \phi'(x)dx = \int_0^x [F(x) - \lambda\phi(x)]dx^2$$

$$\Rightarrow \quad \phi(x) - \phi(0) = \int_0^x [F(\xi) - \lambda\phi(\xi)]d\xi^2$$

$$\Rightarrow \quad \phi(x) = 1 + \int_0^x (x - \xi)[F(\xi) - \lambda\phi(\xi)]d\xi, \text{ since } \phi(0) = 1$$

Which represents a Volterra's integral equation of Second Kind.

Solution Form an integral equation corresponding to the differential equation

$$\frac{d^2y}{dx^2} - \sin x\frac{dy}{dx} + e^x y = x$$

with the initial conditions $y(0) = 1, y'(0) = -1.$

Solution Consider $\frac{d^2y}{dx^2} = \phi(x)$ .....(1)

$$\frac{dy}{dx} = \int_0^x \phi(\xi)d\xi - 1, \text{ since } y'(0) = 1. \quad .....(2)$$

and $$y = \int_0^x (x-\xi)\phi(\xi)d\xi - x + 1, \text{ since } y(0) = 1. \quad .....(3)$$

Substituting the relations (1), (2) and (3) in the given differential equation, we have

$$\phi(x) - \sin x\left[\int_0^x \phi(\xi)d\xi - 1\right] + e^x\left[\int_0^x (x-\xi)\phi(\xi)d\xi - x + 1\right] = x$$

$$\Rightarrow \phi(x) = [x - \sin x + e^x(x-1)] + \int_0^x [\sin x - e^x(x-\xi)]\phi(\xi)d\xi.$$

represents a Volterra's integral equation of Second Kind.

Example 15 Convert differential equation $\frac{d^2\phi}{dx^2} + \lambda\phi = 0$ with the initial conditions $\phi(0) = 0, \phi(l) = 0,$ into Fredholm integral equation of Second Kind. Also, recover the original differential equation from the integral equation you obtain.

Solution The differential equation may be written as

$$\frac{d^2\phi}{dx^2} = -\lambda\phi \quad .....(1)$$

Integrating both the sides with regard to x, we have

$$\int_0^x \frac{d^2\phi}{dx^2}dx = -\lambda\int_0^x \phi dx$$

$$\Rightarrow \left[\frac{d\phi}{dx}\right]_0^x = -\lambda\int_0^x \phi dx$$

$$\Rightarrow \phi'(x) - \phi'(0) = -\lambda\int_0^x \phi(x)dx$$

Consider $\phi'(0) = A$ (constant), then

$$\phi'(x) = A - \lambda\int_0^x \phi(x)dx$$

Integrating both the sides with regard to x, we have

$$[\phi(x)]_0^x = Ax - \lambda\int_0^x \phi(x)dx^2$$

$$\Rightarrow \quad \phi(x) - \phi(0) = Ax - \lambda\int_0^x (x-\xi)\phi(\xi)d\xi$$

$$\Rightarrow \quad \phi(x) = Ax - \lambda\int_0^x (x-\xi)\phi(\xi)d\xi, \text{ since } \phi(0) = 0 \qquad .....(2)$$

since $\phi(l) = 0$, then from equation (2), we have

$$\phi(l) = Al - \lambda\int_0^l (l-\xi)\phi(\xi)d\xi.$$

$$\Rightarrow \quad 0 = Al - \lambda\int_0^l (l-\xi)\phi(\xi)d\xi$$

$$\Rightarrow \quad A = \frac{\lambda}{l}\int_0^l (l-\xi)\phi(\xi)d\xi \qquad .....(3)$$

From the relations (2) and (3), we have

$$\phi(x) = \frac{\lambda}{l}x\int_0^1 (l-\xi)\phi(\xi)d\xi - \lambda\int_0^x (x-\xi)\phi(\xi)d\xi$$

$$\Rightarrow \phi(x) = \int_0^x \frac{\lambda x}{l}(l-\xi)\phi(\xi)d\xi + \int_x^l \frac{\lambda x}{l}(l-\xi)\,\phi(\xi)d\xi - \lambda\int_0^x (x-\xi)\phi(\xi)d\xi.$$

$$= \lambda\int_0^x \left[\frac{x(l-\xi)}{l} - (x-\xi)\right]\phi(\xi)d\xi + \int_x^l \frac{\lambda x(l-\xi)}{l}\phi(\xi)d\xi.$$

$$= \lambda\left[\int_0^x \frac{\xi(l-x)}{l}\phi(\xi)d\xi + \int_x^l \frac{x(l-\xi)}{l}\phi(\xi)d\xi\right]$$

$$= \lambda\int_0^1 K(x,\xi)\phi(\xi)d\xi. \qquad .....(4)$$

where

$$K(x,\xi) = \begin{cases} \dfrac{\xi(l-x)}{l}, & \text{if } 0 \le \xi \le x \\ \dfrac{x(l-\xi)}{l}, & \text{if } x \le \xi \le l \end{cases}$$

which represents Fredholm integral equation of Second Kind.

Converse : The integral equation may be taken as

$$\phi(x)=\int_0^l \frac{\lambda x(l-\xi)}{l}\phi(\xi)d\xi-\int_0^x \lambda(x-\xi)\phi(\xi)\,d\xi. \qquad .....(5)$$

Differentiating both the sides with regards to x, we have

$$\phi'(x)=\frac{d}{dx}\int_0^1 \frac{\lambda(l-\xi)x}{l}\phi(\xi)d\xi-\frac{d}{dx}\int_0^x \lambda(x-\xi)\phi(\xi)d\xi.$$

$$=\int_0^1 \frac{\partial}{\partial x}\left\{\frac{\lambda(l-\xi)x}{l}\phi(\xi)\right\}d\xi-\int_0^x \frac{\partial}{\partial x}\{\lambda x-\xi\}\phi(\xi)\}d\xi$$

$$=\int_0^l \frac{\partial}{\partial x}\left\{\frac{d(l-\xi)x}{l}\phi(\xi)\right\}d\xi-\int_0^x \frac{\partial}{\partial x}\{\lambda(2-\xi)\phi(\xi)\}\,d\xi$$

$$\phi'(x)=\int_0^1 \frac{\lambda(l-\xi)}{l}\phi(\xi)\,d\xi-\int_0^x \lambda\phi(\xi)d\xi$$

Differentiating both the sides with regard to x, we have.

$$\phi''(x)=\frac{d}{dx}\int_0^l \frac{\lambda(l-\xi)}{l}\phi(\xi)d\xi-\frac{d}{dx}\int_0^x \lambda\phi(\xi)d\xi$$

$$=\int_0^l \frac{\partial}{\partial x}\left\{\frac{\lambda(l-\xi)}{l}\phi(\xi)\right\}d\xi-\int_0^x \frac{\partial}{\partial x}\{\lambda\,\phi(\xi)\}d\xi-\lambda\phi(x)$$

$$\Rightarrow \qquad \phi''(x)+\lambda\,\phi(x)=0$$

From the relation (5), we have

$\phi(0)=0$ and $\phi(l)=0$

Which is the required given differential equation.

Example 16 Obtain Fredholm integral equation of Second Kind corresponding to the boundary volume problem $\frac{d^2\phi}{dx^2}+x\phi=1,\ \phi(0)=0,\ \phi(1)=1.$

Also, recover the boundary value problem from the integral equation you obtain.

Solution The given differential equation may be written as $\frac{d^2\phi}{dx^2}=1-x\phi(x)$

Integrating both the sides with regard to x, we have

$$\int_0^x \frac{d^2\phi}{dx^2}dx = \int_0^x dx - \int_0^x x\,\phi(x)dx$$

$$\Rightarrow \quad \phi'(x) - \phi'(0) = x - \int_0^x x\,\phi(x)dx$$

Consider $\phi'(0) = A$, then

$$\phi'(x) = A + x - \int_0^x x\phi(x)dx$$

Integrating both the sides with regard to x, we have

$$[\phi(x)]_0^x = Ax + \frac{x^2}{2} - \int_0^x x\,\phi(x)dx^2$$

$$\Rightarrow \quad \phi(x) - \phi(0) = Ax + \frac{x^2}{2} - \int_0^x (x-\xi)\xi\,\phi(\xi)d\xi$$

Since $\phi(0) = 0$, then

$$\phi(x) = Ax + \frac{x^2}{2} - \int_0^x (x-\xi)\xi\phi(\xi)d\xi$$

$$\phi(x) = Ax + \frac{x^2}{x} - \int_0^x (x-\xi)\xi\,\phi(\xi)d\xi$$

Since $\quad \phi(1) = 1$

$$\Rightarrow \quad 1 = A + \frac{1}{2} - \int_0^1 (1-\xi)\xi\,\phi(\xi)d\xi$$

$$\Rightarrow \quad A = \frac{1}{2} + \int_0^1 (1-\xi)\xi\,\phi(\xi)d\xi$$

$$\Rightarrow \quad \phi(x) = \frac{1}{2}x(1+x) + \int_0^1 x(1-\xi)\xi\,\phi(\xi)d\xi - \int_0^x (x-\xi)\xi\,\phi(\xi)d\xi$$

$$= \frac{1}{2}x(1+x) + \int_0^x x(1-\xi)\xi\phi(\xi)d\xi + \int_x^1 x(1-\xi)\xi\phi(\xi)d\xi$$

$$-\int_0^x (x-\xi)\xi\phi(\xi)d\xi$$

$$= \frac{1}{2}x(1+x) + \int_0^x \xi^2(1-x)\phi(\xi)d\xi + \int_x^1 x\xi(1-\xi)\phi(\xi)d\xi$$

$$\Rightarrow \quad \phi(x) = \frac{1}{2}x(1+x) + \int_0^1 K(x,\xi)\phi(\xi)d\xi \quad .....(1)$$

where

$$K(x,\xi) = \begin{cases} \xi^2(1-x), & 0 \le \xi \le x \\ x\xi(1-\xi), & x \le \xi \le 1 \end{cases}$$

Which represents Fredholm integral equation of Second Kind.

Convers : The integral equation may be taken as.

$$\phi(x) = \frac{1}{2}x(1+x) + \int_0^1 x(1-\xi)\xi\,\phi(\xi)d\xi - \int_0^x (x-\xi)\xi\,\phi(\xi)d\xi \quad .....(2)$$

Differentiating both the sides with regard to x, we get

$$\phi'(x) = \frac{1}{2}(1+2x) + \frac{d}{dx}\int_0^1 x(1-\xi)\xi\,\phi(\xi)d\xi - \frac{d}{dx}\int_0^x (x-\xi)\xi\,\phi(\xi)d\xi$$

$$= \frac{1}{2}(1+2x) + \int_0^1 \frac{\partial}{\partial x}\{x(1-\xi)\xi\,\phi(\xi)\}d\xi - \int_0^x \frac{\partial}{\partial x}\{(x-\xi)\xi\,\phi(\xi)\}d\xi$$

$$= \frac{1}{2}(1+2x) + \int_0^1 \frac{\partial}{\partial x}\{x(1-\xi)\xi\,\phi(\xi)\}d\xi \int_0^x \frac{\partial}{\partial x}\{x-\xi)\xi\,\phi(\xi)\}d\xi$$

$$= \frac{1}{2}(1+2x) + \int_0^1 (1-\xi)\xi\,\phi(\xi)d\xi - \int_0^x \xi\,\phi(\xi)d\xi$$

Differentiating both the sides with regard to x, we have

$$\phi''(x) = 1 + \frac{d}{dx}\int_0^1 (1-\xi)\xi\,\phi(\xi)d\xi - \frac{d}{dx}\int_0^x \xi\phi(\xi)d\xi$$

$$= 1 + \int_0^1 \frac{\partial}{\partial x}\{(1-\xi)\xi\,\phi(\xi)\}d\xi - \int_0^x \frac{\partial}{\partial x}\{\xi\phi(\xi)\}d\xi - x\phi(x)$$

$$\Rightarrow \quad \phi''(x) + x\phi(x) = 1$$

From equation (2), we have

$\phi(0) = 0,\ \phi(1) = 1$

Which is required differential equation.

Example 17 Obtain Fredholm integral equation of Second Kind corresponding to the boundary value problems.

$$\frac{d^2\phi}{dx^2}+\lambda\phi = x;\ \phi(0)=0, \phi'(1)=0.$$

Also, recover the boundary value problem from the integral equation you obtain.

Solution The differential equation may be written as $\frac{d^2\phi}{dx^2}=x-\lambda\phi$

Integrating both the sides, we have

$$\int_0^x \frac{d^2\phi}{dx^2}dx = \int_0^x x\,dx - \lambda\int_0^x \phi\,dx$$

$$\Rightarrow \quad \phi'(x)-\phi'(0)=\frac{x^2}{2}-\lambda\int_0^x \phi(\xi)d\xi$$

Let $\phi'(0)=A,$ then we have

$$\phi'(x)=A+\frac{x^2}{2}-\lambda\int_0^x \phi(\xi)d\xi$$

Integrating both the sides with regard to x, we have

$$\phi(x)-\phi(0)=Ax+\frac{1}{6}x^3-\lambda\int_0^x \phi(\xi)d\xi^2$$

$$\Rightarrow \quad \phi(x)=Ax+\frac{1}{6}x^3-\lambda\int_0^x (x-\xi)\phi(\xi)d\xi, \text{ since } \phi(0)=0$$

Since $\phi'(0)=1$

$$\Rightarrow \quad 0=A+\frac{1}{2}-\lambda\int_0^1 \phi(\xi)d\xi$$

$$\Rightarrow \quad A=-\frac{1}{2}+\lambda\int_0^1 \phi(\xi)d\xi$$

$$\Rightarrow \quad \phi(x)=-\frac{1}{2}x+\frac{1}{6}x^3+\lambda\int_0^1 x\phi(\xi)d\xi-\lambda\int_0^x (x-\xi)\phi(\xi)d\xi$$

$$=-\frac{1}{2}x+\frac{1}{6}x^3+\lambda\int_0^x x\phi(\xi)d\xi+\lambda\int_x^1 x\phi(\xi)d\xi-\lambda\int_0^x (x-\xi)\phi(\xi)d\xi$$

$$= \frac{-1}{2}x + \frac{1}{6}x^3 + \lambda \left[\int_0^x \xi \phi(\xi) d\xi + \int_x^1 x\phi(\xi) d\xi \right]$$

$$\Rightarrow \quad \phi(x) = -\frac{1}{2}x + \frac{1}{6}x^3 + \lambda \int_0^1 K(x,\xi)\phi(\xi) d\xi$$

$$\phi(x) = -\frac{1}{2}x + \frac{1}{6}x^3 + \lambda \int_0^1 K(x_1\xi)\phi(\xi) d\xi$$

Where

$$K(x,\xi) = \begin{cases} \xi, & x > \xi \\ x, & x < \xi \end{cases}$$

Which represents Fredholm integral equation of Second Kind.

Converse : The integral equation may be taken as

$$\phi(x) = -\frac{1}{2}x + \frac{1}{6}x^3 + \lambda \int_0^1 x\phi(\xi) d\xi - \lambda \int_0^x (x-\xi)\phi(\xi) d\xi \qquad .....(1)$$

Differentiating with regard to x, we have

$$\phi'(x) = -\frac{1}{2} + \frac{1}{2}x^2 + \lambda \frac{d}{dx}\int_0^1 x\phi(\xi) d\xi - \lambda \frac{d}{dx}\int_0^x (x-\xi)\phi(\xi) d\xi$$

$$= -\frac{1}{2} + \frac{1}{2}x^2 + \lambda \int_0^1 \phi(\xi) d\xi - \lambda \int_0^x \phi(\xi) d\xi \qquad .....(2)$$

Differentiating again, we have

$$\phi''(x) = x + \lambda \frac{d}{dx}\int_0^1 \phi(\xi) d\xi - \lambda \frac{d}{dx}\int_0^x \phi(\xi) d\xi$$

$$= x + \lambda \int_0^1 \frac{\partial}{\partial x}\phi(\xi) d\xi - \lambda \int_0^x \frac{\partial}{\partial x}\phi(\xi) d\xi - \lambda \phi(x)$$

$$\Rightarrow \quad \phi''(x) = x - \lambda \phi(x)$$

$$\Rightarrow \quad \phi''(x) + \lambda \phi(x) = x$$

From the relations (1) and (2), we have

$$\phi(0) = 0 \text{ and } \phi'(1) = 0$$

Which is the required differential equation.

Practice Problems

1. Verify whether the given functions, $\phi(x)$ are solutions of the corresponding Fredholm integral equations:

   (i) $\phi(x)=e^x; \phi(x)=1-\int_0^1 \sin x\xi\, \phi(\xi)d\xi$

   (ii) $\phi(x)=\cos x; \phi(x)=\sin x+\int_0^\pi (x^2+\xi)\cos\xi\, \phi(\xi)d\xi$

   (iii) $\phi(x)=xe^{-x}; \phi(x)=(x-1)e^{-x}+4\int_0^\pi e^{-(x+\xi)}\phi(\xi)d\xi$

   (iv) $\phi(x)=1-\dfrac{2\sin x}{1-\dfrac{\pi}{2}}; \phi(x)=1+\int_0^\pi \cos(x+\xi)\,\phi(\xi)d\xi$

2. Verify whether the given functions, $\phi(x)$ are solutions of the corresponding Volterra integral equations :

   (i) $\phi(x)=3; x^3=\int_0^x (x-\xi)^2\phi(\xi)d\xi$

   (ii) $\phi(x)=1-x; \int_0^x e^{(x-\xi)}\phi(\xi)d\xi=x$

   (iii) $\phi(x)=x-\dfrac{1}{6}x^3; \phi(x)=x-\int_0^x \sin\text{h}(x-\xi)\ \phi(\xi)d\xi$

3. Show that the function $\phi(x)=\dfrac{1}{\pi\sqrt{x}}$ is a solution of the integral equation

   $\int_0^x \dfrac{\phi(\xi)}{\sqrt{x-\xi}}d\xi=1.$

4. Show that the function $\phi(x)=\sin\left(\dfrac{\pi x}{2}\right)$ is a solution of the equation

   $\phi(x)=\dfrac{1}{2}x+\dfrac{\pi^2}{4}\int_0^1 K(x,\xi)\phi(\xi)d\xi$

Where $K(x,\xi)=\begin{cases}\frac{1}{2}x(2-\xi), & 0\le x\le \xi\\ \frac{1}{2}\xi(2-x), & \xi\le x\le 1\end{cases}$

5. **Form an integral equation corresponding to the differential equation**

$$\frac{d^2y}{dx^2}-5\frac{dy}{dx}+6y=0$$

with the initial conditions.

$$y(0)=0, y'(0)=-1$$

Ans.: $\phi(x)=6x-5+\int_0^x(5-6x+6\xi)\phi(\xi)d\xi$

6. **Form the integral equations corresponding to the following differential equations with the initial conditions:**

(i) $\frac{d^2y}{dx^2}+y=\cos x; y(0)=0, y'(0)=1.$

Ans.: $\phi(x)=\cos x-x-\int_0^x(x-\xi)\phi(\xi)d\xi$

(ii) $\frac{d^3y}{dx^3}-2xy=0; y(0)=\frac{1}{2}, y'(0)=1=y''(0).$

Ans.: $\phi(x)=x(x+1)^2+\int_0^x x(x-\xi)^2\,\phi(\xi)d\xi$

7. **Form the integral equations corresponding to the following differential equations with the initial conditions :**

(i) $\frac{dy}{dx}-y=0; y(0)=1.$ **Ans.:** $\phi(x)=1+\int_0^x\phi(\xi)d\xi$

(ii) $\frac{d^2y}{dx^2}+(1+x^2)y=\cos x; y(0)=0, y'(0)=2.$

Ans.: $\phi(x)=\cos x-2x(1+x^2)-\int_0^x(1+x^2)(x-\xi)\phi(\xi)d\xi$

(iii) $\frac{d^2y}{dx^2}+y=\cos x; y(0)=0=y'(0).$

Ans.: $\phi(x)=\cos x-\int_0^x(x-\xi)\phi(\xi)d\xi$

8. Reduce the differential equation

$$\phi''(x) - 3\phi'(x) + 2\phi(x) = 4\sin x$$

with the conditions

$$\phi(0) = 1, \phi'(0) = -2$$

into a non-homogeneous Volterra's integral equation of Second Kind. Conversely, derive the original differential equation with the initial conditions from the integral equation you obtain.

Ans.: $\phi(x) = 1 - x - 4\sin x + \int_0^x [3 - 2(x-\xi)]\phi(\xi)d\xi$

9. Convert the differential equation

$$\frac{d^2\theta}{dx^2} - 2x\frac{d\theta}{dx} - 3\phi = 0;\ \phi(0) = 0,\ \phi'(0) = 0$$

$$\frac{d^2\phi}{dx^2} - 2x\frac{d\theta}{dx} - 3\phi = 0; \phi(0), \phi'(0) = 0$$

to Volterra's integral equation of Second Kind. Conversely, derive the original differential equation with the initial conditions from the integral equation obtained.

Ans.: $\phi(x) = \int_0^x (x+\xi)\phi(\xi)d\xi$

Solution of Volterra's Integral equation of Second Kind by the method of successive approximation :

A Volterra integral equation of Second Kind

$$\phi(x) = F(x) + \lambda \int_0^x K(x,\xi)\phi(\xi)d\xi$$

has one and only one solution given by the relation

$$\phi(x) = F(x) + \lambda \int_0^x R(x,\xi;\lambda)F(\xi)d\xi$$

where the resolvent Kernel $R(x,\xi;\lambda)$ is the sum of the series

$$R(x,\xi;\lambda) = K(x,\xi) + \sum_{v=1}^{\infty} \lambda^n K_v\ (x,\xi)$$

convergent for all values of λ.

Consider the Volterra integral equation of Second Kind

$$\phi(x) = F(x) + \lambda\int_0^x K(x,\xi)\phi(\xi)d\xi \qquad .....(1)$$

Where the Kernel $K(x,\xi)$ is a continuous function for $0 \le x \le a$, $0 \le \xi \le a$, and the function $F(x)$ is continuous for $0 \le x \le a$. Consider an infinite power series in ascending powers of λ as

$$\phi(x) = \phi_0(x) + \lambda\phi_1(x) + \lambda^2\phi_2(x) + ... + \lambda^n\phi_n(x) + ... \qquad .....(2)$$

Let the series is a solution of the integral equation (1), then

$$\phi_0(x) + \lambda\phi_1(x) + \lambda^2\phi_2(x) + ... + \lambda^n\phi_n(x) + ...$$

$$= F(x) + \lambda\int_0^x K(x,\xi)\,[\phi_0(\xi) + \lambda\phi_1(\xi) + \lambda^2\phi_2(\xi) + ... + \lambda^n\phi_n(\xi)]d\xi \qquad .....(3)$$

Equating the coefficients of like power of λ, we get

$$\phi_0(x) = F(x)$$

$$\phi_1(x) = \int_0^x K(x,\xi)\phi_0(\xi)d\xi$$

$$\phi_2(x) = \int_0^x K(x,\xi)\phi_0(\xi)d\xi$$

$$\phi_n(x) = \int_0^x K(x,\xi)\phi_{n-1}(\xi)d\xi \qquad .....(4)$$

Thus in yields a method for a successive approximation of the function $\phi_n(x)$. It may be shown that the series (2) converges uniformly in x and λ and $x \in [0,a]$ under these assumption with regard to $F(x)$ and $K(x,\xi)$, its sum is a unique solution of the equation (1).

Further, from equation (4),

$$\phi_1(x) = \int_0^x K(x,\xi)F(\xi)d\xi$$

$$\phi_2(x) = \int_0^x K(x,\xi)\left\{\int_0^\xi K(\xi,\xi_1)F(\xi_1)d\xi_1\right\}d\xi$$

Here $\xi_1 = 0, \xi_1 = \xi;\ \xi = 0, \xi = x$

By integrating the order of integration, we get

$$\phi_2(x) = \int_0^x F(\xi_1)d\xi_1 \left\{\int_{\xi_1}^x K(x,\xi)K(\xi,\xi_1)d\xi\right\}$$

$$\Rightarrow \quad \phi_2(x) = \int_0^x K_2(x,\xi_1)F(\xi_1)d\xi_1 \qquad \text{.....(5)}$$

where $K_2(x,\xi) = \int_{\xi_1}^x K(x,\xi)K(\xi,\xi_1)d\xi$ .....(6)

In general, we get

$$\phi_n(x) = \int_0^x K_n(x,\xi)F(\xi)d\xi, \quad n = 1,2,3,..... \qquad \text{.....(7)}$$

The function $K_n(x,\xi)$ are called **iterated Kernels**, which can reading be shown that

$$K_1(x,\xi) = K(x,\xi)$$

and $K_2(x,\xi), K_3(x,\xi)$ etc. are defined recursively by the formulas

$$K_{n+1}(x,\xi) = \int_\xi^x (x,t)K_n(t,\xi)dt, \qquad n = 1,2,3 \qquad \text{.....(8)}$$

The relation (2) which represents the solution of the integral equation (1) can therefore be written as

$$\phi(x) = F(x) + \sum_{v=1}^{\infty} \lambda^v \int_0^x K_v(x,\xi)\ F(\xi)d\xi \qquad \text{.....(9)}$$

where $$R(x,\xi;\lambda) = K_1(x,\xi) + \lambda K_2(x,\xi) + \lambda^2 K_3(x,\xi) + ... + \lambda^{n-1}K_n(x,\xi) + \lambda^n K_{n+1}(x,\xi) + ... \qquad \text{.....(10)}$$

$$\Rightarrow \quad R(x,\xi;\lambda) = \sum_{v=1}^{\infty} \lambda^v K_{v+1}(x,\xi)$$

The function $R(x,\xi;\lambda)$ is called the **resolvent Kernel or reciprocal Kernel** of the integral equation (1).

The resolvent Kernel $R(x,\xi;\lambda)$ satisfies the functional equation

$$R(x,\xi;\lambda) = K(x,\xi) + \lambda \int_\xi^x K(x,t)\ R(t,\xi;\lambda)dt$$

Thus the solution of equation(1) reduces to

$$\phi(x) = F(x) + \lambda \int_0^x R(x,\xi;\lambda)\ F(\xi)d\xi$$

Example 18 Find the resolvent Kernel of the Volterra integral equation with Kernel : $K(x,\xi)=1$.

Solution We know that $K_1(x,\xi) = K(x,\xi) = 1$.

by the iterated Kernel, we have

$$K_2(x,\xi) = \int_\xi^x K(x,t)K_1(t,\xi)dt$$

$$= \int_\xi^x dt = [t]_\xi^x = x - \xi$$

$$K_3(x,\xi) = \int_\xi^x K(x,t)K_2(t,\xi)dt$$

$$= \int_\xi^x 1.(t-\xi)dt = \frac{1}{2!}(x-\xi)^2$$

$$K_4(x,\xi) = \int_\xi^x K(x,t)K_3(t,\xi)dt$$

$$= \int_\xi^x 1.\frac{(t-\xi)^2}{2}dt = \frac{1}{3!}(x-\xi)^3$$

$$K_\nu(x,\xi) = \int_\xi^x 1.K_{\nu-1}(t,\xi)dt$$

$$= \int_\xi^x 1.\frac{(t-\xi)^{\nu-2}}{(\nu-2)!}dt$$

$$= \frac{1}{(\nu-1)!}(x-\xi)^{\nu-1}$$

Hence the resolvent Kernel is

$$R(x,\xi;\lambda) = \sum_{\nu=0}^{\infty} \lambda^\nu K_{\nu+1}(x,\xi)$$

$$= \sum_{\nu=0}^{\infty} \frac{\lambda^{\nu}(x-\xi)^{\nu}}{\nu!}$$

$$= 1 + \frac{\lambda}{1!}(x-\xi) + \frac{\lambda^2}{2!}(x-\xi)^2 + ...$$

$$= e^{\lambda(x-\xi)}$$

Example 19 Find the resolvent Kernel of the volterra integral equations with the following Kernels:

(i) $K(x,\xi) = \begin{cases} e^{x-\xi}, & x > \xi \\ 0, & \xi > x \end{cases}$

(ii) $K(x,\xi) = \dfrac{2+\cos x}{2+\cos \xi}$

Solution (i) We know that

$$K_1(x,\xi) = K(x,\xi) = e^{x-\xi}$$

By the iterated Kernel, we have

$$K_2(x,\xi) = \int_{\xi}^{x} K(x,t)K_1(t,\xi)dt$$

$$= \int_{\xi}^{x} e^{x-t}.e^{t-\xi}dt$$

$$= \int_{\xi}^{x} e^{x-\xi}dt = e^{x-\xi}[t]_{\xi}^{x} = (x-\xi)e^{x-\xi}, x > \xi$$

$$K_3(x,\xi) = \int_{\xi}^{x} K(x,t)K_2(t,\xi)dt$$

$$= \int_{\xi}^{x} e^{x-t}.(t-\xi)e^{t-\xi}dt$$

$$= \int_{\xi}^{x} (t-\xi)e^{x-\xi}dt = \frac{1}{2!}(x-\xi)^2 e^{x-\xi}, x > \xi$$

$$K_4(x,\xi) = \int_{\xi}^{x} K(x,t)K_3(t,\xi)dt$$

$$= \int_{\xi}^{x} e^{x-t}.\frac{1}{2!}(t-\xi)^2 e^{t-\xi}dt$$

$$= \int_{\xi}^{x} \frac{1}{2!}(t-\xi)^2 e^{x-\xi} dt = \frac{1}{3!}(x-\xi)^3 e^{x-\xi}, x > \xi$$

$$K_\nu(x,\xi) = \int_{\xi}^{x} e^{x-1}.K_{\nu-1}(t,\xi)dt$$

$$= \int_{\xi}^{x} e^{x-\xi}.\frac{(t-\xi)^{\nu-2}}{(\nu-2)!} = \frac{1}{(\nu-1)!}(x-\xi)^{\nu-1}e^{x-\xi}, x > \xi$$

Hence the resolvent Kernel is

$$R(x,\xi;\lambda) = \sum_{\nu=0}^{\infty} \lambda^\nu K_{\nu+1}(x,\xi)$$

$$= \sum_{\nu=0}^{\infty} \lambda^\nu \frac{(x-\xi)^\nu}{\nu!} \cdot e^{x-\xi}$$

$$= e^{x-\xi}\left[1 + \frac{\lambda(x-\xi)}{1!} + \frac{\lambda^2(x-\xi)^2}{2!} + ...\right]$$

$$= e^{x-\xi}.e^{\lambda(x-\xi)}$$

$$= e^{(1+\lambda)(x-\xi)}, \ x > \xi$$

(ii) We know that

$$K_1(x,\xi) = K(x,\xi) = \frac{2+\cos x}{2+\cos\xi}$$

By the iterated Kernels, we have

$$K_2(x,\xi) = \int_{\xi}^{x} K(x,t)K_1(t,\xi)dt$$

$$= \int_{\xi}^{x} \frac{2+\cos x}{2+\cos t}.\frac{2+\cos t}{2+\cos\xi}dt$$

$$= \int_{\xi}^{x} \frac{2+\cos x}{2+\cos\xi}dt = (x-\xi)\frac{2+\cos x}{2+\cos\xi}$$

$$K_3(x,\xi) = \int_{\xi}^{x} \frac{2+\cos x}{2+\cos t}.(t-\xi)\frac{2+\cos t}{2+\cos\xi}dt$$

$$= \int_{\xi}^{x} \frac{2+\cos x}{2+\cos \xi}(t-\xi)dt = \frac{1}{2!}(x-\xi)^2 \frac{2+\cos x}{2+\cos \xi}$$

$$K_4(x,\xi) = \int_{\xi}^{x} \frac{2+\cos x}{2+\cos t}.\frac{1}{2!}(t-\xi)^2 \frac{2+\cos t}{2+\cos \xi}dt$$

$$= \int_{\xi}^{x} \frac{1}{2!}\frac{2+\cos x}{2+\cos \xi}(t-\xi)^2 dt = \frac{1}{3!}(x-\xi)^3 \frac{2+\cos x}{2+\cos \xi}$$

----- ----- ----- -----

$$K_v(x,\xi) = \int_{\xi}^{x} \frac{2+\cos x}{2+\cos t}\cdot\frac{1}{(v-2)!}(t-\xi)^{v-2} \frac{2+\cos t}{2+\cos \xi}dt$$

$$= \frac{1}{(v-1)!}(x-\xi)^{v-1}.\frac{2+\cos x}{2+\cos \xi}$$

Hence the resolvent Kernel is

$$R(x,\xi;\lambda) = \sum_{v=0}^{\infty} \lambda^v K_{v+1}(x,\xi)$$

$$= \sum_{v=0}^{\infty} \lambda^v \frac{(x-\xi)^v}{v!} \frac{2+\cos x}{2+\cos \xi}$$

$$= \frac{2+\cos x}{2+\cos \xi}e^{\lambda(x-\xi)}$$

Example 20 With the aid of the resolvent Kernel find the solution of the integral equation

(i) $\phi(x) = x + \int_0^x (\xi - x)\phi(\xi)d\xi$

(ii) $\phi(x) = 1 + \int_0^x (\xi - x)\phi(\xi)d\xi$

Solutions

(i) Here $K_1(x,\xi) = K(x,\xi) = \xi - x$

by the iterated Kernels, we have

$$K_2(x,\xi) = \int_{\xi}^{x} K(x,t)\cdot K_1(t,\xi)dt$$

$$= \int_{\xi}^{x} (t-x)(\xi-t)dt - \frac{1}{3!}(\xi-t)^3$$

$$K_3(x,\xi) \quad = \int_\xi^x K(x,t)K_2(t,\xi)dt$$

$$= \int_\xi^x (t-x)\left\{\frac{-1}{3!}(\xi-t)^3\right\}dt = \frac{1}{5!}(\xi-x)^5$$

----- ----- ----- -----

$$K_v(x,\xi) \quad = (-1)^{v-1}\frac{1}{2^v-1}(\xi-x)^{2v-1}, v=1,2,3...$$

The resolvent Kernel is

$$R(x,\xi;\lambda) = \sum_{v=0}^{\infty}\lambda^v K_{v+1}(x,\xi)$$

$$= \sum_{v=1}^{\infty}\lambda^{v-1}K_v(x,\xi)$$

$$= \sum_{v=1}^{\infty}K_v(x,\xi); \lambda = 1$$

$$= \sum_{v=1}^{\infty}(-1)^{v-1}\frac{1}{(2v-1)!}(\xi-1)^{2v-1}$$

$$= (\xi-x) - \frac{1}{3!}(\xi-x)^3 + \frac{1}{5!}(\xi-x)^5 \ldots\ldots\ldots\ldots$$

$$\Rightarrow \qquad R(x,\xi;\lambda) = \sin(\xi-x)$$

The solution of the integral equation is

$$\phi(x) = F(x) + \lambda\int_0^x R(x,\xi;\lambda)F(\xi)d\xi$$

$$\Rightarrow \phi(x) \quad = x + \int_0^x \xi\sin(\xi-x)d\xi$$

$$\Rightarrow \phi(x) \quad = \sin x$$

(ii) The solution of the integral equation is

$$\phi(x) \ = F(x) + \lambda\int_0^x R(x,\xi;\lambda)F(\xi)d\xi$$

$$= 1 + \int_0^x \sin(\xi-x)d\xi$$

$$= 1 - [\cos(\xi - x]_0^x$$

$$= 1 - 1 + \cos x$$

$$\Rightarrow \quad \phi(x) = \cos x$$

Example 21 Solve the integral equation $\phi(x) = 1 + \int_0^x \phi(\xi)d\xi$

Solution Assume

$$\phi(x) = \phi_0(x) + \lambda\,\phi_1(x) + \lambda^2\,\phi_2(x) + ... + \lambda^n\,\phi_n(x) + ...$$

Here $\quad F(x) = 1, K(x,\xi) = 1, \lambda = 1$

Equation the coefficients of like powers of λ,

$$\phi_0(x) \quad = F(x) = 1$$

$$\phi_1(x) \quad = \int_0^x K(x,\xi)\phi_0(\xi)d\xi = \int_0^x d\xi = x$$

$$\phi_2(x) \quad = \int_0^x K(x,\xi)\phi_1(\xi)d\xi = \int_0^x \xi\,d\xi = \frac{x^2}{2}$$

----- ----- ----- -----

----- ----- ----- -----

$$\therefore \quad \phi(x) = 1 + x + \frac{1}{2!}x^2 + ...$$

$$= e^x$$

Example 22 Solve the integral equations

(i) $\phi(x) = (1+x) + \lambda\int_0^x (x-\xi)\phi(\xi)d\xi$

(ii) $\phi(x) = 1 + \int_0^x x\xi\,\phi(\xi)d\xi$

Solution

(i) Here $K_1(x,\xi) = K(x,\xi) = x - \xi$

By the iterated Kernels, we have

$$K_2(x,\xi) = \int_\xi^x K(x,t)K_1(t,\xi)dt$$

$$= \int_\xi^x (x-t)(t-\xi)dt = \frac{1}{3!}(x-\xi)^3$$

$$K_3(x,\xi) = \int_\xi^x K(x,t)K_2(t,\xi)dt$$

$$= \int_\xi^x (x-t)\frac{(t-\xi)^3}{3!}dt = \frac{1}{5!}(x-\xi)^5$$

----- ----- ----- -----

and so on.

Thus the solution of the integral equation is given as

$$\phi(x) = F(x) + \sum_{\nu=1}^{\infty}\lambda^\nu \int_0^x K_\nu(x,\xi)\, F(\xi)d\xi$$

$$\Rightarrow \phi(x) = (1+x) + \lambda \int_0^x K_1(x,\xi)(1+\xi)d\xi +$$

$$\lambda^2 \int_0^x K_2(x,\xi)(1+\xi)d\xi + \lambda^3 \int_0^x K_3(x,\xi)(1+\xi)d\xi + ...$$

$$= (1+x) + \lambda \int_0^x (x-\xi)\,(1+\xi)d\xi + \lambda^2 \int_0^x \frac{1}{3!}(x-\xi)^3\,(1+\xi)d\xi.$$

$$+\lambda^3 \int_0^x \frac{1}{5!}(x-\xi)^5(1+\xi)d\xi + ...$$

$$= (1+x) + \lambda\left(\frac{x^2}{2} + \frac{x^3}{6}\right) + \frac{\lambda^2}{6}\left(\frac{x^4}{4} + \frac{x^5}{20}\right) + \frac{\lambda^3}{120}\left(\frac{x^6}{6} + \frac{x^7}{42}\right) + ...$$

$$= (1+x) + \lambda\left(\frac{x^2}{2!} + \frac{x^3}{3!}\right) + \lambda^2\left(\frac{x^4}{4!} + \frac{x^5}{5!}\right) + \lambda^3\left(\frac{x^6}{6!} + \frac{x^7}{7!}\right) + ...$$

$$\phi(x) = 1 + x + \frac{x^2}{2!} + \frac{x^3}{3!} + \frac{x^4}{4!} + \frac{x^5}{5!} + ... \text{if } \lambda = 1$$

$$\Rightarrow \quad \phi(x) = e^x$$

(ii) Here $K_1(x,\xi) = K(x,\xi) = x\xi$

By the iterated Kernels, we have

$$K_2(x,\xi) = \int_\xi^x xt.t\xi\, dt = \frac{1}{3}x\xi(x^3 - \xi^3) = \frac{1}{3}(x^4\xi - x\xi^4)$$

$$K_3(x,\xi) = \int_\xi^x xt.\frac{1}{3}(t^4\xi - t\xi^4)dt = \frac{1}{18}(x^7\xi - 2x^4\xi^4 + x\xi^7)$$

$$K_4(x,\xi) = \int_\xi^x xt.\frac{1}{18}(t^7\xi - 2t^4\xi^4 + t\xi^7)dt$$

$$= \frac{1}{162}(x^{10}\xi - 3x^7\xi^4 + 3x^4\xi^7 - x\xi^{10})$$

----- ----- ----- -----

and so on.

Thus the solution of the integral equation is given by

$$\phi(x) = F(x) + \sum_{\nu=1}^{\infty}\lambda^\nu \int_0^x K_\nu(x,\xi)F(\xi)d\xi$$

$$= 1 + \sum_{\nu=1}^{\infty}\int_0^x K_\nu(x,\xi)d\xi$$

$$= 1 + \int_0^x K_1(x,\xi)d\xi + \int_0^x K_2(x,\xi)d\xi + \int_0^x K_3(x,\xi)d\xi + \int_0^x K_4(x,\xi)d\xi + ...$$

$$= 1 + \int_0^x x\xi d\xi + \frac{1}{3}\int_0^x (x^4\xi - x\xi^4)d\xi + \frac{1}{18}\int_0^x (x^7\xi - 2x^4\xi^4 + x\xi^7)d\xi$$

$$+ \frac{1}{162}\int_0^x (x^{10}\xi - 3x^7\xi^4 + 3x^4\xi^7 - x\xi^{10})d\xi + ...$$

$$= 1 + x\left(\frac{\xi^2}{2}\right)_0^x + \frac{1}{3}\left\{x^4\left(\frac{\xi^2}{2}\right)_0^x - x\left(\frac{\xi^5}{5}\right)_0^x\right\}_0^x$$

$$+ \frac{1}{18}\left\{x^7\left(\frac{\xi^2}{2}\right)_0^x - 2x^4\left(\frac{\xi^5}{5}\right)_0^x + x\left(\frac{\xi^8}{8}\right)_0^x\right\}$$

$$+ \frac{1}{162}\left\{x^{10}\left(\frac{\xi^2}{2}\right)_0^x - 3x^7\left(\frac{\xi^5}{5}\right)_0^x + 3x^4\left(\frac{\xi^8}{8}\right)_0^x - x\left(\frac{\xi^{11}}{11}\right)_0^x\right\} + ...$$

$$\Rightarrow \qquad \phi(x) = 1 + \frac{x^3}{2} + \frac{x^6}{2.5} + \frac{x^9}{2.5.8} + ...$$

Example 23 With the aid of the resolvent Kernel, find the solution of the integral equation

$$\phi(x) = e^{x^2} + \int_0^x e^{x^2-\xi^2} \phi(\xi) d\xi$$

Solution Here $K_1(x,\xi) = K(x,\xi) = e^{x^2-\xi^x}$

By the iterated Kernels, we have

$$K_2(x,\xi) = \int_\xi^x K(x,t)\, K_1(t,\xi) dt$$

$$= \int_\xi^x e^{x^2-t^2} . e^{t^2-\xi^2} dt = (x-\xi)e^{x^2-\xi^2}$$

$$K_3(x,\xi) = \int_\xi^x e^{x^2-t^2} .(t-\xi)e^{t^2-\xi^2} dt = \frac{(x-\xi)^2}{2!} e^{x^2-\xi^2}$$

----- ----- ----- -----

$$K_\nu(x,\xi) = \frac{(x-\xi)^{\nu-1}}{(\nu-1)!} e^{x^2-\xi^2}$$

Hence, the resolvent Kernerl is

$$R(x,\xi;\lambda) = \sum_{\nu=1}^{\infty} \lambda^{\nu-1} K_\nu(x,\xi)$$

$$= K_1(x,\xi) + \lambda K_2(x,\xi) + \lambda^2 K_3(x,\xi) + \ldots..$$

$$= e^{x^2-\xi^2}\left[1 + \frac{(x-\xi)}{1!} + \frac{(x-\xi)^2}{2!} + \ldots..\right]; \lambda = 1$$

$$\Rightarrow \quad R(x,\xi;\lambda) = e^{x^2-\xi^2} . e^{x-\xi}$$

Thus the solution of the integral equation is given by

$$\phi(x) = e^{x^2} + \int_0^x e^{x^2-\xi^2}\, e^{x-\xi}\, e^{\xi^2} d\xi$$

$$= e^{x^2} + e^{x^2+x} \int_0^x e^{-\xi} d\xi$$

$$= e^{x^2} + e^{x^2+x}(1-e^{-x}) = e^{x^2+x}$$

Example 24 With the aid of resolvent Kernel, find the solution of the following integral equations:

(i) $\phi(x) = \sin x + 2\int_0^x e^{x-\xi} \phi(\xi) d\xi$

(ii) $\phi(x) = e^x \sin x + \int_0^x \frac{2+\cos x}{2+\cos\xi}.\phi(\xi)d\xi$

Solution

(i) The resolvent Kernel of $e^{x-\xi}$ for $\lambda = 2$ is given by

$R(x,\xi;\lambda) = e^{3(x-\xi)}$

(see example 19 (i))

Hence the solution of the integral equation is given by

$$\begin{aligned}\phi(x) &= \sin x + 2\int_0^x e^{3(x-\xi)} \sin\xi.d\xi \\ &= \sin x + 2e^{3x}\int_0^x e^{-3\xi}\sin\xi d\xi \\ &= \sin x + 2e^{3x}\left[\frac{1}{10}e^{-3\xi}(-3\sin\xi - \cos\xi)\right]_0^x \\ &= \sin x - \frac{1}{5}e^{3x}\left[e^{-3x}(3\sin x + \cos x) - 1\right]\end{aligned}$$

$$\Rightarrow \quad \phi(x) = \frac{1}{5}e^{3x} + \frac{2}{5}\sin x - \frac{1}{5}\cos x$$

(ii) The resolvent Kernel of $\frac{2+\cos x}{2+\cos\xi}$ for $\lambda = 1$ is given by

$$R(x,\xi;1) = \frac{2+cosx}{c+\cos\xi}e^{x-\xi}$$

(seen example 19(ii))

Thus the solution of integral equation is given by

$$\begin{aligned}\phi(x) &= e^x \sin x + \int_0^x \frac{2+\cos x}{2+\cos\xi}e^{x-\xi}\cdot e^{\xi}\sin\xi d\xi \\ &= e^x\sin x + e^x(2+\cos x)\int_0^x \frac{\sin\xi}{2+\cos\xi}d\xi\end{aligned}$$

$$\Rightarrow \quad \phi(x) = e^x\sin x + e^x(2+\cos x)\log\left(\frac{3}{2+\cos x}\right)$$

Determination of some resolvent Kernels : Consider that the Kernel $K(x,\xi)$ is a polynomial of degree $(n-1)$ in ξ such that it may be expressed in the form

$$K(x,\xi) = a_0(x) + a_1(x)(x-\xi) + a_2(x)\frac{1}{2!}$$

$$(x-\xi)^2 + \ldots + \frac{a_{n-1}(x)}{(n-1)!}(x-\xi)^{n-1} + \ldots \quad \ldots(1)$$

where $a_0(x), a_1(x), \ldots a_{n-1}(x)$ are continuous in the interval [0, *a*].

Let the auxiliary function be

$$\phi(x,\xi;\lambda) = \frac{1}{(n-1)!}(x-\xi)^{n-1} + \lambda\int_{\xi}^{x}(t,\xi;\lambda)\frac{(x-t)^{n-1}}{(n-1)!}dt \quad \ldots(2)$$

with the conditions

$$\phi = \frac{d\phi}{dx} = \ldots = \frac{d^{n-2}\phi}{dx^{n-2}} = 0 \text{ at } x=\xi \text{ and } \frac{d^{n-1}\phi}{dx^{n-1}} = 1 \text{ and } x=\xi \quad \ldots(3)$$

In addition, we have

$$R(x,\xi;\lambda) = \frac{1}{\lambda}\frac{d^n}{dx^n}\phi(x,\xi;\lambda) \quad \ldots(4)$$

Since the resolvent Kernel satisfies the functional equation

$$R(x,\xi;\lambda) = K(x,\xi) + \lambda\int_{\xi}^{x} K(x,t)R(t,\xi;\lambda)dt \quad \ldots(5)$$

From equation (4) and (5), we have

$$\frac{d^n}{dx^n}\phi(x,\xi;\lambda) = \lambda K(x,\xi) + \lambda\int_{\xi}^{x} K(x,t)\frac{d^n}{dt^n}\phi(t,\xi;\lambda)dt$$

$$= \lambda K(x,\xi) + \lambda$$

$$\left[K(x,t)\frac{d^{n-1}\phi}{dt^{n-1}} - \frac{\partial K(x,t)}{\partial t}\frac{d^{n-2}\phi}{dt^{n-2}} + \ldots + \frac{d^{n-1}K(x,t)}{dt^{n-1}}\phi\right]_{\xi}^{x} \quad \ldots(6)$$

Using (1) and (3), the relation (6) reduces to

$$D\phi = \frac{d^n\phi}{dx^n} - \lambda\left[a_0(x)\frac{d^{n-1}\phi}{dx^{n-1}} + a_1(x)\frac{d^{n-2}\phi}{dx^{n-2}} + \ldots + a_{n-1}(x)\phi\right] = 0 \quad \ldots(7)$$

The function $\phi(x,\xi;\lambda)$ is therefore the integral of the linear equation $D\phi = 0$, which satisfies the cauchy conditions.

Thus, we have an expression for the resolvent Kernel as

$$R(x,\xi;\lambda) = \frac{1}{\lambda}\frac{d^n}{dx^n}\phi(x,\xi;\lambda)$$

Note: If the Kernel $R(x,\xi)$ is a polynomial of degree $(n-1)$ in x then the resolvent Kernel be

$$R(x,\xi;\lambda) = -\frac{1}{\lambda}\frac{d^n}{d\xi^n}\phi(\xi,x;\lambda)$$

Example 25 Find the resolvent Kernels of integral equations with the following Kernels; $(\lambda = 1)$

(i) $K(x,\xi) = 2x$

(ii) $K(x,\xi) = 2x - (x-\xi)$

(iii) $K(x,\xi) = -2 + 3(x-\xi)$

Solution

(i) Here $K(x,\xi) = 2x; \lambda = 1$

Comparing with the relation $K(x,\xi)$

$$= a_0(x) + a_1(x)(x-\xi) + a_2(x)\frac{1}{2!}(x-\xi)^2 + ...$$

we have $a_0(x) = 2x$ and all other $a_\nu(x) = 0$

Then the differential equation.

$$\frac{d^n\phi}{dx^n} - \lambda\left[a_0(x)\frac{d^{n-1}\phi}{dx^{n-1}} + a_1(x)\frac{d^{n-2}\phi}{dx^{n-2}} + ... + a_{n-1}(x)\phi\right] = 0$$

Reduces to

$$\frac{d\phi}{dx} - 2x\phi = 0;\ \phi = 1 \text{ at } x = \xi \qquad .....(1)$$

$$\Rightarrow \phi = A(\xi)e^{x^2} \qquad .....(2)$$

From (1) and (2), we have

$$\phi(x,\xi;1)=e^{x^2-\xi^2}$$

Thus the resolvent Kernel is given by

$$R(x,\xi;1) = \frac{1}{\lambda}\frac{d}{dx}\phi(x_1\xi;1)$$

$$=\frac{d}{dx}e^{x^2-\xi^2}$$

$$=2xe^{x^2-\xi^2}$$

(ii) Here $K(x,\xi) = 2-(x-\xi);\ \lambda=1$

comparing with the relation, we have.

$a_0(x)=2,\ a_1(x)=-1$ and all other $a_v(x)=0$

The differential equation reduces to

$$\frac{d^2\phi}{dx^2}-2\frac{d\phi}{dx}+\phi=0 \qquad .....(1)$$

with the conditions $\phi=0$ at $x=\xi,\ \frac{d\phi}{dx}=1$ at $x=\xi$ .....(2)

The solution of equation (1)

$$[A(\xi)+B(\xi)x]e^x \qquad .....(3)$$

From (2) and (3), we obtain

$$\phi(x,\xi;1)=(x-\xi)e^{x-\xi}$$

Thus the resolvent Kernel is given by

$$R(x,\xi;1) = \frac{1}{\lambda}\frac{d^2}{dx^2}\phi(x,\xi;1)$$

$$=\frac{d^2}{dx^2}(x-\xi)e^{x-\xi}$$

$$=(x-\xi+2)e^{x-\xi}$$

(iii) Here $K(x,\xi) = -2+3(x-\xi); \lambda = 1$

Comparing with the relation, we have

$a_0(x) = -2, a_1(x) = 3$ and all other $a_v(x) = 0$

The differential equation reduces to

$$\frac{d^2\phi}{dx^2} + 2\frac{d\phi}{dx} - 3\phi = 0 \quad .....(1)$$

with the conditions $\phi = 0$ at $x = \xi$, $\frac{d\phi}{dx} = 1$ at $x = \xi$ .....(2)

The solution of equation (1) is

$$\phi = A(\xi)e^{-3x} + B(\xi)e^{x} \quad .....(3)$$

From (2) and (3), we obtain

$$\phi(x,\xi;1) = \frac{1}{4}e^{x-\xi} - \frac{1}{4}e^{-3(x-\xi)}$$

Hence the resolvent Kernel is given by

$$R(x,\xi;\lambda) = \frac{1}{\lambda}\frac{d^2}{dx^2}\phi(x,\xi;1)$$

$$= \frac{d^2}{dx^2}\left[\frac{1}{4}e^{x-\xi} - \frac{1}{4}e^{-3(x-\xi)}\right]$$

$$= \frac{1}{4}e^{x-\xi} - \frac{9}{4}e^{-3(x-\xi)}$$

Example 26 Solve the Volterra integral equation ;

$$\phi(x) = (1-2x-4x^2) + \int_0^x [3+6(x-\xi)-4(x-\xi)^2]\phi(\xi)d\xi$$

Solution Here $F(x) = 1-2x-4x^2; \lambda = 1$

$K(x,\xi) = 3+6(x-\xi)-4(x-\xi)^2$

Comparing with the relation, we have

$a_0(x) = 3, a_1(x) = 6, a_2(x) = -8$

The differential equation reduces to

$$\frac{d^3\phi}{dx^3}-3\frac{d^2\phi}{dx^2}-6\frac{d\phi}{dx}+8\phi=0 \qquad .....(1)$$

With the conditions

$$\phi=0,\frac{d\phi}{dx}=0 \text{ at } x=\xi \text{ and } \frac{d^2\phi}{dx^2}=1 \text{ at } x=\xi \qquad .....(2)$$

The solution of equation (1) is given by

$$\phi(x,\xi;1)=A(\xi)e^{x}+B(\xi)e^{-2x}+c(\xi)e^{4x} \qquad .....(3)$$

From (2) and (3), we obtain

$$\phi(x,\xi;1)=-\frac{1}{9}e^{x-\xi}+\frac{1}{18}e^{-2(x-\xi)}+\frac{1}{18}e^{4(x-\xi)}$$

Thus the resolvent Kernel is given by

$$\begin{aligned}R(x,\xi;1) &=\frac{1}{\lambda}\frac{d^3}{dx^3}\phi(x,\xi;1)\\ &=\frac{d^3}{dx^3}\left[-\frac{1}{9}e^{x-\xi}+\frac{1}{18}e^{-2(x-\xi)}+\frac{1}{18}e^{4(x-\xi)}\right]\\ &=-\frac{1}{9}\left[x^{x-\xi}+4e^{-2(x-\xi)}-32e^{4(x-\xi)}\right] \qquad .....(4)\end{aligned}$$

Thus the solution of the integral equation is given by

$$\phi(x)=1-2x-4x^2-\frac{1}{9}\int_0^x\left[e^{x-\xi}+4e^{-2(x-\xi)}-32e^{4(x-\xi)}\right](1-2\xi-4\xi^2)d\xi$$

$$=1-2x-4x^2-\frac{1}{9}[e^{x-\xi}(8+10\xi+4\xi^2)+e^{-2(x-\xi)}(4\xi-8\xi^2)+e^{4(x-\xi)}(-32\xi-32\xi^2)]_0^x$$

$$\Rightarrow \phi(x)=1-2x-4x^2-1+2x+4x^2+e^x=e^x$$

Example 27 Solve the linear integral equation

$$\phi(x)=(\cos x-x-2)+\int_0^x(\xi-x)\phi(\xi)d\xi$$

Solution Here $F(x)=\cos x-x-2;\lambda=1$

$$K(x,\xi)=\xi-x$$

Comparing with the relation, we have

$b_1(\xi)=1$ and all other $b_v(\xi)=0$

The differential equation reduces to

$$\frac{d^2\phi}{d\xi^2}+\phi=0 \qquad \text{.....(1)}$$

With the conditions $\phi=0$ at $\xi=x$ and $\frac{d\phi}{d\xi}=1$ at $\xi=x$ .....(2)

The solution of equation (1) is given by

$$\phi(x,\xi;1)=A(x)\cos\xi+B(x)\sin\xi \qquad \text{.....(3)}$$

From (2) and (3), we obtain.

$$\phi(x,\xi;1)=\sin(\xi-x)$$

Thus the resolvent Kernel is given by

$$R(x,\xi;1)=-\frac{1}{\lambda}\frac{d^2}{d\xi^2}\phi(x,\xi;1)=\sin(\xi-x)$$

Thus the solution of integral equation is given by

$$\begin{aligned}\phi(x) &= (\cos x-x-2)+\int_0^x \sin(\xi-x)(\cos\xi-\xi-2)d\xi\\ &=(\cos x-x-2)+\cos x\left[-\frac{1}{4}\cos 2\xi+\xi\cos\xi-\sin\xi+2\cos\xi\right]_0^x\\ &\quad -\sin x\left[\frac{\xi}{2}+\frac{1}{4}\sin 2\xi-\xi\sin\xi-\cos\xi-2\sin\xi\right]_0^x\\ &=\cos x-x+2+\cos x\left[-\frac{1}{4}\cos 2x+x\cos x+2\cos x-\frac{7}{4}\right]\\ &\quad -\sin x\left[\frac{1}{4}\sin 2x-x\sin x-2\sin x+1+\frac{x}{2}\right]\end{aligned}$$

$$\Rightarrow \phi(x)=-\cos x-\sin x-\frac{x}{2}\sin x.$$

Example 28 Solve the volterra integral equation of Second Kind, by using the method of successive approximation:

(i) $\phi(x) = x - \int_0^x (x-\xi)\phi(\xi)\,d\xi$, with $\phi_0(x) = 0$

(ii) $\phi(x) = 1 + \int_0^x \phi(\xi)d\xi$, with $\phi_0(x) = 0$

(iii) $\phi(x) = (1+x) - \int_0^x \phi(\xi)d\xi$, with $\phi_0(x) = 1$

(iv) $\phi(x) = (1+x) + \int_0^x (x-\xi)\phi(\xi)d\xi$, $\phi_0(x) = 1$

Solution

(i) The integral equation is given as

$$\phi(x) = x - \int_0^x (x-\xi)\phi(\xi)d\xi, \phi_0(x) = 0$$

Here $F(x) = x$, $K(x,\xi) = x - \xi$, $\lambda = -1$

then $\phi_v(x) = F(x) + \lambda\int_0^x K(x,\xi)\phi_{v-1}(\xi)d\xi$

$$\Rightarrow \phi_v(x) = x - \int_0^x (x-\xi)\phi_{v-1}(\xi)d\xi$$

Putting $K_3(x_1\xi)$ we have

$$\phi_1(x) = x - \int_0^x (x-\xi)\phi_0(\xi)d\xi = x$$

$$\phi_2(x) = x - \int_0^x (x-\xi)\phi_1(\xi)d\xi$$

$$= x - \int_0^x (x\xi - \xi^2)d\xi = x - \frac{x^3}{3!}$$

$$\phi_3(x) = x - \int_0^x (x-\xi)\phi_2(\xi)d\xi$$

$$= x - \int_0^x (x-\xi)\left(\xi - \frac{\xi^3}{3!}\right)d\xi = x - \frac{x^3}{3!} + \frac{x^5}{5!}$$

----- ----- ----- -----

In general

$$\phi_v(x) = x - \frac{x^3}{3!} + \frac{x^5}{5!} \ldots\ldots + (-1)^{v-1} \frac{x^{2v-1}}{(2v-1)!}$$

Hence the solution of the integral equation is given by

$$\phi(x) = \lim_{v \to \infty} \phi_v(x)$$

$$= x - \frac{x^3}{3!} + \frac{x^5}{5!} - \ldots\ldots + (-1)^{v-1} \frac{x^{2v-1}}{(2v-1)!}$$

$\Rightarrow \phi(x) = \sin x$

(ii) The integral equation is given as

$$\phi(x) = 1 + \int_0^x \phi(\xi) d\xi, \text{ with } \phi_0(x) = 0$$

Here $F(x) = 1,\ K(x, \xi) = 1,\ \lambda = 1$

The vth order approximation is given by

$$\phi_v(x) = F(x) + \lambda \int_0^x K(x, \xi) \phi_{v-1}(\xi) d\xi$$

$$= 1 + \int_0^x \phi_{v-1}(\xi) d\xi$$

Putting $v = 1, 2, 3, \ldots\ldots$, we have

$$\phi_1(x) = 1 + \int_0^x \phi_0(\xi) d\xi = 1$$

$$\phi_2(x) = 1 + \int_0^x \phi_1(\xi) d\xi = 1 + x$$

$$\phi_3(x) = 1 + \int_0^x \phi_2(\xi) d\xi$$

$$= 1 + \int_0^x (1 + \xi) d\xi = 1 + x + \frac{x^2}{2!}$$

In general, we have

$$\phi_v(x) = 1 + x + \frac{x^2}{2!} + \ldots + \frac{x^{v-1}}{(v-1)!} + \ldots$$

Hence the solution of integral equation is given by

$$\phi(x) = \lim_{v \to \infty} \phi_v(x)$$

$$= 1 + x + \frac{x^2}{2!} + ... + \frac{x^{v-1}}{(v-1)!} = e^x$$

(iii) The integral equation is given as

$$\phi(x) = (1+x) - \int_0^x \phi(\xi)d\xi, \text{ with } \phi_0(x) = 1$$

Here $F(x) = 1 + x$, $K(x,\xi) = 1$, $\lambda = -1$

The vth order approximation is given by

$$\phi_v(x) = F(x) + \lambda \int_0^x K(x,\xi)\phi_{v-1}(\xi)d\xi$$

$$= (1+x) - \int_0^x \phi_{v-1}(\xi)d\xi$$

Putting $v = 1, 2, 3.....$, we have

$$\phi_1(x) = (1+x) - \int_0^x \phi_0(\xi)d\xi$$

$$= (1+x) - \int_0^x d\xi = 1$$

$$\phi_2(x) = (1+x) - \int_0^x \phi_1(\xi)d\xi$$

$$= (1+x) - \int_0^x d\xi = 1$$

$$\phi_3(x) = (1+x) - \int_0^x \phi_2(\xi)d\xi = 1$$

----- ----- ----- -----

In general,

$$\phi_v(x) = (1+x) - \int_0^x \phi_{v-1}(\xi)d\xi = 1$$

Hence the solution of integral equation is given by

$$\phi(x) = \lim_{v\to\infty} \phi_v(x) = 1$$

(iv) The integral equation is given as

$$\phi(x) = (1+x) + \int_0^x (x-\xi)\phi(\xi)d\xi, \text{ with } \phi_0(x) = 1$$

Here $F(x) = 1 + x$, $K(x,\xi) = x - \xi$, $\lambda = -1$

The vth order approximation is given by

$$\phi_v(x) = F(x) + \lambda\int_0^x K(x,\xi)\phi_{v-1}(\xi)d\xi$$

$$= (1+x) + \int_0^x (x-\xi)\phi_{v-1}(\xi)d\xi$$

Putting $v = 1,2,3\ldots\ldots$, we have

$$\phi_1(x) = (1+x) + \int_0^x (x-\xi)\phi_0(\xi)d\xi$$

$$= (1+x) + \int_0^x (x-\xi)d\xi = 1 + x + \frac{x^2}{2!}$$

$$\phi_2(x) = (1+x) + \int_0^x (x-\xi)\phi_1(\xi)d\xi$$

$$= (1+x) + \int_0^x (x-\xi)(1+\xi+\frac{\xi^2}{2})d\xi$$

$$= (1+x) + \left[\left(x\xi - \frac{\xi^2}{2}\right) + \left(\frac{x\xi^2}{2} - \frac{\xi^3}{3}\right) + \left(\frac{x\xi^3}{6} - \frac{\xi^4}{8}\right)\right]_0^x$$

$$= (1+x) + \left(x^2 - \frac{x^2}{2}\right) + \left(\frac{x^3}{2} - \frac{x^3}{3}\right) + \left(\frac{x^4}{6} - \frac{x^4}{8}\right)$$

$$\Rightarrow \phi_2(x) = 1 + x + \frac{x^2}{2!} + \frac{x^3}{3!} + \frac{x^v}{4!}$$

In general, we have

$$\phi_v(x) = 1 + x + \frac{x^2}{2!} + \frac{x^3}{3!} + \ldots\ldots + \frac{x^v}{v!}$$

Hence the solution of integral equation is given by

$$\phi(x) = \lim_{v\to\infty} \phi_v(x)$$

$$= 1 + x + \frac{x^2}{2!} + \frac{x^3}{3!} + \ldots\ldots + \frac{x^v}{v!} = e^x$$

Solution of the Fredholm Integral Equation by the Method of Successive Approximation

Consider the Fredholm integral equation

$$\phi(x) = F(x) + \lambda \int_a^b K(x,\xi)\phi(\xi)d\xi \qquad .....(1)$$

Where $F(x)$ is continuous in $a \le x \le b$ and $K(x,\xi)$ is real and continuous in the rectangle $a \le x \le b$ and $a \le \xi \le b$.

Consider the solution of the integral equation (1) is in the form of a powers in λ as.

$$\phi(x) = \phi_0(x) + \lambda\phi_1(x) + \lambda^2\phi_2(x) + ... + \lambda^n\phi_n(x)..... \qquad .....(2)$$

then equation (1) becomes

$$\phi_0(x) + \lambda\phi_1(x) + \lambda^2\phi_2(x) + ... + \lambda^n\phi_n(x).....$$

$$= F(x) + \lambda \int_a^b K(x,\xi)\{\phi_0(\xi) + \lambda\phi_1(\xi) + \lambda^2\phi_2(\xi) + ... + \lambda^n\phi_n(\xi)\}d\xi \qquad .....(3)$$

Equating the coefficients of like powers of λ, we have

$$\phi_0(x) = F(x)$$

$$\phi_1(x) = \int_a^b K(x,\xi)\phi_0(\xi)d\xi = \int_a^b K(x,\xi)F(\xi)d\xi$$

$$\phi_2(x) = \int_a^b K(x,\xi)\phi_1(\xi)d\xi = \int_a^b K_2(x,\xi)F(\xi)d\xi$$

----- ----- ----- -----

$$\phi_n(x) = \int_a^b K(x,\xi)\phi_{n-1}(\xi)d\xi = \int_a^b K_n(x,\xi)F(\xi)d\xi \qquad .....(4)$$

Where $K_1(x,\xi) = K(x,\xi)$

$$K_2(x,\xi) = \int_a^b K(x,t)K_1(t,\xi)dt$$

----- ----- ----- -----

In general, we have

$$K_n(x,\xi) = \int_a^b K(x,t)K_{n-1}(t,\xi)dt \qquad .....(5)$$

The function $K_n(x,\xi)$ are called the **iterated Kernel**

$$K_n(x,\xi) = \int_a^b K_m(x,t)K_{n-m}(t-\xi)d\xi, m < n \qquad .....(6)$$

Consider the series

$$R(x,\xi;\lambda) = K_1(x,\xi) + \lambda K_2(x,\xi) + ... + \lambda^{n-1} K_n(x,\xi) + ...$$

$$\Rightarrow \quad R(x,\xi;\lambda) = \sum_{\nu=0}^{\infty} \lambda^{\nu} K_{\nu+1}(x,\xi)$$

The function $R(x,\xi;\lambda)$ is called the **resolvent Kernel**.

Thus the solution of equation (1) reduces to

$$\phi(x) = F(x) + \lambda \int_a^b R(x,\xi;\lambda)\, F(\xi)\, d\xi$$

Example 29 Find the iterated Kernel for the following :

(i) $K(x,\xi) = x - \xi$, if $a = 0,\ b = 1$

(ii) $K(x,\xi) = x + \sin\xi;\ a = -\pi, b = \pi$

(iiii) $K(x,\xi) = e^x \cos\xi;\ a = 0,\ b = \pi$

Solution

(i) We know that

$$K_1(x,\xi) = K(x,\xi) = x - \xi$$

By the iterated Kernel, we have

$$K_2(x,\xi) = \int_0^1 K(x,t) K_1(t,\xi)\, dt$$

$$= \int_0^1 (x-t)(t-\xi)\, dt = \frac{x+\xi}{2} - x\xi - \frac{1}{3}$$

$$K_3(x,\xi) = \int_0^1 K(x,t) K_2(t,\xi)\, dt$$

$$= \int_0^1 (x-t)\left(\frac{t+\xi}{2} - t\xi - \frac{1}{3}\right) dt = (-1)\frac{x-\xi}{12}$$

$$K_4(x,\xi) = \int_0^1 K(x,t) K_3(t,\xi)\, dt$$

$$= \frac{-1}{12}\int_0^1 (x-t)(t-\xi)\, dt = \frac{-1}{12}\left(\frac{x+\xi}{2} - x\xi - \frac{1}{3}\right)$$

----- ----- ----- -----

and so on.

It follows that iterated Kernels are of the form :

(a) If $n=2\nu-1$ then $K_{2\nu-1}(x,\xi)=\frac{(-1)^{\nu}}{12^{\nu-1}}(x-\xi)$

(b) If $n=2\nu$ then $K_{2\nu}(x,\xi)=\frac{(-1)^{\nu-1}}{12^{\nu-1}}\left(x+\xi-x\xi-\frac{1}{3}\right)$

(ii) We know that

$K_1(x,\xi)=K(x,\xi)=x+\sin\xi$

By the iterated Kernel, we have

$$K_2(x,\xi) \quad = \int_{-\pi}^{\pi}(x+\sin t)(t+\sin\xi)dt = 2\pi(1+x\sin\xi)$$

$$K_3(x,\xi) \quad =2\pi\int_{-\pi}^{\pi}(x+\sin t)(1+t\sin\xi)dt=(2\pi)^2(x+\sin\xi)$$

$$K_4(x,\xi) \quad ==(2\pi)^2\int_{-\pi}^{\pi}(x+\sin t)(t+\sin\xi)dt=$$

$$(2\pi)^3(1+x\sin\xi)$$

$$K_5(x,\xi) \quad =(2\pi)^3\int_{-\pi}^{\pi}(x+\sin t)(1+t\sin\xi)dt=(2\pi)^4(x+\sin\xi)$$

----- ----- ----- -----

as so on.

It follows that the iterated Kenrels are of the form :

$$K_{2\nu-1}(x,\xi) = (2\pi)^{2\nu-2}(x+\sin\xi)$$

and $\quad K_{2\nu}(x,\xi) = (2\pi)^{2\nu-1}(1+x\sin\xi)$

(iii) We know that $K_1(x,\xi)=K(x,\xi)=e^x\cos\xi$

By the iterated Kernels, we have

$$K_2(x,\xi) \quad =\int_0^{\pi}e^x\cos t.e^t\cos\xi dt=-\frac{e^x\cos\xi}{2}(1+e^{\pi})$$

$$K_3(x,\xi) \quad =\int_0^{\pi}e^x\cos t\left\{-\left(\frac{1+e^{\pi}}{2}\right)\right\}e^t\cos\xi\,dt$$

$$=\left(\frac{1+e^{\pi}}{2}\right)^2 e^x\cos\xi$$

----- ----- ----- -----

and so on.

Hence the iterated Kernel reduces

$$K_\nu(x,\xi) = (-1)^{\nu-1}\left(\frac{1+e^\pi}{2}\right)^{\nu-1} e^x \cos\xi$$

Example 30 Find the resolvent Kernels for the following Kernels:

(i) $K(x,\xi) = \sin(x-2\xi);\ 0\le x\le 2\pi,\ \ 0\le\xi\le 2\pi.$

(ii) $K(x,\xi) = (1+x)(1-\xi);\ a=-1,\ b=0$

(iii) $K(x,\xi) = e^{x+\xi};\ a=0,\ b=1$

Solution

(i) We know that

$$K_1(x,\xi) = K_1(x,\xi) = \sin(x-2\xi)$$

$$K_2(x,\xi) = \int_0^{2\pi} \sin(x-2t).\sin(t-2\xi)dt$$

$$= \frac{1}{2}\int_0^{2\pi} [\cos(x+2\xi-3t)-\cos(x-2\xi-t)]dt = 0$$

$$K_3(x,\xi) = \int_0^{2\pi} K(x,t)K_2(x,t)dt = 0$$

$$K_4(x,\xi) = 0$$

Hence the resolvent Kernel is equal to the Kernel itself :

i.e. $R(x,\xi;\lambda) = \sin(x-2\xi)$

(ii) We know that

$$K_1(x,\xi) = K(x,\xi) = (1+x)(1-\xi)$$

By the iterated Kernels, we have

$$K_2(x,\xi) = \int_{-1}^{0}(1+x)(1-t)(1+t)(1-\xi)dt = \frac{2}{3}(1+x)(1-\xi)$$

$$K_3(x,\xi) = \frac{2}{3}\int_{-1}^{0}(1+x)(1-t)(1+t)(1-\xi)dt = \left(\frac{2}{3}\right)^2(1+x)(1-\xi)$$

----- ----- ----- -----

and so on.

Thus in general, we have

$$K_v(x,\xi) = \left(\frac{2}{3}\right)^{v-1} (1+x)(1-\xi)$$

Hence the resolvent Kernel is given by

$$R(x,\xi;\lambda) = \sum_{v=1}^{\infty} \lambda^{v-1} K_v(x,\xi)$$

$$= (1+x)(1-\xi) \sum_{v=1}^{\infty} \left(\frac{2\lambda}{3}\right)^{v-1}$$

$$\text{or } R(x,\xi;\lambda) = \frac{3(1+x)(1-\xi)}{3-2\lambda}, \text{ where } |\lambda| < \frac{3}{2} \left[\because |\lambda| < \frac{1}{p(b-a)}\right]$$

(iii) We know that

$$K_1(x,\xi) = K(x,\xi) = e^{x+\xi}$$

By the iterated Kernels, we have

$$K_2(x,\xi) = \int_0^1 e^{x+t}.e^{t+\xi}\,dt = e^{x+\xi}\left(\frac{e^2-1}{2}\right)$$

$$K_3(x,\xi) = \frac{e^2-1}{2}\int_0^1 e^{x+t}.e^{t+\xi}\,dt = \left(\frac{e^2-1}{2}\right)^2 e^{(x+\xi)}$$

In general, we have

$$K_v(x,\xi) = \left(\frac{e^2-1}{2}\right)^{v-1} e^{x+\xi}$$

Thus the resolvent Kernel is given by

$$R(x,\xi;\lambda) = \sum_{v=1}^{\infty} \lambda^{v-1} K_v(x,\xi)$$

$$e^{x+\xi} \sum_{v=1}^{\infty} \lambda^{v-1} \left(\frac{e^2-1}{2}\right)^{v-1}$$

or $R(x,\xi;\lambda) = \dfrac{2}{2-\lambda(e^2-1)} e^{x+\xi}$, where $|\lambda| < \dfrac{2}{e^2-1}$

Example 31 Solve the following integral equations :

(i) $\phi(x) = \dfrac{5x}{6} + \dfrac{1}{2}\int_0^1 x\xi\,\phi(\xi)d\xi$

(ii) $\phi(x) = 1 + \lambda\int_0^{\pi} \sin(x+\xi)\phi(\xi)d\xi$

Solution (i) Here $F(x) = \dfrac{5x}{6}$, $K(x,\xi) = x\xi, \lambda = \dfrac{1}{2}$

We know that $K(x,\xi) = K(x,\xi) = x\xi$

By the iterated Kernels, we have

$$K_2(x,\xi) = \int_0^1 K(x_1 t)\,K_1(t_1\xi)dt$$

$$= \int_0^1 xt.t\xi\,dt = \frac{1}{3}x\xi$$

$$K_3(x,\xi) = \frac{1}{3}\int_0^1 xt.t\xi\,dt = \left(\frac{1}{3}\right)^2 x\xi$$

In general, $K_n(x,\xi) = \left(\dfrac{1}{3}\right)^{n-1} x\xi$

Hence the resolvent Kernel is given by

$$R(x_1\xi;\lambda) = x\xi\sum_{n=1}^{\infty} \lambda^{n-1}\left(\frac{1}{3}\right)^{n-1} = \frac{3}{3-\lambda}x\xi$$

or $R(x_1\xi;\lambda) = \dfrac{6}{5}x\xi$

Thus the solution of the integral equation is given by.

$$\phi(x) = F(x) + \lambda\int_a^b R(x,\xi;\lambda)F(\xi)d\xi$$

$$= \frac{5x}{6} + \frac{1}{2}\int_0^1 \frac{6}{5}x\xi.\frac{5}{6}\xi\,d\xi$$

$$= \frac{5x}{6} + \frac{x}{6} = x$$

(ii) Here $F(x)=1,\ K(x,\xi)=\sin(x+\xi),\ \lambda=\lambda$

Where know that $K_1(x,\xi) = K(x,\xi) = \sin(x+\xi)$

By the iterated Kernels, we have

$$K_2(x,\xi) = \int_0^{\pi} \sin(x+t).\sin(t+\xi)dt = \frac{\pi}{2}\cos(x-\xi)$$

$$K_3(x,\xi) = \frac{\pi}{2}\int_0^{\pi} \sin(x+t)\cos(t-\xi)dt = \left(\frac{\pi}{2}\right)^2 \sin(x+\xi)$$

Proceeding in this way, we have

$$K_4(x,\xi) = \left(\frac{\pi}{2}\right)^3 \cos(x-\xi),\ K_6(x,\xi) = \left(\frac{\pi}{2}\right)^5 \cos(x-\xi)$$

$$K_5(x,\xi) = \left(\frac{\pi}{2}\right)^4 \sin(x+\xi) \qquad K_7(x,\xi) = \left(\frac{\pi}{2}\right)^6 \sin(x+\xi)$$

Hence the resolvent Kernel is given by

$$R(x,\xi;\lambda) = \sum_{v=1}^{\infty} \lambda^{v-1} K_v(x,\xi)$$

or $R(x,\xi;\lambda) = K_1(x,\xi) + \lambda K_2(x,\xi) + \lambda^2 K_3(x,\xi) + \lambda^3 K_4(x,\xi) + ...$

$$= \sin(x+\xi)\left[1+\left(\frac{\lambda\pi}{2}\right)^2+\left(\frac{\lambda\pi}{2}\right)^4+.....\right]+$$

$$\frac{\lambda\pi}{2}\cos(x-\xi)\left[1+\left(\frac{\lambda\pi}{2}\right)^2+\left(\frac{\lambda\pi}{4}\right)^4+.....\right]$$

$$= \left[\sin(x+\xi)+\frac{\lambda\pi}{2}\cos(x-\xi)\right]\left[1+\left(\frac{\lambda\pi}{2}\right)^2+\left(\frac{\lambda\pi}{2}\right)^4+.....\right]$$

$$= \left[\sin(x+\xi)+\frac{\lambda\pi}{2}\cos(x-\xi)\right]\left(\frac{4}{4-\lambda^2\pi^2}\right), |\lambda| < \frac{2}{\pi}$$

Thus the solution of integral equation is given by

$$\phi(x) = F(x)+\lambda \int_0^{\pi} R(x,\xi;\lambda)\, F(\xi)d\xi$$

$$= 1+\lambda\int_0^{\pi}\left(\frac{4}{4-\lambda^2\pi^2}\right)\left[\sin(x+\xi)+\frac{\lambda\pi}{2}\cos(x-\xi)\right]d\xi\,.$$

$$\phi(x) = 1+\frac{4\lambda}{4-\lambda^2\pi^2}(2\cos x+\lambda\pi\sin x);\ |\lambda|<\frac{2}{\pi}.$$

Example 32 Solve the integral equations :

(i) $\phi(x) = \left(\frac{3}{2}e^x-\frac{1}{2}xe^x-\frac{1}{2}\right)+\frac{1}{2}\int_0^1 \xi\,\phi(\xi)d\xi$

(ii) $\phi(x) = \left(e^x-\frac{1}{2}e+\frac{1}{2}\right)+\frac{1}{2}\int_0^1 \phi(\xi)d\xi.$

(iii) $\phi(x) = f(x)+\lambda\int_0^1 e^{x-\xi}\phi(\xi)d\xi$

Solution

(i) Here $F(x) = \frac{3}{2}e^x-\frac{1}{2}xe^x-\frac{1}{2},\ K(x,\xi)=\xi,\ \lambda=\frac{1}{2}$

We know that $K_1(x,\xi)=K(x,\xi)=\xi$

By the iterated Kernels, we have

$$K_2(x,\xi) = \int_0^1 t\xi\, dt = \frac{1}{2}\xi$$

$$K_3(x,\xi) = \frac{1}{2}\int_0^1 t\xi dt = \frac{1}{2^2}$$

In general $K_n(x,\xi) = \frac{1}{2^{n-1}}\cdot\xi$

Hence the resolvent Kernel is given by

$$R(x,\xi;\lambda)=\sum_{n=1}^{\infty}\lambda^{n-1}\frac{1}{2^{n-1}}\xi = \frac{2\xi}{2-\lambda}$$

or $\qquad R(x,\xi;\lambda) = \frac{4}{3}\xi;\ \lambda = \frac{1}{2}$

Thus the solution of the integral equation is given by

$$\phi(x) = \left(\frac{3}{2}e^x - \frac{1}{2}xe^x - \frac{1}{2}\right) + \frac{1}{2}\int_0^1 \frac{4}{3}\xi\left(\frac{3}{2}e^\xi - \frac{1}{2}\xi e^\xi - \frac{1}{2}\right)d\xi$$

$$= \left(\frac{3}{2}e^x - \frac{1}{2}xe^x - \frac{1}{2}\right) + \frac{2}{3}\int_0^1\left(\frac{3}{2}e^\xi - \frac{1}{2}\xi e^\xi - \frac{1}{2}\right)\xi\ d\xi$$

$$= \left(\frac{3}{2}e^x - \frac{1}{2}xe^x - \frac{1}{2}\right) + \frac{2}{3}\left[\frac{3}{2}(\xi e^\xi - e^\xi) - \frac{1}{2}(\xi^2 e^\xi - 2\xi e^\xi + 2e^\xi) - \frac{\xi^2}{4}\right]_0^1$$

$$\phi(x) = \frac{3}{2}e^x - \frac{1}{2}xe^x - \frac{1}{3}e + 1.$$

(ii) Here $F(x) = e^x - \frac{1}{2}e + \frac{1}{2},\ K(x,\xi) = 1,\ \lambda = \frac{1}{2}$

We know that

$$K_1(x,\xi) = K(x,\xi) = 1$$

By the iterated Kernels, we have

$$K_2(x,\xi) = \int_0^1 dt = 1$$

$$K_3(x,\xi) = \int_0^1 dt = 1$$

In general $K_n(x,\xi) = 1$

Hence the resolvent Kernel is given by

$$R(x,\xi;\lambda) = \sum_{n=1}^{\infty}\lambda^{n-1}.1 = 2;\ \lambda = \frac{1}{2}$$

Thus the solution of the integral equation is given by

$$\phi(x) = \left(e^x - \frac{1}{2}e + \frac{1}{2}\right) + \frac{1}{2}\int_0^1 2\left(e^\xi - \frac{1}{2}e + \frac{1}{2}\right)d\xi$$

$$= \left(e^x - \frac{1}{2}e + \frac{1}{2}\right) + \left[e^\xi - \frac{1}{2}e\xi + \frac{1}{2}\xi\right]_0^1$$

$$= e^x - \frac{1}{2}e + \frac{1}{2} + e - \frac{1}{2}e + \frac{1}{2} - 1 = e^x$$

(iii) Here $K_1(x,\xi) = K(x,\xi) = e^{x-\xi}$

By the iterated Kernels, we have

$$K_2(x,\xi) = \int_0^1 e^{x-t}.e^{t-\xi}dt = e^{x-\xi}$$

Proceeding in this way, we find that all the Kernels coincide with $K(x,\xi)$. Thus the resolvent Kernel is given by

$$R(x,\xi;\lambda) = K(x,\xi)(1+\lambda+\lambda^2+.....) = \frac{e^{x-\xi}}{1-\lambda},\quad \lambda \neq 1$$

Hence, the solution of the integral equation is given by

$$\phi(x) = f(x) + \lambda\int_0^1 \frac{e^{x-\xi}}{1-\lambda}f(\xi)d\xi$$

$$= f(x) - \frac{\lambda}{1-\lambda}\int_0^1 e^{x-\xi} f(\xi)d\xi$$

Example 33 Solve the following integral equations:

(i) $\phi(x) = x + \int_0^{1/2}\phi(\xi)d\xi$ (ii) $\phi(x) = \left(\sin x - \frac{x}{4}\right) + \frac{1}{4}\int_0^{\pi/2}\xi x\phi(\xi)d\xi$

Solution

(i) Here $K(x,\xi) = 1, F(x) = x, \lambda = 1$

We know that

$$K_1(x,\xi) = K(x,\xi) = 1$$

By the iterated Kernels, we have

$$K_2(x,\xi) = \int_0^{1/2} dt = \frac{1}{2}$$

$$K_3(x,\xi) = \frac{1}{2}\int_0^{1/2} dt = \left(\frac{1}{2}\right)^2$$

In general,

$$K_n(x,\xi) = \left(\frac{1}{2}\right)^{n-1}$$

Hence the resolvent Kernel is given by

$$R(x,\xi;\lambda) = \sum_{n=1}^{\infty} \lambda^{n-1}\left(\frac{1}{2}\right)^{n-1} = \frac{2}{2-\lambda} = 2 \text{ as } \lambda = 1$$

Thus the solution of integral equation is given by.

$$\phi(x) = F(x) + \lambda\int_0^{1/2} R(x,\xi;\lambda)F(\xi)d\xi$$

$$= x + \int_0^{1/2} 2.\xi d\xi = x + \frac{1}{4}$$

(ii) Here $F(x) = \sin x - \frac{x}{4},\ K(x,\xi) = x\xi,\ \lambda = \frac{1}{4}$

We know that

$$K_1(x,\xi) \quad = K(x,\xi) = \xi x$$

By the iterated Kernels, we have

$$K_2(x,\xi) \ = \int_0^{\pi/2} tx.\xi t\, dt = \frac{1}{3}\left(\frac{\pi}{2}\right)^3 x\xi$$

$$K_3(x,\xi) \ = \frac{1}{3}\left(\frac{\pi}{2}\right)^3 \int_0^{\pi/2} tx.\xi t\, dt = \left(\frac{1}{3}\right)^2\left(\frac{\pi}{2}\right)^6 x\xi$$

In general

$$K_n(x,\xi) \ = \left[\frac{1}{3}\left(\frac{\pi}{3}\right)^3\right]^{n-1} x\xi$$

Thus the resolvent Kernel is given by

$$R(x_1\xi;\lambda) = \sum_{n=1}^{\infty} \lambda^{n-1}\left[\frac{1}{3}\left(\frac{\pi}{2}\right)^3\right]^{n-1} x\xi$$

$$= \frac{1}{1-\dfrac{\lambda\pi^3}{24}} x\xi = \frac{96}{96-\pi^3} x\xi \text{ as } \lambda = \frac{1}{4}$$

Hence the solution of integral equation is given by

$$\phi(x)=\left(\sin-\frac{x}{4}\right)+\frac{1}{4}\int_0^{\pi/2}\frac{96}{96-\pi^3}.x\xi.\left(\sin\xi-\frac{\xi}{4}\right)d\xi$$

$$=\sin-\frac{x}{4}+\frac{1}{4}.\frac{96x}{96-\pi^3}\int_0^{\pi/2}\xi\left(\sin\xi-\frac{\xi}{4}\right)d\xi$$

$$\Rightarrow \phi(x)=\sin x-\frac{x}{4}+\frac{24x}{96-\pi^3}.\frac{96-\pi^3}{96}=\sin x$$

Example 34 Solve the Fredholm integral equation of the Second Kind, by the method of successive approximation to the third order.

(i) $\phi(x)=2x+\lambda\int_0^1(x+\xi)\phi(\xi)\,d\xi,\ \phi_0(x)=1$

(ii) $\phi(x)=1+\lambda\int_0^1(x+\xi)\phi(\xi)d\xi,\ \phi_0(x)=1$

(iii) $\phi(x)=1+\lambda\int_0^1 x\xi\ \phi(\xi)\,d\xi,\ \phi_0(x)=1$

Solution

(i) The integral equation is given as

$$\phi(x)=2x+\lambda\int_0^1(x+\xi)\phi(\xi)d\xi,\ \phi_0(x)=1 \qquad \text{.....(1)}$$

Consider the nth order approximation, we have

$$\phi_n(x)=2x+\lambda\int_0^1(x+\xi)\phi_{n-1}(\xi)d\xi \qquad \text{.....(2)}$$

Putting $n=1,2,3.....$ in equation (2), we have

$$\phi_1(x)=2x+\lambda\int_0^1(x+\xi)\phi_0(\xi)d\xi=2x+\lambda\int_0^1(x+\xi)d\xi$$

$$=2x+\lambda\left(x+\frac{1}{2}\right) \qquad \text{.....(3)}$$

$$\phi_2(x)=2x+\lambda\int_0^1(x+\xi)\phi_1(\xi)d\xi$$

$$=2x+\lambda\int_0^1(x+\xi)\left[2\xi+\lambda\left(\xi+\frac{1}{2}\right)\right]d\xi$$

$$=2x+\lambda\int_0^1(x+\xi)\left[\frac{1}{2}\lambda+\xi(2+\lambda)\right]d\xi$$

$$\Rightarrow \phi_2(x) = 2x + \lambda\left(x + \frac{2}{3}\right) + \lambda^2\left(x + \frac{7}{12}\right) \quad \text{.....(4)}$$

$$\phi_3(x) = 2x + \lambda\int_0^1 (x+\xi)\phi_2(\xi)d\xi$$

$$= 2x + \lambda\int_0^1 (x+\xi)\left[2\xi + \lambda\left(\xi + \frac{2}{3}\right) + \lambda^2\left(\xi + \frac{7}{12}\right)\right]d\xi$$

$$= 2x + \lambda\int_0^1 (x+\xi)\left(\frac{2\lambda}{3} + \frac{7\lambda^2}{12} + \xi(2+\lambda+\lambda^2)\right)d\xi$$

$$\Rightarrow \phi_3(x) = 2x + \lambda\left(x + \frac{2}{3}\right) + \lambda^2\left(\frac{7}{6}x + \frac{2}{3}\right) + \lambda^3\left(\frac{13}{12}x + \frac{5}{8}\right)$$

(ii) The integral equation is given as

$$\phi(x) = 1 + \lambda\int_0^1 (x+\xi)\phi(\xi)d\xi, \phi_0(x) = 1 \quad \text{.....(1)}$$

Consider n[th] order approximation, we have

$$\phi_n(x) = 1 + \lambda\int_0^1 (x+\xi)\phi_{n-1}(\xi)d\xi, \phi_0(x) = 1 \quad \text{.....(2)}$$

Putting n = 1, 2, 3 ….. in the equation (2), we have

$$\phi_1(x) = 1 + \lambda\int_0^1 (x+\xi)\phi_0(\xi)d\xi$$

$$= 1 + \lambda\int_0^1 (x+\xi)\,d\xi = 1 + \lambda\left(x + \frac{1}{2}\right)$$

$$\phi_2(x) = 1 + \lambda\int_0^1 (x+\xi)\phi_1(\xi)d\xi$$

$$= 1 + \lambda\int_0^1 (x+\xi)\left[1 + \lambda\left(\xi + \frac{1}{2}\right)\right]d\xi$$

$$= 1 + \lambda\left(x + \frac{1}{2}\right) + \lambda^2\left(x + \frac{7}{12}\right)$$

$$\phi_3(x) = 1 + \lambda\int_0^1 (x+\xi)\phi_2(\xi)d\xi$$

$$= 1 + \lambda\int_0^1 (x+\xi)\left[1 + \lambda\left(\xi + \frac{1}{2}\right) + \lambda^2\left(\xi^2 + \frac{7}{12}\right)\right]d\xi$$

$$=1+\lambda\int_0^1\left[x\left(1+\frac{\lambda}{2}+\frac{7\lambda^2}{12}\right)+\xi\left(1+\frac{\lambda}{2}+\frac{7\lambda^2}{12}+x\lambda+x\lambda^2\right)+\lambda(1+\lambda)\xi^2\right]d\xi$$

$$\Rightarrow \phi_3(x)=1+\lambda\left(x+\frac{1}{2}\right)+\lambda^2\left(x+\frac{7}{12}\right)+\lambda^3\left(\frac{13}{12}x+\frac{5}{8}\right)$$

(iii) The integral equation is given as

$$\phi(x)=1+\lambda\int_0^1 x\xi\,\phi(\xi)d\xi,\phi_0(x)=1 \qquad \text{.....(1)}$$

Consider the nth order approximation, we have

$$\phi_n(x)=1+\lambda\int_0^1 x\xi\,\phi_{n-1}(\xi)d\xi,\phi_0(x)=1 \qquad \text{.....(2)}$$

Putting n = 1,2,3 ….. in equation (2), we have

$$\phi_1(x)=1+\lambda\int_0^1 x\xi\,\phi_0(\xi)d\xi$$

$$=1+\lambda\int_0^1 x\xi\,d\xi=1+\frac{\lambda x}{2}$$

$$\phi_2(x)=1+\lambda\int_0^1 x\xi\,\phi_1(\xi)d\xi$$

$$=1+\lambda\int_0^1 x\xi\left[1+\frac{\lambda\xi}{2}\right]d\xi$$

$$=1+\lambda x\int_0^1\left(\xi+\frac{\lambda}{2}\xi^2\right)d\xi=1+\frac{\lambda x}{2}+\frac{\lambda^2 x}{6}$$

$$\phi_3(x)=1+\lambda\int_0^1 x\xi\,\phi_2(\xi)d\xi$$

$$=1+\lambda\int_0^1 x\xi\left[1+\frac{\lambda\xi}{2}+\frac{\lambda^2\xi}{6}\right]d\xi$$

$$=1+\frac{\lambda x}{2}+\frac{\lambda^2 x}{6}+\frac{\lambda^3 x}{18}$$

Practice Problems

1. Find the resolvent Kernels of the volterra integral equation with the following Kernels :

(i) $K(x,\xi)=e^{x^2-\xi^2}$ (ii) $K(x,\xi)=e^{x-\xi}$

(iii) $K(x,\xi)=x-\xi$ (iv) $K(x,\xi)=\dfrac{\cosh x}{\cosh \xi}$

[Ans:

(i) $e^{\lambda(x-\xi)}e^{x^2-\xi^2}$ (ii) $e^{\lambda(x-\xi)}e^{x-\xi}$

(iii) $\dfrac{1}{\sqrt{h}}\sinh\sqrt{\lambda}(x-\xi)$ (iv) $\dfrac{\cosh x}{\cosh \xi}e^{\lambda(x-\xi)}$

2. Solve the integral equation : $\phi(x)=(29+3x)+\int_0^x[5-6(x-\xi)]\phi(\xi)d\xi$

[Ans: $93e^{3x}-64e^{2x}$ **]**

3. With the aid of the resolvent Kernel, find the solution of the integral equation

$$\phi(x)=1+x^2+\int_0^x\frac{1+x^2}{1+\xi^2}\phi(\xi)d\xi$$

[Ans : $e^x(1+x^2)$**]**

4. So that $R(x,\xi;\lambda)=K(x,\xi)+\lambda\int_\xi^x K(x,t)R(t,\xi;\lambda)dt$ is the solution of a non-homogeneous volterra integral equation of Second Kind.

$$\phi(x)=f(x)+\lambda\int_0^x K(x,\xi)\phi(\xi)d\xi$$

Where $R(x,\xi;\lambda)$ is the resolvent kernel of the equation.

5. With the aid of resolvent kernel, find the solution of the integral equation.

(i) $\phi(x)=e^x+\int_0^x e^{x-\xi}\phi(\xi)d\xi$

[Ans: $\phi(x)=e^{2x}$ **]**

(ii) $\phi(x) = x3^x - \int_0^x 3^{x-\xi}\phi(\xi)d\xi$

[Ans: $\phi(x) = 3^x(1-e^{-x})$ **]**

(iii) $\phi(x) = e^{x^2+2x} + 2\int_0^x e^{x^2-\xi^2}\,\phi(\xi)d\xi$

[Ans: $\phi(x) = e^{x^2+2x}(1+2x)$ **]**

(iv) $\phi(x) = f(x) + \int_0^x e^{x-\xi}\,\phi(\xi)d\xi$

[Ans: $\phi(x) = f(x) + \int_0^x e^{2(x-\xi)} f(\xi)d\xi$ **]**

6. **Solve the volterra integral equation of Second Kind, by using the method of successive approximation:**

$$\phi(x) = 1 + \int_0^x (x-\xi)\phi(\xi)d\xi, \text{ with } \phi_0(x) = 0.$$

[Ans: $\phi(x) = \cosh x$ **]**

7. **Using the method of successive approximations, solve the volterra integral equation.**

$$\phi(x) = 1 + x + \int_0^x (x-\xi)\phi(\xi)d\xi, \text{ with } \phi_0(x) = 1 + x$$

[Ans: $\phi(x) = e^x$ **]**

8. **Find the first two iterated Kernels of the Kernel**

$K(x,\xi) = (x-\xi)^2;\ a = -1, b = 1.$

[Ans: $K_2(x,\xi) = \frac{2}{3}(x+\xi)^2 + 2x^2\xi^2 + \frac{4}{3}x\xi + \frac{2}{5}$,

$K_3(x,\xi) = \frac{56}{45}(x^2+\xi^2)\frac{8}{3}x^2\xi^2 - \frac{32}{9}x\xi + \frac{8}{15}$ **]**

9. **Find the iterated Kernel of the following Kernels :**

(i) $K(x,\xi) = xe^{\xi}; a = 0, b = 1$

[Ans: $K_n(x,\xi) = xe^{\xi}$ **]**

(ii) $K(x,\xi) = e^{|x-\xi|}; a = 0, b = 1$ for $n = 2$

$$\textbf{[Ans: } K_2(x,\xi) = \frac{1}{2}\left[e^{x+\xi} + e^{2-x-\xi}\right] + (\xi - x - 1)e^{\xi - x}, 0 \le x \le \xi \textbf{]:}$$

(iii) $K(x,\xi) = \sin(x-\xi);\ a = 0, b = \pi/2$ for n =2, 3.

$$\textbf{[Ans : } K_2(x,\xi) = \frac{1}{2}\sin(x+\xi) - \frac{\pi}{4}\cos(x-\xi),\ K_3(x,\xi) = \frac{4-\pi^2}{16}\sin(x-\xi)\textbf{]}$$

10. Find the resolvent Kernels of the following Kernels :

(i) $K(x,\xi) = 1;\ a = 0, b = 1$

$$\textbf{[Ans: } R(x,\xi;\lambda) = 2\textbf{]}$$

(ii) $K(x,\xi) = x\xi;\ a = -1, b = 1$

$$\textbf{[Ans : } R(x,\xi;\lambda) = \frac{3x\xi}{3-2\lambda};\ |\lambda| < \frac{3}{2}\textbf{]}$$

(iii) $K(x,\xi) = x^2\xi^2;\ a = -1, b = 1$

$$\textbf{[Ans: } R(x,\xi;\lambda) = \frac{5x^2\xi^2}{5-2\lambda};\ |\lambda| < \frac{5}{2}\textbf{]}$$

(iv) $K(x,\xi) = \sin x\cos\xi;\ a = 0, b = \pi/2$

$$\textbf{[Ans : } R(x,\xi;\lambda) = \frac{2\sin x\cos\xi}{2-\lambda};\ |\lambda| < 2\textbf{]}$$

11. Solve the following integral equations:

(i) $\phi(x) = x + \lambda\int_0^1 x\xi\,\phi(\xi)d\xi$

$$\textbf{[Ans : } \phi(x) = \frac{3x}{3-\lambda}, \lambda \ne 3\textbf{]}$$

(ii) $\phi(x) = \frac{5x}{6} - \frac{1}{9} + \frac{1}{3}\int_0^1 (x+\xi)\phi(\xi)d\xi$

$$\textbf{[Ans: } \phi(x) = x\textbf{]}$$

12. Solve the integral equation $\phi(x)=1+\lambda\int_0^1 (1-3x\xi)\phi(\xi)d\xi$

For what values of λ the solution does not exist?

[**Ans** : $\phi(x)=\dfrac{4+2\lambda(2-3x)}{4-\lambda^2}$; $|\lambda|<2$. It odes not exist for $|\lambda|>2$.]

13. Show that the resolvent Kernel of the equation

$$\phi(x)=F(x+\mu)+\int_0^a R(x,\xi;\lambda)\phi(\xi)d\xi \text{ is equal of } R(x,\xi;\lambda+\mu)$$

14. Using the method of successive approximation to solve the Fredholm integral equation

$$y(x)=\sin x+\lambda\int_0^{2\pi} \cos(x+t)\,y(t)\,dt$$

Integral Equation with Fredholm's Determinant

Consider the fredholm integral equation :

$$\phi(x)=F(x)+\lambda\int_a^b K(x,\xi)\phi(\xi)d\xi \qquad .....(1)$$

We consider the partition of the interval (a,b) into n equal parts by the points :

$$a=\xi_0,\xi_1,\xi_2,.....,\xi_{n-1},\xi_n=b$$

Where $\xi_0=a,\xi_1=a+h,\xi_2=a+2h,.....,\xi_n=a+nh$

and $$h=\xi_{n+1}-\xi_n=\frac{b-a}{n} \qquad .....(2)$$

Replace the definite integral in (1) by the sum, corresponding to the points of division, we have

$$\phi(x) = F(x)+\lambda h\sum_{v=1}^{n}K(x,\xi_v)\phi(\xi_v) \qquad .....(3)$$

$$\phi(x)-\lambda h[K(x,\xi_1)\phi(\xi_1)+K(x,\xi_2)\phi(\xi_2)+.....+$$
$$K(x,\xi_n)\phi(\xi_n)]=F(x).$$

Since the equation (3) holds for every x, it must be satisfied at its n points of division $x=\xi_1,\xi_2,.....,\xi_n$. Thus we obtain a set of n linear equations with n unknown values of the function $\phi(\xi_1),\phi(\xi_2),.....\phi(\xi_n)$:

$$\phi(\xi_1)-\lambda h[K(\xi_1,\xi_1)\phi(\xi_1)+K(\xi_1,\xi_2)$$
$$\phi(\xi_2)+.....+K(\xi_1,\xi_n)\phi(\xi_n)]=F(\xi_1)$$
$$\phi(\xi_2)-\lambda h[K(\xi_2,\xi_1)\phi(\xi_1)+K(\xi_2,\xi_2)\phi(\xi_2)+.....+$$
$$K(\xi_2,\xi_n)\phi(\xi_n)=F(\xi_2)$$
$$..... \quad ..... \quad ..... \quad .....$$
$$\phi(\xi_n)-\lambda h[K(\xi_n,\xi_1)\phi(\xi_1)+K(\xi_n,\xi_2)\phi(\xi_2)+.....+$$
$$K(\xi_n,\xi_n)\phi(\xi_n)]=F(\xi_n) \qquad .....(4)$$

With the notations $F(\xi_i)=F_i,\phi(\xi_i)=\phi_i,K(\xi_i,\xi_j)=K_{ij}$

The system of equations (4) reduces to

$$(1-\lambda h\,K_{11})\phi_1-\lambda h\,K_{12}\,\phi_2-\lambda h\,K_{13}\,\phi_3-.....-\lambda h\;K_{1n}\phi_n=F_1$$
$$-\lambda h\,K_{21}\phi_1+(1-\lambda hK_{22})\phi_2-\lambda h\,K_{23}\,\phi_3.....\lambda h\;K_{2n}\phi_n=F_2$$
$$-\lambda h\,K_{31}\,\phi_1-\lambda h\,K_{32}\,\phi_2+(1-\lambda h\,K_{33})\phi_3-.....-\lambda h\,\phi_{3n}\,\phi_n=F_3$$
$$..... \quad ..... \quad ..... \quad .....$$
$$-\lambda h\,K_{n1}\phi_1-\lambda h\,K_{n2}\phi_2-\lambda h\,K_{n3}\phi_3-.....+(1-\lambda hK_{nn})\phi_n=F_n \qquad .....(5)$$

The system of equations (5) may be expressed in the form of determinants :

$$D_n(\lambda)=\begin{vmatrix} 1-\lambda hK_{11} & -\lambda hK_{12} & -\lambda hK_{13} & ..... & -\lambda hK_{1n} \\ -\lambda hK_{21} & (1-\lambda hK_{22}) & -\lambda hK_{23} & ..... & -\lambda hK_{2n} \\ -\lambda hK_{31} & -\lambda hK_{32} & (1-\lambda hK_{33}) & ..... & -\lambda hK_{3n} \\ ..... & ..... & ..... & ..... & ..... \\ -\lambda hK_{n1} & -\lambda hK_{n2} & -\lambda hK_{n3} & & (1-\lambda hK_{nn}) \end{vmatrix} \qquad .....(6)$$

Provided that $D_n(\lambda) \neq 0$.

$$D_n(\lambda) = 1 - \lambda h \sum_{v=1}^{n} K_{vv} + \frac{(-\lambda h)^2}{2!} \sum_{r,s=1}^{n} \begin{vmatrix} K_{rr} & K_{rs} \\ K_{sr} & K_{ss} \end{vmatrix} + \frac{(-\lambda h)^3}{3!} \sum_{r,s,t=1}^{n} \begin{vmatrix} K_{rr} & K_{rs} & K_{rt} \\ K_{sr} & K_{ss} & K_{st} \\ K_{tr} & K_{ts} & K_{tt} \end{vmatrix} + \ldots + \frac{(-\lambda h)^n}{n!} \sum_{\lambda_1,\lambda_2,\ldots,\lambda_n=1}^{n} \begin{vmatrix} K_{\lambda_1\lambda_1} & K_{\lambda_1\lambda_2} & \ldots & K_{\lambda_1\lambda_n} \\ K_{\lambda_2\lambda_1} & K_{\lambda_2\lambda_2} & \ldots & K_{\lambda_2\lambda_n} \\ \ldots & \ldots & \ldots & \ldots \\ K_{\lambda_n\lambda_1} & K_{\lambda_n\lambda_2} & \ldots & K_{\lambda_n\lambda_n} \end{vmatrix} \quad \ldots(7)$$

Let $h \to 0$ and $n \to \infty$, then each terms of sum (7) reduces to some single, double or triple integral etc. Thus, we have

$$D(\lambda) = 1 - \lambda \int_a^b K(\xi_1, \xi_1) d\xi_1 + \frac{\lambda^2}{2!} \int_a^b \int_a^b \begin{vmatrix} K(\xi_1,\xi_1) & K(\xi_1,\xi_2) \\ K(\xi_2,\xi_1) & K(\xi_2,\xi_2) \end{vmatrix} d\xi_1 d\xi_2$$

$$-\frac{\lambda^3}{3!} \int_a^b \int_a^b \int_a^b \begin{vmatrix} K(\xi_1,\xi_1) & K(\xi_1,\xi_2) & K(\xi_1,\xi_3) \\ K(\xi_2,\xi_1) & K(\xi_2,\xi_2) & K(\xi_2,\xi_3) \\ K(\xi_3,\xi_1) & K(\xi_3,\xi_2) & K(\xi_3,\xi_3) \end{vmatrix} d\xi_1 d\xi_2 d\xi_3 + \ldots \quad \ldots(8)$$

Where $D(\lambda)$ is called the **Fredholm's determinant.**

Equation (8) may be written an

$$D(\lambda) = 1 + \sum_{m=1}^{\infty} \frac{(-\lambda)^m}{m!} \int_a^b \int_a^b \ldots \int_a^b K\begin{pmatrix} \xi_1, & \xi_2, & \ldots & \xi_m \\ \xi_1, & \xi_2, & \ldots & \xi_m \end{pmatrix} d\xi_1, d\xi_2 \ldots d\xi_m \quad \ldots(9)$$

$$\Rightarrow D(\lambda) = 1 + \sum_{m=1}^{\infty} \frac{(-1)^m \lambda^m}{m!} Cm. \quad \ldots(10)$$

Where

$$K\begin{pmatrix} \xi_1, \xi_2, \ldots \xi_m \\ \xi_1, \xi_2 \ldots \xi_m \end{pmatrix} = \begin{vmatrix} K(\xi_1,\xi_1) & K(\xi_1,\xi_2) & \ldots & K(\xi_1,\xi_m) \\ K(\xi_2,\xi_1) & K(\xi_2,\xi_2) & \ldots & K(\xi_2,\xi_m) \\ \ldots & \ldots & & \ldots \\ K(\xi_m,\xi_1) & K(\xi_m,\xi_2) & \ldots & K(\xi_m,\xi_m) \end{vmatrix}$$

This is called Fred holm's first series.

$$C_m = \int_a^b .....\int_a^b \int_a^b K\begin{pmatrix} \xi_1, \xi_2.....\xi_m \\ \xi_1, \xi_2.....\xi_m \end{pmatrix} d\xi_1, d\xi_2.....d\xi_m$$

Now, the solution of the Fred holm integral equation is

$$\phi(x) = F(x) + \lambda \int_a^b R(x,\xi;\lambda) F(\xi) d\xi$$

Where $$R(x,\xi;\lambda) = \frac{D(x,\xi;\lambda)}{D(\lambda)} \quad .....(11)$$

And is called the **resolvent Kernel** and the function $D(x,\xi;\lambda)$ is said to be **Fredholm minor**.

We know that

$$R(x,\xi;\lambda) = K(x,\xi) + \lambda \int_a^b K(x,\xi_1) R(\xi,\xi;\lambda) d\xi_1$$

$$\Rightarrow \quad \frac{D(x,\xi;\lambda)}{D(\lambda)} = K(x,\xi) + \lambda \int_a^b K(x,\xi_1) \frac{D(\xi_1,\xi;\lambda)}{D(\lambda)} d\xi_1$$

$$D(x,\xi;\lambda) = K(x,\xi) D(\lambda) + \lambda \int_a^b K(x,\xi_1) D(\xi_1,\xi;\lambda) d\xi_1 \quad .....(12)$$

The solution of the equation in the form of series is given by

$$D(x,\xi;\lambda) = B_0(x,\xi) + \sum_{m=1}^{\infty} (-1)^m \frac{\lambda^m}{m!} B_m(x,\xi) \quad .....(13)$$

From the relations (13) and (10), the equation (12) reduces

$$B_0(x,\xi) + \sum_{m=1}^{\infty} \frac{(-1)^m \lambda^m}{m!} B_m(x,\xi) = K(x,\xi) \left\{ 1 + \sum_{m=1}^{\infty} \frac{(-1)^m \lambda^m}{m!} C_m \right\}$$

$$+ \lambda \int_a^b K(x,\xi_1) \left\{ B_0(\xi_1,\xi) + \sum_{m=1}^{\infty} \frac{(-1)^m \lambda^m}{m!} B_m(\xi_1,\xi) \right\} d\xi_1$$

Substituting $\lambda = 0 \Rightarrow B_0(x,\xi) = K(x,\xi)$

To determine the coefficients $B_m(x,\xi)$ equating the coefficients of λ^m, we have

$$\frac{(-1)^m}{m!} B_m(x,\xi) = \frac{(-1)^m}{m!} C_m K(x,\xi) + \frac{(-1)^{m-1}}{(m-1)!} \int_a^b K(x,\xi_1) B_{m-1}(\xi_1,\xi) d\xi_1$$

$$\Rightarrow \qquad B_m(x,\xi) = C_m K(x,\xi) - m\int_a^b K(x,\xi_1) B_{m-1}(\xi_1,\xi)\,d\xi_1 \qquad .....(14)$$

Which is a **recursive relation.**

Thus for $m = 1$, we have

$$\begin{aligned} B_1(x,\xi) &= C_1 K(x,\xi) - \int_a^b K(x,\xi_1) B_0(\xi_1,\xi)\,d\xi_1 \\ &= K(x,\xi)\int_a^b K(\xi_1,\xi_1)\,d\xi_1 - \int_a^b K(x,\xi_1) K(\xi_1,\xi)\,d\xi_1 \\ &= \int_a^b \begin{vmatrix} K(x,\xi) & K(x,\xi_1) \\ K(\xi_1,\xi) & K(\xi,\xi_1) \end{vmatrix} d\xi_1 = \int_a^b K\begin{pmatrix} x, \xi_1 \\ \xi, \xi_1 \end{pmatrix} d\xi_1 \end{aligned}$$

In general,

$$B_m(x,\xi) = \int_a^b \int_a^b .....\int_a^b K\begin{pmatrix} x, \xi_1, \xi_2, ...\xi_m \\ \xi, \xi_1, \xi_2, ...\xi_m \end{pmatrix} d\xi_1\, d\xi_2 .....d\xi_m$$

This is called Fredholm Second Series.

Example 35 **Using Fredholm determinants, find the resolvent Kernel of the following Kernels :**

(i) $K(x,\xi) = xe^{\xi}; a = 0, b = 1$

(ii) $K(x,\xi) = 1 + 3x\xi; 0 \le x \le 1, 0 \le \xi \le 1$

(iii) $K(x,\xi) = x^2\xi - x\xi^2; 0 \le x \le 1, 0 \le \xi \le 1$

(iv) $K(x,\xi) = \sin x - \sin\xi; 0 \le x \le 2\pi, 0 \le \xi \le 2\pi$

(v) $K(x,\xi) = 2x - \xi; 0 \le x \le 1, 0 \le \xi \le 1$

(vi) $K(x,\xi) = \sin x \cos\xi; 0 \le x \le 2\pi, 0 \le \xi \le 2\pi$

Solution

(i) Here $B_0(x,\xi) = K(x,\xi) = xe^{\xi}$

$$B_1(x,\xi) = \int_0^1 \begin{vmatrix} K(x,\xi) & K(x,\xi_1) \\ K(\xi_1,\xi) & K(\xi_1,\xi_1) \end{vmatrix} d\xi_1 = \int_0^1 \begin{vmatrix} xe^{\xi} & xe^{\xi_1} \\ \xi_1 e^{\xi} & \xi_1 e^{\xi_1} \end{vmatrix} d\xi_1 = 0$$

$$B_2(x,\xi) = \int_0^1\int_0^1 \begin{vmatrix} K(x,\xi) & K(x,\xi_1) & K(x,\xi_2) \\ K(\xi_1,\xi) & K(\xi_1,\xi_1) & K(\xi_1,\xi_2) \\ K(\xi_2,\xi) & K(\xi_2,\xi_1) & K(\xi_2,\xi_2) \end{vmatrix} d\xi_1 d\xi_2$$

$$= \int_0^1 \int_0^1 \begin{vmatrix} xe^{\xi} & xe^{\xi_1} & xe^{\xi_2} \\ \xi_1 e^{\xi} & \xi_1 e^{\xi_1} & \xi_1 e^{\xi_2} \\ \xi_2 e^{\xi} & \xi_2 e^{\xi_1} & \xi_2 e^{\xi_2} \end{vmatrix} d\xi_1 d\xi_2 = 0$$

It follows that all subsequent $B_m(x,\xi) = 0$ for $m \geq 3$

$$\text{Also } C_1 = \int_0^1 K(\xi_1, \xi_1) d\xi_1 = \int_0^1 \xi_1 e \xi d\xi_1 = 1$$

$$C_2 = \int_0^1 \int_0^1 \begin{vmatrix} K(\xi_1, \xi_1) & K(\xi_1, \xi_2) \\ K(\xi_2, \xi_1) & K(\xi_2, \xi_2) \end{vmatrix} d\xi_1 d\xi_2$$

$$= \int_0^1 \int_0^1 \begin{vmatrix} \xi_1 e^{\xi_1} & \xi_1 e^{\xi_2} \\ \xi_2 e^{\xi_1} & \xi_2 e^{\xi_2} \end{vmatrix} d\xi_1 d\xi_2 = 0$$

$\Rightarrow$ That all subsequent $C_m = 0$

Hence, we have

$$D(x, \xi; \lambda) = K(x, \xi) + \sum_{m=1}^{\infty} \frac{(-1)^m \lambda^m}{m!} B_m(x, \xi) = xe^{\xi}$$

$$\text{and } D(\lambda) = 1 + \sum_{m=1}^{\infty} \frac{(-1)^m}{m!} C_m = 1 - \lambda$$

Thus the Fredholm resolvent Kernel is given by

$$R(x, \xi; \lambda) = \frac{D(x, \xi; \lambda)}{D(\lambda)} = \frac{xe^{\xi}}{1 - \lambda}$$

(ii) Here $B_0(x, \xi) = K(x, \xi) = 1 + 3x\xi$

$$B_1(x, \xi) = \int_0^1 \begin{vmatrix} 1+3x\xi & 1+3x\xi_1 \\ 1+3\xi_1\xi & 1+3\xi_1^2 \end{vmatrix} d\xi_1 = -\left[\frac{3(x+\xi)}{2} - 3x\xi - 1 \right]$$

$$B_2(x, \xi) = \int_0^1 \int_0^1 \begin{vmatrix} 1+3x\xi & 1+3x\xi_1 & 1+3x\xi_2 \\ 1+3\xi_1\xi & 1+3\xi_1^2 & 1+3\xi_1\xi_2 \\ 1+3\xi_2\xi & 1+3\xi_2\xi_1 & 1+3\xi_2^2 \end{vmatrix} \xi_1 d\xi_2 = 0$$

It follows that all other $B_m(x, \xi) = 0$

Also $C_1 = \int_0^1 (1+3\xi_1^2) d\xi_1 = 2$

$$C_2 = \int_0^1 \int_0^1 \begin{vmatrix} 1+3\xi_1^2 & 1+3\xi_1\xi_2 \\ 1+3\xi_2\xi_1 & 1+3\xi_2^2 \end{vmatrix} d\xi_1 d\xi_2 = \frac{1}{2}$$

$\Rightarrow$ That all subsequent $C_m = 0$

$$\text{Hence } D(x,\xi;\lambda) = K(x,\xi) + \sum_{m=1}^{\infty} \frac{(-1)^m \lambda^m}{m!} B_m(x,\xi)$$

$$= (1+3x\xi) + \lambda\left[\frac{3(x+\xi)}{2} - 3x\xi - 1\right]$$

$$\text{And } D(\lambda) = 1 + \sum_{m=1}^{\infty} \frac{(-1)^m \lambda^m}{m!} C_m = 1 - 2\lambda + \frac{1}{4}\lambda^2$$

Thus the Fredholm resolvent Kernel is given by

$$R(x,\xi;\lambda) = \frac{D(x,\xi;\lambda)}{D(\lambda)} = \frac{1+3x\xi+\lambda\left[\dfrac{3(x+\xi)}{2} - 3x\xi - 1\right]}{1-2\lambda+\dfrac{1}{4}\lambda^2}$$

(iii) Here $B_0(x,\xi) = K(x,\xi) = x^2\xi - x\xi^2$

$$B_1(x,\xi) = \int_0^1 \begin{vmatrix} x^2\xi - x\xi^2 & x^2\xi_1 - x\xi_1^2 \\ \xi_1^2\xi - \xi_1\xi^2 & 0 \end{vmatrix} d\xi_1$$

$$= -x\xi\left(\frac{x+\xi}{4} - \frac{x\xi}{3} - \frac{1}{5}\right)$$

$$B_2(x,\xi) = B_3(x,\xi) = 0$$

It follows that all subsequent $B_m(x,\xi) = 0$

Also $C_1 = \int_0^1 0\, d\xi_1 = 0$

$$C_2 = \int_0^1 \int_0^1 \begin{vmatrix} 0 & \xi_1^2\xi_2 - \xi_1\xi_2^2 \\ \xi_1^2\xi_1 - \xi_2\xi_1^2 & 0 \end{vmatrix} d\xi_1 d\xi_2 = \frac{1}{120}$$

$$C_3 = C_4 = 0 \; C_3 = C_4 = 0$$

It follows that all subsequent $C_m = 0$.

Hence $D(x,\xi;\lambda) = K(x,\xi) + \sum_{m=1}^{\infty} \frac{(-1)^m \lambda^m}{m!} B_m(x_1\xi)$

$$= x^2\xi - x\xi^2 + x\xi\left(\frac{x+\xi}{4} - \frac{x\xi}{3} - \frac{1}{5}\right)\lambda$$

and $D(\lambda) = 1 + \sum_{m=1}^{\infty} \frac{(-1)^m \lambda^m}{m!} C_m = 1 + \frac{\lambda^2}{240}$

Thus the Fredholm resolvent Kernel is given by

$$R(x,\xi;\lambda) = \frac{(x,\xi;\lambda)}{D(\lambda)} = \frac{x^2\xi - x\xi^2 + x\xi\left(\frac{x+\xi}{4} - \frac{x\xi}{3} - \frac{1}{5}\right)\lambda}{1 + \frac{\lambda^2}{240}}$$

(iv) Here $B_0(x,\xi) = K(x,\xi) = \sin x - \sin \xi$.

$$B_1(x,\xi) = \int_0^{2\pi} \begin{vmatrix} \sin x - \sin \xi & \sin x - \sin \xi_1 \\ \sin \xi_1 - \sin \xi & 0 \end{vmatrix} d\xi_1 = \pi(1 + 2\sin x \sin \xi)$$

$B_2(x,\xi)$

$$= \int_0^{2\pi}\int_0^{2\pi} \begin{vmatrix} \sin x - \sin \xi & \sin x - \sin \xi_1 & \sin x - \sin \xi_2 \\ \sin \xi_1 - \sin \xi & 0 & \sin \xi_1 - \sin \xi_2 \\ \sin \xi_2 - \sin \xi & \sin \xi_2 - \sin \xi_1 & 0 \end{vmatrix} d\xi_1\, d\xi_2 = 0$$

It follows that all subsequent $B_m(x,\xi) = 0$

Also $C_1 = \int_0^{2\pi} (\sin \xi_1 - \sin \xi_1) d\xi_1 = 0$

$$C_2 = \int_0^{2\pi}\int_0^{2\pi} \begin{vmatrix} 0 & \sin \xi_1 - \sin \xi_2 \\ \sin \xi_2 - \sin \xi_1 & 0 \end{vmatrix} d\xi_1\, d\xi_2 = 4\pi^2$$

$C_3 = 0 \Rightarrow$ that all other $C_m = 0$

Hence $D(x,\xi;\lambda) = K(x,\xi) + \sum_{m=1}^{\infty} \frac{(-1)^m \lambda^m}{m!} B_m(x,\xi)$

$$= \sin x - \sin \xi - \pi(1 + 2\sin x \sin \xi)$$

And $D(\lambda) = 1 + \sum_{m=1}^{\infty} \frac{(-1)^m \lambda^m}{m!} C_m = 1 + 2\lambda^2 \pi^2$

Hence the Fredholm resolvent Kernel is given by

$$R(x,\xi;\lambda) = \frac{D(x,\xi;\lambda)}{D(\lambda)} = \frac{\sin x - \sin\xi - \pi(1 + 2\sin x \sin\xi)}{1 + 2\lambda^2 \pi^2}$$

(v) Here $B_0(x,\xi) = K(x,\xi) = 2x - \xi$

$$B_1(x,\xi) = \int_0^1 \begin{vmatrix} 2x-\xi & 2x-\xi_1 \\ 2\xi_1 - \xi & \xi_1 \end{vmatrix} d\xi_1 = -\left[(x + \xi - 2x\xi) - \frac{2}{3}\right]$$

$$B_2(x,\xi) = \int_0^1 \int_0^1 \begin{vmatrix} 2x-\xi & 2x-\xi_1 & 2x-\xi_2 \\ 2\xi_1 - \xi & \xi_1 & 2\xi_1 - \xi_2 \\ 2\xi_2 - \xi & 2\xi_2 - \xi_1 & \xi_2 \end{vmatrix} d\xi_1 d\xi_2 = 0$$

It follows that all subsequent $B_m(x,\xi) = 0$

Also $C_1 = \int_0^1 \xi_1 d\xi_1 = \frac{1}{2}$

$$C_2 = \int_0^1 \int_0^1 \begin{vmatrix} \xi_1 & 2\xi_1 - \xi_2 \\ 2\xi_2 - \xi_1 & \xi_2 \end{vmatrix} d\xi_1\, d\xi_2 = \frac{1}{3}$$

$$C_3 = \int_0^1 \int_0^1 \int_0^1 \begin{vmatrix} \xi_1 & 2\xi_1 - \xi_2 & 2\xi_1 - \xi_3 \\ 2\xi_2 - \xi_1 & \xi_2 & 2\xi_2 - \xi_3 \\ 2\xi_3 - \xi_1 & 2\xi_3 - \xi_2 & \xi_3 \end{vmatrix} d\xi_1 d\xi_2 d\xi_3 = 0$$

$\Rightarrow$ That all other $C_m = 0$

Hence $D(x,\xi;\lambda) = K(x,\xi) + \sum_{m=1}^{\infty} \frac{(-1)^m \lambda^m}{m!} B_m(x,\xi)$

$$= 2x - \xi + \lambda\left[x + \xi - 2x\xi - \frac{2}{3}\right]$$

and $D(\lambda) = 1 + \sum_{m=1}^{\infty} \frac{(-1)^m \lambda^m}{m!} C_m = 1 - \frac{\lambda}{2} + \frac{\lambda^2}{6}$

Thus the Fredholm resolvent Kernel is given by.

$$R(x_1\xi;\lambda)=\frac{D(x,\xi;\lambda)}{D(\lambda)}=\frac{2x-\xi+\lambda\left[x+\xi-2x\xi-\frac{2}{3}\right]}{1-\frac{\lambda}{2}+\frac{\lambda^2}{6}}$$

(vi) Here $B_0(x,\xi)=K(x,\xi)=\sin x\cos\xi$.

$$B_1(x,\xi)=\int_0^{2\pi}\begin{vmatrix}\sin x\cos\xi & \sin x\cos\xi_1\\ \sin\xi_1\cos\xi & \sin\xi_1\cos\xi_1\end{vmatrix}d\xi_1=0$$

$$B_2(x,\xi)=0$$

It follows that all subsequent $B_m(x,\xi)=0$

Also $C_1=\int_0^{2\pi}\sin\xi_1\cos\xi_1 d\xi_1=0$

$C_2=0$

$\Rightarrow$ That all other $C_m=0$.

Hence $D(x,\xi;\lambda)=K(x,\xi)+\sum_{m=1}^{\infty}\frac{(-1)^m\lambda^m}{m!}B_m(x,\xi)=\sin x\cos\xi$

and $D(\lambda)=1+\sum_{m=1}^{\infty}\frac{(-1)^m\lambda^m}{m!}C_m=1$

Thus the Fredholm resolvent Kernel is given by

$$R(x,\xi;\lambda)=\frac{D(x,\xi;\lambda)}{D(\lambda)}=\sin x\cos\xi.$$

Example 36 Find the resolvent Kernel of the following Kernels :

(i) $K(x,\xi)=x-2\xi;\ 0\le x\le 1,\ 0\le\xi\le 1$

(ii) $K(x,\xi)=x+\xi+1;\ -1\le x\le 1,\ -1\le\xi\le 1$

Solution

(i) Here $B_0(x,\xi)=K(x,\xi)=x-2\xi$.

By the recursive relations.

$$B_m(x,\xi)=C_mK(x,\xi)-m\int_a^b K(x,\xi_1)B_{m-1}(\xi_1,\xi)d\xi_1 \qquad .....(1)$$

and $C_m = \int_a^b B_{m-1}(\xi_1, \xi_1)d\xi_1$ .....(2)

From the relation (2), we have

$$C_1 = \int_a^b B_0(\xi_1, \xi_2)d\xi_1 = \int_0^1 (-\xi_1)d\xi_1 = -\frac{1}{2}$$

From the relation (1), we have

$$B_1(x,\xi) = C_1 K(x,\xi) - \int_a^b K(x,\xi_1) B_0(\xi_1,\xi)d\xi_1$$

$$= -\frac{1}{2}(x-2\xi) - \int_0^1 (x-2\xi_1)(\xi_1 - 2\xi)d\xi_1$$

or $B_1(x,\xi) = -x - \xi + 2x\xi + \dfrac{2}{3}$

Also $C_2 = \int_0^1 B_1(\xi_1, \xi_1)d\xi_1$

$$= \int_0^1 \left(-2\xi_1 + 2\xi_1^2 + \frac{2}{3}\right)d\xi_1 = \frac{1}{3}$$

And $B_2(x,\xi) = C_2 K(x,\xi) - 2\int_0^1 K(x,\xi_1)B_1(\xi,\xi)d\xi_1$

$$= \frac{x-2\xi}{2} - 2\int_0^1 (x-2\xi_1)\left(-\xi_1 - \xi + 2\xi_1\xi + \frac{2}{3}\right)d\xi_1 = 0$$

$$\Rightarrow C_3 = C_4 = ... = 0 \text{ and } B_3(x,\xi) = B_4(x,\xi) = ... = 0$$

Hence $D(x,\xi;\lambda) = K(x,\xi) + \sum_{m=1}^{\infty} \dfrac{(-1)^m \lambda^m}{m!} B_m(x,\xi)$

$$= x - 2\xi + \left(x + \xi - 2x\xi - \frac{2}{3}\right)\lambda$$

and $D(\lambda) = 1 + \sum_{m=1}^{\infty} \dfrac{(-1)^m \lambda^m}{m!} C_m = 1 + \dfrac{\lambda}{2} + \dfrac{\lambda^2}{6}$

thus the resolvent Kernel is given by

$$R(x,\xi;\lambda) = \frac{D(x,\xi;\lambda)}{D(\lambda)} = \frac{x - 2\xi + \left(x + \xi - 2x\xi - \frac{2}{3}\right)\lambda}{1 + \frac{\lambda}{2} + \frac{\lambda^2}{6}}$$

(ii) Here $B_0(x,\xi)=K(x,\xi)=x+\xi+1$

By the recursive relations :

$$B_m(x,\xi)=C_mK(x,\xi)-m\int_a^b K(x,\xi)B_{m-1}(\xi_1,\xi)d\xi_1 \qquad .....(1)$$

and $C_m=\int_a^b B_{m-1}(\xi_1,\xi_1)d\xi_1$

From the relation (2), we have

$$C_1=\int_{-1}^{1} B_0(\xi_1,\xi_1)d\xi_1=\int_{-1}^{1}(2\xi_1+1)d\xi_1=2$$

From the relation (1), we have

$$B_1(x,\xi)=C_1K(x,\xi)-\int_a^b K(x,\xi_1)B_0(\xi_1,\xi)d\xi_1$$

$$=2(x+\xi+1)-\int_{-1}^{1}(x+\xi_1+1)(\xi_1+\xi+1)d\xi_1=-2\left(x\xi+\frac{1}{3}\right)$$

Further, we get

$$C_2=-2\int_{-1}^{1}\left(\xi_1^2+\frac{1}{3}\right)d\xi_1=-\frac{8}{3}$$

and

$$B_2(x,\xi)=-\frac{8}{3}(x+\xi+1)+4\int_{-1}^{1}(x+\xi_1+1)\left(\xi_1\xi+\frac{1}{3}\right)d\xi_1=0$$

$\Rightarrow\ C_3=C_4=.....=0$ and $B_3(x,\xi)=B_4(x,\xi)=.....=0$

Hence $D(x,\xi;\lambda)=K(x,\xi)+\sum_{m=1}^{\infty}\frac{(-1)^m\lambda^m}{m!}B_m(x,\xi)$

$$=(x+\xi+1)+2\lambda\left(x\xi+\frac{1}{3}\right)$$

and $D(\lambda)=1+\sum_{m=1}^{\infty}\frac{(-1)^m\lambda^m}{m!}C_m=1-2\lambda-\frac{4}{3}\lambda^2$

Thus the resolvent Kernel is given by

$$R(x,\xi;\lambda)=\frac{D(x,\xi;\lambda)}{D(\lambda)}=\frac{(x+\xi+1)+2\lambda\left(x\xi+\frac{1}{3}\right)}{1-2\lambda-\frac{4}{3}\lambda^2}$$

Example 37 Solve the following integral equations:

(i) $\phi(x) = \sec^2 x + \lambda \int_0^1 \phi(\xi) d\xi$

(ii) $\phi(x) = e^x + \lambda \int_0^{10} x\xi \, \phi(\xi) d\xi$

(iii) $\phi(x) = e^x + \lambda \int_0^1 2e^x e^\xi \, \phi(\xi) d\xi$

(iv) $\phi(x) = \sin x + \lambda \int_4^{10} x \phi(\xi) d\xi$

Solution :

(i) The integral equation is given as

$\phi(x) = \sec^2 x + \lambda \int_0^1 \phi(\xi) d\xi$

Here $B_0(x, \xi) = K(x, \xi) = 1$

and $B_1(x, \xi) = B_2(x, \xi) = ... = 0$

also $C_2 = \int_0^1 K(\xi_1, \xi_1) d\xi_1 = \int_0^1 d\xi_1 = 1$

and $C_2 = C_2 = ... = 0$

Hence $D(x, \xi; \lambda) = K(x, \xi) + \sum_{m=1}^{\infty} \frac{(-1)^m \lambda^m}{m!} B_m(x, \xi) = 1$

and $D(\lambda) = 1 + \sum_{m=1}^{\infty} \frac{(-1)^m \lambda^m}{m!} C_m = 1 - \lambda$

Thus resolvent Kernel is given by

$$R(x, \xi; \lambda) = \frac{D(x, \xi; \lambda)}{D(\lambda)} = \frac{1}{1 - \lambda}$$

Hence the solution of the integral equation is given by

$$\phi(x) = F(x) + \lambda \int_0^1 R(x, \xi; \lambda) F(\xi) d\xi$$

$$= \sec^2 x + \lambda \int_0^1 \frac{1}{1 - \lambda} . \sec^2 \xi \, d\xi$$

or $\phi(x) = \sec^2 x + \frac{\lambda}{1 - \lambda} \tan 1.$

(ii) The integral equation is given as

$$\phi(x) = e^x + \lambda\int_0^{10} x\xi\,\phi(\xi)d\xi$$

Here $B_0(x,\xi) = K(x,\xi) = x\xi$

$$B_1(x,\xi) = \int_0^{10}\begin{vmatrix} x\xi & x\xi_1 \\ \xi_1\xi & \xi_1^2 \end{vmatrix} d\xi_1 = 0$$

and $B_2(x,\xi) = B_3(x,\xi) = ... = 0$

also $C_1 = \int_0^{10} K(\xi_1,\xi_1)d\xi_1 = \int_0^{10} \xi_1^2 d\xi = \dfrac{10^3}{3}$

$$C_2 = \int_0^{10}\int_0^{10}\begin{vmatrix} \xi_1^2 & \xi_1\xi_2 \\ \xi_2\xi_1 & \xi_2^2 \end{vmatrix} d\xi_1 d\xi_2 = 0$$

or $C_3 = C_4 = ... = 0$

Hence $D(x,\xi;\lambda) = K(x,\xi) + \sum_{m=1}^{\infty}\dfrac{(-1)^m\lambda^m}{m!}B_m(x,\xi) = x\xi$

and $D(\lambda) = 1 + \sum_{m=1}^{\infty}\dfrac{(-1)^m\lambda^m}{m!}C_m = 1 - \dfrac{10^3}{3}\lambda$

Thus resolvent Kernel is obtained as

$$R(x,\xi;\lambda) = \frac{D(x,\xi;\lambda)}{D(\lambda)} = \frac{x\xi}{1-\dfrac{10^3}{3}\lambda}$$

Hence the solution of the integral equation is given by

$$\phi(x) = e^x + \frac{\lambda}{1-\dfrac{10^3}{3}\lambda}\int_0^{10} x\xi.e^{\xi}\,d\xi$$

or $$\phi(x) = e^x + \frac{\lambda x}{1-\dfrac{10^3}{3}\lambda}(9e^{10}+1)$$

(iii) The integral equation is given as

$$\phi(x) = e^x + \lambda\int_0^1 2e^x e^\xi \phi(\xi)d\xi$$

Here $B_0(x,\xi) = K(x,\xi) = 2e^x e^\xi$

$$B_1(x,\xi) = \int_0^1 \begin{vmatrix} 2e^x e^\xi & 2e^x e^{\xi_1} \\ 2e^{\xi_1} e^\xi & 2e^{2\xi_1} \end{vmatrix} d\xi_1 = 0$$

$$B_2(x,\xi) = B_3(x,\xi) = ... = 0$$

and $C_1 = \int_0^1 2e^{2\xi_1} d\xi_1 = e^2 - 1$

$$C_2 = \int_0^1\int_0^1 \begin{vmatrix} 2e^{2\xi_1} & 2e^{\xi_1} e^{\xi_2} \\ 2e^{\xi_2} e^{\xi_1} & 2e^{2\xi_2} \end{vmatrix} d\xi_1 d\xi_2 = 0$$

or $C_3 = C_4 = ... = 0$

Thus $D(x,\xi;\lambda) = K(x,\xi) + \sum_{m=1}^{\infty} \frac{(-1)^m \lambda^m}{m!} B_m(x_1\xi) = 2e^x e^\xi$

and $D(\lambda) = 1 + \sum_{m=1}^{\infty} \frac{(-1)^m \lambda^m}{m!} C_m = 1 - (e^2 - 1)\lambda$

Then the resolvent Kernel is obtained as

$$R(x,\xi;\lambda) = \frac{D(x,\xi;\lambda)}{D(\lambda)} = \frac{2e^x e^\xi}{1-(e^2-1)\lambda}$$

Thus the solution of the integral equation is given by

$$\phi(x) = e^x + \frac{2\lambda e^x}{1-(e^2-1)\lambda}\int_0^1 e^\xi e^\xi d\xi$$

$$= e^x + \frac{\lambda(e^2-1)e^x}{1-(e^2-1)\lambda}$$

or $\phi(x) = \dfrac{e^x}{1-(e^2-1)\lambda}$

(iv) The integral equation is given as

$$\phi(x) = \sin x + \lambda\int_4^{10} x\phi(\xi)d\xi$$

Here $B_0(x,\xi) = K(x,\xi) = x$

$$B_1(x,\xi) = \int_4^{10}\begin{vmatrix} x & x \\ \xi_1 & \xi_1 \end{vmatrix} d\xi_1 = 0$$

$$B_2(x,\xi) = B_3(x,\xi) = ... = 0$$

and $C_1 = \int_4^{10} \xi_1 d\xi = 42$

$$C_2 = \int_4^{10}\int_4^{10}\begin{vmatrix} \xi_1 & \xi_1 \\ \xi_2 & \xi_2 \end{vmatrix} d\xi_1 d\xi_2 = 0$$

$$C_3 = C_4 = ... = 0$$

Hence $D(x,\xi;\lambda) = K(x,\xi) + \sum_{m=1}^{\infty}\frac{(-1)^m \lambda^m}{m!}B_m(x,\xi) = x$

and $D(\lambda) = 1 + \sum_{m=1}^{\infty}\frac{(-1)^m \lambda^m}{m!}C_m = 1 - 42\lambda$

The resolvent Kernel is obtained as

$$R(x,\xi;\lambda) = \frac{D(x,\xi;\lambda)}{D(\lambda)} = \frac{x}{1-42\lambda}$$

Thus the solution of integral equation is given by

$$\phi(x) = \sin x + \frac{\lambda x}{1-42\lambda}\int_4^{10}\sin\xi\, d\xi$$

$$= \sin x + \frac{2\lambda x}{1-42\lambda}\sin 7.\sin 3.$$

Integral Equation with Degenerate (Separable) Kernels

A Kernel K(x, ξ) of Fredholm integral equation of the Second Kind is said to be separable (or degenerate) if it can be expressed as the sum of a finite number of terms, each of which is the product of a function of x alone and a function of ξ alone, i.e., it is the form.

$$K(x,\xi)=\sum_{n=1}^{m} f_n(x)g_n(\xi) \qquad .....(1)$$

Consider the Fredholm integral equation

$$\phi(x)=F(x)+\lambda\int_a^b K(x,\xi)\phi(\xi)d\xi \qquad .....(2)$$

Substituting (1) in (2), we get

$$\phi(x)=F(x)+\lambda\int_a^b\left[\sum_{n=1}^{m} f_n(x)g_n(\xi)\right]\phi(\xi)d\xi$$

$$=F(x)+\lambda\sum_{n=1}^{m} f_n(x)\left\{\int_a^b g_n(\xi)\phi(\xi)d\xi\right\} \qquad .....(3)$$

Evidently $\int_a^b g_n(\xi)\phi(\xi)\xi=C_n$ (say), is a constant.

Then (3) becomes

$$\phi(x)=F(x)+\lambda\sum_{n=1}^{m} C_n f_n(x) \qquad .....(4)$$

This is a solution of (2) in which m constants. $C_1, C_2 .....C_m$ are to be determined.

Now multiplying both sides of (4) by $g_k(x)$ and integrating from a to b, we obtain

$$\int_a^b \phi(x)g_k(x)dx=\lambda\int_a^b\sum_{n=1}^{m} C_n f_n(x)g_k(x)dx+\int_a^b F(x)g_k(x)dx$$

writing $\int_a^b g_k(x)f_n(x)dx=\alpha_{kn}$ and $\int_a^b F(x)g_k(x)dx=\beta_k$

The above equation becomes

$$C_k=\beta_k+\lambda\sum_{n=1}^{m} C_n\cdot\alpha_{kn}$$

$$=\beta_k+\lambda(C_1\alpha_{k1}+C_2\alpha_{k2}+.....+C_m\alpha_{km})$$

Taking $K=1,2,.....m$, we get

$$(1-\lambda\alpha_{11})C_1-\lambda\alpha_{12}C_2-.....-\lambda\alpha_{1m}C_m=\beta_1$$

$$-\lambda\alpha_{21}C_1+(1-\lambda\alpha_{22})C_2-.....-\lambda\alpha_{2m}C_m=\beta_2$$

....

.....

$$-\lambda\alpha_{m1}C_1 - \lambda\alpha_{m2}C_2 - ..... + (1-\lambda\alpha_{mm})C_m = \beta_m \qquad .....(5)$$

Equation (5) will have a unique solution if the determinant Δ of the coefficients $C_1, C_2.....C_m$ is not zero.

Now the following cases arise :

I. When $F(x) = 0$ then (2) is said to be a homogeneous integral equation and all β's are zero.

(i) If $\Delta \neq 0$, the only solution of (4) is the trivial solution $C_1 = C_2 = ... = C_m = 0$. then $\phi(x) = 0$ is the trivial solution of (2).

(ii) If $\Delta \neq 0$, at least one of the C's can be assigned any value. Then (4) gives infinitely many solutions. The value of λ are the **eigen values**. Any non trivial solution of the homogeneous integral equation for a certain value of λ is called the corresponding **eigen function**.

II. When $F(x) \neq 0$. Let us assume that

$$\int_a^b g_m(x)F(x)dx = 0, \text{ so that } B_m = 0.$$

(i) If $\Delta = 0$, then only of (2) is the trivial solution $C_1 = C_2 = ...C_m = 0$. Then $\phi(x) = F(x)$ is desired solution of (2).

(ii) (ii) If $\Delta = 0$, at least one of the C's can be given any value then (u) gives infinitely many solutions.

III. When at least one of the B's $\neq 0$

(i) If $\Delta \neq 0$, then equation (5) give a unique solution. Hence there is a unique solution of (2).

(ii) If $\Delta \neq 0$, then (5) will be inconsistent. i.e. either there is not solution or infinitely many solutions of (2) exist.

Example 38 Solve the following integral equations with degenerate Kernels :

(i) $\phi(x) = e^x + \lambda\int_0^1 2e^x e^\xi \phi(\xi)d\xi$

(ii) $\phi(x) = x + \lambda\int_0^1 (x\xi^2 + x^2\xi)\phi(\xi)d\xi$

Solution

(i) $$\phi(x) = e^x + \lambda \int_0^1 2e^x e^\xi \phi(\xi) d\xi \qquad .....(1)$$

or $$\phi(x) = e^x + 2\lambda e^x \int_0^1 e^\xi \phi(\xi) d\xi$$

or $$\phi(x) = e^x + 2\lambda e^x C \qquad .....(2)$$

where $$C = \int_0^1 e^\xi \phi(\xi) d\xi \qquad .....(3)$$

From the equation (2), the relation (3) reduces to

$$C = \int_0^1 e^\xi (1 + 2\lambda C) e^\xi d^\xi$$

$$C = (1 + 2\lambda C) \int_0^1 e^{2\xi} d\xi$$

$$C = \frac{1}{2}(1 + 2\lambda C)(e^2 - 1)$$

or $$C = \frac{1}{2} \frac{e^2 - 1}{1 - \lambda(e^2 - 1)}, \lambda \neq \frac{1}{e^2 - 1}$$

Thus the solution of the integral equation is given by

$$\phi(x) = e^x + 2\lambda e^x \frac{e^2 - 1}{2[1 - \lambda(e^x - 1)]}, \text{ from (2)}$$

or $$\phi(x) = \left[1 + \frac{\lambda(e^2 - 1)}{1 - \lambda(e^2 - 1)}\right] e^x = \frac{e^x}{1 - \lambda(e^2 - 1)}$$

(ii) $$\phi(x) = x + \lambda \int_0^1 (x\xi^2 + x^2\xi) \phi(\xi) d\xi \qquad .....(1)$$

or $$\phi(x) = x + \lambda \left[x \int_0^1 \xi^2 \phi(\xi) d\xi + x^2 \int_0^1 \xi \phi(\xi) d\xi \right]$$

or $$\phi(x) = x + \lambda(C_1 x + C_2 x^2) \qquad .....(2)$$

where $$C_1 = \int_0^1 \xi^2 \phi(\xi) d\xi \text{ and } C_2 = \int_0^1 \xi \phi(\xi) d\xi \qquad .....(3)$$

from equation (2), the relation (3) reduces to

$$C_1 = \int_0^1 \xi^2[\xi + \lambda(C_1\xi + C_2\xi^2)]d\xi$$

and $$C_2 = \int_0^1 \xi[\xi + \lambda(C_1\xi + C_2\xi^2)]d\xi$$

or $$\left(1 - \lambda\int_0^1 \xi^3\, d\xi\right)C_1 - \left(\lambda\int_0^1 \xi^4 d\xi\right)C_2 = \int_0^1 \xi^3 d\xi$$

and $$\left(-\lambda\int_0^1 \xi^2 d\xi\right)C_1 + \left(1 - \lambda\int_0^1 \xi^3 d\xi\right)C_2 = \int_0^1 \xi^2 d\xi$$

or $$\left(1 - \frac{\lambda}{4}\right)C_1 - \frac{\lambda}{5}C_2 = \frac{1}{4}$$

and $$-\frac{\lambda}{3}C_1 + \left(1 - \frac{\lambda}{4}\right)C_2 = \frac{1}{3}$$

The determinant of the eigne values will be

$$D(\lambda) = \begin{vmatrix} 1 - \frac{\lambda}{4} & -\frac{\lambda}{5} \\ -\frac{\lambda}{3} & \left(1 - \frac{\lambda}{4}\right) \end{vmatrix} \neq 0 \Rightarrow 240 - 120\lambda - \lambda^2 \neq 0$$

$$\Rightarrow \frac{C_1}{\frac{\lambda}{15} + \frac{1}{14}\left(1 - \frac{\lambda}{4}\right)} = \frac{C_2}{\frac{1}{3}\left(1 - \frac{\lambda}{4}\right) + \frac{\lambda}{12}} = \frac{1}{\left(1 - \frac{\lambda}{4}\right)^2 - \frac{\lambda^2}{15}}$$

$$\Rightarrow C_1 = \frac{60 + \lambda}{240 - 120\lambda - \lambda^2},\ C_2 = \frac{80}{240 - 120\lambda - \lambda^2}$$

Substituting $\phi(x)$ in equation (2), the solution of integral equation becomes

$$= \frac{\lambda}{(1 - \lambda\pi)\left(1 - \frac{1}{2}\lambda\pi\right) + 4\lambda^2}$$

$$= x + \lambda\left[\frac{60 + \lambda}{240 - 120\lambda - \lambda^2}x + \frac{80x^2}{240 - 120\lambda - \lambda^2}\right]$$

or $\phi(x) = \dfrac{(240-60\lambda)x+80\lambda x^2}{240-120\lambda-\lambda^2}$

Example 39 Solve the following integral equations :

(i) $\phi(x) = x+\lambda\int_{-\pi}^{\pi}(x\cos\xi+\xi^2\sin x+\cos x\sin\xi)\phi(\xi)d\xi$

(ii) $\phi(x) = \cos x+\lambda\int_0^{\pi}\sin(x-\xi)\phi(\xi)d\xi$

(iii) $\phi(x) = x+\lambda\int_0^1(1+x+\xi)\phi(\xi)d\xi$

(iv) $\phi(x) = 1+\int_0^1(1+e^{e+\xi})\phi(\xi)d\xi$

Solution

(i) $\phi(x) = x+\lambda\int_{-\pi}^{\pi}(x\cos\xi+\xi^2\sin x+\cos x\sin\xi)\phi(\xi)d\xi$

Writing the integral equation in the following ferm :

$$\phi(x) = x+\lambda\left[x\int_{-\pi}^{\pi}\cos\xi\,\phi(\xi)d\xi+\sin x\int_{-\pi}^{\pi}\xi^2\,\phi(\xi)d\xi+\cos x\int_{-\pi}^{\pi}\sin\xi\,\phi(\xi)d\xi\right] \quad .....(1)$$

or $\phi(x) = x+\lambda[C_1x+C_2\sin x+C_3\cos x]$ .....(2)

where $C_1 = \int_{-\pi}^{\pi}\cos\xi\,\phi(\xi)d\xi,\ C_2 = \int_{-\pi}^{\pi}\xi^2\phi(\xi)d\xi,$

$C_3 = \int_{-\pi}^{\pi}\sin\xi\,\phi(\xi)d\xi$ .....(3)

From equation (2), the relations (3) becomes

$$C_1 = \int_{-\pi}^{\pi}\cos\xi[\xi+C_1\lambda\xi+C_2\lambda\sin\xi+C_3\lambda\cos\xi]d\xi$$

$$C_2 = \int_{-\pi}^{\pi}\xi^2[\xi+C_1\lambda\xi+C_2\lambda\sin\xi+C_3\lambda\cos\xi]d\xi$$

$$C_3 = \int_{-\pi}^{\pi}\sin\xi[\xi+C_1\lambda\xi+C_2\lambda\sin\xi+C_3\lambda\cos\xi]d\xi$$

or $C_1\left[1-\lambda\int_{-\pi}^{\pi}\xi\cos\xi\,d\xi\right]-C_2\ \lambda\int_{-\pi}^{\pi}\sin\xi\cos\xi\,d\xi-C_3\lambda$

$$\int_{-\pi}^{\pi}\cos^2\xi\,d\xi=\int_{-\pi}^{\pi}\xi\cos\xi\,d\xi$$

$$-C_1\lambda\int_{-\pi}^{\pi}\xi^3 d\xi+C_2\left[1-\lambda\int_{-\pi}^{\pi}\xi^2\sin\xi d\xi\right]-C_3\lambda\int_{-\pi}^{\pi}\xi^2\cos\xi\,d\xi=\int_{-\pi}^{\pi}\xi^3 d\xi$$

$$-C_1\lambda\int_{-\pi}^{\pi}\xi\sin\xi d\xi-C_2\lambda\int_{-\pi}^{\pi}\sin^2\xi d\xi+C_3\left[\int_{-\pi}^{\pi}\cos\xi\sin\xi\,d\xi\right]=\int_{-\pi}^{\pi}\xi\sin\xi\,d\xi$$

or $C_1+OC_2-\lambda\pi C_3=0$

$OC_1+C_2+4\lambda\pi C_3=0$

$-2\lambda\pi C_1-\lambda\pi C_2+C_3=2\pi$

The determinant of this system is

$$D(\lambda)=\begin{vmatrix}1 & 0 & -\lambda\pi\\ 0 & 1 & 4\lambda\pi\\ -2\lambda\pi & -\lambda\pi & 1\end{vmatrix}=1+2\lambda^2\pi^2\neq 0$$

Thus the system has a unique solution.

$$C_1=\frac{2\lambda\pi^2}{1+2\lambda^2\pi^2},\quad C_2=\frac{-8\lambda\pi^2}{1+2\lambda^2\pi^2},\quad C_3=\frac{2\pi}{1+2\lambda^2\pi^2}$$

Substituting C_1, C_2, C_3 in equation (2), the solution of integral equation as follows :

$$\phi(x)=x+\frac{2\lambda\pi}{1+2\lambda^2\pi^2}\ [\lambda\pi x-4\lambda\pi\sin x+\cos x].$$

(ii) $\phi(x)=\cos x+\lambda\int_0^{\pi}\sin(x-\xi)\phi(\xi)d\xi$

Writing the integral equation in the following from :

$$\phi(x)=\cos x+\lambda\left[\sin x\int_0^{\pi}\cos\xi\,\phi(\xi)d\xi-\cos x\int_0^{\pi}\sin\xi\,\phi(\xi)d\xi\right]$$

or $\phi(x)=\cos x+\lambda[\sin x\,C_1-\cos x\,C_2]$ .....(1)

where $C_1=\int_0^{\pi}\cos\xi\,\phi(\xi)d\xi,\ C_2=\int_0^{\pi}\sin\xi\,\phi(\xi)d\xi$ .....(2)

From equation (1), relation (2) reduces to

$$C_1 = \int_0^\pi \cos\xi[\cos\xi + \lambda C_1 \sin\xi - \lambda C_2 \cos\xi]d\xi$$

$$C_2 = \int_0^\pi \sin\xi[\cos\xi + \lambda C_1 \sin\xi - \lambda C_2 \cos\xi]d\xi$$

or $C_1\left[1-\lambda\int_0^\pi \cos\xi\sin\xi\,d\xi\right] + C_2\lambda\int_0^\pi \cos^2\xi d\xi = \int_0^\pi \cos^2\xi\,d\xi$

$$-C_1\lambda\int_0^\pi \sin^2\xi\,d\xi + C_2\left[1+\lambda\int_0^\pi \sin\xi\cos\xi\,d\xi\right] = \int_0^\pi \sin\xi\cos\xi\,d^x$$

or $C_1 + \frac{1}{2}C_2\lambda\pi = \frac{\pi}{2}$

$$-\frac{1}{2}C_1\lambda\pi + C_2 = 0$$

The determinant of this system is

$$D(\lambda) = \begin{vmatrix} 1 & \frac{1}{2}\pi\lambda \\ -\frac{1}{2}\pi\lambda & 1 \end{vmatrix} = 1 + \frac{1}{4}\lambda^2\pi^2 \neq 0$$

Thus the system has a unique solution.

$$C_1 = \frac{2\pi}{4+\lambda^2\pi^2},\quad C_2 = \frac{\lambda\pi^2}{4+\lambda^2\pi^2}$$

Substituting C_1, C_2 in (1), the solution of integral equation becomes

$$\phi(x) = \cos x + \frac{\lambda}{4+\lambda^2\pi^2}[2\pi\sin x - \lambda\pi^2\cos x]$$

or $\phi(x) = \frac{2}{4+\lambda^2\pi^2}[2\cos x + \pi\lambda\sin x]$

(iii) $\phi(x) = x + \lambda\int_0^1 (1+x+\xi)\phi(\xi)d\xi$

Writing the integral equation in the following form :

$$\phi(x) = x + \lambda\left[(1+x)\int_0^1 \phi(\xi)d\xi + \int_0^1 \xi\phi(\xi)d\xi\right]$$

$$\phi(x) = x + \lambda[(1+x)C_1 + C_2] \qquad \text{.....(1)}$$

where $C_1 = \int_0^1 \phi(\xi)d\xi,\quad C_2 = \int_0^1 \xi\phi(\xi)d\xi$ .....(2)

Substituting (1) into (2), we have

$$C_1 = \int_0^1 [\xi + \lambda(1+\xi)C_1 + \lambda C_2]d\xi\ C_1$$

$$C_2 = \int_0^1 \xi[\xi + \lambda(1+\xi)C_1 + \lambda C_2]d\xi$$

or $\left[1-\lambda\int_0^1 (1+\xi)d\xi\right]C_1 - \lambda C_2\int_0^1 d\xi = \int_0^1 \xi d\xi$

$$-\lambda C_1\int_0^1 (1+\xi)\xi d\xi + \left[1-\lambda\int_0^1 \xi d\xi\right]C_2 = \int_0^1 \xi^2 d\xi$$

or $\left(1-\dfrac{3\lambda}{2}\right)C_1 - \lambda C_2 = \dfrac{1}{2}$

$$-\frac{5\lambda}{6}C_1 + \left(1-\frac{\lambda}{2}\right)C_2 = \frac{1}{3}$$

The determinant of this system is

$$D(\lambda) = \begin{vmatrix} 1-\dfrac{3\lambda}{2} & -\lambda \\ -\dfrac{5\lambda}{6} & 1-\dfrac{\lambda}{2} \end{vmatrix} = 12 - 24\lambda - \lambda^2 \neq 0$$

Thus the system has a unique solution.

$$C_1 = \frac{6+\lambda}{12-24\lambda-\lambda^2},\ C_2 = \frac{4-\lambda}{12-24\lambda-\lambda^2}$$

Substituting C_1, C_2 in (1), the solution of integral equation becomes

$$\phi(x) = x + \frac{\lambda}{12-24\lambda-\lambda^2}[10 + (6+\lambda)x].$$

(iv) $\phi(x) = 1 + \int_0^1 (1+e^{x+\xi})\phi(\xi)d\xi$

Writing the integral equation in the following form :

$$\phi(x) = 1 + \int_0^1 \phi(\xi)d\xi + e^x \int_0^1 e^\xi \phi(\xi)d\xi$$

or $\quad \phi(x) = 1 + C_1 + C_2 e^x \qquad .....(1)$

where $C_1 = \int_0^1 \phi(\xi)d\xi, \; C_2 = \int_0^1 e^\xi \phi(\xi)d\xi \qquad .....(2)$

From the equation (1), the relation (2) reduces to

$$C_1 = \int_0^1 (1 + C_1 + C_2 e^\xi)d\xi \Rightarrow C_1 = 1 + C_1 + C_2(e-1) \Rightarrow C_2 = -\frac{1}{e-1}$$

$$C_2 = \int_0^1 e^\xi (1 + C_1 + C_2 e^\xi)d\xi$$

or $\quad C_2 = (1 + C_1)(e-1) + \frac{C_2}{2}(e^2 - 1) \Rightarrow C_1 = \frac{4e - e^2 - 5}{2(e-1)^2}$

Substituting C_1, C_2 in equation (1), the solution of integral equation becomes

$$\phi(x) = 1 - \frac{e^2 - 4e + 5}{2(e-1)^2} - \frac{1}{e-1}e^x$$

or $\quad \phi(x) = \frac{(e^2 - 3) - 2(e-1)e^x}{2(e-1)^2}.$

Practice Problems

1. Using Fredholm determinants, find the resolvent Kernel of the following Kernels :

(i) $K(x, \xi) = 4x\xi - x^2;\; 0 \le x \le 1,\; 0 \le \xi \le 1.$

Ans. $\dfrac{(4x\xi - x^2) - \left(2x^2\xi - \frac{4}{3}x^2 + x - \frac{4}{3}x\xi\right)\lambda}{1 - \lambda + \frac{\lambda^2}{18}}$

(ii) $K(x, \xi) = 1;\; a = 0, b = 1$

Ans. $\dfrac{1}{1-\lambda}$

(iii) $K(x, \xi) = 2e^x e^\xi;\; a = 0, b = 1$

Ans. $\dfrac{2e^x e^\xi}{1 - (e^2 - 1)\lambda}$

2. Using the method of degenerated Kernels, determine the solution of the following integral equations :

(i) $\phi(x) = \cos + \lambda\int_0^{2\pi}(\sin x\cos\xi - \sin 2x\cos 2\xi + \sin 3x\cos 3\xi)\,\phi(\xi)d\xi$

Ans. $\phi(x) = \cos x + \pi\lambda\sin x$

(ii) $\phi(x) = \cos x + \lambda\int_0^{\pi}\sin x\,\phi(\xi)d\xi$

Ans. $\phi(x) = \cos x$

(iii) $\phi(x) = x + \lambda\int_0^{\pi}(1 + \sin x\sin\xi)\,\phi(\xi)d\xi$

Ans. $\phi(x) = \dfrac{\lambda}{(1-\lambda\pi)\left(1-\frac{1}{2}\lambda\pi\right)+4\lambda^2}\left[\left\{2\lambda\pi+\frac{1}{2}\pi^2\left(1-\frac{\lambda\pi}{2}\right)\right\}+\pi(1-2\lambda\pi)\sin x\right]$

(iv) $\phi(x) = (2x - \pi) + 4\int_0^{\pi/2}\sin^2 x\,\phi(\xi)d\xi$

Ans. $\phi(x) = 2x - \pi + \dfrac{\pi^2\sin x}{\pi - 1}$

(v) $\phi(x) = f(x) + \lambda\int_0^{2\pi}\sin x\cos\xi\,\phi(\xi)d\xi$

Ans. $\phi(x) = f(x) + \lambda\int_0^{2\pi}\sin x\cos\xi\,d\xi$

(vi) $\phi(x) = f(x) + \lambda\int_{-1}^{1}(x\xi + x^2\xi^2)\phi(\xi)d\xi$

Ans. $\phi(x) = f(x) + \lambda\int_{-1}^{1}\left(\dfrac{3x\xi}{3-2\lambda} + \dfrac{5x^2\xi^2}{5-2\lambda}\right) + (\xi)d\xi$

(vii) $\phi(x) = \dfrac{6}{5}(1-4x) + \lambda\int_0^{1}(x\log\xi - \xi\log x)\,\phi(\xi)d\xi$

Ans. $\dfrac{6}{5}(1-4x) + \dfrac{48}{48+29\lambda^2}\left[2\lambda^2 x + \left(\lambda + \dfrac{\lambda^2}{4}\right)\log x\right]$

(viii) $\phi(x)\ \tan x + \int_{-1}^{1}e^{\sin^{-1}x}\phi(\xi)d\xi$

Ans. $\phi(x) = \tan x$

Construction of Green's Function

Consider a differential equation of order n :

$$L(4) = P_0(x)u^n + P_1(x)u^{n-1} + P_2(x)u^{n-2} + ... + P_n(x)u = 0 \qquad .....(1)$$

where the functions $P_0(x), P_1(x), P_2(x), ...P_n(x)$ are continuous n $[a, b]$, $P_0(x) \neq 0$ and the boundary conditions.

$$V_k(u) = \alpha_k 4(a) + \alpha_k{}^1 u^1(a) + \alpha_k{}^2 u^{11}(a) + ... + \alpha_k{}^{n-1} u^{n-1}(a) + \beta_K u(b) + \beta_K{}^1 u^1(b)$$
$$+ \beta_K{}^2 u^{11}(b) + ... + \beta_K{}^{n-1} u^{n-1}(b), \; K = 1, 2, ...n. \qquad .....(2)$$

where $V_1, V_2, ...V_n$ in $u(a), u^1(a).....u^{n-1}(a), 4(b), 4^1(b), ...u^{n-1}(b)$ are linearly independent.

The boundary value problem (1), (2) contains only a trivial solution $u(x) = 0$.

Green's Function of the boundary value problem (1), (2) is the function $C_1(x_1\xi)$ constructed for any ξ, $a < \xi < b$ satisfying the following properties:

I. $C_1(x, \xi)$ is continuous in x for fixed ξ and has continuous derivatives with regard to x up to order $(n-2)$.

II. Its $(n-1)^{th}$ derivative with regard to x and at a point $x = \xi$ has a discontinuity i.e.

$$\left\{\frac{\partial^{n-1}}{\partial x^{n-1}} C_1(x_1\xi)\right\}_{x=\xi+0} - \left\{\frac{\partial^{n-1}}{\partial x^{n-1}} C_1(x_1\xi)\right\}_{x=\xi-0} = -\frac{1}{P_0(\xi)} \qquad .....(3)$$

III. In each of the interval $[a, \xi]$ and $[\xi, b]$ the function $C_1(x_1\xi)$ considered as a function of x, is a solution of equation (1) i.e.,

$L(G) = 0$

IV. The function $G(x_1\xi)$ satisfies the boundary conditions (2).

$V_k(G) = 0, \; K = 1, 2, ...n.$

Consider $u_1(x), u_2(x), ..., u_n(x)$ be linearly independent solution of the equation d $L(u) = 0$. From the condition III, the unknown Green's Function $G(x_1\xi)$ must have the representation on the interval $[a_1\xi)$ and $(\xi, b]$:

$$G(x_1\xi) = a_1u_1(x) + a_2u_2(x) + ... + a_nu_n(x), \; a \leq x \leq \xi$$

and $G(x_1\xi) = b_1u_1(x) + b_2u_2(x) + ... + b_nu_n(x), \ \xi \le x \le b$

where $a_1, a_2, ..., a_n, b_1, b_2, ..., b_n$ are some function of ξ.

From the condition I. The continuity of the function $G(x_1\xi)$ and of its first $(n-2)$ derivatives w.r.t. x at the point $x = \xi$ yields.

$$[b_1u(\xi)+b_2u_2(\xi)+.....+b_nu_n/\xi]-[a_1u_1(\xi)+.....+a_nu_n(\xi)]=0$$

$$[b_1u_1^1(\xi)+b_2u_2^1(\xi)+.....+b_nu_n^1(\xi)-[a_1u_1^1(\xi)+.....+a_nu_n^1(\xi)]=0$$

$$..... \qquad ..... \qquad ..... \qquad .....$$

$$[b_1u_1^{n-2}(\xi)+b_2u_2^{n-2}(\xi)+.....+b_nu_n^{n-2}(\xi)]-[a_nu_1^{n-2}(\xi)+.....+a_nu_n^{n-2}(\xi)]=0$$

Also $[b_1u^{b-1}(\xi)+b_2u_2^{n-1}(\xi)+.....+b_nu_n^{n-1}(\xi)]-$

$$[a_nu_1^{n-1}(\xi)+a_nu_2^{n-1}(\xi)+.....+a_nu_n^{n-1}(\xi)]=-\frac{1}{p_0(\xi)}$$

Assume $C_K(\xi)=b_K(\xi)-a_K(\xi), K=1,2,.....n.$

Then the systems are

$$c_1u_1(\xi)+c_2u_2(\xi)+.....+c_nu_n(\xi)=0$$

$$c_1u_1^1(\xi)+c_2u_2^1(\xi)+.....+c_nu_n^1(\xi)=0$$

$$..... \qquad ..... \qquad ..... \qquad .....$$

$$c_1u_1^{n-2}(\xi)+c_2u_2^{n-2}(\xi)+.....+c_nu_n^{n-2}(\xi)=0$$

and $c_1u_1^{n-1}(\xi)+c_2u_2^{n-1}(\xi)+.....+c_nu_n^{n-1}(\xi)=-\dfrac{1}{p_0(\xi)}$ .....(4)

The determinant of the system is equal to the value of the Wronskian $W(u_1,u_2,.....,u_n)$ at the point $x = \xi$ and is therefore different from zero.

From the boundary condition (2), we have

$$V_k(u)=A_k(u)+B_k(u)$$

Using the condition IV, we have

$$V_k(u)=a_1A_k(u_1)+a_2A_k(u_2)+.....+a_nA_k(u_n)+b_1B_k(u_1)+b_2B_k(u_2)+.....+b_nB_k(u_n)=0$$

Which is a linear system and hence it is different from zero :

$$W[u_1(\xi),u_1(\xi),....u_n(\xi)]=\begin{vmatrix} u_1(\xi) & u_2(\xi) & ..... & u_n(\xi) \\ u_1^1(\xi) & u_2^1(\xi) & ..... & u_n^1(\xi) \\ ..... & ..... & ..... & ..... \\ u_1^{n-2} & u_2^{n-2}(\xi) & ..... & u_2^{n-2}(\xi) \end{vmatrix} \neq 0$$

The system of equations contains a unique solution.

I. **If the boundary value problems (1), (2) is self adjoint. Then Green's function is symmetric i.e.,**

$$G_1(x_1\xi)=g(\xi_1x)$$

II. **It at one of the extremities of an interval $[a,b]$ the coefficient of the derivative vanishes i.e., $P_0(a)=0$, then the natural boundary condition for the boundedness of the solution $x=a$ is imposed, and at the other extremity the ordinary boundary conditions is specified.**

Example 40 Construct Green's function for the homogeneous boundary value problem.

$$\frac{d^4u}{dx^4}=0$$

With the conditions. $u(0)=u^1(0)=0$

$$u(1)=u^1(1)=0$$

Solution The differential equation is given as

$$\frac{d^4u}{dx^4}=0 \qquad .....(1)$$

Its general solution is of the form

$$u(x)=A+Bx+Cx^2+Dx^3$$

Where $u_1(x)=1,\ u_2(x)=2,\ u_3(x)=x^2,\ u_4(x)=x^3$ .....(2)

The boundary conditions given the relations for determining the constants A, B, C, D.

$$u(0)=0 \Rightarrow A=0$$

$u^1(0)=0 \Rightarrow B=0$

$u(1) = 0 \Rightarrow A + B + C + D = 0$

$u^1(1)=0 \Rightarrow B+2C+3D=0$

$\Rightarrow \qquad A = B = C = D = 0$

Thus the boundary value problem has only a zero solution $u(x) = 0$, and hence we can construct a unique Green's function for it.

Consider the function $G(x_1\xi)$ as :

$$G(x_1\xi) = a_1 + a_2x + a_3x^2 + a_4x^3, \qquad 0 \le x \le \xi$$

$$b_1 + b_2x + b_3x^2 + b_4x^3, \quad \xi \le x \le 1 \qquad .....(3)$$

Where $a_1, a_2, a_3, a_4, b_1, b_2, b_3, b_4$ are the unknown function of ξ.

Consider $c_k(\xi) = b_k(\xi) - a_k(\xi), \ K = 1, 2, ...$ .....(4)

The system of linear equations for determining the function $c_k(\xi)$ become

$$c_1 + c_2\xi + c_3\xi^2 + c_4\xi^3 = 0$$

$$c_2 + 2c_2\xi + 3c_4\xi^2 = 0$$

$$2c_2 + 6c_4\xi = 0$$

$$6c_4 = 1$$

$$\Rightarrow c_4(\xi) = \frac{1}{6}, c_3(\xi) = \frac{-1}{2}\xi, c_2(\xi) = \frac{1}{2}\xi^2, c_1(\xi) = \frac{-1}{6}\xi^3 \qquad .....(5)$$

From the property IV, it must satisfy the boundary conditions :

$G(0,\xi) = 0, \ G^1(0,\xi) = 0$

$G(1,\xi) = 0, \ G^1(1,\xi) = 0$

The relation (3) reduces to

$$G(0_1\xi)=0 \Rightarrow a_1=0$$

$$G^1(0_1\xi)=0 \Rightarrow a_2=0$$

$$G(0_1\xi)=0 \Rightarrow b_2+b_2+b_3+b_4=0$$

$$G^1(1_1\xi)=0 \Rightarrow b_2+2b_2+2b_3+3b_4=0$$

From equations (4), (5) and (6), we have

$$C_1=b_1-a_1 \Rightarrow b_1=-\frac{1}{6}\xi^3$$

$$c_2=b_2-a_2 \Rightarrow b_2=-\frac{1}{2}\xi^2$$

$$b_3+b_4 \Rightarrow \frac{1}{6}\xi^3-\frac{1}{2}\xi^2$$

and $2b_3+3b_4=-\frac{1}{2}\xi^2$

$$\Rightarrow b_3=\frac{1}{2}\xi^3-\xi^2,\ b_4=\frac{1}{2}\xi^2-\frac{1}{3}\xi^3$$

and $C_3=b_3-a_3 \Rightarrow a_3=\frac{1}{2}\xi^3-\xi^2+\frac{1}{2}\xi$

$$c_4=b_4-a_4 \Rightarrow a_4=\frac{1}{2}\xi^2-\frac{1}{3}\xi^3-\frac{1}{6}$$

From relation (B), the Green's function $G(x_1\xi)$ is obtained as

$$G(x_1\xi)=\begin{cases}\left(\frac{1}{2}\xi-\xi^2+\frac{1}{2}\xi^3\right)x^2-\left(\frac{1}{6}-\frac{1}{2}\xi^2+\frac{1}{3}\xi^3\right)x^3, & 0\le x\le\xi\\ -\frac{1}{6}\xi^3+\frac{1}{2}\xi^2x+\left(\frac{1}{2}\xi^3-\xi^2\right)x^2+\left(\frac{1}{2}\xi^2-\frac{1}{3}\xi^3\right)x^2, & \xi\le x\le 1\end{cases}$$

The expression $G(x_1\xi)$ may be transformed to

$$G(x_1\xi)=\left(\frac{1}{2}x-x^2+\frac{1}{2}x^3\right)\xi^2-\left(\frac{1}{6}-\frac{1}{2}x^2+\frac{1}{3}x^3\right)\xi^3,\ \xi\le x\le 1$$

$$\Rightarrow G(x_1\xi) = G(\xi, x)$$

i.e., Green's function is symmetric.

Example 41 Construct Green's function for the equation

$$x\frac{d^2u}{dx^2}+\frac{du}{dx}=0$$

With the conditions $u(x)$ is bounded as $x \to 0$

$$u(1)=\mu\mu^1(1), \mu\neq 0$$

Solution The differential equation is given by

$$x\frac{d^2u}{dx^2}+\frac{du}{dx}=u \qquad \text{.....(1)}$$

$$\left(\frac{d^2u/dx^2}{du/dx}\right)dx = -\frac{1}{x}dx$$

Integrating, $\log\frac{du}{dx} = -\log x + \log A$

or $\frac{du}{dx} = \frac{A}{x}$

Integrating again, we get

$$u(x) = A\log x + B \qquad \text{.....(2)}$$

The conditions $u(x)$ is bounded as $x \to 0$ and $u(1)=\mu u^4(x), \mu\neq 0$ has only a trivial solution $u(x) = 0$,

Hence we can construct a Green's Function $G(x_1\xi)$.

Consider the function $G(x_1\xi)$ as:

$$G(x_1\xi) \begin{cases} a_1 + a_2\log x & 0<x\leq\xi \\ b_1 + b_2\log x & \xi\leq x\leq 1 \end{cases} \qquad \text{.....(3)}$$

where a_1, a_2, b_1, b_2 are unknown function of ξ.

Consider $C_k = b_k(\xi) - a_k(\xi)$, $K = 1, 2, 3, ...$

From the continuity for $x=\xi$, we obtain

$$b_1+b_2\log\xi-a_1-a_2\log\xi=0$$

The derivative of G has a discontinuity of magnitude of the point $x=\xi$.

where $P_0(\xi)$=coefficient of highest order derivative in the given differential equation (1).

$$\left(\frac{\partial G}{\partial x}\right)_{x=\xi+0}-\left(\frac{\partial G}{\partial^{x}}\right)_{x=\xi-0}=\frac{-1}{\xi}$$

$$\Rightarrow b_2\frac{1}{\xi}-a_2\frac{1}{\xi}=\frac{-1}{\xi}$$

$$\Rightarrow b_2-a_2=-1 \qquad .....(4)$$

Putting $c_1=b_1-a_1,\ c_2=b_2-a_2$ .....(5)

$$\Rightarrow c_1+c_2\log\xi=0$$

From equation (4) and (5) we have

$$c_1=\log\xi, c_2=-1 \qquad .....(6)$$

Also $G(x,\xi)=\mu G^1(x_1\xi)$

$$\Rightarrow b_1=\mu b_2$$

From equations (4), (5), (6) and (7), we obtain

$$a_1=-(\mu+\log\xi), a_2=0, b_1=-\mu, b_2=-1$$

Substituting the values of a_1, a_2, b_1, b_2 in the relation (3), the Green's function is obtained as

$$G(x_1\xi)=\begin{cases}-(\mu+\log\xi),\ 0<x\le\xi\\ -(\mu+\log x),\ \xi\le x\le 1\end{cases}$$

Example 42 Construct the Green's function for the boundary value problem

$$\frac{d^2u}{dx^2}+\mu^2u=0$$

With the conditions $u(0), u(l) = 0$.

Solution The differential equation is given as

$$\frac{d^2u}{dx^2}+\mu^2u=0 \qquad \text{.....(1)}$$

The general solution is

$$u(x) = A\cos\mu x + B\sin\mu x \qquad \text{.....(2)}$$

Since $u(0) = 0 \Rightarrow A = 0$

and $u(l) = 0 \Rightarrow B\sin\mu x = 0 \Rightarrow B = 0$

Thus the boundary value problem contains only a trivial solution $u(x) = 0$. Hence we can construct a Green's function.

Consider the Green's function as :

$$G(x_1\xi) = \begin{cases} a_1\cos\mu x + a_2\sin\mu x, & 0 \le x < \xi \\ b_1\cos\mu x + b_2\sin\mu x, & \xi < x \le 1 \end{cases} \qquad \text{.....(3)}$$

where a_1, a_2, b_1, b_2 are unknown function of ξ.

The Green's function $G(x_1\xi)$ must satisfy the following properties :

I. $G(x_1\xi)$ is continuous at $x = \xi$, i.e.1

$$b_1\cos\mu\xi + b_2\sin\mu\xi - a_1\cos\mu\xi - a_2\sin\mu\xi = 0$$

$$\text{or } (b_1 - a_1)\cos\mu\xi + (b_2 - a_2)\sin\mu\xi = 0 \qquad \text{.....(4)}$$

II. The derivatives of G has a discontinuity, i.e.

$$\left(\frac{\partial G}{\partial x}\right)_{x=\xi+0} = \left(\frac{\partial G}{\partial x}\right)_{x=\xi-0} = -1.$$

$$\mu(-b_1\sin\mu\xi + b_2\cos\mu\xi) - \mu(-a_1\sin\mu\xi + a_2\cos\mu\xi) = -1$$

$$\Rightarrow -(b_1 - a_1)\sin\mu\xi + (b_2 - a_2)\cos\mu\xi = -\frac{1}{\mu} \qquad \text{.....(5)}$$

III. The Green's function must satisfy the boundary conditions :

$$G(0_1\xi)=0 \Rightarrow a_1 = 0$$

$$G(l_1\xi)=0 \Rightarrow b_1\cos\mu l + b_2\sin\mu l = 0 \qquad .....(6)$$

Consider $c_K = b_K = a_K$

The equations (4) and (5) may be written as

$$c_1\cos\mu\xi + c_2\sin\mu\xi = 0$$

$$-c_1\sin\mu\xi + c_2\cos\mu\xi = \frac{-1}{\mu}$$

$$\Rightarrow \frac{c_1}{\frac{1}{\mu}\sin\mu\xi} = \frac{c_2}{-\frac{1}{\mu}\cos\mu\xi} = \frac{1}{1}$$

$$\Rightarrow c_1 = \frac{1}{\mu}\sin\mu\xi, c_2 = -\frac{1}{\mu}\cos\mu\xi$$

So $b_1 - a_1 = \frac{1}{\mu}\sin\mu\xi, b_2 - a_2 = \frac{-1}{\mu}\cos\mu\xi$

Since $a_1 = 0, b_1 = \frac{1}{\mu}\sin\mu\xi$

From equation (6), we have

$$b_2 = -\frac{\sin\mu\xi\cos\mu l}{\mu\sin\mu l}$$

$$a_2 = -\frac{\sin\mu(\xi - l)}{\mu\sin\mu l}$$

Therefore, the relation (3) gives.

$$G(x_1\xi) = \begin{cases} -\dfrac{\sin\mu(\xi-1)\sin\mu x}{\mu\sin\mu l}, 0 \le x < \xi \\ -\dfrac{\sin\mu\xi\ \sin\mu(x-l)}{\mu\sin\mu l}, \xi \le x \le l \end{cases}$$

Example 43 Construct the Green's function for the boundary value problem

$$\frac{d^2u}{dx^2}=0$$

With the conditions $u(0)=u^1(1)$ and $u^1(0)=u(1)$.

Solution The differential equation is given as

$$\frac{d^2u}{dx^2}=0 \qquad .....(1)$$

The general solution of equation (1) is

$$u(x) = Ax+B \qquad .....(2)$$

Since $u(0) = u^1(1) \Rightarrow B = A$

and $u^1(0) = u(1) \Rightarrow B = A$

Thus the boundary value problem contains only a trivial solution $u(x) \equiv 0$, hence we can construct a Green's function for it.

Consider a Green's function as :

$$G(x_1\xi) = \begin{cases} a_1x+a_2, & 0 \le x < \xi \\ b_1x+b_2, & \xi < x \le x \end{cases} \qquad .....(3)$$

where a_1, a_2, b_1, b_2 are unknown function of ξ. The Green's function must satisfy the following properties:

$$b_1\xi+b_2-a_1\xi-a_2=0$$

$$\Rightarrow (b_1-a_2)\xi+(b_2-a_2)=0 \qquad .....(4)$$

II. The derivative of G has a discontinuity at $x=\xi$, i.e.

$$\left(\frac{\partial G}{\partial x}\right)_{x=\xi+0} - \left(\frac{\partial G}{\partial x}\right)_{x=\xi-0} = -1$$

$$\Rightarrow b_1-a_1=-1 \qquad .....(5)$$

III. **The Green's function must satisfy the boundary conditions :**

$$G(0,\xi) = G^1(l,\xi) \Rightarrow a_2 = b_1 \quad .....(6)$$

$$G(0,\xi) = G(l,\xi) \Rightarrow a_1 = b_1 + b_2$$

Consider $c_K = b_K - a_K \quad .....(7)$

$$\Rightarrow c_1 = b_1 - a_1, c_2 = b_2 - a_2$$

From equations (4), (5), (6) and (7), we have

$$c_1 = -1, c_2 = -\xi$$

$$a_1 = -\xi + 2, a_2 = -\xi + 1, b_1 = -\xi + 1, b_2 = 1$$

Hence the relation (3), the Green's function is obtained :

$$G(x,\xi) = \begin{cases} (-\xi+2)x + (-\xi+1), & 0 \le x < \xi \\ (-\xi+1)x + 1, & \xi < x \le 1 \end{cases}$$

Example 44 **Construct the Green's function for the boundary value problem**

$$\frac{d^3u}{dx^3} = 0$$

With the boundary conditions $u(0) = u^1(1) = 0$ **and** $u^1(0) = u(1)$.

Solution **The differential equation is given as**

$$\frac{d^3u}{dx^3} = 0 \quad .....(1)$$

The general solution for the differential equation (1) is

$$u(x) = A + Bx + Cx^2 \quad .....(2)$$

Since $u(0) = 0 \Rightarrow A = 0$

$$u^1(1) = B + 2C = 0$$

$$u^1(0) = u(1) \Rightarrow B = A + B + C$$

$$\Rightarrow A = B = C = 0$$

Thus the boundary value problem has a trivial solution $u(x)=0$.

Hence we can construct a Green's function for it.

Consider a Green's function as :

$$G(x_1\xi)=\begin{cases} a_1+a_2x+a_3x^2, & 0\le x<\xi \\ b_1+b_2x+b_3x^2, & \xi<x\le 1 \end{cases} \qquad .....(3)$$

Where a_1,a_2,a_3,b_1,b_2,b_3 are unknown functions of ξ.

Consider $c_K=b_K-a_K, K=1,2,3,.....$

The system of linear equation for determining the function c_K becomes

$$c_1+c_2\xi+c_3\xi^2=0$$

$$c_2+2c_3\xi=0$$

$$2c_3=1$$

$$\Rightarrow c_3=\frac{1}{2}, c_2=-\xi, c_1=\frac{1}{2}\xi^2 \qquad .....(4)$$

From the property IV, the Green's function must satisfy the boundary Conditions :

$$G(0,\xi)\quad =0\Rightarrow a_1=0$$

$$G^1(1,\xi)\quad =0\Rightarrow b_2+2b_3=0$$

$$G^1(0,\xi)=G(l,\xi)\Rightarrow a_2=b_1+b_2+b_3$$

From the realtion (4) and (5), we have

$$c_1=b_1-a_1\Rightarrow b_1-a_1=\frac{1}{2}\xi^2$$

$$c_2=b_2-a_2\Rightarrow b_2-a_2=-\xi$$

$$c_3=b_3-a_3\Rightarrow b_3-a_3=\frac{1}{2}$$

$$\Rightarrow a_1 = 0, b_1 = \frac{1}{2}\xi^2, a_2 = \xi^2 - \xi, b_2 = \xi^2 - 2\xi, b_3 = \xi - \frac{1}{2}\xi^2,$$

$$a_3 = \xi - \frac{1}{2}\xi^2 - \frac{1}{2}$$

Hence the relation (3), the Green's function $a(x_1\xi)$ is obtained as

$$c_1(x_1\xi) = \begin{cases} (\xi^2 - \xi)x + \left(\xi - \frac{1}{2}\xi^2 - \frac{1}{2}\right)x^2, & 0 \le x < \xi \\ \frac{1}{2}\xi^2(\xi^2 - 2\xi)x + \left(\xi - \frac{1}{2}\xi^2\right)x^2, & \xi < x \le 1 \end{cases}$$

$$\text{or } c_1(x_1\xi) = \begin{cases} \frac{1}{2}x(\xi - 1)[x - x\xi + 2\xi], & 0 \le x < \xi \\ \frac{1}{2}\xi[x(1-x)(\xi - 2) + \xi], & \xi < x \le 1 \end{cases}$$

To Solve the boundary value problem: The solution of the given boundary value problem is given by

$$u(x) = \int_a^b G(x_1\xi)F(\xi)d\xi$$

where $G(x_1\xi)$ known as the Green's function.

Example 45 Using Green's function, solve the boundary value problem

$$\frac{d^2u}{dx^2} + u = x$$

With the boundary conditions.

$$u(0) = 0 \text{ and } u\left(\frac{\pi}{2}\right) = 0.$$

Solution The boundary value problem is given as

$$\frac{d^2u}{dx^2} + u = x \qquad \text{.....(1)}$$

Consider the corresponding homogeneous boundary value problem :

$$\frac{d^2u}{dx^2}+u=0 \qquad .....(2)$$

The general solution of equation (2) is given by

$$u(x) = A\cos x + B\sin x \qquad .....(3)$$

Since $u(0) = 0 \Rightarrow A = 0$

$$u\left(\frac{\pi}{2}\right) = 0 \Rightarrow B = 0$$

Thus the boundary conditions contain only a trivial solution $u(x) = 0$.

Hence the Green's function exists and is

$$G(x_1\xi) = \begin{cases} a_1\cos x + a_2\sin x, & 0 \le x < \xi \\ b_1\cos x + b_2\sin x, & \xi < x \le 1 \end{cases} \qquad .....(4)$$

where a_1, a_2, b_1, b_2 are the unknown functions of ξ. The Green's function $G(x_1\xi)$ must satisfy the following properties :

I. $G(x_1\xi)$ is continuous of $x = \xi$, i.e.

$$b_1\cos\xi + b_2\sin\xi = a_1\cos\xi + b_1\sin\xi$$

$$\Rightarrow (b_1 - a_1)\cos\xi + (b_2 - a_2)\sin\xi = 0$$

$$\Rightarrow c_1\cos\xi + c_2\sin\xi = 0 \qquad .....(5)$$

II. The derivative of $G(x_1\xi)$ has a discontinuity at $x = \xi$, i.e.

$$\left(\frac{\partial c_1}{\partial x}\right)_{x=\xi+\partial} - \left(\frac{\partial c_1}{\partial x}\right)_{x=\xi-0} = -1$$

$$-b_1\sin\xi + b_2\cos\xi - (-a_1\sin\xi + a_2\cos\xi) = -1$$

$$\Rightarrow c_1\sin\xi + c_2\cos\xi = -1 \qquad .....(6)$$

III. **The function $G(x_1\xi)$ must satisfy the boundary conditions :**

$$G(0_1\xi) = 0 \Rightarrow a_1 = 0$$

$$G\left(\frac{\pi}{2},\xi\right) = 0 \Rightarrow b_2 = 0.$$

From equation (5) and (6), we have

$$c_1 = \sin\xi, c_2 = -\cos\xi$$

Consider $c_K = b_K = a_k$

Then $c_1 = b_1 - a_1 \Rightarrow b_1 - a_1 = \sin\xi$

$$c_2 = b_2 - a_2 \Rightarrow b_2 - a_2 = \cos\xi$$

From the relation (4), the Green's function becomes.

$$G(x_1\xi) = \begin{cases} \cos\xi\sin x, & 0 \le x < \xi \\ \sin\xi\cos x, & \xi < x \le 1 \end{cases} \qquad .....(7)$$

Therefore, the solution of the given boundary value problem (1) is given by

$$u(x) = \int_0^{\pi/2} G(x_1\xi)F(\xi)d\xi$$

$$\Rightarrow u(x) = \int_0^x \xi\sin\xi\cos x d\xi + \int_x^{\pi/2} \xi\cos\xi\sin x d\xi$$

$$= \cos x\{-x\cos x + \sin x\} + \sin x\left\{\frac{\pi}{2} - x\sin x - \cos x\right\}$$

$$\Rightarrow u(x) = -x + \frac{\pi}{2}\sin x.$$

Example 46 **Solve the boundary value problem using Green's function :**

$$\frac{d^2u}{dx^2} + u = x^2;\ u(0) = u\left(\frac{\pi}{2}\right) = 0.$$

Solution **The solution of the given boundary value problem is given by**

$$u(x) = \int_0^{\pi/2} G(x_1\xi)\, f(\xi)d\xi$$

$$u(x) = \int_0^x \xi^2 \sin\xi \cos x d\xi + \int_x^{\pi/2} \xi^2 \cos\xi \sin x \, dx \text{ (see example 45)}$$

$$= \frac{\pi^2}{4}\sin x - x^2 - 2(\cos x + \sin x - 1)$$

Or $u(x) = -\left[2\cos x + \sin x\left(2 - \frac{\pi^2}{4}\right) + x^2 - 2\right].$

Example 47 Solve the boundary value problem using Green's function.

$$\frac{d^2u}{dx^2} - u = -2e^x$$

With boundary conditions : $u(0) = u^1(0), u(l) + u^1(l) = 0.$

Solution The differential equation is given as

$$\frac{d^2u}{dx^2} - u = -2e^{-x} \quad \text{.....(1)}$$

Consider the corresponding homogeneous boundary value problem

$$\frac{d^2u}{dx^2} - u = 0 \quad \text{.....(2)}$$

The general solution of equation (2) is given by

$$u(x) = Ae^x + Be^{-x} \quad \text{.....(3)}$$

Since $u(0) = u^1(0) \Rightarrow A + B = A - B \Rightarrow B = 0$

$$u(l) + u^1(l) = 0 \Rightarrow A = 0$$

i.e., $u(x) = 0$. Thus the Green's function exists and is

$$c_1(x_1\xi) = \begin{cases} a_1e^x + a_2e^{-x}, & 0 \le x\xi \\ b_1e^x + b_2e^{-x}, & \xi < x \le l \end{cases} \quad \text{.....(4)}$$

where a_1, a_2, b_1, b_2 are unknown functions of ξ. The Green's function $G(x_1\xi)$ must satisfy the following properties :

I. $c_1(x_1\xi)$ is continuous at $x = \xi$, i.e.

$$b_1e^{\xi} + b_2e^{-\xi} = a_1e^{\xi} + a_2e^{-\xi}$$

$$\Rightarrow c_1e^{\xi} + c_2e^{-\xi} = 0$$

II. The derivative of $c_1(x_1\xi)$ has a discontinuity at $x = \xi$, i.e.

$$\left(\frac{\partial c_1}{\partial x}\right)_{x=\xi+0} - \left(\frac{\partial c_1}{\partial x}\right)_{x=\xi-0} = -1.$$

$$\Rightarrow (b_1e^{\xi} - b_ee^{-\xi}) - (a_1e^{\xi} - a_2e^{-\xi}) = -1$$

$$\Rightarrow (b_1 - a_1)e^{\xi} - (b_2 - a_2)e^{-\xi} = -1$$

$$\Rightarrow c_1e^{\xi} - c_2e^{-\xi} = -1 \qquad \text{.....(6)}$$

III. The function $c_1(x_1\xi)$ must satisfy the boundary conditions :

$$c_1(0_1\xi) = c_1^1(0_1\xi) \Rightarrow a_1e^{\xi} + a_2e^{-\xi} = a_1e^{\xi} - a_2e^{-\xi} \Rightarrow a_2 = 0 \qquad \text{.....(7)}$$

and $c_1(l_1\xi) + c_1^1(\lambda, \xi) = 0 \Rightarrow b_1e^{l} + b_2e^{-l} + b_1e^{l} - b_2e^{-l} = 0 \Rightarrow b_1 = 0$

From equations (5) and (6), we have

$$c_1 = -\frac{1}{2}e^{-\xi},\ c_2 = \frac{1}{2}e^{\xi}$$

Consider $c_K = b_K - a_K$

$$c_1 = b_1 - a_1 \Rightarrow b_1 - a_1 = -\frac{1}{2}e^{-\xi} \qquad \text{.....(8)}$$

$$c_2 = b_2 - a_2 \Rightarrow b_2 - a_2 = \frac{1}{2}e^{\xi}$$

From (7) and (8), we have

$$a_1 = \frac{1}{2}e^{-\xi},\ a_2 = 0,\ b_1 = 0,\ b_2 = \frac{1}{2}e^{\xi}$$

Thus, from relation (C_1), the Green's function becomes

$$c_1(x_1\xi) = \begin{cases} \frac{1}{2}e^{x-\xi},\ 0 \le x < \xi \\ \frac{1}{2}e^{-(x-\xi)},\ \xi < x \le l \end{cases} \qquad .....(9)$$

Therefore, the solution of boundary value problem of equation (1) is given by

$$u(x) = \int_0^l c_1(x_1\xi)F(\xi)d\xi.$$

$$\text{or } u(x) = \int_0^x \frac{1}{2}e^{-(x-\xi)}(-2e^{\xi})d\xi + \int_x^l \frac{1}{2}e^{(x-\xi)}(-2e^{\xi})d\xi$$

$$= -e^{-x}\int_0^x e^{2\xi}d\xi - e^x\int_x^l d\xi$$

$$= -\frac{1}{2}e^x(e^{2x}-1) - e^x(l-x)$$

$$\Rightarrow \qquad u(x) = -\frac{1}{2}(e^x - e^{-x}) - e^x(l-x).$$

Construction of green's function when the boundary value problem contains a parameter : Consider a differential equation of order n

$$L(u) - \lambda^u = h(x) \qquad .....(1)$$

With boundary conditions $V_K(u) = 0, K = 1, 2, ...n.$ .....(2)

Where $h(x)$ is a given continuous function of x and λ is some non-zero numerical parameter.

For $h(x) = 0,$ the equation (1) reduces to homogeneous boundary value problem

$$L(u) = \lambda u$$

$$V_K(u) = 0, \quad K = 1, 2, \ldots n. \qquad \ldots(3)$$

The values of λ for which the boundary value problem has non-trivial solution $c_1(x)$.

If the boundary value problem

$$L(u) = 0$$

$$V_K(u) = 0, \; K = 1, 2, \ldots n. \qquad \ldots(4)$$

Contains the Green's function $c_1(x_1\xi)$, then the boundary value problem (1) and (2) is equivalent to the Fredholm integral equation equation.

$$u(x) = \lambda \int_a^b c_1(x_1\xi) u(\xi) d\xi + F(x) \qquad \ldots(5)$$

where $F(x) \int_a^b c_1(x_1\xi) h(\xi) d\xi.$

Example 48 Reduce the boundary value problem

$$\frac{d^2u}{dx^2} + \lambda u = x; u(0) = u\left(\frac{\pi}{2}\right) = 0.$$

to an integral equation using Green's function.

Solution Consider the associated boundary value problem

$$\frac{d^2u}{dx^2} = 0 \qquad \ldots(1)$$

The general solution is given by

$$u(x) = Ax + B$$

since $u(0) = 0 \Rightarrow A = 0$

$$u\left(\frac{\pi}{2}\right) = 0 \Rightarrow A = 0$$

$\therefore \; u(x) = 0.$ Therefore the Green's function $c_1(x_1\xi)$ exists and is

$$c_1(x_1\xi) = \begin{cases} a_1x + a_2, & 0 \le x < \xi \\ b_1x + b_2, & \xi < x \le \dfrac{\pi}{2} \end{cases} \quad .....(2)$$

The Green's function must satisfy the following properties :

I. $c_1(x_1\xi)$ is continuous at $x = \xi$, i.e.

$$b_1\xi + b_2 = a_2\xi + a_2$$

$$\Rightarrow (b_1 - a_1)\xi + (b_2 - a_2) = 0$$

II. The derivative $c_1(x_1\xi)$ has a discontinuity at $x = \xi$, i.e.

$$\left(\frac{\partial c_1}{\partial^x}\right)_{x=\xi+0} - \left(\frac{\partial c_1}{\partial^x}\right)_{x=\xi-0} = -1$$

$$\Rightarrow b_1 - a_1 = -1$$

III. The Green's Function $c_1(x_1\xi)$ must satisfy the boundary conditions :

$$c_1(0_1\xi) = 0 \Rightarrow a_2 = 0$$

$$c_1\left(\frac{\pi}{2}, \xi\right) = 0 \Rightarrow b_1\frac{\pi}{2} + b_2 = 0 \quad .....(5)$$

Solving equations (3), (4) and (5), we have

$$a_1 = 1 - \frac{2\xi}{\pi}, a_2 = 0, b_1 = \frac{-2\xi}{\pi}, b_2 = \xi$$

Substituting the values in the relation (2), the Green's function is obtained.

$$u_1(x_1\xi) = \begin{cases} \left(1 - \dfrac{2\xi}{\pi}\right)x, & 0 \le x < \xi \\ \left(1 - \dfrac{2x}{\pi}\right)\xi, & \xi < x \le \dfrac{\pi}{2} \end{cases}$$

Since $c_1(x_1\xi)$ is the Green's function of the associated boundary value problem the given boundary value problem (1) may be reduced to an integral equation :

$$u(x) = \int_0^{\pi/2} c_1(x_1\xi)\{-\lambda c_1(\xi)+\xi\}d\xi$$

$$= -\lambda\int_0^{\pi/2} c_1(x_1\xi)c(\xi)d\xi + \int_0^{\pi/2} c_1(x_1\xi)\xi\, d\xi \qquad .....(6)$$

Where $F(x) = \int_0^{\pi/2} c_1(x_1\xi)\xi\, d\xi$

$$= \int_0^{x}\left(1-\frac{2x}{\pi}\right)\xi^2\, d\xi + \int_x^{\pi/2}\left(1-\frac{2\xi}{\pi}\right)x\xi\, d\xi$$

$$= \frac{1}{3}\left(1-\frac{2x}{\pi}\right)x^3 + x\left(-\frac{x^2}{2}+\frac{\pi^2}{24}+\frac{2x^3}{3\pi}\right)$$

$$= \frac{1}{3}x^3 - \frac{2}{3\pi}x^4 + \frac{\pi^2 x}{24} - \frac{1}{2}x^3 + \frac{2}{3\pi}x^4$$

$$= \frac{\pi^3 x}{24} - \frac{x^3}{6}$$

Thus, the given boundary value problem has been reduced to an integral equation.

$$u(x) + \lambda\int_0^{\pi/2} c_1(x_1\xi)u(\xi)d\xi = \frac{\pi^2}{24}x - \frac{1}{6}x^3$$

Example 49 Reduce the boundary value problem

$$\frac{d^2u}{dx^2} + \frac{\pi^2}{u}u = \lambda c_1 + \cos\frac{\pi x}{2}$$

$u(-1) = u(1)$ and $u^1(-1) = u^1(1)$

to an integral equation.

Solution Consider the boundary value problem is

$$\frac{d^2x}{dx^2} + \frac{\pi^2}{c_1}c_1 = 0 \qquad .....(1)$$

The general solution is

$$c_1(x) = A\cos\frac{\pi}{2}x + B\sin\frac{\pi}{2}x \qquad .....(2)$$

since $\quad u(-1) = u(1) \Rightarrow B = 0$

$$u^1(-1) = u^1(1) \Rightarrow A = 0$$

Thus the boundary value problem contains only a trivial solution $u(x) = 0$. therefore Green's function $c_1(x_1\xi)$ exists and is

$$c_1(x_1\xi) = \begin{cases} a_1\cos\frac{\pi x}{2} + a_2\sin\frac{\pi x}{2}, & -1 \le x < \xi \\ b_1\cos\frac{\pi x}{2} + b_2\sin\frac{\pi x}{2}, & \xi < x \le 1 \end{cases} \qquad .....(3)$$

The Green's function must satisfy the following properties:

I. $c_1(x_1\xi)$ is continuous at $x = \xi$, i.e.

$$b_1\cos\frac{\pi\xi}{2} + b_2\sin\frac{\pi\xi}{2} = a_1\cos\frac{\pi\xi}{2} + a_2\sin\frac{\pi\xi}{2}$$

$$\Rightarrow (b_1 - a_1)\cos\frac{\pi\xi}{2} + (b_2 - a_2)\sin\frac{\pi\xi}{2} = 0 \qquad .....(4)$$

II. The derivative has a discontinuity at $x = \xi$, i.e.

$$\left(\frac{\partial c_1}{\partial x}\right)_{x=\xi+0} - \left(\frac{\partial c_1}{\partial x}\right)_{x=\xi-0} = -1$$

$$-\frac{\pi}{2}(b_1 - a_1)\sin\frac{\pi\xi}{2} + \frac{\pi}{2}(b_2 - a_2)\cos\frac{\pi\xi}{2} = -1 \qquad .....(5)$$

III. The function $c_1(x_1\xi)$ must satisfy the boundary conditions :

$$c_1(-1,\xi)\, c_1(1,\xi) \Rightarrow -a_2 = b_2 \Rightarrow a_2 + b_2 = 0$$

$$c_1^1(-1,\xi) = c_1^1(1,\xi) \Rightarrow \frac{\pi}{2}a_1 = \frac{-\pi}{2}b_1 \Rightarrow a_1 + b_1 = 0 \qquad .....(6)$$

Solving equations (4), (5) and (6), we have

$$a_1 = \frac{-1}{\pi}\sin\frac{\pi\xi}{2},\ a_2 = \frac{1}{\pi}\cos\frac{\pi\xi}{2},\ b_1 = \frac{1}{\pi}\sin\frac{\pi\xi}{2},\ b_2 = \frac{-1}{\pi}\cos\frac{\pi\xi}{2}$$

Substituting the values in relation (3), we have

$$c_1(x_1\xi) = \begin{cases} -\frac{1}{\pi}\sin\frac{\pi\xi}{2}\cos\frac{\pi x}{2} + \frac{1}{\pi}\cos\frac{\pi\xi}{2}\sin\frac{\pi x}{2}, & -1 \le x < \xi \\ \frac{1}{\pi}\sin\frac{\pi\xi}{2}\cos\frac{\pi x}{2} - \frac{1}{\pi}\cos\frac{\pi\xi}{2}\sin\frac{\pi x}{2}, & \xi < x \le 1 \end{cases}$$

$$\Rightarrow c_1(x_1\xi) = \begin{cases} \frac{1}{\pi}\sin\frac{\pi}{2}(x-\xi), & -1 \le x < \xi \\ \frac{1}{\pi}\sin\frac{\pi}{2}(\xi - x), & \xi < x \le 1 \end{cases} \quad \text{.....(7)}$$

Since $c_1(x_1\xi)$ is the Green's function of the associated boundary value problem then the given boundary value problem may reduced to an integral equation :

$$c_1(x) = \int_{-1}^{1} c_1(x_1\xi)\left\{\lambda c_1(\xi) + \cos\frac{\pi\xi}{2}\right\} d\xi$$

$$\text{or } c_1(x) = \lambda\int_{-1}^{1} c_1(x_1\xi)\,d\xi + \int_{-1}^{1} c_1(x_1\xi)\cos\frac{\pi\xi}{2}\,d\xi \quad \text{.....(8)}$$

$$\text{where } F(x) = \int_{-1}^{1} c_1(x_1\xi)\cos\frac{\pi\xi}{2}\,d\xi$$

$$= \int_{-1}^{1}\frac{1}{\pi}\sin\frac{\pi}{2}(\xi - x)\cos\frac{\pi\xi}{2}\,d\xi + \int_{x}^{1}\frac{1}{\pi}\sin\frac{\pi}{2}(x-\xi)\cos\frac{\pi\xi}{2}\,d\xi$$

$$= \frac{1}{2\pi}\int_{-1}^{x}\left\{\sin\frac{\pi}{2}(2\xi - x) - \sin\frac{\pi}{2}x\right\}$$

$$d\xi + \frac{1}{2\pi}\int_{x}^{1}\left\{\sin\frac{\pi x}{2} + \sin\frac{\pi}{2}(x - 2\xi)\right\}d\xi$$

$$= \frac{1}{2\pi}\left[-\frac{1}{\pi}\cos\frac{\pi}{2}(2\xi - x) - \xi\sin\frac{\pi x}{2}\right]_{-1}^{x}$$

$$+\frac{1}{2\pi}\left[\xi\sin\frac{\pi x}{2} + \frac{1}{\pi}\cos\frac{\pi}{2}(x-2\xi)\right]_{x}^{1}$$

$$= -\frac{x}{\pi}\sin\frac{\pi x}{2} - \frac{2}{\pi^2 \cos}\cos\frac{\pi x}{2}$$

Thus, the given boundary value problem has been reduced to an integral equation.

$$c_1(x) - \lambda\int_{-1}^{1} c_1(x_1\xi)c_1(\xi)d\xi = -\left(\frac{x}{\pi}\sin\frac{\pi x}{2} + \frac{2}{\pi^2}\cos\frac{\pi x}{2}\right)$$

Example 50 Reduce the Bessel's differential equation

$$x^2\frac{d^2u}{dx^2} + x\frac{du}{dx} + (\lambda x^2 - 1)c_1 = 0$$

With the conditions $c_1(0)=0, c_1(1)=0$ into an integral equation.

Solution The Bessel's equation is given by

$$x^2\frac{d^2u}{dx^2} + x\frac{du}{dx} + (\lambda x^2 - 1)u = 0 \quad .....(1)$$

$$\text{or } x\frac{d^2u}{dx^2} + \frac{du}{dx} + \left(\lambda x - \frac{1}{x}\right)u = 0$$

$$\text{or } \left[\frac{d}{dx}\left(x\frac{d}{dx}\right) + \left(\frac{-1}{x}\right)\right]u + \lambda x u = 0$$

Comparing the equation with the standard equation

$$\text{L}\,\lambda\, r(x)u(x) = F(x) \quad .....(2)$$

$$\left[\frac{d}{dx}\left(p\frac{du}{dx}\right) + q\right]u + \lambda\, r(x)u(x) = F(x)$$

$$\Rightarrow p(x) = x, q(x) = -\frac{1}{x}, r(x) = x, F(x) = 0$$

Consider the Green's function for the boundary value problem

$$\left[\frac{d}{dx}\left(x\frac{d}{dx}\right)+\left(-\frac{1}{x}\right)\right]u=0 \qquad .....(3)$$

with the boundary conditions $u(0)=0, u(1)=0$

We know that the Green's function $G_1(x_1\xi)$ is given by

$$c_1(x_1\xi)=\begin{cases}-\frac{1}{c}u_1(x)u_2(\xi), & x<\xi\\ -\frac{1}{c}u_1(x)u_2(x), & x>\xi\end{cases} \qquad .....(4)$$

where C is given by the Abel's formula.

$$u_1(\xi)u_2^1(\xi)-u_2(\xi)u_1^1(\xi)=\frac{C}{p(\xi)} \qquad .....(5)$$

where $u_1(\xi), u_2(\xi)$ are two independent solutions of equation (3)

Thus $\frac{d}{dx}\left(x\frac{du}{dx}\right)-\frac{u}{x}=0$

or $x^2\frac{d^2u}{dx^2}+x\frac{du}{dx}-4=0 \qquad .....(6)$

The general solution of equation (6) becomes.

$u(x)=Ax+\frac{B}{x}$

The functions $u_1(x)=x$ and $u_2(x)=\frac{1}{x}-x$ are linear independent solutions of equation (6) that satisfy the conditions.

$u(0)=0$ and $u(1)=0$

The Wronskian $W[u_1(x), u_2(x)]$ is given by

$$W[u_1(x), u_2(x)] = \begin{vmatrix} u_1(x) & u_2(x) \\ u_1^1(x) & u_2^1(x) \end{vmatrix}$$

$$= \begin{vmatrix} x & \frac{1}{x} - x \\ 1 & -\frac{1}{x^2} - 1 \end{vmatrix} = \frac{-2}{x} \neq 0$$

which shows that $u_1(x)$ and $u_2(x)$ are linearly independent solution of (3).

Also $u_1(\xi)u_2^1(\xi) - u_2(\xi)u_1^1(\xi) = \dfrac{C}{P(\xi)}$

where $u_1(\xi) = \xi, u_2(\xi) = \dfrac{1}{\xi} - \xi, P(\xi) = \xi$

$\Rightarrow c = -2$

$\therefore$ From relation (u), $G_1(x_1\xi)$ reduces to

$$G_1(x_1\xi) = \frac{x}{2\xi}(1-\xi^2),\ 0 \le x \le \xi$$

$$= \frac{\xi}{2x}(1-x^2),\ \xi \le x \le 1$$

Thus the Bessel's equation reduces to an integral equation

$$u(x) = \lambda \int_0^1 G_1(x_1\xi) r(\xi) u(\xi) - \int_0^1 G_1(x_1\xi) F(\xi) d\xi$$

or $u(x) = \lambda \int_0^1 G_1(x_1\xi) r(\xi) u(\xi) d\xi$

the second integral vanishes as $F(x) = 0 \Rightarrow F(\xi) = 0$.

Practice Problems

1. Construct the Green's function for the boundary value problem

$$u''(x)+\mu^2 u=0$$

with the conditions $u(00 = u(1) = 0$

2. Find the Green's function for the boundary value problem.

$$\frac{d^2u}{dx^2}-u(x)=0$$

with the conditions $u(0) = u(1) = 0.$

3. Construct the green's function for the boundary value problem.

$$x^2\frac{d^2u}{dx^2}+x\frac{du}{dx}-4=0$$

with $u(x)$ is bounded as $x \to 0$ and $u(1) = 0$.

4. Construct the Green's function for the boundary value problem

$$\frac{d^2u}{dx^2}-u=0$$

with the conditions $u(0)=u^1(0)$ and $u(l) + \lambda u'(l) = 0.$

5. Using Green's function, solve the boundary value problem

$$u''(x) - u(x) = x$$

with the conditions $u(0) = u(1) = 0.$

6. Solve the boundary value problem. Using Green's function

$$\frac{d^4u}{dx^4}=1,\ u(0)=u^1(0)=u''(1)=u'''(1)=0.$$

7. Reduce the boundary value problem

$$\frac{d^2u}{dx^2}+xu=1, u(0)=u(1)=0$$

to an integral equation.

8. Reduce the boundary value problem to an integral equation.

$$\frac{d^4u}{dx^4}=\lambda u+1,\ u(0)=0,\ u''(1)=u'''(1)=0$$

9. Reduce the following boundary value problem to integral equations.

(i) $u''+\lambda u=2x+1;\ u(0)=u'(1),\ u'(0)=u(1)$.

(ii) $u''+\lambda u=e^x;\ u(0)=u'(0),\ u'(1)=u^1(1).$

10. Determine the Green's function $c_1(x_1\xi)$ for the differential equation

$$\left[\frac{d}{dx}\left(x\frac{d}{x}\right)-\frac{n^2}{x}\right]u=0$$

With the condition $u(0)=0$ and $u(1)=0$.

Answers

1. $$c_1(x_1\xi)=\begin{cases}\dfrac{\sin\mu(\xi-1)\sin\mu x}{\mu\sin\mu}, & u\le x<\xi \\[2ex] & u\le x<\xi \\[2ex] \dfrac{\sin\mu\xi\sin\mu(x-1)}{\mu\sin\mu}, & \xi<x\le 1\end{cases}$$

2. $$c_1(x_1\xi)=\begin{cases}\dfrac{\sin hx\sin h(\xi-1)}{\sin 1}, & 0\le x<\xi \\[2ex] \dfrac{\sin h\xi\sin h\,x-1)}{\sin h1}, & \xi\le x<1\end{cases}$$

3. $$c_1(x_1\xi)=\begin{cases}\dfrac{1}{2}x\left(\dfrac{1}{\xi^2}-1\right), & 0\le x<\xi \\[2ex] \dfrac{1}{2}\left(\dfrac{1}{x}-x\right), & \xi<x\le 1\end{cases}$$

$$-\frac{1}{2}\left(\frac{1-\lambda}{1+\lambda}\right)e^{x+\xi-2l}+\frac{1}{2}e^{x-\xi},\ 0\le x<\xi$$

4. $c_1(x_1\xi)=$

$$-\frac{1}{2}\left(\frac{1-\lambda}{1+\lambda}\right)e^{x+\xi-2l}+\frac{1}{2}e^{\xi-x},\ \xi<x\le 1$$

5. $u(x)=\dfrac{\sin hx}{\sin h1}-x$

6. $u(x)=\dfrac{1}{24}x^2(x^2-4x+6)$

7. $u(x)=\int_0^1 c_1(x_1\xi)u(\xi)\xi d\xi-\dfrac{1}{2}x(1-x)$

8. $u(x)=\lambda\int_0^1 c_1(x_1\xi)u(\xi)d\xi+\dfrac{1}{24}x^2(x^2-4x+6)$

9. (i) $u(x)=-\lambda\int_0^1 c_1(x_1\xi)u(\xi)d\xi-\dfrac{1}{6}(2x^3+3x^2-17x-5)$

(ii) $u(x)=-\lambda\int_0^1 c_1(x_1\xi)u(\xi)d\xi-e\xi$

10. $c_1(x_1\xi)=$

$$\frac{x^n}{2n\xi^n}(1-\xi^{2n}),\ x<\xi$$

$$\frac{\xi^n}{2nx^n}(1-x^{2n}),\ x>\xi$$

Practice Problems

1. Write a general linear partial differential equation of second order and conditions when it becomes parabolic, elliptic and hyper bolic. Solve $\nabla^2 u = 0$ under the conditions (h=1, k=1):

   $u(0,y)=0$, $u(4,y)=12+y$ for $0 \le y \le 4$,

   $u(x,0)=3x$, $u(x,4)=x^2$ for $0 \le x \le 4$

2. (a) Find the solution of $u_t=u_{xx}$ subject to $u(x,0)=\sin \pi x$, $0 \le x \le 1$, $u(0,t) = u(1,t) = 0$ using Schmidt method.

   (b) Find the Fourier sine transform of $f(x) = \frac{1}{x}$.

   (c) Explain fast Fourier transform and its concept.

3. (a) Use Fourier transform to determine the displacement y(x,t) of an infinite string given that the string is initially at rest and the initial displacement is f(x); $-\infty < x < \infty$.

   (b) Write the definition and inversion formula for Mellin and Hankel transform.

4. (a) Define Fredholm integral equations of first and second kind. When an equation is said to be singular? How do we construct Green function?

   (b) Solve the integral equation:

   $F(t) = a \sin t - z\int_0^t F(u)\cos(t-u)du$

5. (a) Show that y(x) =1 is a solution of the Fredholm integral equation:

   $y(x)+\int_0^1 x(e^{tx}-1)y(t)dt = e^x - x$

   (b) Solve the boundary value problem using Green's function:

   $\frac{d^2y}{dx^2}+u=x^2$, $u(\pi/2) = 0 = u(0)$

6. (a) Prove that a necessary condition for:

   $I = \int_{x_1}^{x_2} f(x,y,y')dx$

   to be an extremum is Euler's equation:

   $\frac{\partial f}{\partial y} - \frac{d}{dx}\left(\frac{\partial f}{\partial y}\right) = 0$

   (b) Prove that the sphere is the solid figure of revolution which for a given surface area has maximum volume.

7. (a) Solve:

$$y'' + y = -x, 0 < x < 1, y(0) = 0 = y(1)$$

by Galerkin's method.

(b) Using Rayleigh-Ritz method, solve the boundary value problem defined by:

$$\frac{\partial^2 y}{\partial x^2} + 2x = 0; y(0) = y(1) = 0.$$

8. (a) Give the concept of calculus of variation and prove that the shortest distance between two point in a plane is straight line.

(b) Find the Mellin transform of $\frac{1}{1+x^2}$.

(c) Find the Hankel transform of $\frac{1}{x}$.

9. (a) Explain the following terms giving examples:
   (i) Wavelet transform
   (ii) Error function
   (iii) Haar transform
   (iv) Hash function

(b) Prove that Hermite Polynomial:

$$H_n(x) = (-1)^n . e^{x^2} \frac{d^n y}{dx^n}\left(e^{-x^2}\right).$$

10. (a) Prove that the vectors (1, 2, 1), (1, 2, 0), (1,-1, 2) form a basis of R^3.

(b) Show that the mapping $T:V_3(R) \to V_2(R)$ defined by f(a,b,c) = f(a,b) is Linear transformation. What is the kernel of this transform.

11. (a) Find DFT of sample sequence x(n)={1,1,2,2,3,3} and compute the corresponding amplitude and phase spectrum.

(b) Use separation of variable technique to solve:

$$3u_x + 2u_y = 0,$$

With $u(x, 0) = 4 e^{-x}$.

12. (a) Solve the Poisson's equation:

$$\frac{\partial^2 u}{\partial x^2} + \frac{\partial^2 u}{\partial y^2} = -10(x^2 + y^2 + 10)$$

over the square with sides x=y=0; x=y=3 with u(x,y)=0 on the boundary and mesh length =1.

(b) Find the deflection u(x,t) of vibrating string of length π and ends fixed, corresponding to zero initial velocity and initial deflection :

$$f(x) = k(\sin x - \sin 2x)$$

13. (a) Explain the following with example:
    (i) Hypothesis
    (ii) Discrete random variables
    (iii) Heavisdes's unit function
    (iv) Stochastic process

    (b) If the variance of the Poisson distribution is 2, find the probabilities for r = 1,2,3,4 from the recurrence relation of the Poisson distribution. Also find P(x≥4).

14. (a) A coin is tossed until the head appears. What is expectation of the number of tosses?

    (b) Derive stochastic matrix for one step transition probabilities.

15. (a) Establish the probability distribution formula for pure death process.

    (b) In a railway marshalling yard goods trains arrive at a rate of 30 trains per day. Assuming that the inter-arrival time follows an exponential distribution and the service time distribution is also exponential with an average of 36 minutes, calculate:
    (i) Expected queue size (line length)
    (ii) Probability that the queue size exceeds 10.

    (c) Use Rayleigh-Ritz method to solve the equation:
    $$\frac{\partial^2 y}{\partial x^2} + y = x, y(0) = y(1) = 1$$

16. (a) Use Galerkin's method to solve the equation:
    $$\frac{\partial^2 y}{\partial x^2} - y + x = 0, y(0) = y(1) = 0$$

    (b) Obtain the steady state difference equation for the queuing model $\{(M|M|1)\langle N|FCFS\rangle\}$ and show that:
    $$P_n = \frac{(1-\rho)}{1-\rho^{N+1}}\rho^n, 0 \leq n \leq N.$$

17. (a) Show that the mapping f:V_3(R)→V_2(R), defined by f(a,b,c) = (a,b) is linear transformation. What is the kernel of this transformation?

    (b) Define linearly dependent and linearly independent sets. Are the vectors (3,2,7), (2,4,1), (1,-2,6) linearly independent or linearly dependent.

18. (a) Prove that Hermite Polynomial:

$H_n(x)=(-1)^n . e^{x^2} \frac{\partial^n y}{\partial x^n}(e^{-x^2})$

(b) Find the solution of the one-dimensional heat equation by variable separable method.

19. (a) Find Fourier sine transform of :

$\left(\frac{e^{-ax}}{x}\right)$

(b) Find the values of u(x,t) satisfying the parabolic equation $\frac{\partial u}{\partial t} = u\frac{\partial^2 y}{\partial x^2}$ and the boundary conditions u(0,t)=0=u(8,t) and u(x,0) = 4x-$\frac{1}{2}x^2$ at the points x=i; i=0,1,2,.......8 and t=$\frac{1}{8}.j$;j=0,1,2,...5.

20. (a) In a certain factory turning out razor blades, there is a small chance of 0.002 for any blade to be defective. The blades are supplied in packets of 10, use Poisson distribution to calculate the approximate number of packets containing no defective, one defective and two defective blades respectively in a consignment of 10000 packets.

(b) In a normal distribution, 31% of the items are under 45 and 8% are over 64. Find the mean and standard deviation of the distribution.

(c) In a normal distribution, 31% of the items are under 45 and 8% are over 64. Find the mean and standard deviation of the distribution.

21. (a) Explain the following with example:

(i) Null hypothesis

(ii) Theory of estimation

(iii) Test of significance

(iv) Discrete Random variable

(b) In a sample of 600 men from a certain city, 450 are found smokers. In another sample of 900 men from another city, 450 are smokers. Do the data indicate that the cities are significantly different with respect to the habit of smoking among men?

22. (a) Test the following transition matrix to see if the Markov chain is regular and ergodic where x is some positive P_{ij} value:

$$P = \begin{bmatrix} 0 & x & x & 0 \\ x & 0 & 0 & x \\ x & 0 & 0 & x \\ 0 & x & x & 0 \end{bmatrix}$$

(b) Define stochastic process and Markov process with examples.

23. (a) Obtain the steady state difference equation for the queuing model {(M/M/1): (N/FC FS)} and show that:

$$P_n = \frac{(1-\rho)}{1-\rho^{N+1}}\rho^n, 0 \le n \le N.$$

(b) In a railway marshalling yard, goods trains arrive at a rate of 30 trains per day. Assuming that the inter-arrival time follows an exponential distribution and the service time distribution is also exponential with an average of 36 minutes. Calculate:

(i) Expected queue size (line length).

(ii) Probability that the queue size exceeds 10.

24. Explain descretization in finite elements method. Use Rayleigh-Ritz method to solve the equation:

$$\frac{\partial^2 y}{\partial x^2}+y=x; \; y(0)=0, \; y(1)=1.$$

25. (a) A black and white TV picture consists of about 2 X 10^6 picture elements with 16 different brightness levels with equal probabilities. If the pictures are repeated at the rate of 32 per second, calculate the average rage of information conveyed by this TV picture source. If SNR is 30 dB, what is the maximum band width required to support the transmission of the resultant video signal.

(b) Find the mutual information and channel capacity of the channel shown below:

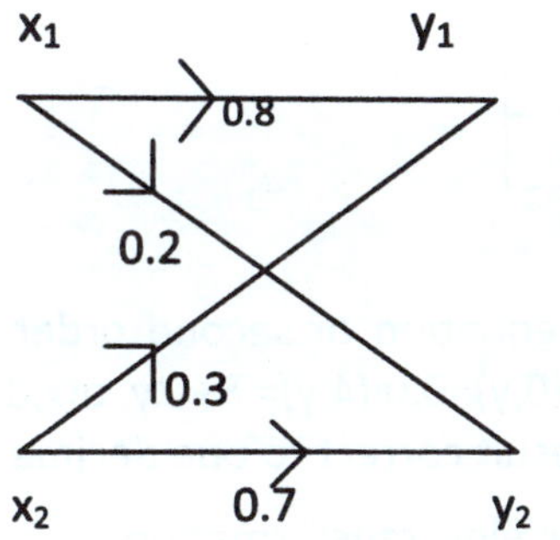

Given $P(x_1) = 0.6$ and $P(x_2) = 0.4$

26. (a) Let e be the (3,8) encoding function $e:B^3 \to B^8$ defined by:

$$\left.\begin{aligned} e(000) &= 00000000 \\ e(001) &= 10111000 \\ e(010) &= 00101101 \\ e(011) &= 10010101 \\ e(100) &= 10100100 \\ e(101) &= 10001001 \\ e(110) &= 00011100 \\ e(111) &= 00110001 \end{aligned}\right\} \textit{code words}$$

and let d be an (8,3) maximum likelihood decoding function associated with e. How many errors will e detect ? Also find that how many errors can (e, d) correct?

(b) What is coding efficiency? Show that the coding efficiency is maximum when P (0) = P(1).

27. (a) Describing the concept of membership, find $\underset{\sim}{A} \cup \underset{\sim}{B}, \underset{\sim}{A} \cap \underset{\sim}{B}, \underset{\sim}{A}-\underset{\sim}{B}$ and $\underset{\sim}{B}-\underset{\sim}{A}$ in the following cases:

(i) $\underset{\sim}{A}=\{((x_1|0.2),(x_2|0.8)),(x_3|0.6)\}$

$\underset{\sim}{B}=\{((x_1|0.4),(x_2|0.6)),(x_3|1.0)\}$

(ii) $\underset{\sim}{A}=\{((x_1|1.0),(x_2|0.0)),(x_3|0.5),(x_4|0.7)\}$

$\underset{\sim}{B}=\{((x_1|0.8),(x_2|1.0)),(x_3|0.0),(x_4|0.6)\}$

(b) If a binary fuzzy relation is given by:

	y_1	y_2
x_1	0.3	0.2
x_2	0	1
x_3	0.6	0.4

Then find the inverse of fuzzy relation.

28.(a) Classify the linear partial differential equation of second order and solve $u_{xx}+u_{yy}=0$ in $0 \le x \le 4$, $0 \le y \le 4$, given that $u(0,y)=0$, $u(4,y)= 8+2y$, $u(x,0)=x^2/2$ and $u(x,4)=x^2$. Take h=k=1 and obtain the result correct to one decimal.

(b) difference between DFT and FFT for wavelet transformation.

29. (a) State and prove Baye's theorem.

(b) Fit a poisson distribution to the following data and calculate the theoretical frequencies:

Deaths	Frequencies
0	122
1	60
2	15
3	2
4	1

Given that $e^{-0.5}$=0.61.

30. (a) Assuming M=3 show that the entropy is maximum when all the messages are equiprobable.

    (b) Determine the DFT of the sequence:

    $$x(t)=\begin{cases}\frac{1}{4}\ for\ 0 \le t \le 2\\ 0\ otherwise\end{cases}$$

31. (a) A class consists of 6 girls and 10 boys. If a committee of 3 is chosen at random from the class find the probability that:

    (i) 3 boys are selected.

    (ii) Exactly two boys are selected

    (iii) at least one boy is selected

    (b) Solve the equation $\nabla u^2 = -10(x^2 + y^2 + 10)$ over the square mesh with sides x=0=y,x=3=y with u=0 on the boundary and mess length =1.

32. Write short notes on any three of the following:
    (a) Fuzzy reasoning
    (b) Shannon's measure of information
    (c) Linear block codes
    (d) Fuzzy fiction and Difuzzification
    (e) Parity checks codes.

33. (a) Find the deflection u(x,t) of the vibrating string of length π and fixed, corresponding to zero initial velocity and initial deflection f

34. (a) Find the deflection u(x,t) of the vibrating string of length π and ends fixed, corresponding to zero initial velocity and initial deflection f(x)=k(sin x – sin 2x), given c^2=1.

    (b) Solve Laplace equation in rectangle with u(0,y)=u(a,y)=u(x,b)=0 and u(x,0)=f(x), see figure:

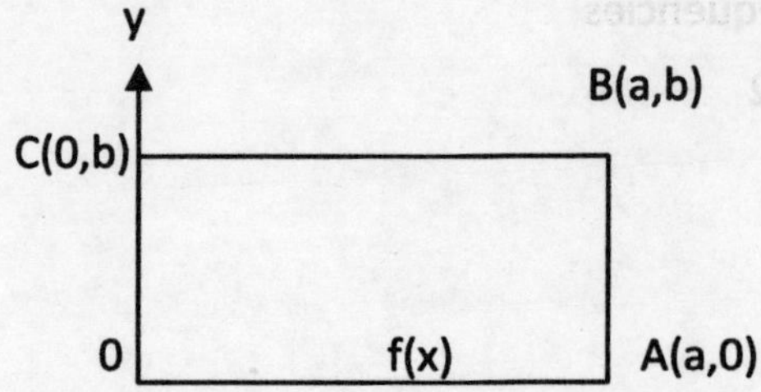

35. (a) Solve $\frac{\partial u}{\partial t} = \frac{\partial^2 u}{\partial x^2}$ subject to initial condition u=sinπx at t=0 for 0≤x≤1 and u=0 at x=0 and x=1 for t>0, by the Gauss-Seidel interrative method.

(b) Write the properties of DFT, WFT and Haar transform.

36. (a) Out of 800 families with 4 children each, how many families would be expected to have (i) 2 boys and 2 girls (ii) at least one boy (iii) no girl (iv) at most two girls? Assume equal probabilities for boys and girls.

(b) A random variable X has the probability function:

X	P(X=x)=P(x)
0	A
1	3a
2	5a
3	7a
4	9a
5	11a
6	13a
7	15a
8	17a

(i) Determine the value of a.

(ii) Find P(X>2), P(X≥6), P(3<X<5)

37. (a) Explain the meaning and importance of sampling distribution.

(b) Define stochastic processes and explain classification.

38. (a) Let:

$$S = \begin{bmatrix} 0.2 & 0.3 & 0.5 \\ 0.4 & 0.4 & 0.2 \\ 0.4 & 0.6 & 0 \end{bmatrix}$$

is a transition matrix. Give the transition matrix for three steps and four steps.

(b) Obtain the steady state difference equations for the queuing model (M/M/1:N/FCFS) in usual notation and solve them for p_0 and p_1.

39. (a) Obtain the distribution of the number in the system in steady state for M/M/S model by considering it as a birth and death process.

(b) What are the roles of α-cuts and strong α-cuts in fuzzy set theory? What is the difference between them?

40. (a) Let A,B be two fuzzy numbers whose membership functions are given by:

$$A(x) = \begin{cases} \frac{x+2}{2} \; for -2 < x \leq 0 \\ \frac{2-x}{2} \; for\; 0 < x < 2 \\ 0 \; otherwise \end{cases}$$

$$B(x) = \begin{cases} \frac{x-2}{2} \; for\; 2 < x \leq 4 \\ \frac{6-x}{2} \; for\; 0 \leq x < 6 \\ 0 \; otherwise \end{cases}$$

Calculate the fuzzy numbers A+B, A-B, B-A, A.B, A/B, Min.(A,B) and Max.(A,B).

41. (a) The failure rate of a certain component is $h(t)=\lambda_0 t$, where $y_0>0$ is a given constant. Determine the reliability R(t) of the component. Repeat for $h(t) =\lambda_0 t^{\frac{1}{2}}$.

(b) Write short notes on the following:

(i) Decision theory

(ii) Goal programming.

42. Let F be a subfield of complex numbers and let T be the function from F^3 to F^3 defined by:

$$T(x_1,x_2,x_3) = (x_1-x_2+2x_3, 2x_1+x_2, -x_1-2x_2+2x_3)$$

(a) Verify that T is a linear transformation.

(b) If (a,b,c) is a vector in F^3, what are the conditions on a,b and c that the vector be in the range of T? what is the rank of T?

(c) What are the conditions on a,b and c that (a,b,c) be in the null space of T? What is the nullity of T?

Or

43. **Define Hash function with example and write the Hermite differential equation and Mathematical form of Hermite polynomial. Check whether the following functions T from R^2 into R^2 are linear transformations or not:**

(a) $T(x_1,x_2) = (1+x_1, x_2)$

(b) $T(x_1,x_2)=(x_2,x_1)$

(c) $T(x_1,x_2) = x_1^2, x_2$

(d) $T(x_1,x_2)=(\sin x_1, x_2)$

(e) $T(x_1,x_2) =(x_1-x_2 , 0)$

44. **The longitudinal displacement in a vibrating elastic bar can be described by the boundary value problem:**

$$\frac{\partial^2 y}{\partial t^2} = c^2 \frac{\partial^2 u}{\partial x^2}, \ 0<x<l,\ t>0;$$

$$\frac{\partial u}{\partial x}(l,t)=0 \quad , t>0$$

$$u(x,0) = x^2(1-\sin x),$$

$$\frac{\partial u}{\partial t}(x,0) = 0,\ 0<x<l.$$

Find the solution of the above boundary value problem by using method of separation of variables.

45. **(a)** **Explain the following with one example in each case where it is applied**

(i) DFT (ii) WFT

(b) **Solve Laplace equation at mesh points:**

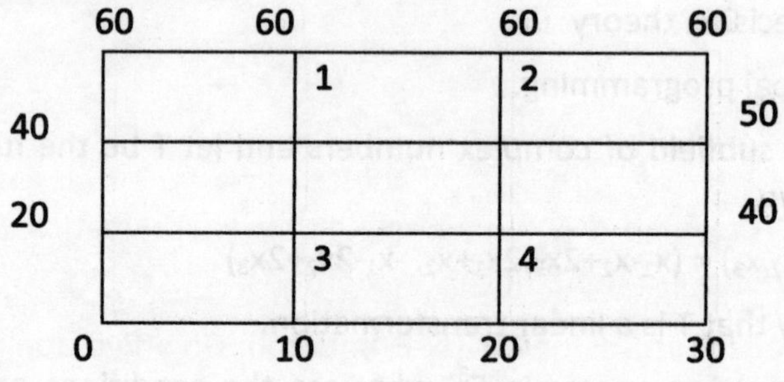

With given conditions.

46. **Define discrete and continuous random variables. Let X have the density function:**

$f(x) = 0.75(1-x^2)$, if $-1\le x\le 1$

$= 0$, otherwise

Find the distribution function. Find the probabilities

$P\left(-\frac{1}{2} < X \leq \frac{1}{2}\right)$ and $P\left(\frac{1}{4} \leq X \leq 2\right)$. Find x s.t.

$P(X \leq x) = 0.95$.

Also compute the probability of obtaining at least two "six" in rolling a fair die 4 times.

47. Define null and alternative hypothesis. Define confidence interval.

If the probability of producing a defective screw is p=0.01, what is the probability that a lot of 100 screws will contain more than 2 defectives?

48. Distinguish between discrete parameter Markov chain and continuous parameter Markov chain. A man either drives his car or takes a train to work each day. Suppose he never takes the train two days in a row, but if he drives to work, then the next day he is just likely to drive again as he is to take the train. Find the transition matrix. Also find the probability that he changes from going by train to driving exactly in four days.

Or

49. A one-person barber shop has six chairs to accommodate people waiting for haircut. Assume customers who arrive when all six chairs are full leave without entering the barber shop. Customer's arrive at the average rate of 3 per hour and spend an average of 15 minutes in the barber shop. Then find:
    (a) The probability a customer can get directly into the barber chair upon arrival.
    (b) Expected number of customers waiting for a haircut.
    (c) Effective arrival rate.
    (d) The time a customer can expect to spend in the barber shop.

50. (a) What is fuzzy membership function? Explain the triangular, trapezoidal and Gaussian membership function with their mathematical form.
    (b) Explain different defuzzification methods.

51. Explain various features of MATLAB in detail. Give five applications where it is used.

52. (a) If f is a mapping from $V_3(F)$ into $V_2(F)$, defined as f(x,y,z) = (y,z), then show that f is linear.
    (b) Show that the set:

    $$W=\{(a,b,c):a-3b+4c=0 \ \forall \ a,b,c \in R\}$$

    is a sub-space of the 3-tuple vector space $V_3(R)$.

53. (a) Prove that:

$$H_n(-x) = (-1)^n H_n(x)$$

(b) Show that:

$$\int_0^\infty e^{-x^2-2ax}dx = \frac{\sqrt{\pi}}{2}e^{a^2}\{1\text{-erf}(a)\}$$

54. Solve $\frac{\partial u}{\partial x} = 2\frac{\partial u}{\partial t} + u$, subject to the condition:

$$u(x,0)=6e^{-3x}$$

55. (a) Solve the wave equation:

$$\frac{\partial^2 u}{\partial t^2} = 16\frac{\partial^2 u}{\partial x^2}, \quad 0 \le x \le 5, \ 0 \le t \le 1.25$$

Subject to the conditions $u(0,t) = u(5,t) = 0$, $t<0$:

$$\left.\begin{aligned} u(x,0) &= x^2(5-x) \\ u_t &= (x,0) = 0 \end{aligned}\right\} 0 \le x \le 5$$

(b) Write notes on the following:

(i) WFT

(ii) Wavelet transform

(iii) Haar transform

56. (a) Two dice are thrown. Find the probability that the sum of the faces is :

(i) 7 or 8

(ii) More than 8

(b) In four tosses of a coin, let x be the number of heads. Calculate the expected value of x. What can you say about the expectation of number of tails?

57. (a) Find the mean and variance of Poisson distribution.

(b) The mean and variance of a Binomial distribution are 4 and 4/3 respectively. Find:

(i) the probability of 2 successes

(ii) the probability of more than two successes

(iii) The probability of more than three successes

58. Derive the expression:

$P_n(t) = P[x(t)=n] = e^{-\lambda t}\frac{(\lambda t)^n}{n!}$ $n = 0,1,2...\infty$ in case of a Poisson process.

59. (a) What is the difference between FEM and DFT?

    (b) Solve the boundary value problem:

    $y'' - y + x = 0$, $(0 \le x \le 1)$, $y(0) = y(1) = 0$ by Rayleigh-Ritz method.

60. (a) Show that the vectors (1,2,0) (0,3,0) and (-1,0,1) are linearly dependent in the vector space V_3 of real numbers R.

    (b) Let $T: R^2 \to R^3$ be defined by:

    $T(x,y) = \{-x+2y, y, -3x+3y\}$

    Find the matrix representation of T with respect to the bases A = {(1,2),(-2,1)} and B = {(-1,0,2),(1,2,3),(1,-1,-1)} of R^2 and R^3 respectively.

61. (a) Give Mathematical formula for {$He_n(x)${Hermite polynomial}} and show that:

    $He_3(x) = x^3 - 3x$

    Also write weight function for Hermite polynomial to be orthogonal.

    (b) A string is stretched and fastened to two points l apart. Motion is started by displacing the string in the form y = a sin $\frac{\pi x}{l}$ from which it is released at time t=0. Show that the displacement of any point at a distance x from one end at time t is given by:

    $$y(x,t) = a \sin \frac{\pi x}{l} \cos \frac{\pi c t}{l}$$

62. (a) Solve $\frac{\partial u}{\partial t} = \frac{\partial^2 u}{\partial x^2}$ subject to initial condition $u = \sin \pi x$ at t=0 for $0 \le x \le 1$ and u=0 at x=0 and x=1 for t>0, by the Gauss-Seidel iterative method method.

    (b) Write the properties of DFT, WFT and Haar Transform.

    (c) A can hit a target 4 times in 5 shots; B 3 times in 4 shots; C twice in 3 shots. They fire a volley. What is the probability that at least two shots hit ?

63. (a) Fit a Poisson distribution to the following and calculate theoretical frequencies:

x	:	0	1	2	3	4
f	:	122	60	15	2	1

    (b) Write short notes on the following:

    (i) Concept of estimation theory

    (ii) Theory of hypotheses

64. (a) Show that every stochastic process $\{x_t, t = 0,1,2,...\}$ with independent increment is a Markov process.

(b) Obtain the distribution of the number in the system in steady state for (M/M/1: ∞/FCFS) model by considering it as birth and death process.

65. (a) Customers arrive at a one man barber shop according to Poisson process with a mean interval time of 12 minutes. Customer spend an average of 10 min. in the barber's shop:

(i) What is the expected number of customers in the barber shop and in the queue ?

(ii) How much time can a customer expect to spend in the barber's shop?

(iii) What is the average time customer spends in the queue?

(b) What are fuzzy sets? Describe the importance of fuzzy sets. Compare it with classical sets.

66. (a) Write five transfer functions and their MATLAB functions.

(b) Define Heaviside's unit function and error function and where they are used.

67. (a) Solve the elliptic equation $u_{xx}+u_{yy}=0$ for the following square mesh with boundary values as shown:

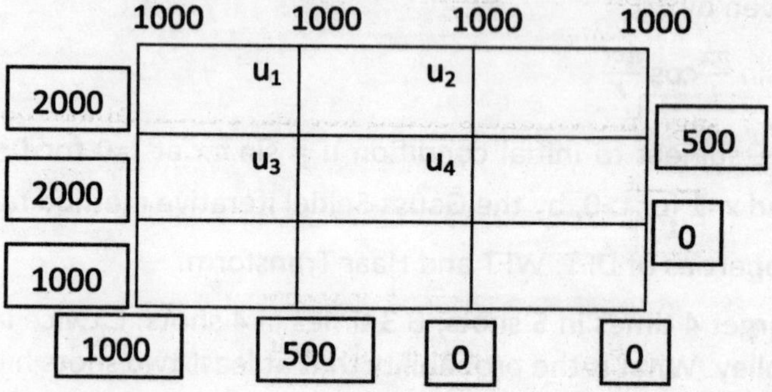

(b) Solve the heat equation $\frac{\partial u}{\partial t}=\frac{\partial^2 y}{\partial x^2}$ subject to the initial condition u=sin πx at t=0 for 0≤x≤1 and u=0 at x=0 and x=1 for t>0 by Gauss-Seidel iterative method.

68. (a) Use the method of separation of variables to solve the equation:

$$\frac{\partial^2 y}{\partial x^2}=\frac{\partial u}{\partial t}$$

Given that u=0 when t→∞ as well as u = 0 at x = 0 and x =l.

(b) A random variable X has the probability function:

X	P(X=x) = P(x)
0	a
1	3a
2	5a
3	7a
4	9a
5	11a
6	13a
7	15a
8	17a

(i) Determine the value of a.

(ii) Find $P(X>2)$, $P(X\geq6)$, $P(3<X<5)$.

69. (a) A person throws two dice, one the common cube and other a regular tetrahedron, the number on the lowest face being taken in the case of tetrahedron. Find the chances of throwing 6 and 10.

(b) A manufacturer of other cotter pins knows that 5% of his product is defective. Pins are sold in boxes of 100. He guarantees that not more than 10 pins will be defective. Determine the probability that a box will fail to meet the guarantee.

70. (a) Define stochastic processes and explain classification of stochastic process.

(b) Derive the difference equations for the queuing model (M/M/1):(∞/FCFS).

71. (a) Let:

$$S = \begin{bmatrix} 0.2 & 0.3 & 0.5 \\ 0.4 & 0.4 & 0.2 \\ 0.4 & 0.6 & 0 \end{bmatrix}$$

is a transition matrix. Give the transition matrix for three steps and four steps.

(b) In a railway marshalling yard, goods trais arrive at a rate of 30 trains per day. Assuming that the inter-arrival time follows an exponential distribution and the service time distribution is also exponential with an average of 36 minutes, calculate:

(i) Expected queue size (line length).

(ii) Probability that the queue size exceeds 10.

72. (a) Define the following terms giving examples:

(i) α-cut of fuzzy set

(ii) Union of two fuzzy sets

(iii) Intersection of two fuzzy sets

(iv) Normal fuzzy set

(b) Describe fuzz if then rule and composition rule of inference for a fuzzy system.

73. (a) Differentiate between DFT and FFT MATLAB tools for wavelet transform.

(b) Variables a,b,c and d have been initialized to the following values: a=2, b=3,c=4, d=3. Evaluate the following MATLAB Assignments:

(i) a*b+c*d

(ii) a*b*d

(iii) a*b+b*d

(iv) a*b+c*d*a

74. (a) Write short notes on the following:

(i) Reliability

(ii) Failure rate

(iii) Decision theory

(iv) Goal programming

(b) The failure rate of a certain component is $h(t) = \lambda_0 t$, $\lambda_0 > 0$ is a given constant. Determine the reliability R(t) of the component. Repeat for $h(t) = \lambda_0 t^{\frac{1}{2}}$.

75. (a) Solve the boundary value problem $\frac{\partial u}{\partial t} = \frac{\partial^2 y}{\partial x^2}$ under the condition u(0,t) = u(1,t)=0 and u(x,0)=sin πx, $0 \leq x \leq 1$, taking h=0.2 and k= 0.02 using Schmidt method.

(b) Discuss the solution of Poisson equation. Solve the partial differential equation $u_{xx}+u_{yy}=8x^2y^2$ for the square mesh of the following fig with u(x,y)= 0 on the boundary and mesh length =1.

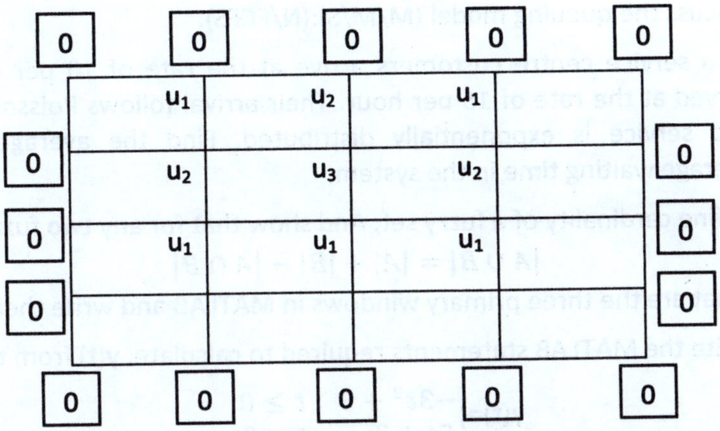

76. (a) Find the solution of $\frac{\partial^2 u}{\partial x^2}=h^2\frac{\partial u}{\partial t}$ for which u(0,t) = u(l,t)= 0, u(x,0) $\frac{\sin \pi x}{l}$ by method of variable separation.

(b) find the solution of two-dimensional heat equation.

77. (a) In a sample of 1000 cases the mean of certain test is 14 and σ=2.5. Assuming the distribution to be normal. Then find:

(i) How many students score between 12 and 15?

(ii) How many score above 18?

(iii) How many score below 8?

(iv) How many score 16?

(b) If the Probability of committing an error of magnitude x is given by:

$y=\frac{h}{\sqrt{\pi}}\ e^{-h^2x^2}$

Compute the probable error from the following data:

m_1=1.305, m_3=1.295, m_5=1.318

m_2=1.301, m_4=1.286, m_6=1.321

78. (a) Obtain the steady state solution of the queuing model (M/M/1)(N/FCFS).

(b) Patients arrive at a clinic according to a Poisson distribution at the rate of 30 patients per hour. The waiting room does not accommodate more than 14 patients. Examination time per patient is exponential with mean rate 20 per hour:

(i) Find the effective arrival rate at the clinic.

(ii) What is the expected waiting time until a patient is discharged from a clinic?

79. (a) Discuss the queuing model (M/M/S):(N/FCFS).

    (b) At a service centre customers arrive at the rate of 10 per hour and are served at the rate of 15 per hour. Their arrival follows Poisson distribution and service is exponentially distributed. Find the average length and average waiting time in the system.

80. (a) Define cardinality of a fuzzy set, And show that for any two fuzzy sets:

$$|A \cup B| = |A| + |B| - |A \cap B|$$

    (b) What are the three primary windows in MATLAB and write their purpose?

81. (a) Write the MATLAB statements required to calculate. y(t) from the equation:

$$y\{t\}=\begin{cases} -3t^2+5 & : t \geq 0 \\ 5t+2 & : t < 0 \end{cases}$$

    for values of t between -9 and 9 in step of 0.5.

    (b) Define the standard operations on fuzzy sets and show that for any two fuzzy sets A and B of X-

$$(A\cap B)' = A' \cup B'$$

82. (a) Define gamma model and find time to failure f(t), probability of failure F(t), probability of working reliability R(t) and mean time to failure MTTF.

    (b) The time to failure density function (PDF)for a system is :

    f(t)=0.01; 0≤t≤100 days

    Find:

    (i) R(t)

    (ii) The hazard rate function

    (iii) The MTTF

    (iv) The standard deviation

    (v) The median time to failure.

83. (a) Show that S={$\alpha_1, \alpha_2, \alpha_3,$} is linearly dependent over field R, where α_1=(1,3,2), α_2=(1,-7,-8), α_3=(2,1,-1).

    (b) Prove the recurrence relation for Hermite polynomial:

    $xH'_n(x) = nH'_{n-1}(x) + nH_n(x)$

84. (a) Solve 3x≡10 (mod 19)and x≡ 19(mod 21) using successive substitution.

    (b) A tightly stretched string of length l with fixed ends is initially in equilibrium position. It is set vibrating by giving each point a velocity $y_0 \sin \frac{3\pi x}{l}$. Find the displacement y(x,t).

84. (a) Solve the elliptic equation $\frac{\partial^2 y}{\partial x^2} + \frac{\partial^2 u}{\partial y^2} = 0$ for the square mesh with boundary values as shown in figure.

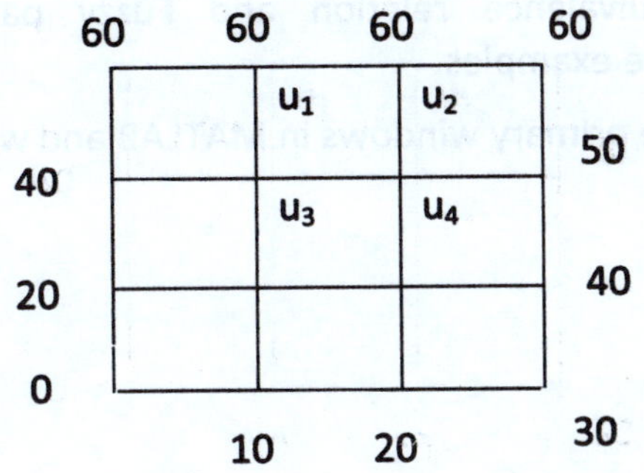

(b) Solve the equation $\frac{\partial u}{\partial t} = \frac{\partial^2 u}{\partial x^2}$ subject to the conditions u(x,0) = sin πx, $0 \leq x \leq 1$; u(0,t)=u(1,t)=0 using Crank-Nicolson method. Carry out computations for two levels taking h=1/3, k=1/36.

85. (a) Explain the following with the example in each case, where it is applied:
   (i) Haar transform
   (ii) Wavelet transform

(b) A box containing 6 red, 4 white and 5 black balls. A person draws 4 balls from the box at random. Find the probability that among the balls drawn there is at least one ball of each colour.

86. (a) If 10% of the bolts produced by a machine are defective, determine the probability that out of 10 bolts chosen at random (i) one (ii) none (iii) at most 2 bolts will be defective.

(b) Explain the following with examples:
   (i) Testing of hypothesis
   (ii) Theory of estimation
   (iii) Discrete random variable

87. (a) Define Stochastic process and Markov process with example.

(b) Test the following transition matrix to see if the Markov chain is regular and ergodic where a is some positive p_{ij} value:

$$P = \begin{bmatrix} 0 & a & a & 0 \\ a & 0 & 0 & a \\ a & 0 & 0 & a \\ 0 & a & a & 0 \end{bmatrix}$$

88. (a) Obtain the steady state solution of (M/M/1)(N/FCFS) system.

(b) Explain Fuzzy set and Fuzzy set operations giving examples.

89. (a) Define Fuzzy equivalence relation and Fuzzy partial order relation considering suitable examples.

(b) What are the three primary windows in MATLAB and write their purpose?